AF678565

Introduction to Tropical Geometry

Introduction to Tropical Geometry

Diane Maclagan
Bernd Sturmfels

Graduate Studies in Mathematics
Volume 161

American Mathematical Society
Providence, Rhode Island

EDITORIAL COMMITTEE

Dan Abramovich
Daniel S. Freed
Rafe Mazzeo (Chair)
Gigliola Staffilani

2010 *Mathematics Subject Classification.* Primary 14T05;
Secondary 05B35, 12J25, 13P10, 14M25, 15A80, 52B20.

For additional information and updates on this book, visit
www.ams.org/bookpages/gsm-161

Library of Congress Cataloging-in-Publication Data

Maclagan, Diane, 1974–
Introduction to tropical geometry / Diane Maclagan, Bernd Sturmfels.
pages cm. — (Graduate studies in mathematics ; volume 161)
Includes bibliographical references and index
ISBN 978-0-8218-5198-2 (alk. paper)
1. Tropical geometry—Study and teaching (Graduate) 2. Geometry, Algebraic—Study and teaching (Graduate) I. Sturmfels, Bernd, 1962– II. Title. III. Tropical geometry.

QA582.M33 2015
516.3′5—dc23

2014036141

Copying and reprinting. Individual readers of this publication, and nonprofit libraries acting for them, are permitted to make fair use of the material, such as to copy select pages for use in teaching or research. Permission is granted to quote brief passages from this publication in reviews, provided the customary acknowledgment of the source is given.

Republication, systematic copying, or multiple reproduction of any material in this publication is permitted only under license from the American Mathematical Society. Permissions to reuse portions of AMS publication content are handled by Copyright Clearance Center's RightsLink® service. For more information, please visit: `http://www.ams.org/rightslink`.

Send requests for translation rights and licensed reprints to `reprint-permission@ams.org`.

Excluded from these provisions is material for which the author holds copyright. In such cases, requests for permission to reuse or reprint material should be addressed directly to the author(s). Copyright ownership is indicated on the copyright page, or on the lower right-hand corner of the first page of each article within proceedings volumes.

© 2015 by the American Mathematical Society. All rights reserved.
The American Mathematical Society retains all rights
except those granted to the United States Government.
Printed in the United States of America.

∞ The paper used in this book is acid-free and falls within the guidelines
established to ensure permanence and durability.
Visit the AMS home page at `http://www.ams.org/`

10 9 8 7 6 5 4 3 2 1 20 19 18 17 16 15

To Saul and Hyungsook

Contents

Preface

Tropical geometry is an exciting new field at the interface between algebraic geometry and combinatorics with connections to many other areas. At its heart it is geometry over the tropical semiring, which is $\mathbb{R} \cup \{\infty\}$ with the usual operations of addition and multiplication replaced by minimum and addition, respectively. This turns polynomials into piecewise-linear functions and replaces an algebraic variety by an object from polyhedral geometry, which can be regarded as a "combinatorial shadow" of the original variety.

In this book we introduce this subject at a level that is accessible to beginners. Tropical geometry has become a large field, and only a small selection of topics can be covered in a first course. We focus on the study of tropical varieties that arise from classical algebraic varieties. Methods from commutative algebra and polyhedral geometry are central to our approach. This necessarily means that many important topics are left out. These include the systematic development of tropical geometry as an intrinsic geometry in its own right, connections to enumerative and real algebraic geometry, connections to mirror symmetry, connections to Berkovich spaces and abstract curves, and the more applied aspects of max-plus algebra. Luckily most of these topics are covered in other recent or forthcoming books, such as [**BCOQ92**], [**But10**], [**Gro11**], [**IMS07**], [**Jos**], [**MR**], and [**PS05**].

Prerequisites. This text is intended to be suitable for a class on tropical geometry for first-year graduate students in mathematics. We have attempted to make the material accessible to readers with a minimal background in algebraic geometry, at the level of the undergraduate text book *Ideals, Varieties, and Algorithms* by Cox, Little, and O'Shea [**CLO07**].

Essential prerequisites for this book are mastery of linear algebra and the material on rings and fields in a first course in abstract algebra. Since tropical geometry draws on many fields of mathematics, some additional background in geometry, topology, or number theory will be beneficial.

Polyhedra and polytopes play a fundamental role in tropical geometry, and some prior exposure to convexity and polyhedral combinatorics may help. For that we recommend Ziegler's book *Lectures on Polytopes* [**Zie95**].

Chapter 1 offers a friendly welcome to our readers. It has no specific prerequisites and is meant to be enjoyable for all. The first three sections in Chapter 2 cover background material in abstract algebra, algebraic geometry, and polyhedral geometry. Enough definitions and examples are given that an enthusiastic reader can fill in any gaps. All students (and their teachers) are strongly urged to explore the exercises for Chapters 1 and 2.

Some of the results and their proofs will demand more mathematical maturity and expertise. Chapters 2 and 3 require some commutative algebra. Combinatorics and multilinear algebra will be useful for studying Chapters 4 and 5. Chapter 6 assumes familiarity with modern algebraic geometry.

Overview. We begin by relearning the arithmetic operations of addition and multiplication. The rest of Chapter 1 offers tapas that can be enjoyed in any order. They show a glimpse of the past, present, and future of tropical geometry and serve as an introduction to the more formal contents of this book. In Chapter 2, the first half covers background material, while the second half develops a version of Gröbner basis theory suitable for algebraic varieties over a field with valuation. The highlights are the construction of the Gröbner complex and the resulting finiteness of tropical bases.

Chapter 3 is the heart of the book. The two main results are the Fundamental Theorem 3.2.3, which characterizes tropical varieties in seemingly different ways, and the Structure Theorem 3.3.5, which says that they are connected balanced polyhedral complexes of the correct dimension. Stable intersections of tropical varieties reveal a hint of intersection theory.

Tropical linear spaces and their parameter spaces, the Grassmannian and the Dressian, appear in Chapter 4. Matroid theory plays a foundational role. Our discussion of complete intersections includes mixed volumes of Newton polytopes and a tropical proof of Bernstein's Theorem for n equations in n variables. We also study the combinatorics of surfaces in 3-space.

Chapter 5 covers spectral theory for tropical matrices, tropical convexity, and determinantal varieties. It also showcases computations with Bergman fans of matroids and other linear spaces. Chapter 6 concerns the connection between tropical varieties and toric varieties. It introduces the tropical approach to degenerations, compactifications, and enumerative geometry.

Teaching possibilities. A one-semester graduate course could be based on Chapters 2 and 3, plus selected topics from the other chapters. One possibility is to start with two or three weeks of motivating examples selected from Chapter 1 before moving on to Chapters 2 and 3. A course for more advanced graduate students could start with Gröbner bases as presented in the second half of Chapter 2, cover Chapter 3 with proofs, and end with a sampling of topics from the later chapters. Students with an interest in combinatorics and computation might gravitate toward Chapters 4 and 5. An advanced course for students specializing in algebraic geometry would focus on Chapters 3 and 6. Covering the entire book would require a full academic year or an exceptionally well-prepared group of participants.

We have attempted to keep the prerequisites low enough to make parts of the book appropriate for self-study by a final-year undergraduate. The sections in Chapter 2 could serve as first introductions to their subject areas. A simple route through Chapter 3 is to focus in detail on the hypersurface case, and to discuss the Fundamental Theorem and Structure Theorem without proofs. The exercises suggest many possibilities for senior thesis projects.

Acknowledgments. We have drawn on the rich and ever-growing literature in tropical geometry when preparing this book. While most direct sources are mentioned, the bibliography is by no means complete. We thank the authors whose work we have drawn on for their inspiration and apologize for any omissions. Readers are encouraged to search the keywords of this book and the MSC code 14T05 to explore this beautiful subject, including the topics missing from this book.

We are grateful to the many readers who have offered mathematical help and comments on drafts of this book during its long incubation. These include Frank Ban, Roberto Barrera, Florian Block, Lucia Caporaso, Dustin Cartwright, Federico Castillo, Andrew Chan, Melody Chan, Angelica Cueto, Jan Draisma, Mareike Dressler, Laura Escobar, Rodrigo Ferreira da Rosa, Gunnar Fløystad, Jennifer Garcia Castilla, Falko Gauss, Noah Giansiracusa, Walter Gubler, Maria Isabel Herrero, June Huh, Florencia Orosz Hunziker, Nathan Ilten, Anders Jensen, Michael Joswig, Dagan Karp, Steven Karp, Sara Lamboglia, Hwangrae Lee, Yoav Len, Bo Lin, Madhusudan Manjunath, Hannah Markwig, Ralph Morrison, Benjamin Nill, Mounir Nisse, Jin Hyung Park, Sam Payne, Nathan Pflueger, Dhruv Ranganathan, Felipe Rincón, Kristin Shaw, Erez Sheiner, David Speyer, Stefan Stadlöder, Jenia Tevelev, Ngoc Mai Tran, Paolo Tripoli, Emanuel Tsukerman, Jan Verschelde, Daping Weng, Annette Werner, Jessie Yang, Josephine Yu, Magdalena Zajaczkowska, and Dylan Zwick. Some of this book was written when the authors were both resident at various mathematics institutes. Particular thanks to MSRI (Berkeley), MPIM (Bonn), and NIMS (Daejeon) for their hospitality. A skeleton of parts of the book was created in the lecture notes the first author wrote for a four-week AARMS (Atlantic Association for Research in the Mathematical Sciences) summer school in 2008. Thanks to Colin Ingalls, Barry Monson, and Hugh Thomas for that opportunity. This project was partially supported by the U.K. Engineering and Physical Sciences Research Council and the U.S. National Science Foundation.

Diane Maclagan
Mathematics Institute
University of Warwick
Coventry, CV4 7AL
United Kingdom

Bernd Sturmfels
Department of Mathematics
University of California
Berkeley, CA 94720
United States

Chapter 1

Tropical Islands

In tropical algebra, the sum of two numbers is their minimum and the product of two numbers is their sum. This algebraic structure is known as the *tropical semiring* or as the min-plus algebra. With minimum replaced by maximum we get the isomorphic max-plus algebra. The adjective "tropical" was coined by French mathematicians, notably Jean-Eric Pin [**Pin98**], to honor their Brazilian colleague Imre Simon [**Sim88**], who pioneered the use of min-plus algebra in optimization theory. There is no deeper meaning to the adjective "tropical". It simply stands for the French view of Brazil.

The origins of algebraic geometry lie in the study of zero sets of systems of multivariate polynomials. These objects are algebraic varieties, and they include familiar examples such as plane curves and surfaces in three-dimensional space. It makes perfect sense to define polynomials and rational functions over the tropical semiring. These functions are piecewise linear. Algebraic varieties can also be defined in the tropical setting. They are now subsets of $\mathbb{R}^n$ that are composed of convex polyhedra. Thus tropical algebraic geometry is a piecewise-linear version of algebraic geometry.

This chapter serves as a friendly welcome to tropical mathematics. We present the basic concepts concerning the tropical semiring, we discuss some of the historical origins of tropical geometry, and we show by way of elementary examples how tropical methods can be used to solve problems in algebra, geometry, and combinatorics. Proofs, precise definitions, and the general theory will be postponed to later chapters. Our primary objective here is to show the reader that the tropical approach is both useful and fun.

The chapter title stands for our view of a day at the beach. The sections are disconnected but island hopping between them should be quick and easy.

1.1. Arithmetic

Our basic object of study is the *tropical semiring* $(\mathbb{R} \cup \{\infty\}, \oplus, \odot)$. As a set this is just the real numbers $\mathbb{R}$, together with an extra element ∞ which represents infinity. In this semiring, the basic arithmetic operations of addition and multiplication of real numbers are redefined as follows:

$$x \oplus y \ := \ \min(x, y) \qquad \text{and} \qquad x \odot y \ := \ x + y.$$

In words, the *tropical sum* of two numbers is their minimum, and the *tropical product* of two numbers is their usual sum. Here are some examples of how to do arithmetic in this exotic number system. The tropical sum of 4 and 9 is 4. The tropical product of 4 and 9 equals 13. We write this as follows:

$$4 \oplus 9 \ = \ 4 \qquad \text{and} \qquad 4 \odot 9 \ = \ 13.$$

Many of the familiar axioms of arithmetic remain valid in tropical mathematics. For instance, both addition and multiplication are *commutative*:

$$x \oplus y \ = \ y \oplus x \qquad \text{and} \qquad x \odot y \ = \ y \odot x.$$

These two arithmetic operations are also associative, and the times operator $\odot$ takes precedence when plus $\oplus$ and times $\odot$ occur in the same expression.

The *distributive law* holds for tropical addition and multiplication:

$$x \odot (y \oplus z) \ = \ x \odot y \ \oplus \ x \odot z.$$

Here is a numerical example to show distributivity:

$$\begin{aligned} 3 \odot (7 \oplus 11) \ &= \ 3 \odot 7 \ = \ 10, \\ 3 \odot 7 \ \oplus \ 3 \odot 11 \ &= \ 10 \oplus 14 \ = \ 10. \end{aligned}$$

Both arithmetic operations have an identity element. Infinity is the *identity element* for addition and zero is the *identity element* for multiplication:

$$x \oplus \infty = x \qquad \text{and} \qquad x \odot 0 = x.$$

We also note the following identities involving the two identity elements:

$$x \odot \infty = \infty \qquad \text{and} \qquad x \oplus 0 = \begin{cases} 0 & \text{if } x \geq 0, \\ x & \text{if } x < 0. \end{cases}$$

Elementary school students prefer tropical arithmetic because the multiplication table is easier to memorize, and even long division becomes easy.

Here is a tropical *addition table* and a tropical *multiplication table*:

$\oplus$	**1**	**2**	**3**	**4**	**5**	**6**	**7**
1	1	1	1	1	1	1	1
2	1	2	2	2	2	2	2
3	1	2	3	3	3	3	3
4	1	2	3	4	4	4	4
5	1	2	3	4	5	5	5
6	1	2	3	4	5	6	6
7	1	2	3	4	5	6	7

$\odot$	**1**	**2**	**3**	**4**	**5**	**6**	**7**
1	2	3	4	5	6	7	8
2	3	4	5	6	7	8	9
3	4	5	6	7	8	9	10
4	5	6	7	8	9	10	11
5	6	7	8	9	10	11	12
6	7	8	9	10	11	12	13
7	8	9	10	11	12	13	14

An essential feature of tropical arithmetic is that there is no subtraction. There is no real number x that we can call "13 minus 4" because the equation $4 \oplus x = 13$ has no solution x at all. Tropical division is defined to be classical subtraction, so $(\mathbb{R} \cup \{\infty\}, \oplus, \odot)$ satisfies all ring axioms (and indeed field axioms) except for the existence of an additive inverse. Such algebraic structures are called *semirings*, whence the name tropical semiring.

It is extremely important to remember that "0" is the multiplicative identity element. For instance, the tropical *Pascal's triangle* looks like this:

$$\begin{array}{ccccccccc} & & & & 0 & & & & \\ & & & 0 & & 0 & & & \\ & & 0 & & 0 & & 0 & & \\ & 0 & & 0 & & 0 & & 0 & \\ 0 & & 0 & & 0 & & 0 & & 0 \\ \cdots & \cdots & & \cdots & & \cdots & & \cdots & \cdots \end{array}$$

The rows of Pascal's triangle are the coefficients appearing in the *Binomial Theorem*. For instance, the third row in the triangle represents the identity

$$\begin{aligned} (x \oplus y)^3 &= (x \oplus y) \odot (x \oplus y) \odot (x \oplus y) \\ &= 0 \odot x^3 \oplus 0 \odot x^2 y \oplus 0 \odot x y^2 \oplus 0 \odot y^3. \end{aligned}$$

Of course, the zero coefficients can be dropped in this identity:

$$(x \oplus y)^3 = x^3 \oplus x^2 y \oplus x y^2 \oplus y^3.$$

Moreover, the *Freshman's Dream* holds for all powers in tropical arithmetic:

$$(x \oplus y)^3 = x^3 \oplus y^3.$$

The validity of the three displayed identities is easily verified by noting that the following equations hold in classical arithmetic for all $x, y \in \mathbb{R}$:

$$3 \cdot \min\{x, y\} = \min\{3x, 2x + y, x + 2y, 3y\} = \min\{3x, 3y\}.$$

The linear algebra operations of adding and multiplying vectors and matrices make sense over the tropical semiring. For instance, the tropical scalar product in $\mathbb{R}^3$ of a row vector with a column vector is the scalar

$$\begin{aligned}(u_1, u_2, u_3) \odot (v_1, v_2, v_3)^{\mathrm{T}} &= u_1 \odot v_1 \oplus u_2 \odot v_2 \oplus u_3 \odot v_3 \\ &= \min\{u_1 + v_1,\, u_2 + v_2,\, u_3 + v_3\}.\end{aligned}$$

Here is the product of a column vector and a row vector of length 3:

$$(u_1, u_2, u_3)^T \odot (v_1, v_2, v_3)$$
$$= \begin{pmatrix} u_1 \odot v_1 & u_1 \odot v_2 & u_1 \odot v_3 \\ u_2 \odot v_1 & u_2 \odot v_2 & u_2 \odot v_3 \\ u_3 \odot v_1 & u_3 \odot v_2 & u_3 \odot v_3 \end{pmatrix} = \begin{pmatrix} u_1 + v_1 & u_1 + v_2 & u_1 + v_3 \\ u_2 + v_1 & u_2 + v_2 & u_2 + v_3 \\ u_3 + v_1 & u_3 + v_2 & u_3 + v_3 \end{pmatrix}.$$

Any matrix which can be expressed as such a product has *tropical rank* 1. See Section 5.3 for three different definitions of the rank of a tropical matrix.

Here are a few more examples of arithmetic with vectors and matrices:

$$2 \odot (3, -7, 6) = (5, -5, 8)\,, \quad (\infty, 0, 1) \odot (0, 1, \infty)^T = 1\,,$$

$$\begin{pmatrix} 3 & 3 \\ 0 & 7 \end{pmatrix} \oplus \begin{pmatrix} 4 & 1 \\ 5 & 2 \end{pmatrix} = \begin{pmatrix} 3 & 1 \\ 0 & 2 \end{pmatrix}, \text{ and } \begin{pmatrix} 3 & 3 \\ 0 & 7 \end{pmatrix} \odot \begin{pmatrix} 4 & 1 \\ 5 & 2 \end{pmatrix} = \begin{pmatrix} 7 & 4 \\ 4 & 1 \end{pmatrix}.$$

Given a $d \times n$-matrix A, we might be interested in computing its image $\{A \odot \mathbf{x} : \mathbf{x} \in \mathbb{R}^n\}$ and in solving the linear systems $A \odot \mathbf{x} = \mathbf{b}$ for various right-hand sides $\mathbf{b}$. We will discuss the relevant geometry in Section 5.2. For an introduction to tropical linear systems and their applications we recommend the books *Synchronization and Linearity* by Baccelli, Cohen, Olsder, and Quadrat [**BCOQ92**] and *Max-linear Systems* by Butkovič [**But10**].

Students of computer science and discrete mathematics may encounter tropical matrix multiplication in algorithms for shortest paths in graphs and networks. The general framework for such algorithms is known as *dynamic programming*. We shall explore this connection in the next section.

Let $x_1, x_2, \ldots, x_n$ be variables which represent elements in the tropical semiring $(\mathbb{R} \cup \{\infty\}, \oplus, \odot)$. A *monomial* is any product of these variables, where repetition is allowed. Throughout this book, we generally allow negative integer exponents. By commutativity, we can sort the product and write monomials in the usual notation, with the variables raised to exponents:

$$x_2 \odot x_1 \odot x_3 \odot x_1 \odot x_4 \odot x_2 \odot x_3 \odot x_2 = x_1^2 x_2^3 x_3^2 x_4.$$

A monomial represents a function from $\mathbb{R}^n$ to $\mathbb{R}$. When evaluating this function in classical arithmetic, what we get is a linear function:

$$x_2 + x_1 + x_3 + x_1 + x_4 + x_2 + x_3 + x_2 = 2x_1 + 3x_2 + 2x_3 + x_4.$$

Remark 1.1.1. Every linear function with integer coefficients arises in this way, so tropical monomials are linear functions with integer coefficients.

A *tropical polynomial* is a finite linear combination of tropical monomials:

$$p(x_1,\ldots,x_n) \quad = \quad a \odot x_1^{i_1} x_2^{i_2} \cdots x_n^{i_n} \ \oplus\ b \odot x_1^{j_1} x_2^{j_2} \cdots x_n^{j_n} \ \oplus\ \cdots .$$

Here the coefficients $a, b, \ldots$ are real numbers and the exponents $i_1, j_1, \ldots$ are integers. Every tropical polynomial represents a function $\mathbb{R}^n \to \mathbb{R}$. When evaluating this function in classical arithmetic, what we get is the minimum of a finite collection of linear functions, namely

$$p(x_1,\ldots,x_n) = \min\bigl(a + i_1x_1 + \cdots + i_nx_n\,,\, b + j_1x_1 + \cdots + j_nx_n\,,\, \ldots\bigr).$$

This function $p : \mathbb{R}^n \to \mathbb{R}$ has the following three important properties:

- p is continuous,
- p is piecewise linear with a finite number of pieces, and
- p is concave: $p\bigl(\frac{1}{2}(\mathbf{x}+\mathbf{y})\bigr) \geq \frac{1}{2}\bigl(p(\mathbf{x}) + p(\mathbf{y})\bigr)$ for all $\mathbf{x}, \mathbf{y} \in \mathbb{R}^n$.

Every function which satisfies these three properties can be represented as the minimum of a finite set of linear functions; see Exercise 1.9(4). We conclude:

Lemma 1.1.2. *The tropical polynomials in n variables $x_1, \ldots, x_n$ are precisely the piecewise-linear concave functions on $\mathbb{R}^n$ with integer coefficients.*

It is instructive to examine tropical polynomials and the functions they define even for polynomials in one variable. For instance, consider the general cubic polynomial in one variable x:

$$p(x) \quad = \quad a \odot x^3 \ \oplus\ b \odot x^2 \ \oplus\ c \odot x \ \oplus\ d. \tag{1.1.1}$$

To graph this function we draw four lines in the (x,y) plane: $y = 3x + a$, $y = 2x + b$, $y = x + c$, and the horizontal line $y = d$. The value of $p(x)$ is the smallest y-value such that (x,y) is on one of these four lines; the graph of $p(x)$ is the lower envelope of the lines. All four lines actually contribute if

$$b - a \ \leq\ c - b\ \leq d - c. \tag{1.1.2}$$

These three values of x are the breakpoints where $p(x)$ fails to be linear, and the cubic has a corresponding factorization into three linear factors:

$$p(x) \quad = \quad a \odot (x \oplus (b-a)) \odot (x \oplus (c-b)) \odot (x \oplus (d-c)).$$

The three breakpoints (1.1.2) of the graph are called the *roots* of the cubic polynomial $p(x)$. The graph and its breakpoints are shown in Figure 1.1.1.

Every tropical polynomial function can be written uniquely as a tropical product of tropical linear functions; the *Fundamental Theorem of Algebra* holds tropically (see Exercise 1.9(2)). In this statement we must underline the word "function". Distinct polynomials can represent the same function $p\colon \mathbb{R} \to \mathbb{R}$. We are not claiming that every polynomial factors into linear functions. What we are claiming is that every polynomial can be replaced

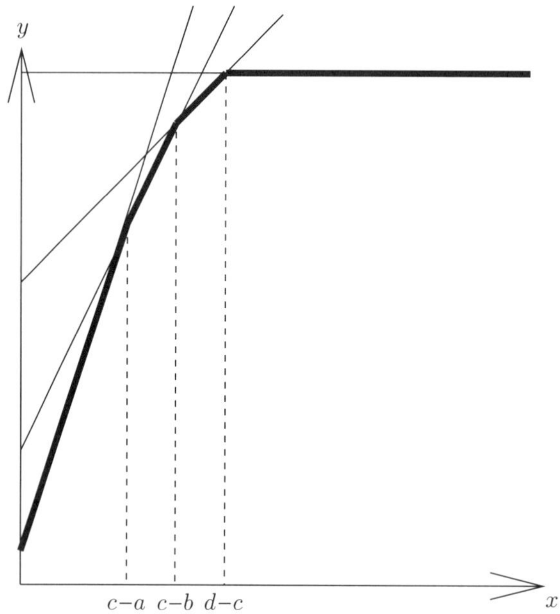

Figure 1.1.1. The graph of a cubic polynomial and its roots.

by an equivalent polynomial, representing the same function, that can be factored into linear factors. Here is an example of a quadratic polynomial function and its unique factorization into linear polynomial functions:

$$x^2 \oplus 17 \odot x \oplus 2 \quad = \quad x^2 \oplus 1 \odot x \oplus 2 \quad = \quad (x \oplus 1)^2.$$

Unique factorization of tropical polynomials holds in one variable, but it no longer holds in two or more variables. What follows is a simple example of a bivariate polynomial that has two distinct irreducible factorizations:

$$(1.1.3) \quad (x \oplus 0) \odot (y \oplus 0) \odot (x \odot y \oplus 0)$$
$$= (x \odot y \oplus x \oplus 0) \odot (x \odot y \oplus y \oplus 0).$$

Here is a geometric way of interpreting this identity.

Definition 1.1.3. Let $f(x, y)$ be a polynomial in two variables, in either classical or tropical arithmetic. Its *Newton polygon* $\mathrm{Newt}(f)$ is defined as the convex hull in $\mathbb{R}^2$ of all points (i, j) such that $x^i y^j$ appears in the expansion of $f(x, y)$. For more information see Definition 2.3.4 and Figure 2.3.5.

The Newton polygon of the polynomial in (1.1.3) is a hexagon. The identity means that the hexagon is the Minkowski sum of three line segments and also the Minkowski sum of two triangles. We refer to (2.3.1) and (2.3.3) for precise definitions of the relevant concepts in arbitrary dimensions.

1.2. Dynamic Programming

To see why tropical arithmetic might be relevant for computer science, let us consider the problem of finding shortest paths in a weighted directed graph. We fix a directed graph G with n nodes that are labeled by $1, 2, \ldots, n$. Every directed edge (i, j) in G has an associated length d_{ij} which is a nonnegative real number. If (i, j) is not an edge of G, then we set $d_{ij} = +\infty$.

We represent the weighted directed graph G by its $n \times n$-adjacency matrix $D_G = \big(d_{ij}\big)$ whose off-diagonal entries are the edge lengths d_{ij}. The diagonal entries of D_G are zero: $d_{ii} = 0$ for all i. The matrix D_G need not be symmetric; it may well happen that $d_{ij} \neq d_{ji}$ for some i, j. However, if G is an undirected graph with edge lengths, then we can represent G as a directed graph with two directed edges (i, j) and (j, i) for each undirected edge $\{i, j\}$. In that special case, D_G is a symmetric matrix, and we can think of $d_{ij} = d_{ji}$ as the distance between node i and node j. For a general directed graph G, the adjacency matrix D_G will not be symmetric.

Consider the $n \times n$-matrix with entries in $\mathbb{R}_{\geq 0} \cup \{\infty\}$ that results from tropically multiplying the given adjacency matrix D_G with itself $n-1$ times:

$$D_G^{\odot(n-1)} \quad = \quad D_G \odot D_G \odot \cdots \odot D_G. \tag{1.2.1}$$

Proposition 1.2.1. *Let G be a weighted directed graph on n nodes with $n \times n$-adjacency matrix D_G. The entry of the matrix $D_G^{\odot(n-1)}$ in row i and column j equals the length of a shortest path from node i to node j in G.*

Proof. Let $d_{ij}^{(r)}$ denote the minimum length of any path from node i to node j which uses at most r edges in G. We have $d_{ij}^{(1)} = d_{ij}$ for any two nodes i and j. Since the edge weights d_{ij} were assumed to be nonnegative, a shortest path from node i to node j visits each node of G at most once. In particular, any shortest path in the directed graph G uses at most $n-1$ directed edges. Hence the length of a shortest path from i to j equals $d_{ij}^{(n-1)}$.

For $r \geq 2$ we have a recursive formula for the length of a shortest path:

$$d_{ij}^{(r)} \quad = \quad \min\big\{d_{ik}^{(r-1)} + d_{kj} \,:\, k = 1, 2, \ldots, n\big\}. \tag{1.2.2}$$

Using tropical arithmetic, this formula can be rewritten as follows:

$$\begin{aligned} d_{ij}^{(r)} \quad &= \quad d_{i1}^{(r-1)} \odot d_{1j} \,\oplus\, d_{i2}^{(r-1)} \odot d_{2j} \,\oplus\, \cdots \,\oplus\, d_{in}^{(r-1)} \odot d_{nj} \\ &= \quad (d_{i1}^{(r-1)}, d_{i2}^{(r-1)}, \ldots, d_{in}^{(r-1)}) \odot (d_{1j}, d_{2j}, \ldots, d_{nj})^T. \end{aligned}$$

From this it follows, by induction on r, that $d_{ij}^{(r)}$ coincides with the entry in row i and column j of the $n \times n$-matrix $D_G^{\odot r}$. Indeed, the right-hand side of the recursive formula is the tropical product of row i of $D_G^{\odot(r-1)}$ and column

j of D_G, which is the (i,j) entry of $D_G^{\odot r}$. In particular, $d_{ij}^{(n-1)}$ coincides with the entry in row i and column j of $D_G^{\odot(n-1)}$. This proves the claim. □

The iterative evaluation of the formula (1.2.2) is *Floyd's algorithm* for finding shortest paths in a weighted digraph. This algorithm and its running time are featured in undergraduate textbooks on discrete mathematics. For us, running that algorithm means performing the matrix multiplication

$$D_G^{\odot r} \quad = \quad D_G^{\odot(r-1)} \odot D_G \qquad \text{for } r = 2, \ldots, n-1.$$

Example 1.2.2. Let G be the weighted directed graph on $n = 4$ nodes, with no loops, that is defined by the adjacency matrix

$$D_G \quad = \quad \begin{pmatrix} 0 & 1 & 3 & 7 \\ 2 & 0 & 1 & 3 \\ 4 & 5 & 0 & 1 \\ 6 & 3 & 1 & 0 \end{pmatrix}.$$

The first and second tropical power of this matrix are

$$D_G^{\odot 2} \quad = \quad \begin{pmatrix} 0 & 1 & 2 & 4 \\ 2 & 0 & 1 & 2 \\ 4 & 4 & 0 & 1 \\ 5 & 3 & 1 & 0 \end{pmatrix} \qquad \text{and} \qquad D_G^{\odot 3} \quad = \quad \begin{pmatrix} 0 & 1 & 2 & 3 \\ 2 & 0 & 1 & 2 \\ 4 & 4 & 0 & 1 \\ 5 & 3 & 1 & 0 \end{pmatrix}.$$

The entries in $D_G^{\odot 3}$ are the lengths of the shortest paths in the digraph G.

The tropical computation mirrors the following matrix computation in ordinary arithmetic. Let ϵ denote an indeterminate that represents a very small positive real number, and let $A_G(\epsilon)$ be the $n \times n$-matrix whose entry in row i and column j is the monomial $\epsilon^{d_{ij}}$. In our example we have

$$A_G(\epsilon) \quad = \quad \begin{pmatrix} 1 & \epsilon^1 & \epsilon^3 & \epsilon^7 \\ \epsilon^2 & 1 & \epsilon^1 & \epsilon^3 \\ \epsilon^4 & \epsilon^5 & 1 & \epsilon^1 \\ \epsilon^6 & \epsilon^3 & \epsilon^1 & 1 \end{pmatrix}.$$

Now, we compute the third power of this matrix in ordinary arithmetic:

$$A_G(\epsilon)^3 = \begin{pmatrix} 1 + 3\epsilon^3 + \cdots & 3\epsilon + \epsilon^4 + \cdots & 3\epsilon^2 + 3\epsilon^3 + \cdots & \epsilon^3 + 6\epsilon^4 + \cdots \\ 3\epsilon^2 + 4\epsilon^5 + \cdots & 1 + 3\epsilon^3 + \cdots & 3\epsilon + \epsilon^3 + \cdots & 3\epsilon^2 + 3\epsilon^3 + \cdots \\ 3\epsilon^4 + 2\epsilon^6 + \cdots & 3\epsilon^4 + 6\epsilon^5 + \cdots & 1 + 3\epsilon^2 + \cdots & 3\epsilon + \epsilon^3 + \cdots \\ 6\epsilon^5 + 3\epsilon^6 + \cdots & 3\epsilon^3 + \epsilon^5 + \cdots & 3\epsilon + \epsilon^3 + \cdots & 1 + 3\epsilon^2 + \cdots \end{pmatrix}.$$

The entry of the classical matrix power $A_G(\epsilon)^3$ in row i and column j is a polynomial in ϵ which represents the lengths of all paths from node i to node j using at most three edges. The lowest exponent appearing in this polynomial is the (i,j)-entry in the tropical matrix power $D_G^{\odot 3}$. ◇

This is a general phenomenon, summarized informally as follows:

$$\text{(1.2.3)} \qquad \text{tropical} \quad = \quad \lim_{\epsilon \to 0} \log_\epsilon\big(\text{classical}(\epsilon)\big).$$

This process of passing from classical arithmetic to tropical arithmetic is referred to as *tropicalization*. Equation (1.2.3) is not a mathematical statement. To make this rigorous we use the algebraic notion of *valuations* which will be developed in our introductory discussion of fields in Section 2.1.

We shall give two more examples of how tropical arithmetic ties in naturally with algorithms in discrete mathematics. The first example concerns the dynamic programming approach to *integer linear programming*. The integer linear programming problem can be stated as follows. Let $A = (a_{ij})$ be a $d \times n$-matrix of nonnegative integers, let $\mathbf{w} = (w_1, \ldots, w_n)$ be a row vector with real entries, and let $\mathbf{b} = (b_1, \ldots, b_d)$ be a column vector with nonnegative integer entries. Our task is to find a nonnegative integer column vector $\mathbf{u} = (u_1, \ldots, u_n)$ which solves the following optimization problem:

$$\text{(1.2.4)} \qquad \text{Minimize } \mathbf{w} \cdot \mathbf{u} \text{ subject to } \mathbf{u} \in \mathbb{N}^n \text{ and } A\mathbf{u} = \mathbf{b}.$$

Let us assume that all columns of the matrix A sum to the same number α and that $b_1 + \cdots + b_d = m\alpha$. This assumption is convenient because it ensures that all feasible solutions $\mathbf{u} \in \mathbb{N}^n$ of (1.2.4) satisfy $u_1 + \cdots + u_n = m$.

We can solve the integer programming problem (1.2.4) using tropical arithmetic as follows. Let $x_1, \ldots, x_d$ be variables and consider the expression

$$\text{(1.2.5)} \quad w_1 \odot x_1^{a_{11}} \odot x_2^{a_{21}} \odot \cdots \odot x_d^{a_{d1}} \;\oplus\; \cdots \;\oplus\; w_n \odot x_1^{a_{1n}} \odot x_2^{a_{2n}} \odot \cdots \odot x_d^{a_{dn}}.$$

Proposition 1.2.3. *The optimal value of* (1.2.4) *is the coefficient of the monomial* $x_1^{b_1} x_2^{b_2} \cdots x_d^{b_d}$ *in the* m*th power of the tropical polynomial* (1.2.5).

The proof of this proposition is not difficult and is similar to that of Proposition 1.2.1. The process of taking the mth power of the tropical polynomial (1.2.5) can be regarded as solving the shortest path problem in a certain graph. This is the dynamic programming approach to (1.2.4). This approach furnishes a polynomial-time algorithm for integer programming in fixed dimension under the assumption that the integers in A are bounded.

Example 1.2.4. Let $d = 2$, let $n = 5$, and consider the instance of (1.2.4) with

$$A = \begin{pmatrix} 4 & 3 & 2 & 1 & 0 \\ 0 & 1 & 2 & 3 & 4 \end{pmatrix}, \quad \mathbf{b} = \begin{pmatrix} 5 \\ 7 \end{pmatrix}, \quad \text{and} \quad \mathbf{w} = (2, 5, 11, 7, 3).$$

Here we have $\alpha = 4$ and $m = 3$. The matrix A and the cost vector $\mathbf{w}$ are encoded by a tropical polynomial as in (1.2.5):

$$p \quad = \quad 2x_1^4 \oplus 5x_1^3 x_2 \oplus 11 x_1^2 x_2^2 \oplus 7 x_1 x_2^3 \oplus 3 x_2^4.$$

The third power of this polynomial, evaluated tropically, is equal to

$$p \odot p \odot p \;=\; 6x_1^{12} \oplus 9x_1^{11}x_2 \oplus 12x_1^{10}x_2^2 \oplus 11x_1^9x_2^3 \oplus 7x_1^8x_2^4 \oplus 10x_1^7x_2^5 \oplus 13x_1^6x_2^6$$
$$\oplus\, 12x_1^5x_2^7 \oplus 8x_1^4x_2^8 \oplus 11x_1^3x_2^9 \oplus 17x_1^2x_2^{10} \oplus 13x_1x_2^{11} \oplus 9x_2^{12}.$$

The coefficient 12 of $x_1^5x_2^7$ in $p \odot p \odot p$ is the optimal value. An optimal solution to this integer programming problem is $\mathbf{u} = (1,0,0,1,1)^{\mathrm{T}}$. ◊

Our final example concerns the notion of the determinant of an $n \times n$-matrix $X = (x_{ij})$. As there is no negation in tropical arithmetic, the *tropical determinant* is the same as the *tropical permanent*, namely, the sum over the diagonal products obtained by taking all $n!$ permutations π of $\{1,2,\ldots,n\}$:

$$\text{(1.2.6)} \qquad \operatorname{trop}\det(X) \quad := \quad \bigoplus_{\pi \in S_n} x_{1\pi(1)} \odot x_{2\pi(2)} \odot \cdots \odot x_{n\pi(n)}.$$

Here S_n is the *symmetric group* of permutations of $\{1,2,\ldots,n\}$. Evaluating the tropical determinant means solving the classical *assignment problem* of combinatorial optimization. Imagine a company that has n jobs and n workers, and each job needs to be assigned to exactly one of the workers. Let x_{ij} be the cost of assigning job i to worker j. The company wishes to find the cheapest assignment $\pi \in S_n$. The optimal total cost is the minimum:

$$\min\bigl\{x_{1\pi(1)} + x_{2\pi(2)} + \cdots + x_{n\pi(n)} \;:\; \pi \in S_n\bigr\}.$$

This number is precisely the tropical determinant of the matrix $Q = (x_{ij})$:

Remark 1.2.5. The tropical determinant solves the assignment problem.

In the assignment problem we seek the minimum over $n!$ quantities. This appears to require exponentially many operations. However, there is a well-known polynomial-time algorithm for solving this problem. It was developed by Harold Kuhn in 1955 who called it the *Hungarian Assignment Method* [**Kuh55**]. This algorithm maintains a price for each job and an (incomplete) assignment of workers and jobs. At each iteration, an unassigned worker is chosen and a shortest augmenting path from this person to the set of jobs is chosen. The total number of arithmetic operations is $O(n^3)$.

In classical arithmetic, the evaluation of determinants and the evaluation of permanents are in different complexity classes. The determinant of an $n \times n$-matrix can be computed in $O(n^3)$ steps, namely by *Gaussian elimination*, while computing the permanent of an $n \times n$-matrix is a fundamentally harder problem. A famous theorem due to Leslie Valiant says that computing the (classical) permanent is *#P-complete*. In tropical arithmetic, computing the permanent is easier, thanks to the Hungarian Assignment Method. We can think of that method as a certain tropicalization of Gaussian elimination.

For an example, consider a 3×3-matrix $A(\epsilon)$ whose entries are polynomials in the unknown ϵ. For each entry we list the term of lowest order:

$$A(\epsilon) \quad = \quad \begin{pmatrix} a_{11}\epsilon^{x_{11}} + \cdots & a_{12}\epsilon^{x_{12}} + \cdots & a_{13}\epsilon^{x_{13}} + \cdots \\ a_{21}\epsilon^{x_{21}} + \cdots & a_{22}\epsilon^{x_{22}} + \cdots & a_{23}\epsilon^{x_{23}} + \cdots \\ a_{31}\epsilon^{x_{31}} + \cdots & a_{32}\epsilon^{x_{32}} + \cdots & a_{33}\epsilon^{x_{33}} + \cdots \end{pmatrix}.$$

Suppose that the a_{ij} are sufficiently general integers, so that no cancellation occurs in the lowest-order coefficient when we expand the determinant of $A(\epsilon)$. Writing X for the 3×3-matrix with entries x_{ij}, we have

$$\det(A(\epsilon)) \quad = \quad \alpha \cdot \epsilon^{\operatorname{trop}\det(X)} + \cdots \qquad \text{for some } \alpha \in \mathbb{R}\backslash\{0\}.$$

Thus the tropical determinant of X can be computed from this expression by taking the logarithm and letting ϵ tend to zero, as suggested by (1.2.3).

The material in this section is closely related to Chapter 2 in the book *Algebraic Statistics for Computational Biology* by Lior Pachter and Bernd Sturmfels [**PS05**]. The connection to computational biology arises because many algorithms in that field (e.g., for sequence alignment and gene prediction) are based on dynamic programming. These algorithms can be interpreted as the evaluation of a tropical polynomial. The book [**PS05**] and the paper [**PS04**] that preceded it argue that the tropical interpretation of dynamic programming algorithms is useful for statistical inference.

Readers who enjoyed this section might like to take a peek at Section 5.1. That section concerns the eigenvalue and eigenvectors of a square matrix.

1.3. Plane Curves

A tropical polynomial function $p : \mathbb{R}^n \to \mathbb{R}$ is given as the minimum of a finite set of linear functions. We define the *hypersurface* $V(p)$ of p to be the set of all points $\mathbf{w} \in \mathbb{R}^n$ at which this minimum is attained at least twice. Equivalently, a point $\mathbf{w} \in \mathbb{R}^n$ lies in $V(p)$ if and only if p is not linear at $\mathbf{w}$.

For instance, let $n = 1$ and fix the univariate tropical polynomial

$$p(x) \quad = \quad a \odot x^3 \,\oplus\, b \odot x^2 \,\oplus\, c \odot x \,\oplus\, d$$

seen in (1.1.1). If the assumption $b - a \leq c - b \leq d - c$ of (1.1.2) holds, then

$$V(p) \quad = \quad \{\, b - a,\, c - b,\, d - c \,\}.$$

Thus the hypersurface $V(p)$ is the set of "roots" of the polynomial $p(x)$.

For an example of a tropical polynomial in many variables consider the determinant function $p = \operatorname{trop}\det$ from (1.2.6). Its hypersurface $V(p)$ consists of all $n \times n$-matrices that are *tropically singular*. A square matrix being tropically singular means that the optimal solution to the assignment problem discussed in the previous section is not unique, so among all $n!$ ways of assigning n workers to n jobs, there are at least two assignments both of which minimize the total cost. For further information see Example 3.1.11.

In this section we study the geometry of a polynomial in two variables:

$$p(x, y) \quad = \quad \bigoplus_{(i,j)} c_{ij} \odot x^i \odot y^j.$$

The corresponding tropical hypersurface $V(p)$ is a *plane tropical curve*. The following proposition summarizes the salient features of such a curve.

Proposition 1.3.1. *The curve $V(p)$ is a finite graph that is embedded in the plane $\mathbb{R}^2$. It has both bounded and unbounded edges, all edge slopes are rational, and this graph satisfies a* balancing condition *around each node.*

This result is a consequence of the Structure Theorem for tropical varieties, which is our Theorem 3.3.5. Balancing refers to the following geometric condition: Consider any node (x, y) of the graph. The edges adjacent to this node lie on lines with rational slopes. Translate (x, y) to the origin $(0, 0)$. In the direction of each edge, now consider the first nonzero lattice vector on that line. *Balancing* at (x, y) means that the sum of these vectors is zero.

In general, it will be necessary to assign a positive integer *multiplicity* to each edge of $V(p)$, in order for this balancing condition to hold. These multiplicities will make informal appearances throughout this chapter. The precise definition will be given, for arbitrary dimensions, in Section 3.4.

Our first example is a *line* in the plane. It is defined by a polynomial

$$p(x, y) \quad = \quad a \odot x \;\oplus\; b \odot y \;\oplus\; c, \qquad \text{where } a, b, c \in \mathbb{R}. \tag{1.3.1}$$

The tropical curve $V(p)$ consists of all points (x, y) where the function

$$p \,:\, \mathbb{R}^2 \to \mathbb{R}\,, \qquad (x, y) \;\mapsto\; \min\big(a + x,\, b + y,\, c\big)$$

is not linear. It consists of three half-rays emanating from the point $(x, y) = (c - a, c - b)$ into the northern, eastern, and southwestern directions.

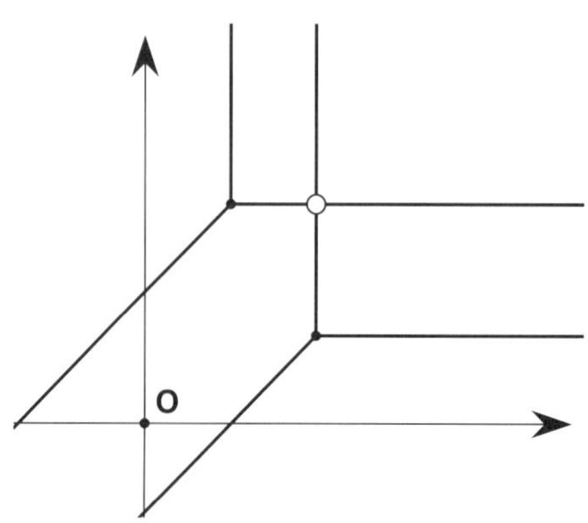

Figure 1.3.1. Two lines in the tropical plane meet in one point.

Two lines in the tropical plane will always meet in one point. This is shown in Figure 1.3.1. When the lines are in special position, it can happen that the set-theoretic intersection is a half-ray. In that case the notion of stable intersection discussed below is used to get a unique intersection point.

Let p be any tropical polynomial in x and y, and consider any term $\gamma \odot x^i \odot y^j$ appearing in p. In classical arithmetic this represents the linear function $(x, y) \mapsto \gamma + ix + jy$. The tropical polynomial function $p : \mathbb{R}^2 \to \mathbb{R}$ is given by the minimum of these linear functions. The graph of p is concave and piecewise linear. It looks like a tent over the plane $\mathbb{R}^2$. The tropical curve $V(p)$ is the set of all points in $\mathbb{R}^2$ at which the function is not linear.

As an example we consider the general quadratic polynomial

$$p(x, y) \quad = \quad a \odot x^2 \;\oplus\; b \odot xy \;\oplus\; c \odot y^2 \;\oplus\; d \odot y \;\oplus\; e \;\oplus\; f \odot x.$$

Suppose that the coefficients $a, b, c, d, e, f \in \mathbb{R}$ satisfy the inequalities

$$b + f < a + d, \qquad d + f < b + e, \qquad b + d < c + f.$$

Then the graph of $p : \mathbb{R}^2 \to \mathbb{R}$ is the lower envelope of six planes in $\mathbb{R}^3$. This is shown in Figure 1.3.2, where each linear piece of the graph is labeled by the corresponding linear function. Below this "tent" lies the tropical quadratic curve $V(p) \subset \mathbb{R}^2$. This curve has four vertices, three bounded edges, and six half-rays (two northern, two eastern, and two southwestern).

If $p(x, y)$ is a tropical polynomial, then its curve $V(p)$ can be constructed from its Newton polygon $\text{Newt}(p)$, which we recall from Definition 1.1.3. Namely, the planar graph dual to $V(p)$ is a subdivision of $\text{Newt}(p)$ into smaller polygons. This subdivision is determined by the coefficients of p. Typically, these smaller polygons are triangles, in which case the subdivision is a *triangulation*. The triangulation is *unimodular* if

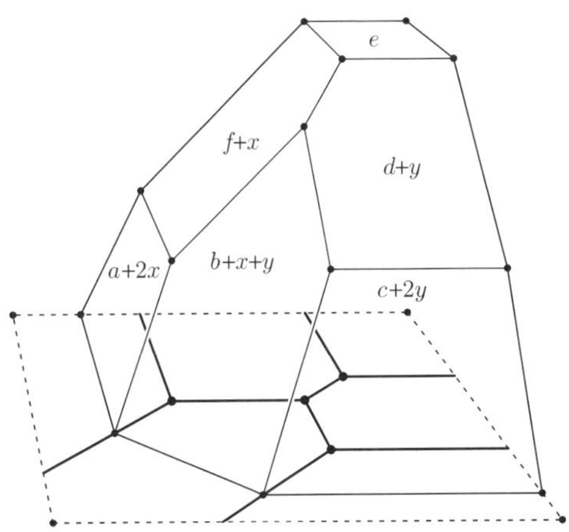

Figure 1.3.2. The graph and the curve defined by a quadratic polynomial.

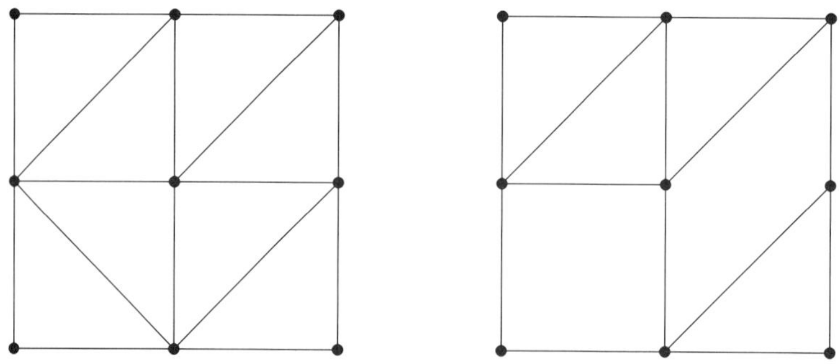

Figure 1.3.3. Two subdivisions of the Newton polygon of a biquadratic curve. Their planar duals are the curves in Figure 1.3.4.

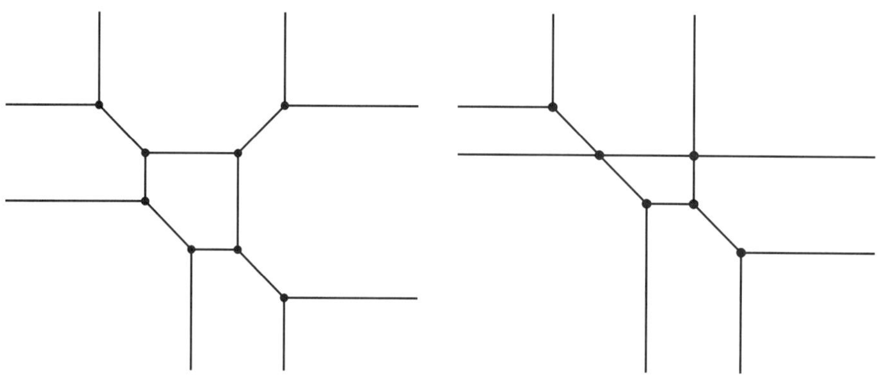

Figure 1.3.4. Two tropical biquadratic curves. The curve on the left is smooth.

each cell is a lattice triangle of unit area 1/2. In this case we call $V(p)$ a *smooth tropical curve*. The adjective "smooth" will be justified in Proposition 4.5.1. For our subdivisions and triangulations in arbitrary dimensions, see Definition 2.3.8.

The unbounded rays of a tropical curve $V(p)$ are perpendicular to the edges of the Newton polygon. For example, if p is a biquadratic polynomial, then $\mathrm{Newt}(p)$ is the square with vertices $(0,0)$, $(0,2)$, $(2,0)$, $(2,2)$. Here, $V(p)$ has two unbounded rays for each of the four edges of the square. Figure 1.3.3 shows two subdivisions. The corresponding tropical curves are shown in Figure 1.3.4. The curve on the left is smooth. It has genus one. The unique cycle corresponds to the interior lattice point of $\mathrm{Newt}(p)$. This is an example of a *tropical elliptic curve*. The curve on the right is not smooth.

If we draw tropical curves in the plane, then we discover that they intersect and interpolate just as algebraic curves do. In particular, we observe

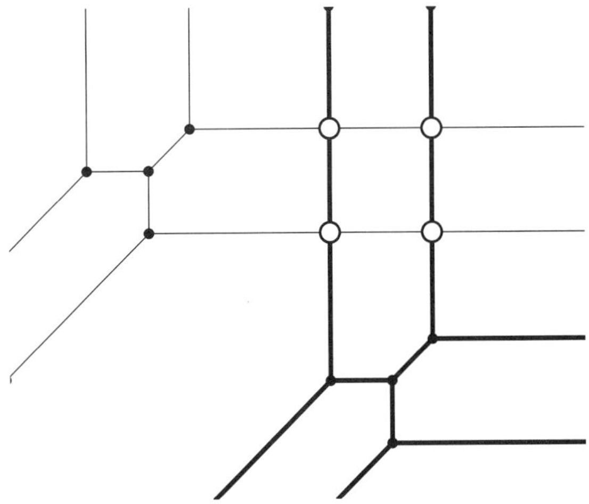

Figure 1.3.5. Bézout's Theorem: Two quadratic curves meet in four points.

the following:

- Two general lines meet in one point (Figure 1.3.1).
- Two general points lie on a unique line.
- A general line and quadric meet in two points (Figure 1.3.6).
- Two general quadrics meet in four points (Figures 1.3.5 and 1.3.7).
- Five general points lie on a unique quadric.

A classical result from algebraic geometry, known as *Bézout's Theorem*, holds in tropical algebraic geometry as well. In order to state this theorem, we need the multiplicities that were mentioned after Proposition 1.3.1. In addition to that, we assign a positive integer to any two lines with distinct rational slopes in $\mathbb{R}^2$. If their primitive direction vectors are $(u_1, u_2) \in \mathbb{Z}^2$ and $(v_1, v_2) \in \mathbb{Z}^2$, respectively, then the intersection multiplicity of the two lines at their unique common point is $|u_1v_2 - u_2v_1|$. We multiply that number with the product of the multiplicities of the two edges determining the lines.

We now focus on tropical curves whose Newton polygons are the standard triangles with vertices $(0,0)$, $(0,d)$, and $(d,0)$. We refer to such a curve as a *curve of degree d*. A curve of degree d has d rays, possibly counting multiplicities, perpendicular to each of the three edges of its Newton triangle. Suppose that C and D are two tropical curves in $\mathbb{R}^2$ that intersect transversally, that is, every common point lies in the relative interior of a unique edge in C and also in D. The multiplicity of that point is the product of the multiplicities of the edges times the intersection multiplicity $|u_1v_2 - u_2v_1|$.

Theorem 1.3.2 (Bézout). *Consider two tropical curves C and D of degree c and d in $\mathbb{R}^2$. If the two curves intersect transversally, then the number of intersection points, counted with multiplicities as above, is equal to cd.*

Just as in classical algebraic geometry, it is possible to remove the restriction "intersect transversally" from the statement of Bézout's Theorem. In fact, the situation is even better here because of the following important phenomenon, which is false in classical geometry. The intersection points depend continuously on the coefficients of the two tropical polynomials. These continuous functions are well defined on the entire space of coefficients, even at locations when the two polynomials are very special.

We explain this for the intersection of two curves C and D of degrees c and d in $\mathbb{R}^2$. Suppose the intersection of C and D is not transverse or not even finite. Pick *any* nearby curves C_ϵ and D_ϵ such that C_ϵ and D_ϵ intersect transversely in finitely many points. Then, according to the refined count of Theorem 1.3.2, the intersection $C_\epsilon \cap D_\epsilon$ is a multiset of cardinality cd.

Theorem 1.3.3 (Stable Intersection Principle). *The limit of the point configuration $C_\epsilon \cap D_\epsilon$ is independent of the choice of perturbations. It is a well-defined multiset of cd points contained in the intersection $C \cap D$.*

Here the limit is taken as ϵ tends to 0. Multiplicities add up when points collide. The limit is a finite configuration of points in $\mathbb{R}^2$ with multiplicities, where the sum of the multiplicities is cd. We call this limit the *stable intersection* of the curves C and D. This is a multiset of points, denoted by

$$C \cap_{\mathrm{st}} D \quad = \quad \lim_{\epsilon \to 0} (C_\epsilon \cap D_\epsilon).$$

Hence we can strengthen the statement of Bézout's Theorem as follows.

Corollary 1.3.4. *Any two curves of degrees c and d in $\mathbb{R}^2$, no matter how special they might be, intersect stably in a well-defined multiset of cd points.*

The Stable Intersection Principle is illustrated in Figures 1.3.6 and 1.3.7. In Figure 1.3.6 we see the intersection of a tropical line with a tropical quadric, moving from general position to special position. In the diagram on the right, the set-theoretic intersection of the two curves is infinite, but the stable intersection is well defined. It consists of two points A and B.

Figure 1.3.7 shows an even more dramatic situation. In that picture, a quadric is intersected stably with itself. For any small perturbation of the coefficients of the two tropical polynomials, we obtain four intersection points near the four nodes of the original quadric. This shows that the stable intersection of a quadric with itself consists precisely of the four nodes.

We refer to Section 3.6 for a thorough treatment of stable intersections.

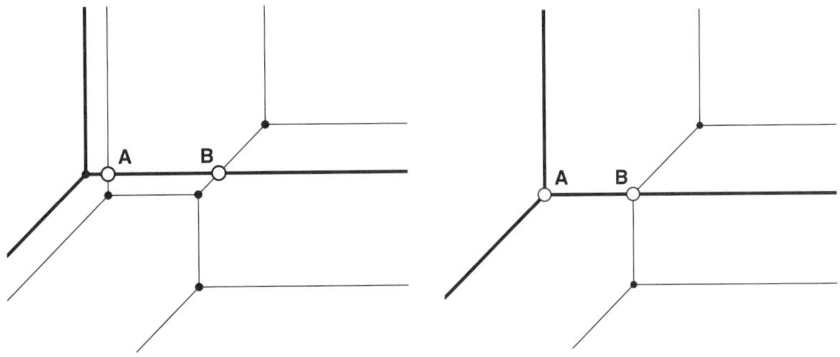

Figure 1.3.6. The stable intersection of a line and a quadric.

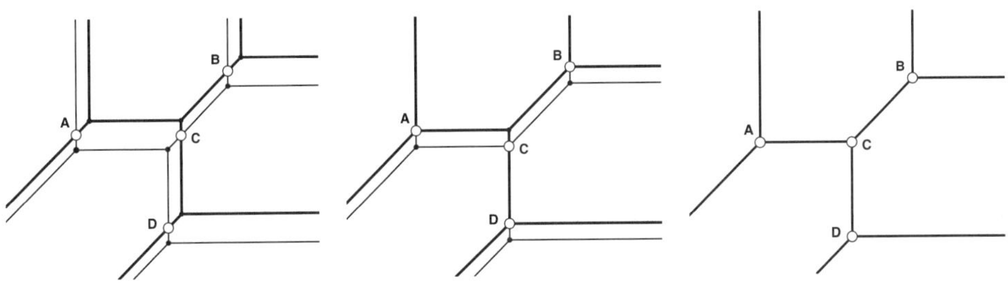

Figure 1.3.7. The stable intersection of a quadric with itself.

1.4. Amoebas and their Tentacles

One early source in tropical algebraic geometry is a 1971 paper on *the logarithmic limit-set of an algebraic variety* by George Bergman [**Ber71**]. With hindsight, the structure introduced by Bergman is the same as the tropical variety arising from a subvariety in a complex algebraic torus $(\mathbb{C}^*)^n$. Here $\mathbb{C}^* = \mathbb{C}\backslash\{0\}$ denotes the multiplicative group of nonzero complex numbers.

The *amoeba* of such a variety is its image under taking the coordinate-wise logarithm of the absolute value of any point on the variety. The term "amoeba" was coined by Gel'fand, Kapranov, and Zelevinsky in their monograph *Discriminants, Resultants, and Multidimensional Determinants* [**GKZ08**]. Bergman's logarithmic limit set arises from the amoeba as the set of all tentacle directions. In this section we discuss these and related topics.

Let I be an ideal in the Laurent polynomial ring $S = \mathbb{C}[x_1^{\pm 1}, x_2^{\pm 1}, \ldots, x_n^{\pm 1}]$. Its algebraic variety is the common zero set of all Laurent polynomials in I:

$$V(I) \;=\; \big\{ \mathbf{z} \in (\mathbb{C}^*)^n \;:\; f(\mathbf{z}) = 0 \text{ for all } f \in I \big\}.$$

Note that this is well defined because $0 \notin \mathbb{C}^*$. The *amoeba* of the ideal I is the subset of $\mathbb{R}^n$ defined as image of the coordinate-wise logarithm map:

$$\mathcal{A}(I) = \big\{ \big(\log(|z_1|), \log(|z_2|), \ldots, \log(|z_n|)\big) \in \mathbb{R}^n \,:\, \mathbf{z} = (z_1, \ldots, z_n) \in V(I) \big\}.$$

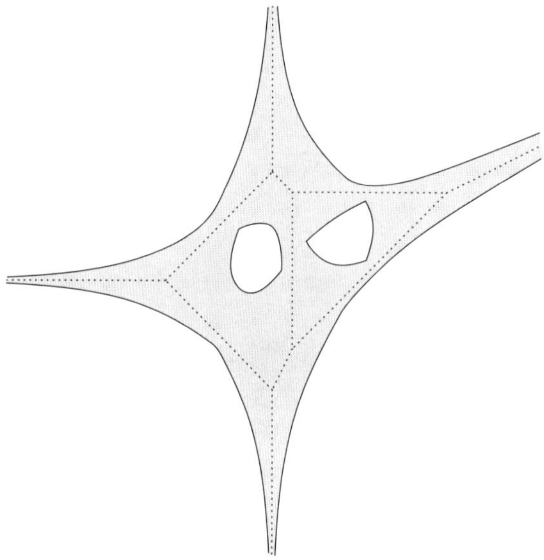

Figure 1.4.1. The amoeba of a plane curve and its spine.

If $n = 1$ and I is a proper ideal in $S = \mathbb{C}[x, x^{-1}]$, then I is a principal ideal. It is generated by a single polynomial $f(x)$ that factors over $\mathbb{C}$:

$$f(x) = (u_1 + iv_1 - x)(u_2 + iv_2 - x) \cdots (u_m + iv_m - x).$$

Here $u_1, v_1, \ldots, u_m, v_m \in \mathbb{R}$ are the real and imaginary parts of the various roots of $f(x)$, and the amoeba is the following set of at most m real numbers:

$$\mathcal{A}(I) = \mathcal{A}(f) = \left\{ \log\left(\sqrt{u_1^2 + v_1^2}\right), \log\left(\sqrt{u_2^2 + v_2^2}\right), \ldots, \log\left(\sqrt{u_m^2 + v_m^2}\right) \right\}.$$

It is instructive to draw some amoebas for $n = 2$. Let $I = \langle f(x_1, x_2) \rangle$ be the ideal of a curve $\{f(x_1, x_2) = 0\}$ in $(\mathbb{C}^*)^2$. The amoeba $\mathcal{A}(f)$ of that curve is a closed subset of $\mathbb{R}^2$ whose boundary is described by analytic functions. It has finitely many tentacles that emanate toward infinity, and the directions of these tentacles are precisely the directions perpendicular to the edges of the Newton polygon $\mathrm{Newt}(f)$. The complement $\mathbb{R}^2 \backslash \mathcal{A}(f)$ of the amoeba is a finite union of open convex subsets of the plane $\mathbb{R}^2$.

We refer to work of Passare and his collaborators [**PR04, PT05**] for foundational results on amoebas of hypersurfaces in $(\mathbb{C}^*)^n$, and to the article by Theobald [**The02**] for methods for computing and drawing amoebas. An interesting Nullstellensatz for amoebas was established by Purbhoo [**Pur08**].

Example 1.4.1. Figure 1.4.1 shows the complex amoeba of the curve

$$f(z, w) = 1 + 5zw + w^2 - z^3 + 3z^2w - z^2w^2.$$

Note the two bounded convex components in the complement of

$$\mathcal{A}(f) \;=\; \big\{\, (\log(|z|), \log(|w|)) \in \mathbb{R}^2 \;:\; z, w \in \mathbb{C}^* \text{ and } f(z,w) = 0 \,\big\}.$$

They correspond to the two interior lattice points of the Newton polygon of f. The tentacles of the amoeba converge to four rays in $\mathbb{R}^2$. Up to sign, the union of these rays is the plane curve $V(p)$ defined by the tropical polynomial

$$p \;=\; \operatorname{trop}(f) \;=\; 0 \oplus u \odot v \oplus v^2 \oplus u^3 \oplus u^2 \odot v \oplus u^2 \odot v^2.$$

This expression is the tropicalization of f, to be defined formally in (2.4.1). All coefficients of p are zero because the coefficients of f are real numbers.

Note that in our definition of amoeba (and in Figure 1.4.1) the max-convention was used. (Mikael Passare always preferred max-convention because of his son's first name: it is Max and not Min). Inside the amoeba of Figure 1.4.1, we see the curve defined by a tropical polynomial of the form

$$q \;=\; c_1 \oplus c_2 \odot u \odot v \oplus c_3 \odot v^2 \oplus c_4 \odot u^3 \oplus c_5 \odot u^2 \odot v \oplus c_6 \odot u^2 \odot v^2.$$

The tropical curve $V(q)$ is a canonical deformation retract of $-\mathcal{A}(f)$. It is known as the *spine* of the amoeba. The coefficients c_i are defined below. ◊

There are three different ways in which tropical varieties arise from amoebas. We associate the name of a mathematician with each of them.

The Passare Construction: Every complex hypersurface amoeba $\mathcal{A}(f)$ has a *spine* which is a canonical tropical hypersurface contained in $\mathcal{A}(f)$. Suppose $f = f(z,w)$ is a polynomial in two variables. Then its *Ronkin function* is

$$N_f(u,v) \;=\; \frac{1}{(2\pi i)^n} \int_{\operatorname{Log}^{-1}(u,v)} \log|f(z,w)| \frac{dz}{z} \wedge \frac{dw}{w}.$$

Passare and Rullgård [**PR04**] showed that this function is convex and that it is linear on each connected component of the complement of $\mathcal{A}(f)$. Let $q(u,v)$ denote the negated maximum of these affine-linear functions, one for each component in the amoeba complement. Then $q(u,v)$ is a tropical polynomial function (a piecewise-linear concave function) which satisfies $N_f(u,v) \geq -q(u,v)$ for all $(u,v) \in \mathbb{R}^2$. Its tropical curve $V(q)$ is the spine.

The Maslov Construction: Tropical varieties arise as limits of amoebas as one changes the base of the logarithm and makes it either very large or very small. This limit process is also known as *Maslov dequantization*, and it can be made precise as follows. Given $h > 0$, we redefine arithmetic as follows:

$$x \oplus_h y \;=\; h \cdot \log\Big(\exp(\frac{x}{h}) \,+\, \exp(\frac{y}{h}) \Big) \qquad \text{and} \qquad x \odot_h y \;=\; x + y.$$

This is what happens to ordinary addition and multiplication of positive real numbers under the coordinate transformation $\mathbb{R}_+ \to \mathbb{R}, x \mapsto h \cdot \log(x)$.

We now consider a polynomial $f_h(z,w)$ whose coefficients are rational functions of the parameter h. For each $h > 0$, we take the amoeba $\mathcal{A}_h(f_h)$ of f_h with respect to scaled logarithm map $(z,w) \mapsto h \cdot \big(\log(|z|), \log(|w|)\big)$. The limit in the Hausdorff topology of the set $-\mathcal{A}_h(f_h)$ as $h \to 0+$ is a tropical hypersurface $V(q)$. For details see [**Mik04**]. The coefficients of the tropical polynomials q are the orders (of poles or zeros) of the coefficients at $h = 0$. This process can be thought of as a sequence of amoebas converging to their spine, but it is different from the construction using Ronkin functions.

The Bergman Construction: Our third connection between amoebas on tropical varieties arises by examining their tentacles. Here we disregard the interior structure of $\mathcal{A}(f)$, such as the bounded convex regions in the complement. We focus only on the asymptotic directions. This makes sense for any subvariety of $(\mathbb{C}^*)^n$, so our input now is an ideal $I \subset S$ as above.

We denote the unit sphere by $\mathbb{S}^{n-1} = \{\mathbf{x} \in \mathbb{R}^n : ||\mathbf{x}|| = 1\}$. For any real number $M > 0$, we consider the following set:

$$\mathcal{A}_M(I) \;=\; -\frac{1}{M}\mathcal{A}(I) \cap \mathbb{S}^{n-1}.$$

The *logarithmic limit set* $\mathcal{A}_\infty(I)$ is the set of points $\mathbf{v}$ on the sphere $\mathbb{S}^{n-1}$ such that there exists a sequence of points $\mathbf{v}_M \in \mathcal{A}_M(I)$ converging to $\mathbf{v}$:

$$\lim_{M\to\infty} \mathbf{v}_M \;=\; \mathbf{v}.$$

We next exhibit the relationship to the *tropical variety* $\mathrm{trop}(V(I))$ of I. Here $\mathrm{trop}(V(I))$ is defined to be the intersection of the tropical hypersurfaces $V(p)$ where $p = \mathrm{trop}(f)$ is the tropicalization of any polynomial $f \in I$.

Theorem 1.4.2. *The tropical variety of I coincides with the cone over the logarithmic limit set $\mathcal{A}_\infty(I)$, i.e., a nonzero vector $\mathbf{w} \in \mathbb{R}^n$ lies in $\mathrm{trop}(V(I))$ if and only if the corresponding unit vector $\frac{1}{||\mathbf{w}||}\mathbf{w}$ lies in $\mathcal{A}_\infty(I)$.*

When the ideal I is principal, this appears in [**MR01**]. For a proof of Theorem 1.4.2 and connections of amoebas to Berkovich spaces, see [**Jon14**].

The tropical variety $\mathrm{trop}(V(I))$ is the principal actor in this book. It will be studied in great detail in Chapter 3. We shall see in Corollary 3.5.5 that $\mathrm{trop}(V(I))$ has the structure of a polyhedral fan, and we shall establish various properties of that fan. Theorem 1.4.2 and the fan property of $\mathrm{trop}(V(I))$ imply that $\mathcal{A}_\infty(I)$ is a spherical polyhedral complex in $\mathbb{S}^{n-1}$.

It is interesting to see the motivation behind the paper [**Ber71**]. Bergman introduced tropical varieties in order to prove a conjecture of Zalessky concerning the multiplicative action of $\mathrm{GL}(n,\mathbb{Z})$ on the Laurent polynomial ring S. Here, an integer matrix $g = (g_{ij})$ acts on S as the ring homomorphism that maps each variable x_i to the Laurent monomial $\prod_{j=1}^n x_j^{g_{ij}}$.

If I is a proper ideal in S, then we consider its stabilizer subgroup:

$$\mathrm{Stab}(I) \;=\; \{\, g \in \mathrm{GL}(n,\mathbb{Z}) \,:\, gI = I \,\}.$$

The following result from [**Ber71**, Theorem 1] answers Zalessky's question:

Corollary 1.4.3. *The stabilizer* $\mathrm{Stab}(I) \subset \mathrm{GL}(n,\mathbb{Z})$ *of a proper ideal* $I \subset S$ *has a subgroup of finite index that stabilizes a nontrivial sublattice of* $\mathbb{Z}^n$.

Proof. The tropical variety of $V(I)$ has the structure of a proper polyhedral fan in $\mathbb{R}^n$. Let $\mathcal{U}$ be the finite set of linear subspaces of $\mathbb{R}^n$ that are spanned by the maximal cones in $V(I)$. While the fan structure is not unique, the set $\mathcal{U}$ of linear subspaces of $\mathbb{R}^n$ is uniquely determined by I. The set $\mathcal{U}$ does not change under refinement or coarsening of the fan structure on $\mathrm{trop}(V(I))$.

The group $\mathrm{Stab}(I)$ acts by linear transformations on $\mathbb{R}^n$, and it leaves the tropical variety of I invariant. This implies that it acts by permutations on the finite set $\mathcal{U}$ of subspaces in $\mathbb{R}^n$. Fix one particular subspace $U \in \mathcal{U}$, and let G be the subgroup of all elements $g \in \mathrm{Stab}(I)$ that fix U. Then G has finite index in $\mathrm{Stab}(I)$ and it stabilizes the sublattice $U \cap \mathbb{Z}^n$ of $\mathbb{Z}^n$. □

A counterpart to the amoeba $\mathcal{A}(I)$ is the *co-amoeba*, which records the phases of the coordinates of all points in a complex variety $V(I)$. An analogue of Bergman's logarithmic limit set for co-amoebas is the *phase limit set* of $V(I)$. See [**NS13**] for recent results and references on these topics.

1.5. Implicitization

An algebraic variety can be represented either as the image of a rational map or as the zero set of some multivariate polynomials. The latter representation exists for all algebraic varieties while the former representation requires that the variety be *unirational*, which is a very special property in algebraic geometry. The transition between two representations is a basic problem in computer algebra. *Implicitization* is the problem of passing from the first representation to the second, that is, given a rational map Φ, one seeks to determine the prime ideal of all polynomials that vanish on the image of Φ.

In this section we examine the simplest instance, namely, we consider the case of a plane curve in $\mathbb{C}^2$ that is given by a rational parameterization:

$$\Phi : \mathbb{C} \to \mathbb{C}^2 ,\;\; t \mapsto (\phi_1(t), \phi_2(t)). \tag{1.5.1}$$

To make the map Φ actually well defined, here we tacitly assume that the poles of ϕ_1 and ϕ_2 have been removed from the domain $\mathbb{C}$. The implicitization problem is to compute the unique (up to scaling) irreducible polynomial $f(x,y)$ vanishing on the curve $\mathrm{Image}(\Phi) = \{(\phi_1(t), \phi_2(t)) \in \mathbb{C}^2 : t \in \mathbb{C}\}$.

Example 1.5.1. Consider the plane curve defined parametrically by

$$\Phi(t) \;=\; \left(\frac{t^3+4t^2+4t}{t^2-1},\; \frac{t^3-t^2-t+1}{t^2}\right).$$

The implicit equation of this curve equals

$$f(x,y) \;=\; x^3y^2 - x^2y^3 - 5x^2y^2 - 2x^2y - 4xy^2 - 33xy + 16y^2 + 72y + 81.$$

This irreducible polynomial vanishes on all points $(x,y) = \Phi(t)$ for $t \in \mathbb{C}$. ◊

Two standard methods used in computer algebra for solving implicitization problems are Gröbner bases and resultants. These methods are explained in the textbook by Cox, Little, and O'Shea [**CLO07**]. Specifically, the desired polynomial $f(x,y)$ equals the Sylvester resultant of the numerator of $x - \phi_1(t)$ and the numerator of $y - \phi_2(t)$ with respect to the variable t. For instance, the implicit equation in Example 1.5.1 is easily found by

$$f(x,y) \;=\; \mathrm{resultant}_t\big(\, t^3+4t^2+4t-(t^2-1)x\,,\; t^3-t^2-t+1-t^2y\big).$$

For larger problems in higher dimensions, Gröbner bases and resultants often do not perform well enough or do not give enough geometric insight. This is where the approach to implicitization using tropical geometry comes in. We shall explain the basic idea behind this approach for rational plane curves.

Suppose we are given the parameterization Φ, and wish to compute the implicit equation $f(x,y)$. Tropical geometry allows us to compute the Newton polygon $\mathrm{Newt}(f)$ first, directly from the parameterization Φ, without knowing $f(x,y)$. This is the content of Theorem 1.5.2. Once the Newton polygon $\mathrm{Newt}(f)$ is known, we recover the desired polynomial $f(x,y)$ by a linear algebra computation. The next paragraph explains that computation.

Pretend that the polynomial $f(x,y)$ in Example 1.5.1 is unknown and impossible to compute using resultants or Gröbner bases. Suppose further that we are given its Newton polygon. According to Definition 1.1.3, this is

$$(1.5.2) \qquad \mathrm{Newt}(f) \;=\; \mathrm{conv}\left\{\begin{pmatrix}0\\0\end{pmatrix}, \begin{pmatrix}2\\1\end{pmatrix}, \begin{pmatrix}3\\2\end{pmatrix}, \begin{pmatrix}2\\3\end{pmatrix}, \begin{pmatrix}0\\2\end{pmatrix}\right\}.$$

This pentagon has four additional lattice points in its interior, so $\mathrm{Newt}(f)$ contains precisely nine lattice points. That information reveals

$$f(x,y) \;=\; c_1x^3y^2 + c_2x^2y^3 + c_3x^2y^2 + c_4x^2y + c_5xy^2 + c_6xy + c_7y^2 + c_8y + c_9,$$

where the coefficients $c_1, c_2, \ldots, c_9$ are unknown parameters. At this point we can set up a linear system of equations as follows. For any choice of complex number τ, the equation $f(\phi_1(\tau), \phi_2(\tau)) = 0$ holds. This equation translates into one linear equation for the nine unknowns c_i. Eight of such linear equations will determine the coefficients uniquely (up to scaling). For instance, in our example, if we take $\tau = \pm 2, \pm 3, \pm 4, \pm 5$, then we get eight linear equations which stipulate that the vector $(c_1, c_2, c_3, c_4, c_5, c_6, c_7, c_8, c_9)^{\mathrm{T}}$ lies in the kernel of the following 8×9-matrix of rational numbers

$$\begin{array}{c} \tau \\ -5 \\ -4 \\ -3 \\ -2 \\ 2 \\ 3 \\ 4 \\ 5 \end{array}
\begin{array}{c}
\begin{array}{ccccccccc} x^3y^2 & x^2y^3 & x^2y^2 & x^2y & xy^2 & xy & y^2 & y & 1 \end{array} \\
\begin{pmatrix}
-\frac{2187}{10} & -\frac{419904}{625} & \frac{2916}{25} & -\frac{81}{4} & -\frac{7776}{125} & \frac{54}{5} & \frac{20736}{625} & -\frac{144}{25} & 1 \\
-\frac{80}{3} & -\frac{1875}{16} & 25 & -\frac{16}{3} & -\frac{375}{16} & 5 & \frac{5625}{256} & -\frac{75}{16} & 1 \\
-\frac{2}{3} & -\frac{512}{81} & \frac{16}{9} & -\frac{1}{2} & -\frac{128}{27} & \frac{4}{3} & \frac{1024}{81} & -\frac{32}{9} & 1 \\
0 & 0 & 0 & 0 & 0 & 0 & \frac{81}{16} & -\frac{9}{4} & 1 \\
\frac{2048}{3} & 48 & 64 & \frac{256}{3} & 6 & 8 & \frac{9}{16} & \frac{3}{4} & 1 \\
\frac{15625}{6} & \frac{40000}{81} & \frac{2500}{9} & \frac{625}{4} & \frac{800}{27} & \frac{50}{3} & \frac{256}{81} & \frac{16}{9} & 1 \\
\frac{34992}{5} & \frac{32805}{16} & 729 & \frac{1296}{5} & \frac{1215}{16} & 27 & \frac{2025}{256} & \frac{45}{16} & 1 \\
\frac{235298}{15} & \frac{3687936}{625} & \frac{38416}{25} & \frac{2401}{6} & \frac{18816}{125} & \frac{196}{5} & \frac{9216}{625} & \frac{96}{25} & 1
\end{pmatrix}
\end{array}.$$

This matrix has rank 8, so its kernel is one dimensional. Any generator of that kernel translates into (a scalar multiple of) the desired polynomial $f(x, y)$.

While the implicit equation $f(x, y)$ of a parametric curve can always be recovered from its Newton polygon by solving linear equations, the relevant matrices tend to be dense and ill conditioned. It is a nontrivial challenge to recover the coefficients numerically when $f(x, y)$ has thousands of terms.

By contrast, some mathematicians can rightfully consider the implicitization problem to be solved once the Newton polygon has been found. Thus, in what follows, we study the following alternative version of implicitization:

Tropical implicitization problem: Given two rational functions $\phi_1(t)$ and $\phi_2(t)$, compute the Newton polygon $\mathrm{Newt}(f)$ of the implicit equation $f(x, y)$.

We shall present the solution to the tropical implicitization problem for plane curves. By the Fundamental Theorem of Algebra, the two given rational functions are products of linear factors over the complex numbers $\mathbb{C}$:

$$\begin{aligned} \phi_1(t) &= (t - \alpha_1)^{u_1} (t - \alpha_2)^{u_2} \cdots (t - \alpha_m)^{u_m}, \\ \phi_2(t) &= (t - \alpha_1)^{v_1} (t - \alpha_2)^{v_2} \cdots (t - \alpha_m)^{v_m}. \end{aligned} \tag{1.5.3}$$

Here the α_i are the zeros and poles of either of the two functions ϕ_1 and ϕ_2. It may occur that u_i is zero while v_i is nonzero or vice versa.

For what follows we do not need the algebraic numbers α_i but only the exponents u_i and v_j occurring in the factorizations. These can be found by symbolic computation. For instance, it suffices to factor $\phi_1(t)$ and $\phi_2(t)$ over their field of definition, say, the rational numbers $\mathbb{Q}$. No field extensions or floating point computations are needed to obtain the integers u_i and v_j.

We abbreviate $u_0 = -u_1 - u_2 - \cdots - u_m$ and $v_0 = -v_1 - v_2 - \cdots - v_m$, and we consider the following collection of $m+1$ integer vectors in the plane:

$$\begin{pmatrix} u_0 \\ v_0 \end{pmatrix}, \begin{pmatrix} u_1 \\ v_1 \end{pmatrix}, \begin{pmatrix} u_2 \\ v_2 \end{pmatrix}, \ldots, \begin{pmatrix} u_m \\ v_m \end{pmatrix}. \tag{1.5.4}$$

We consider the rays spanned by these $m+1$ vectors. Each ray has a natural multiplicity, namely the sum of the lattice lengths of all vectors $(u_i, v_i)^T$ lying on that ray. Since the vectors in (1.5.4) sum to zero, this configuration of rays satisfies the balancing condition: it is a tropical curve in the plane $\mathbb{R}^2$.

The following result can be derived from the Fundamental Theorem 3.2.3. We will ask for a proof in Exercise 5.6(26). A higher-dimensional generalization of Theorem 1.5.2 is presented in Theorem 5.5.1. As stated, Theorem 1.5.2 and Corollary 1.5.3 need the hypothesis that the map Φ is one-to-one. Otherwise, one first divides (1.5.4) by the degree of Φ.

Theorem 1.5.2. *The tropical curve* $V(\mathrm{trop}(f))$ *defined by the unknown polynomial* f *equals the tropical curve determined by the vectors in* (1.5.4).

The Newton polygon $\mathrm{Newt}(f)$ can be recovered from the tropical curve $V(f)$ as follows. The first step is to rotate our vectors by 90 degrees:

$$\begin{pmatrix} v_0 \\ -u_0 \end{pmatrix}, \begin{pmatrix} v_1 \\ -u_1 \end{pmatrix}, \begin{pmatrix} v_2 \\ -u_2 \end{pmatrix}, \ldots, \begin{pmatrix} v_m \\ -u_m \end{pmatrix}. \tag{1.5.5}$$

Since these vectors sum to zero, there exists a convex polygon P whose edges are translates of these vectors. We construct P by sorting the vectors by increasing slope and then simply concatenating them. The polygon P is unique up to translation. Hence there exists a unique translate P^+ of the polygon P that lies in the nonnegative orthant $\mathbb{R}^2_{\geq 0}$ and that has nonempty intersection with both the x-axis and the y-axis. The latter requirements are necessary (and sufficient) for a lattice polygon to arise as the Newton polygon of an irreducible polynomial in $\mathbb{C}[x, y]$. We conclude:

Corollary 1.5.3. *The polygon* P^+ *equals the Newton polygon* $\mathrm{Newt}(f)$ *of the defining irreducible polynomial of the parameterized curve* $\mathrm{Image}(\Phi)$.

This solves the tropical implicitization problem for plane curves over $\mathbb{C}$. We illustrate this solution for our running example.

Example 1.5.4. Write the map of Example 1.5.1 in factored form (1.5.3):

$$\begin{aligned} \phi_1(t) &= (t-1)^{\mathbf{-1}}\, t^{\mathbf{1}}\, (t+1)^{\mathbf{-1}}\, (t+2)^{\mathbf{2}}, \\ \phi_2(t) &= (t-1)^{\mathbf{2}}\, t^{\mathbf{-2}}\, (t+1)^{\mathbf{1}}\, (t+2)^{\mathbf{0}}. \end{aligned}$$

The derived configuration of five vectors as in (1.5.4) equals

$$\begin{pmatrix} -1 \\ -1 \end{pmatrix}, \begin{pmatrix} -1 \\ 2 \end{pmatrix}, \begin{pmatrix} 1 \\ -2 \end{pmatrix}, \begin{pmatrix} -1 \\ 1 \end{pmatrix}, \begin{pmatrix} 2 \\ 0 \end{pmatrix}.$$

We form their rotations as in (1.5.5), and we order them by increasing slope:

$$\begin{pmatrix} 2 \\ 1 \end{pmatrix}, \begin{pmatrix} 1 \\ 1 \end{pmatrix}, \begin{pmatrix} -1 \\ 1 \end{pmatrix}, \begin{pmatrix} -2 \\ -1 \end{pmatrix}, \begin{pmatrix} 0 \\ -2 \end{pmatrix}.$$

We concatenate these vectors starting at the origin. The resulting edges all remain in the nonnegative orthant. The result is the pentagon P^+ in Corollary 1.5.3. As predicted, it coincides with the pentagon in (1.5.2). $\Diamond$

The technique of tropical implicitization can be used, in principle, to compute the tropicalization of any parametrically presented algebraic variety. The details are more complicated than the simple curve case discussed here. A proper treatment requires toric geometry and concepts from resolution of singularities. The proof of Theorem 6.5.16 demonstrates this point. For further reading on tropical implicitization, we refer to [**STY07, SY08**].

1.6. Group Theory

One of the origins in tropical geometry is the work of Bieri, Groves, Strebel, and Neumann in group theory [**BG84, BS80, BNS87**]. Starting in the late 1970s, these authors associate polyhedral fans to certain classes of discrete groups, and they establish remarkable results concerning generators, relations and higher cohomology of these groups in terms of their fans. This part of our tropical island is more secluded and offers breathtaking vistas.

We begin with an easy illustrative example. Fix a nonzero real number ξ, and let G_ξ denote the group generated by the two invertible 2×2-matrices

$$A = \begin{pmatrix} 1 & 1 \\ 0 & 1 \end{pmatrix} \quad \text{and} \quad X = \begin{pmatrix} 1 & 0 \\ 0 & \xi \end{pmatrix}. \tag{1.6.1}$$

What relations do these two generators satisfy? In particular, is the group G_ξ finitely presented? Does this property depend on the number ξ?

To answer these questions, we explore some basic computations such as

$$X^u A^c X^{-u} X^v A^d X^{-v} = \begin{pmatrix} 1 & c\,\xi^{-u} + d\,\xi^{-v} \\ 0 & 1 \end{pmatrix}. \tag{1.6.2}$$

Here u, v, c, and d can be arbitrary integers. This identity shows that the two matrices $X^u A^c X^{-u}$ and $X^v A^d X^{-v}$ commute, and this commutation relation is a valid relation among the two generators of G_ξ. If the number ξ is not algebraic over $\mathbb{Q}$, then the set of all such commutation relations constitutes a complete presentation of G_ξ, and in this case the group G_ξ is never finitely presented. On the other hand, if ξ is an algebraic number, then additional relations can be derived from the irreducible minimal polynomial $f \in \mathbb{Z}[x]$ of ξ. To show how this works, we consider the explicit example $\xi = \sqrt{2} + \sqrt{3}$.

The minimal polynomial of this algebraic number is $f(x) = x^4 - 10x^2 + 1$. This polynomial translates into the matrix identity

$$(1.6.3) \qquad (X^{-4}A^1X^4)\cdot(X^{-2}A^{-10}X^2)\cdot(X^0A^1X^0) \;=\; \begin{pmatrix}1 & 0\\ 0 & 1\end{pmatrix}.$$

The left-hand side gives the word $X^{-4}AX^2A^{-10}X^2A$ in the generators A and X. That word is a relation in G_ξ. Our question is whether the group of all such relations is finitely generated. It turns out that the answer is affirmative for $\xi = \sqrt{2}+\sqrt{3}$, and we shall list the generators in Example 1.6.10.

In general, finite presentation is characterized by the following result:

Theorem 1.6.1. *The group $G_\xi = \langle A, X\rangle$ is finitely presented if and only if either the real number ξ or its reciprocal $1/\xi$ is an algebraic integer over $\mathbb{Q}$.*

The condition that either ξ or $1/\xi$ is an algebraic integer says that either the highest term or the lowest term of $f(x)$ has coefficient $+1$ or -1. This is equivalent to saying that either the highest or the lowest term of the minimal polynomial $f(x)$ is a unit in $\mathbb{Z}[x, x^{-1}]$. It is precisely this condition on leading terms that underlies the tropical thread in geometric group theory.

Bieri and Strebel introduced tropical varieties over $\mathbb{Z}$ in their 1980 paper on metabelian groups [**BS80**]. Later work with Neumann [**BNS87**] extended their construction to a wider class of discrete groups. In what follows we restrict ourselves to metabelian groups whose corresponding module is cyclic. This special case suffices in order to explain the general idea and to shed light on the mystery of why Theorem 1.6.1 might be true.

We begin with some commutative algebra definitions. Consider the Laurent polynomial ring $S = \mathbb{Z}[x_1^{\pm 1}, \ldots, x_n^{\pm 1}]$ over the integers $\mathbb{Z}$. The units in S are the *monomials* $\pm x^{\mathbf{a}} = \pm x_1^{a_1}\cdots x_n^{a_n}$ where $\mathbf{a} = (a_1, \ldots, a_n)$ runs over $\mathbb{Z}^n$. For $f \in S$ and $\mathbf{w} \in \mathbb{R}^n$, the *initial form* $\mathrm{in}_{\mathbf{w}}(f)$ is the sum of all terms in f whose $\mathbf{w}$-weight is minimal. If I is a proper ideal in S, then the *initial ideal* $\mathrm{in}_{\mathbf{w}}(I)$ is the ideal generated by all initial forms $\mathrm{in}_{\mathbf{w}}(f)$ where f runs over I. Computing $\mathrm{in}_{\mathbf{w}}(I)$ from a generating set of I requires *Gröbner bases over the integers*. The relevant algorithm for computing $\mathrm{in}_{\mathbf{w}}(I)$ from I is implemented in computer algebra systems such as `Macaulay2` and `Magma`.

The *tropical variety* (over $\mathbb{Z}$) of the ideal I is the following subset of $\mathbb{R}^n$:

$$\mathrm{trop}_{\mathbb{Z}}(I) \;=\; \{\mathbf{w} \in \mathbb{R}^n : \mathrm{in}_{\mathbf{w}}(I) \neq S\}.$$

This tropical variety contains the tropical variety over the field $\mathbb{Q}$ as a subset:

$$\mathrm{trop}_{\mathbb{Z}}(I) \;\supseteq\; \mathrm{trop}_{\mathbb{Q}}(I).$$

This containment is strict in general. For example, if $n = 2$ and $I = \langle x_1 + x_2 + 3\rangle$, then $\mathrm{trop}_{\mathbb{Q}}(I)$ is the tropical line (1.3.1), which has three rays. However, $\text{trop}_{\mathbb{Z}}(I)$ also contains the positive quadrant because 3 is not a unit in $\mathbb{Z}$.

We write $R = S/I$ for the quotient $\mathbb{Z}$-algebra, and, by mild abuse of notation, we write R^* for the multiplicative group generated by the images of the monomials. It follows from the results to be proved later in Chapter 3 that the complex variety of the ideal I is finite if and only if $\text{trop}_{\mathbb{Q}}(I) = \{\mathbf{0}\}$. Here we state the analogous result for tropical varieties over the integers.

Theorem 1.6.2 (Bieri and Strebel). *The $\mathbb{Z}$-algebra $R = S/I$ is finitely generated as a $\mathbb{Z}$-module if and only if*

$$\text{trop}_{\mathbb{Z}}(I) = \{\mathbf{0}\}. \tag{1.6.4}$$

Proof. See [**BS80**, Theorem 2.4]. □

This raises the questions of how to test this criterion in practice and, if (1.6.4) holds, how to determine a finite set of monomials $\mathcal{U} \subset R^*$ that generate the abelian group R^*. It turns out that this can be done in `Macaulay2`.

Example 1.6.3. Fix integers m and n where $|m| > 1$. Consider the ideal

$$J = \langle ms^{-1}t^{-1} + s^{-1} + t^{-1} + n + st,\ mst + s + t + n + s^{-1}t^{-1} \rangle \subset \mathbb{Z}[s^{\pm 1}, t^{\pm 1}].$$

This ideal is a variation on Example 43 in Strebel's exposition [**Str84**]. The condition (1.6.4) is satisfied. To find a generating set $\mathcal{U}$, we can run the following four lines of `Macaulay2` code for various fixed values of m and n:

```
R = ZZ[s,t,S,T];              m = 7; n = 13;
J = ideal(m*S*T+S+T+n+s*t,m*s*t+s+t+n+S*T,s*S-1,t*T-1);
toString leadTerm J
toString basis(R/J)
```

The output of this script is the same for all m and n, namely,

$$\mathcal{U} = \{1, s, st^{-1}, t, s^{-1}, s^{-1}t^{-1}, t^{-1}, t^{-2}\}. \tag{1.6.5}$$

For a proof that $\mathbb{Z}\mathcal{U} = R/J$, it suffices to show that the initial ideal of J with respect to the reverse lexicographic term order is generated by

```
(m^2-1)*S*T,t*T,m*s*T,S^2,t*S,s*S,t^2,s*t,s^2,T^3,S*T^2,s*T^2.
```

This proof amounts to computing a Gröbner basis over the integers $\mathbb{Z}$. ◇

The integral tropical variety $\text{trop}_{\mathbb{Z}}(I)$ is of interest even in the case $n = 1$.

Example 1.6.4. Suppose that ξ is an algebraic number over $\mathbb{Q}$ and I is the prime ideal of all Laurent polynomials $f(x)$ in $\mathbb{Z}[x, x^{-1}]$ such that $f(\xi) = 0$.

There are four possible cases of what the integral tropical variety can be:

- If ξ and $1/\xi$ are both algebraic integers, then $\text{trop}_{\mathbb{Z}}(I) = \{0\}$.
- If ξ is an algebraic integer but $1/\xi$ is not, then $\text{trop}_{\mathbb{Z}}(I) = \mathbb{R}_{\geq 0}$.
- If $1/\xi$ is an algebraic integer but ξ is not, then $\text{trop}_{\mathbb{Z}}(I) = \mathbb{R}_{\leq 0}$.
- If neither ξ nor $1/\xi$ are algebraic integers, then $\text{trop}_{\mathbb{Z}}(I) = \mathbb{R}$.

Examples of numbers for the first, third, and last cases are $\xi = \sqrt{2} + \sqrt{3}$, $\xi = \frac{1}{\sqrt{2}} + \frac{1}{\sqrt{3}}$, and $\xi = \sqrt{2} + \frac{1}{\sqrt{3}}$, respectively. In particular, we see from Theorem 1.6.1 that G_ξ is finitely presented if and only if $\text{trop}_{\mathbb{Z}}(I) \neq \mathbb{R}$. ◊

We now come to the punchline of this section, namely, the extension of Example 1.6.4 to $n \geq 2$ variables. Let I be any ideal in $S = \mathbb{Z}[x_1^{\pm 1}, \ldots, x_n^{\pm 1}]$, and let $R = S/I$. We associate with I the following group of $2{\times}2$-matrices:

$$G_I = \begin{pmatrix} 1 & R \\ 0 & R^* \end{pmatrix}.$$

This is a *metabelian group*, which means that the commutator subgroup of G_I is abelian. The elements of G_I are $\left(\begin{smallmatrix} 1 & f \\ 0 & m \end{smallmatrix}\right)$, where f is a Laurent polynomial and m is a Laurent monomial, but both are considered modulo I. The following result generalizes Theorem 1.6.1 to higher dimensions:

Theorem 1.6.5 (Bieri and Strebel). *The metabelian group G_I is finitely presented if and only if the integer tropical variety* $\text{trop}_{\mathbb{Z}}(I)$ *contains no line.*

This was the main theorem in the remarkable 1980 paper by Bieri and Strebel [**BS80**, Theorem A]. It predates the 1984 paper by Bieri and Groves [**BG84**], which has been cited by tropical geometers for its resolution of problems left open in Bergman's 1971 paper [**Ber71**] on the logarithmic limit set.

In what follows we aim to shed some light on the presentation of the metabelian group G_I. We begin with the observation that G_I is always finitely generated, namely, by a natural set of $n{+}1$ matrices over $R = S/I$:

Lemma 1.6.6. *The metabelian group G_I is generated by the $n+1$ matrices*

$$A = \begin{pmatrix} 1 & 1 \\ 0 & 1 \end{pmatrix} \quad \textit{and} \quad X_i = \begin{pmatrix} 1 & 0 \\ 0 & x_i \end{pmatrix} \quad \textit{for } i = 1, 2, \ldots, n.$$

For $n = 1$, these are the two generators in (1.6.1), and $G_I = G_\xi$ if $I = \langle f(x) \rangle$ is the principal ideal generated by the minimal polynomial of ξ.

We now examine the relations among the $n + 1$ generators in Lemma 1.6.6. Let us first assume that $I = \langle 0 \rangle$ is the zero ideal, so that $R = S$. The matrices X_i and X_j commute, so the commutator $[X_i, X_j] = X_i X_j X_i^{-1} X_j^{-1}$

is the 2×2-identity matrix. Next we consider the action of the group R^* on G_I by conjugation. For any monomial $m = x^{\mathbf{u}}$ we have $X^{\mathbf{u}} = \left(\begin{smallmatrix} 1 & 0 \\ 0 & x^{\mathbf{u}} \end{smallmatrix}\right)$, and the product $A^m = X^{-\mathbf{u}} A X^{\mathbf{u}}$ is equal to $\left(\begin{smallmatrix} 1 & m \\ 0 & 1 \end{smallmatrix}\right)$. Likewise, we have $A^{-m} = X^{-\mathbf{u}} A^{-1} X^{\mathbf{u}} = \left(\begin{smallmatrix} 1 & -m \\ 0 & 1 \end{smallmatrix}\right)$, so the same identity holds for monomials whose coefficient is -1. In particular, for any monomial m in S, the two matrices A and A^m commute. Hence, in the group $G_{\langle 0 \rangle}$ we have

$$(1.6.6) \qquad [X_i, X_j] = [A, A^m] = 1 \text{ for } 1 \le i < j \le n \text{ and monomials } m \in S^*.$$

Lemma 1.6.7. *The relations* (1.6.6) *define a presentation of the group* $G_{\langle 0 \rangle}$.

We next extend this to all ideals I. For any $f \in S$, consider the matrix

$$A^f = \begin{pmatrix} 1 & f \\ 0 & 1 \end{pmatrix}.$$

Matrix products such as (1.6.2) show that A^f lies in $G_{\langle 0 \rangle}$.

Proposition 1.6.8. *For any ideal I in S, the group G_I has the presentation*

$$(1.6.7) \qquad [X_i, X_j] = [A, A^m] = A^f = 1,$$

where m runs over monomials, f runs over I, and $1 \le i < j \le n$.

This presentation is infinite. We wish to know whether (1.6.7) can be replaced by a finite subset. Is the group G_I finitely presented? To answer this, we first note that the conjugation action satisfies the relations

$$A^f A^g = A^g A^f = A^{f+g} \quad \text{and} \quad (A^f)^g = (A^g)^f = A^{fg} \ \text{ for } f, g \in S.$$

This shows that it suffices to take f from any finite generating set of the ideal I. So, the question is whether there exists a finite subset $\mathcal{U} \subset \mathbb{Z}^n$ such that the monomials $m = \pm x^{\mathbf{u}}$ with $\mathbf{u} \in \mathcal{U}$ suffice in the presentation (1.6.7).

Theorem 1.6.5 offers a criterion for testing whether such a finite set $\mathcal{U}$ exists. For instances in which the answer is affirmative, we can use the techniques in [**BS80**, §3] to construct an explicit generating set $\mathcal{U}$. These techniques are quite delicate and have not yet been developed into an actual algorithm. In what follows we outline a proposal for how to approach this.

The first step is to compute the integral tropical variety $\mathrm{trop}_{\mathbb{Z}}(I)$ from the given generators of I. We would replace I by its homogenization and compute the Gröbner fan. If K is a field, then the Gröbner fan of $I \subset K[x_0, x_1, \ldots, x_n]$ is a polyhedral fan in $\mathbb{R}^{n+1}$ such that the initial ideal $\mathrm{in}_{\mathbf{w}}(I)$ is constant as $\mathbf{w}$ ranges over the relative interior of any cone. See Corollary 2.5.12. However, here we need an extension to $K = \mathbb{Z}$, and this theory has yet to be developed. Gröbner fans over $\mathbb{Z}$ will be finer than those over $K = \mathbb{Q}$. For example, if $I = \langle 2x_1, x_1x_2 - x_1x_3 \rangle$, then the Gröbner fan over $\mathbb{Q}$ consists of a single cone, while the Gröbner fan over $\mathbb{Z}$ has a wall on the plane $\{w_2 = w_3\}$.

In the course of computing the Gröbner fan of I, we would obtain generators for every initial ideal $\mathrm{in}_{\mathbf{w}}(I)$. From these we would derive a finite generating set $\mathcal{B}$ of I with the property that, for every $\mathbf{w} \in \mathbb{R}^n$, either $\mathrm{in}_{\mathbf{w}}(I)$ is a proper ideal in S or the finite set $\{\mathrm{in}_{\mathbf{w}}(f) : f \in \mathcal{B}\}$ contains a unit. A subset $\mathcal{B}$ of the ideal I that enjoys this property is a *tropical basis* over $\mathbb{Z}$. Every Laurent polynomial in a tropical basis $\mathcal{B}$ can be scaled by a unit, so we can always assume that the relevant leading monomial is the constant 1.

Suppose now that I is an ideal in S which satisfies the condition of Theorem 1.6.5 and that we have computed a tropical basis $\mathcal{B}$ for I. Then

(1.6.8) For all $\mathbf{w} \in \mathbb{R}^n$ there is $f \in \mathcal{B}$ with $\mathrm{in}_{\mathbf{w}}(f) = 1$ or $\mathrm{in}_{-\mathbf{w}}(f) = 1$.

For each Laurent polynomial f in the tropical basis $\mathcal{B}$, let support(f) denote the set of all vectors $\mathbf{a} \in \mathbb{Z}^n$ such that the monomial $x^{\mathbf{a}}$ appears with nonzero coefficient in f. We define the Newton polytope of the tropical basis $\mathcal{B}$ as the convex hull of the union of these support sets for all f in $\mathcal{B}$:

$$\mathrm{Newt}(\mathcal{B}) \quad := \quad \mathrm{conv}\Big(\bigcup_{f\in\mathcal{B}} \mathrm{support}(f)\Big).$$

By examining the proof technique used in [**BS80**, §3.5], one can derive the following explicit version of the "if" direction in the Bieri–Strebel Theorem:

Theorem 1.6.9. *Fix a tropical basis $\mathcal{B}$ satisfying* (1.6.8) *for the ideal I. Then the metabelian group G_I is presented by the relations* (1.6.6)*, where f runs over the elements in the tropical basis $\mathcal{B}$ and $m = x^{\mathbf{u}}$ runs over the set* $\mathrm{Newt}(\mathcal{B}) \cap \mathbb{Z}^n$ *of lattice points $\mathbf{u}$ in the Newton polytope the tropical basis.*

Example 1.6.10. Let $n = 1$, and let I be the prime ideal of $\xi = \sqrt{2} + \sqrt{3}$. The singleton $\mathcal{B} = \{x^4 - 10x^2 + 1\}$ is a tropical basis of I satisfying (1.6.8). Then the group $G_\xi = G_I$ is presented by five relations. The first relation is the word in (1.6.3), and the other four required relations are the words

$$[A, A^{x^i}] = AX^{-i}AX^iA^{-1}X^{-i}A^{-1}X^i \quad \text{for } i = 1,2,3,4. \qquad \diamondsuit$$

Example 1.6.11. Consider the group in [**Str84**, Example 43]. Let $S = \mathbb{Z}[s^{\pm1}, t^{\pm1}]$ with $I = \langle f \rangle$ generated by the polynomial in Example 1.6.3:

$$f(s,t) \quad = \quad ms^{-1}t^{-1} + s^{-1} + t^{-1} + n + st.$$

The tropical variety $\mathrm{trop}_{\mathbb{Z}}(I)$ contains no line. A minimal tropical basis satisfying the condition (1.6.8) consists of three Laurent polynomials:

$$\mathcal{B} \;=\; \big\{ s^{-1}t^{-1}f(s,t)\,,\, sf(s,t),\, tf(s,t) \,\big\}.$$

The corresponding polytope $\mathrm{Newt}(\mathcal{B})$ is a planar convex 7-gon that has 14 lattice points, corresponding to the 14 Laurent monomials:

$$m \;=\; s^2t,\ st^2,\ st,\ s,\ s/t,\ t,\ 1,\ 1/t,\ t/s,\ 1/s,\ 1/st,\ 1/st^2,\ 1/s^2t,\ 1/s^2t^2.$$

The metabelian group G_I has three generators A, X_1, X_2. The description in Theorem 1.6.9 gives a presentation with $17 = 3 + 14$ relations. ◊

In this section we saw a connection between Gröbner bases over $\mathbb{Z}$ and tropical geometry. The beautiful group theory results by Bieri, Groves, Neumann, and Strebel suggest that further research on this topic is desirable.

1.7. Curve Counting

The breakthrough that brought tropical methods to the attention of geometers was the work of Mikhalkin [**Mik05**] on Gromov–Witten invariants of the plane. These invariants count the number of complex algebraic curves of a given degree and genus passing through a given number of points. Mikhalkin proved that complex curves can be replaced by tropical curves, and he then derived a combinatorial formula for the count in the tropical case. We already saw the first case of this result in Section 1.3: there is a unique tropical line (degree 1 curve) through two general points in $\mathbb{R}^2$. The objective of this section is to present the basic ideas and the main result.

We begin by reviewing some classical facts about curves in the complex projective plane $\mathbb{P}^2$. If C is a smooth curve of degree d in $\mathbb{P}^2$, then its *genus* is the number of handles of the Riemann surface of C. That genus equals

$$g(C) \;=\; \frac{1}{2}(d-1)(d-2).$$

Moreover, that same number counts the lattice points in the interior of the Newton polygon of the general curve of degree d. That Newton polygon is the triangle with vertices $(d,0,0)$, $(0,d,0)$, and $(0,0,d)$. In symbols,

$$g(C) \;=\; \#\left(\mathrm{int}(\mathrm{Newt}(C)) \cap \mathbb{Z}^3\right).$$

The set of all curves of degree d forms a projective space of dimension

$$\binom{d+2}{2} - 1 \;=\; \frac{1}{2}(d-1)(d-2) + 3d - 1. \tag{1.7.1}$$

As the $\binom{d+2}{2}$ coefficients of its defining polynomial vary, the curve C may acquire one or more singular points. The simplest type of singularity is a *node*. Each time the curve acquires a node, the genus drops by one. Thus for a singular curve C_{sing} with ν nodes and no other singularities, the genus is

$$g(C_{\mathrm{sing}}) \;=\; \frac{1}{2}(d-1)(d-2) - \nu. \tag{1.7.2}$$

We are interested in the following problem of enumerative geometry: *What is the number $N_{g,d}$ of irreducible curves of genus g and degree d that pass through $g + 3d - 1$ general points in the complex projective plane $\mathbb{P}^2$?*

This question makes sense because the moduli space of curves of degree d and genus g is expected to have dimension $g+3d-1$, by (1.7.1) and (1.7.2), since acquiring a node poses a codimension-1 condition on the curve. Thus we expect the number $N_{g,d}$ of curves satisfying all constraints to be finite. Gromov–Witten theory offers the tools for proving this finiteness result.

The numbers $N_{g,d}$ are called *Gromov–Witten invariants* of the plane $\mathbb{P}^2$. Their study has been a topic of considerable interest among geometers.

Example 1.7.1. The simplest Gromov–Witten invariants are $N_{0,1} = 1$ and $N_{0,2} = 1$. This translates into saying that a unique line passes through two points and that a unique quadric passes through five points. We also have $N_{1,3} = 1$, which says that a unique cubic passes through nine points. ♢

Example 1.7.2. The first nontrivial number is $N_{0,3} = 12$, and we wish to explain this in some detail. It concerns curves defined by cubic polynomials

$$f = c_0x^3+c_1x^2y+c_2x^2z+c_3xy^2+c_4xyz+c_5xz^2+c_6y^3+c_7y^2z+c_8yz^2+c_9z^3.$$

For general coefficients $c_0, \ldots, c_9$, the curve $\{f = 0\}$ is smooth of genus $g = 1$. The curve becomes rational, i.e., the genus drops to $g = 0$, precisely when it has a singular point. This happens if and only if the *discriminant* of f vanishes. The discriminant $\Delta(f)$ is a homogeneous polynomial of degree 12 in the ten unknown coefficients $c_0, c_1 \ldots, c_9$. It is a sum of 2040 monomials:

$$\Delta(f) = 19683c_0^4c_6^4c_9^4 - 26244c_0^4c_6^3c_7c_8c_9^3 + \cdots - c_2^2c_3c_4^4c_5^3c_6^2. \tag{1.7.3}$$

The study of discriminants and resultants is the topic of the book by Gel'fand, Kapranov, and Zelevinsky [**GKZ08**], which contains many formulas for computing them. Here is a simple determinantal formula for (1.7.3). The Hessian H of the quadrics $\frac{\partial f}{\partial x}$, $\frac{\partial f}{\partial y}$, and $\frac{\partial f}{\partial z}$ is a polynomial of degree 3. Form the 6×6-matrix $M(f)$ whose entries are the coefficients of the six quadrics $\frac{\partial f}{\partial x}, \frac{\partial f}{\partial y}, \frac{\partial f}{\partial z}, \frac{\partial H}{\partial x}, \frac{\partial H}{\partial y}$, and $\frac{\partial H}{\partial z}$. Then the discriminant (1.7.3) equals $\Delta(f) = \det(M(f))$.

Now, suppose the cubic $\{f = 0\}$ is required to pass through eight given points in $\mathbb{P}^2$. This translates into eight linear equations in $c_0, c_1, \ldots, c_9$. Combining the eight linear equations with the degree 12 equation $\Delta(f) = 0$, we obtain a system of equations that has 12 solutions in $\mathbb{P}^9$. These solutions are the coefficient vectors of the $N_{0,3} = 12$ rational cubics that we seek. ♢

Example 1.7.3. Quartic curves in the plane $\mathbb{P}^2$ can have genus 3, 2, 1, or 0. The Gromov–Witten numbers corresponding to these four cases are

$$N_{3,4} = 1, \;\; N_{2,4} = 27, \;\; N_{1,4} = 225, \;\text{ and }\; N_{0,4} = 620.$$

Here 27 is the degree of the discriminant of a ternary quartic. The last entry means that there are 620 rational quartics through 11 general points. ♢

The result of Mikhalkin [**Mik05**] can be stated informally as follows:

Theorem 1.7.4. *The Gromov–Witten numbers $N_{g,d}$ can be found tropically.*

The following discussion is aimed at making precise what this means. We consider tropical curves of degree d in $\mathbb{R}^2$. Each such curve C is the planar dual graph to a regular subdivision of the triangle with vertices $(0,0)$, $(0,d)$, and $(d,0)$. We say that the curve C is *smooth* if this subdivision consists of d^2 triangles each having unit area $1/2$. Equivalently, the tropical curve C is smooth if it has d^2 vertices. These vertices are necessarily trivalent.

We already encountered smoothness of tropical curves in Section 1.3. Proposition 4.5.1 explains this for hypersurfaces in arbitrary dimensions. The next property for plane tropical curves is more inclusive than "smooth".

A tropical curve C is called *simple* if each vertex is either trivalent or is locally the intersection of two line segments. Equivalently, C is simple if the corresponding subdivision consists only of triangles and parallelograms. Here the triangles are allowed to have large area. Let $t(C)$ be the number of trivalent vertices, and let $r(C)$ be the number of unbounded edges of C.

We define the *genus* of a simple tropical curve C by the formula

$$g(C) \;=\; \frac{1}{2}t(C) - \frac{1}{2}r(C) + 1. \tag{1.7.4}$$

It is instructive to check that this definition makes sense for smooth tropical curves. Indeed, if C is smooth, then $t(C) = d^2$ and $r(C) = 3d$, and we recover the formula for the genus of a smooth classical complex curve:

$$g(C) \;=\; \frac{1}{2}d^2 - \frac{1}{2}3d + 1 \;=\; \frac{1}{2}(d-1)(d-2).$$

We finally define the *contribution* of a simple curve C as the product of the normalized areas of all triangles in the corresponding subdivision. Thus, in computing the contribution of C, we disregard the "nodal singularities", which correspond to 4-valent crossings. We just multiply positive integers attached to the trivalent vertices. The contribution of a trivalent vertex equals $w_1 w_2 |\det(\mathbf{u}_1, \mathbf{u}_2)|$, where w_1, w_2, w_3 are the weights of the adjacent edges and $\mathbf{u}_1, \mathbf{u}_2, \mathbf{u}_3$ are their primitive edge directions. That formula is independent of the choice made because of the balancing condition $w_1\mathbf{u}_1 + w_2\mathbf{u}_2 + w_3\mathbf{u}_3 = 0$. If the curve is smooth, then its contribution equals 1.

Here now is the precise statement of what was meant in Theorem 1.7.4:

Theorem 1.7.5 (Mikhalkin's Correspondence Principle). *The number of simple tropical curves of degree d and genus g that pass through $g + 3d - 1$ general points in $\mathbb{R}^2$, where each curve is counted with its contribution, equals the Gromov–Witten number $N_{g,d}$ of the complex projective plane $\mathbb{P}^2$.*

The proof of Theorem 1.7.5 given by Mikhalkin in [**Mik05**] uses methods from complex geometry, specifically, deformations of J-holomorphic curves. Subsequently, Gathmann and Markwig [**GM07a, GM07b**] developed an algebraic approach. See also the work of Tyomkin [**Tyo12**]. Mikhalkin's Correspondence Principle led to the systematic development of tropical moduli spaces and tropical intersection theory on such spaces.

We close with one more example of what can be done with tropical curves in enumerative geometry. The Gromov–Witten invariants $N_{0,d}$ for rational curves (genus $g = 0$) satisfy the following remarkable recursion:

$$(1.7.5) \quad N_{0,d} \;=\; \sum_{\substack{d_1+d_2=d \\ d_1,d_2>0}} \left(d_1^2 d_2^2 \binom{3d-4}{3d_1-2} \;-\; d_1^3 d_2 \binom{3d-4}{3d_1-1} \right) N_{0,d_1} N_{0,d_2}.$$

This equation is due to Kontsevich, who derived it from the WDVV equations, named after the theoretical physicists Witten, Dijkgraaf, Verlinde, and Verlinde, which express the associativity of quantum cohomology of $\mathbb{P}^2$.

Using Mikhalkin's Correspondence Principle, Gathmann and Markwig [**GM08**] gave a proof of this formula using tropical methods. Namely, they establish the combinatorial result that simple tropical curves of degree d and genus 0 passing through $3d-1$ points satisfy the Kontsevich relations (1.7.5).

Students wishing to learn the foundations of tropical geometry as it pertains to the topic of this section are referred to the text by Mikhalkin and Rau [**MR**]. The present book does not contain a proof of Theorem 1.7.5. We do not focus on metric graphs, curves, and their moduli. Instead we study embedded tropical varieties that are derived from polynomial ideals.

1.8. Compactifications

Many of the advanced tools of algebraic geometry, such as intersection theory, are custom tailored for varieties that are compact, such as complex projective varieties. Yet, in concrete problems, the given spaces are often not compact. In such a case one first needs to replace the given variety X by a nice compact variety $\overline{X}$ that contains X as dense subset. Here the emphasis lies on the adjective "nice" because the advanced tools will not work or will give incorrect answers if the boundary $\overline{X}\backslash X$ is not good enough.

We begin by considering a nonsingular curve X in the n-dimensional complex torus $(\mathbb{C}^*)^n$. The curve X is not compact, and we wish to add a finite set of points to X so as to get a smooth compactification $\overline{X}$ of X.

From a geometric point of view, it is clear what must be done. Identifying the complex plane $\mathbb{C}$ with $\mathbb{R}^2$, the curve X becomes a surface. More precisely, X is a noncompact Riemann surface. It is an orientable smooth compact surface of some genus g with a certain number m of points removed.

The problem is to identify the m missing points and to fill them back in. What is the algebraic procedure that accomplishes this geometric process?

To illustrate the algebraic complications, we begin with a plane curve

$$X \;=\; \{\,(x,y)\in(\mathbb{C}^*)^2 \,:\, f(x,y)=0\,\}.$$

Our smoothness hypothesis says that the Laurent polynomial equations

$$f(x,y) \;=\; \frac{\partial f}{\partial x}(x,y) \;=\; \frac{\partial f}{\partial y}(x,y) \;=\; 0 \tag{1.8.1}$$

have no common solutions (x,y) in the algebraic torus $(\mathbb{C}^*)^2$. A first attempt at compactifying X is to replace $f(x,y)$ with the homogeneous polynomial

$$f^{\mathrm{hom}}(x,y,z) \;=\; z^N\cdot f\bigl(\frac{x}{z},\frac{y}{z}\bigr).$$

Here N is the smallest integer such that this expression is a polynomial. This homogeneous polynomial defines a curve in the complex projective plane $\mathbb{P}^2$:

$$X^{\mathrm{hom}} \;=\; \bigl\{\,(x:y:z)\in\mathbb{P}^2 \,:\, f^{\mathrm{hom}}(x,y,z)=0\,\bigr\}.$$

This curve is a compactification of X but it is usually not what we want.

Example 1.8.1. Let X be the curve in $(\mathbb{C}^*)^2$ defined by the polynomial

$$f(x,y) \;=\; c_1 + c_2xy + c_3x^2y + c_4x^3y + c_5x^3y^2. \tag{1.8.2}$$

Here c_1, c_2, c_3, c_4, c_5 are any complex numbers that satisfy

$$c_2c_3^4 - 8c_2^2c_3^2c_4 + 16c_2^3c_4^2 - c_1c_3^3c_5 + 36c_1c_2c_3c_4c_5 - 27c_1^2c_4c_5^2 \;\neq\; 0. \tag{1.8.3}$$

This condition ensures that the given noncompact curve X is smooth. The discriminant polynomial in (1.8.3) is computed by eliminating x and y from (1.8.1). The homogenization of the polynomial $f(x,y)$ equals

$$f^{\mathrm{hom}}(x,y,z) \;=\; c_1z^5 + c_2xyz^3 + c_3x^2yz^2 + c_4x^3yz + c_5x^3y^2.$$

The corresponding projective curve X^{hom} in $\mathbb{P}^2$ is compact but it is not smooth. The boundary we have added to compactify consists of two points

$$X^{\mathrm{hom}}\backslash X \;=\; \bigl\{\,(1:0:0),\,(0:1:0)\,\bigr\}.$$

Both of these points are singular on the compact curve X^{hom}. Their respective multiplicities are 2 and 3. In this context, *multiplicities* refers to the lowest degrees seen in $f^{\mathrm{hom}}(1,y,z)$ and $f^{\mathrm{hom}}(x,1,z)$, respectively.

Another thing one might try is the closure of our curve $X\subset(\mathbb{C}^*)^2$ in the product of two projective lines $\mathbb{P}^1\times\mathbb{P}^1$. Then the ambient coordinates are $\bigl((x_0:x_1),(y_0:y_1)\bigr)$, and our polynomial is replaced by its *bihomogenization*

$$x_0^3y_0^2f\bigl(\frac{x_1}{x_0},\frac{y_1}{y_0}\bigr) \;=\; c_1x_0^3y_0^2 + c_2x_1y_1x_0^2y_0 + c_3x_1^2y_1x_0y_0 + c_4x_1^3y_1y_0 + c_5x_1^3y_1^2.$$

The compactification X^{bihom} of X is the zero set of this polynomial in $\mathbb{P}^1 \times \mathbb{P}^1$. Now, the boundary we have added to compactify consists of three points

$$X^{\text{bihom}} \backslash X \;=\; \{((1:0),(0:1)),\ ((0:1),(1:0)),\ ((0:1),(c_5,-c_4))\}.$$

The compactification X^{bihom} is better than X^{hom} but is still mildly singular. The first point above is singular, of multiplicity 2, but the last two points are smooth on X^{bihom}. They correctly fill in two of the holes in X. ◊

The solution to our problem offered by tropical geometry is to replace a given noncompact variety $X \subset (\mathbb{C}^*)^n$ by a *tropical compactification* X^{trop}. Each such compactification of X is characterized by a polyhedral fan in $\mathbb{R}^n$ whose support is the tropical variety corresponding to X. When the dimension or codimension is small, there is a unique coarsest fan structure. This includes all curves and all hypersurfaces. In these cases we obtain a canonical tropical compactification. However, in general, picking a tropical compactification requires making choices, and X^{trop} will depend on these choices. See Example 3.5.4 for a concrete illustration.

Tropical compactifications were introduced by Jenia Tevelev in [**Tev07**]. The geometric foundation for his construction is the theory of *toric varieties.* In Chapter 6, we shall explain the relationship between toric varieties and tropical geometry. In Section 6.4 we shall see the precise definition of tropical compactifications X^{trop} of a variety $X \subset (\mathbb{C}^*)^n$, and we shall prove its key geometric properties. In what follows, we keep the discussion informal and entirely elementary, and we simply go over a few examples.

Example 1.8.2. Let X be the plane complex curve in (1.8.2). Its tropical compactification X^{trop} is a smooth elliptic curve, i.e., it is a Riemann surface of genus $g = 1$. The boundary $X^{\text{trop}} \backslash X$ consists of $m = 4$ points. Unlike the extra points in the bad compactifications X^{hom} and X^{bihom} in Example 1.8.1, these four new points are smooth on X^{trop}. This confirms that the complex curve X is a real torus with $m = 4$ points removed.

The tropical compactification of a plane curve is derived from its Newton polygon, here the quadrilateral $\text{Newt}(f) = \text{conv}\{(0,0),(1,1),(3,2),(3,1)\}$. The genus g of X is the number of interior lattice points of $\text{Newt}(f)$.

The tropical curve is the union of the inner normal rays to the four edges of this quadrilateral. In other words, $\text{trop}(X)$ consists of the four rays spanned by $(1,-1)$, $(1,-2)$, $(-1,0)$, and $(-1,3)$. Each ray has multiplicity one because the edges of $\text{Newt}(f)$ have lattice length 1. This shows that $m = 4$ points need to be added to X to get X^{trop}. The directions of the rays specify how these new points should be glued into X in order to make them smooth in X^{trop}. Algebraically, this process can be described by replacing the given polynomial f by a certain homogeneous polynomial f^{trop}, but the homogenization process is now more tricky. One uses *homogeneous coordinates*, on the toric surface given by Newt(f). These generate the Cox homogeneous coordinate ring, to be defined in Section 6.1. Here, it suffices to think of the homogeneous coordinates we know for $\mathbb{P}^2$ and for $\mathbb{P}^1 \times \mathbb{P}^1$. $\Diamond$

The example of plane curves has two natural generalizations in $(\mathbb{C}^*)^n$, $n \geq 3$, namely curves and hypersurfaces. We briefly discuss both of these.

If X is a curve in $(\mathbb{C}^*)^n$, then the geometry is still easy. All we will do is fill in m missing points in a punctured Riemann surface of genus g. However, the algebra is more complicated than in Example 1.8.2. The curve X is given by an ideal $I \subset \mathbb{C}[x_1^{\pm 1}, \ldots, x_n^{\pm 1}]$. Our primary challenge is to determine the number m from I. The number m is the sum of the multiplicities of the rays in the tropicalization of X. The tropical curve trop(X) is a finite union of rays in $\mathbb{R}^n$ but it is generally impossible to find these rays from (the Newton polytopes of) the given generators of I. To understand how trop(X) arises from I, one needs the concepts pertaining to Gröbner bases and initial ideals, which will be introduced in Chapters 2 and 3. In practice, the software `Gfan`, due to Anders Jensen [**Jen**], can be used to compute the tropical curve trop(X) and the multiplicity of each of its rays.

If X is a hypersurface in $(\mathbb{C}^*)^n$, then the roles are reversed. The algebra is still easy but the geometry is more complicated now than in Example 1.8.2. Let $f = f(x_1, \ldots, x_n)$ be the polynomial that defines X. We compute its Newton polytope Newt(f) $\subset \mathbb{R}^n$, as introduced in Definition 2.3.4.

The tropical compactification X^{trop} has one boundary divisor for each facet of Newt(f). These boundary divisors are varieties of dimension $n-2$. They get glued to the $(n-1)$-dimensional variety X in order to create the compact $(n-1)$-dimensional variety X^{trop}. The precise nature of this gluing is determined by the ray normal to the facet. What is different from the curve case is that the boundary divisors are themselves nontrivial hypersurfaces in X, and they are no longer pairwise disjoint. In fact, describing their intersection pattern in $X^{\text{trop}} \backslash X$ is an essential part of the construction. The relevant combinatorics is encoded in the facial structure of the polytope Newt(f), and we record this data in the tropical hypersurface.

Tropical geometry furnishes such a compactification for any subvariety X of the algebraic torus $(\mathbb{C}^*)^n$. Starting from an ideal $I \subset \mathbb{C}[x_1^{\pm 1}, \ldots, x_n^{\pm 1}]$ with $X = V(I)$, we can compute the tropical variety trop(X). For small examples this can be done by hand, but for larger examples we use software such as `Gfan` for that computation. The output is a polyhedral fan Δ in $\mathbb{R}^n$ whose support $|\Delta|$ equals trop(X). That fan determines a tropical compactification $X^{\text{trop}}(\Delta)$ of the variety X. Now, this compactification may not be quite nice enough, so one sometimes has to replace the fan Δ by a refinement Δ'. This induces a map $X^{\text{trop}}(\Delta') \to X^{\text{trop}}(\Delta)$. For example, Δ may not be a simplicial fan, and we could take Δ' to be a smooth fan that triangulates Δ. Further, we may want to require the flatness condition in Definition 6.4.13.

Let us consider the case when X is an irreducible surface in $(\mathbb{C}^*)^n$. In any compactification $\overline{X}$ of X, the boundary $\overline{X}\backslash X$ is a finite union of irreducible curves. What is desired is that these curves are smooth and that they intersect each other transversally. If this holds, then the boundary $\overline{X}\backslash X$ has *normal crossings*. The tropical compactifications of a surface X usually have the normal crossing property. Here the tropical variety trop(X) supports a two-dimensional fan in $\mathbb{R}^n$. Such a fan has a unique coarsest fan structure. We identify the tropical surface trop(X) with that coarsest fan Δ, and we abbreviate $X^{\text{trop}} = X^{\text{trop}}(\Delta)$. The rays in the fan trop(X) correspond to the irreducible curves in $\overline{X}\backslash X$, and two such curves intersect if and only if the corresponding rays span a two-dimensional cone. Since the fan trop(X) is two dimensional, it has no cones of dimension ≥ 3. Hence, the intersection of any three of the irreducible curves in $\overline{X}\backslash X$ is empty.

Example 1.8.3. Let I be the ideal minimally generated by three linear polynomials $a_1x_1 + a_2x_2 + a_3x_3 + a_4x_4 + a_5x_5 + a_6$ in $\mathbb{C}[x_1^{\pm 1}, x_2^{\pm 1}, x_3^{\pm 1}, x_4^{\pm 1}, x_5^{\pm 1}]$. Its variety X is a noncompact surface in $(\mathbb{C}^*)^5$. If we took the variety of I in affine space $\mathbb{C}^5$, then this would simply be an affine plane $\mathbb{C}^2$. But the torus $(\mathbb{C}^*)^5$ is obtained from $\mathbb{C}^5$ by removing the hyperplanes $\{x_i = 0\}$. Hence our noncompact surface X equals the affine plane $\mathbb{C}^2$ with five lines removed. Equivalently, X is the complex projective plane $\mathbb{P}^2$ with six lines removed.

If the three linear generators of I have random coefficients, then the six lines form a normal crossing arrangement in $\mathbb{P}^2$ and the tropical compactification simply fills the six lines back in, so that $X^{\text{trop}} = \mathbb{P}^2$. Here, trop$(X)$ consists of six rays and the 15 two-dimensional cones spanned by any two of the rays. Five of the rays are spanned by the standard basis vectors $\mathbf{e}_1, \mathbf{e}_2, \mathbf{e}_3, \mathbf{e}_4, \mathbf{e}_5$ of $\mathbb{R}^5$, and the sixth ray is spanned by $-\mathbf{e}_1 - \mathbf{e}_2 - \mathbf{e}_3 - \mathbf{e}_4 - \mathbf{e}_5$.

The situation is more interesting if the generators of I are special, e.g.,

$$I \quad = \quad \langle\, x_1 + x_2 - 1,\; x_3 + x_4 - 1,\; x_1 + x_3 + x_5 - 1 \,\rangle. \tag{1.8.4}$$

For this particular ideal, the configuration of six lines in $\mathbb{P}^2$ has four triples of lines that meet in one point. Two of these special intersection points are

$$\{x_1 = x_4 = x_5 = 0,\ x_2 = x_3 = 1\} \text{ and } \{x_2 = x_3 = x_5 = 0,\ x_1 = x_4 = 1\}.$$

The other two points lie on the line at infinity, where they are determined by $\{x_1 = x_2 = 0\}$ and $\{x_3 = x_4 = 0\}$, respectively. The tropical compactification is constructed by blowing up these four special points. This process replaces each triple intersection point with a new line that meets the three old lines transversally at three distinct points. Thus X^{trop} is a compact surface whose boundary $X^{\text{trop}}\backslash X$ consists of ten lines, namely, the six old lines that had been removed from $\mathbb{P}^2$ plus the four new lines from blowing up. Now, no three lines intersect, so the boundary $X^{\text{trop}}\backslash X$ is normal crossing. There are 15 pairwise intersection points, three on each of the four new lines, and three old intersection points. The latter are determined by $\{x_1 = x_3 = 0\}$, $\{x_2 = x_4 = 0\}$ and by intersecting $\{x_5 = 0\}$ with the line at infinity.

The combinatorics of this situation is encoded in the tropical plane trop(X). It consists of 15 two-dimensional cones which are spanned by ten rays. The rays correspond to the ten lines. The rays are spanned by

$$\mathbf{e}_1\,,\ \mathbf{e}_2\,,\ \mathbf{e}_3\,,\ \mathbf{e}_4\,,\ \mathbf{e}_5\,,\ -\mathbf{e}_1-\mathbf{e}_2-\mathbf{e}_3-\mathbf{e}_4-\mathbf{e}_5,$$
$$\mathbf{e}_1+\mathbf{e}_4+\mathbf{e}_5,\ \mathbf{e}_2+\mathbf{e}_3+\mathbf{e}_5,\ -\mathbf{e}_3-\mathbf{e}_4-\mathbf{e}_5,\ -\mathbf{e}_1-\mathbf{e}_2-\mathbf{e}_5.$$

The tropical plane trop(X) is the cone over the *Petersen graph*, shown in Example 4.1.12. The ten vertices of the Petersen graph correspond to the ten lines in $X^{\text{trop}}\backslash X$, and the 15 edges of the Petersen graph correspond to the pairs of lines that intersect on the tropical compactification X^{trop}. ◇

The previous example shows that tropical compactifications are nontrivial and interesting even for linear ideals I. Since linear ideals cut out linear spaces, we refer to trop(X) as a *tropical linear space*. The combinatorics of tropical linear spaces is governed by the theory of *matroids*. This will be explained in Chapter 4. In the linear case, the open variety $X \subset (\mathbb{C}^*)^n$ is the complement of an arrangement of $n+1$ hyperplanes in a projective space, and the tropical compactification X^{trop} was already known before the advent of tropical geometry. It is essentially equivalent to the *wonderful compactifications* of a hyperplane arrangement complement due to De Concini and Procesi. This was shown in [**FS05**, Theorem 6.1].

1.9. Exercises

(1) Consider the 2×2-matrices $A = \left(\begin{smallmatrix} 2 & 3 \\ 5 & 9 \end{smallmatrix}\right)$ and $B = \left(\begin{smallmatrix} 9 & 5 \\ 3 & 2 \end{smallmatrix}\right)$. Compute $A \odot B$ and $A \oplus B$ tropically. Also compute $A \oplus A^2 \oplus \cdots \oplus A^{1000}$.

(2) Formulate and prove the Fundamental Theorem of Algebra in the tropical setting. Why is the tropical semiring "algebraically closed"?

(3) Find all roots of the quintic $x^5 \oplus 1{\odot}x^4 \oplus 3{\odot}x^3 \oplus 6{\odot}x^2 \oplus 10{\odot}x \oplus 15$.

(4) Let $p : \mathbb{R}^n \to \mathbb{R}$ be a function that is continuous, concave, and piecewise linear with finitely many pieces that are linear functions with integer coefficients. Show that p can be represented by a tropical polynomial in $x_1, \ldots, x_n$.

(5) Prove the following generalization of Proposition 1.2.1. Let $B \in (\mathbb{R} \cup \{\infty\})^{n \times n}$ be a matrix, and let G be the associated weighted directed graph as in Section 1.2. We now allow negative edge weights,

and G may have loops. Assume that G has no negative cost circuit, so there is no path from a vertex to itself in G for which the sum of the edge weights is negative. Consider the matrix

$$B^+ \;=\; B \oplus B^2 \oplus B^3 \oplus \cdots \oplus B^n.$$

Show that B^+_{ij} is the length of the shortest path from i to j. What goes wrong if G has a negative cost circuit?

(6) Prove Proposition 1.2.3. This concerns the tropical interpretation of the dynamic programming method for integer programming.

(7) Let $D = (d_{ij})$ be a symmetric $n\times n$-matrix with zeros on the diagonal and positive off-diagonal entries. We say that D represents a *metric space* if the triangle inequalities $d_{ik} \leq d_{ij} + d_{jk}$ hold for all indices i, j, k. Show that D represents a metric space if and only if the matrix equation $D \odot D = D$ holds.

(8) The tropical 3×3-determinant is a piecewise-linear real-valued function $\mathbb{R}^{3\times 3} \to \mathbb{R}$ on the nine-dimensional space of 3×3-matrices. Describe all the regions of linearity of this function and their boundaries. What does it mean for a matrix to be tropically singular?

(9) How many combinatorial types of quadratic curves are there?

(10) Prove that the stable self-intersection of a plane curve equals its set of vertices. What does this mean for classical algebraic geometry?

(11) Given five general points in $\mathbb{R}^2$, there exists a unique tropical quadric passing through these points. Compute and draw the quadratic curve through the points $(0,5)$, $(1,0)$, $(4,2)$, $(7,3)$, $(9,4)$.

(12) For any multiset of five points in the plane there is a unique tropical quadric passing through them. Argue how stable intersections can be used to get uniqueness for configurations in special position.

(13) A tropical cubic curve in $\mathbb{R}^2$ is *smooth* if it has precisely nine nodes. Prove that every smooth cubic curve has a unique bounded region, and that this region can have either three, four, five, six, seven, eight, or nine edges. Draw examples for all seven cases.

(14) Install Anders Jensen's software `Gfan` [**Jen**] on your computer. Download the manual and try running a few examples.

(15) Find explicit tropical biquadratic polynomials whose curves look like those shown in Figure 1.3.4.

(16) The amoeba of a curve of degree 4 in the plane $\mathbb{C}^2$ can have either 0, 1, 2, or 3 bounded convex regions in its complement. Construct explicit examples for all four cases.

(17) Determine the logarithmic limit set $\mathcal{A}_\infty(I)$ for the line given by the ideal $I = \langle x_1 + x_2 + 1 \rangle$ in $S = \mathbb{C}[x_1^{\pm 1}, x_2^{\pm 1}]$. Verify Theorem 1.4.2 for this example. How would the picture look for a line in 3-space?

(18) Consider the plane curve given by the parameterization

$$x = (t-1)^{13}t^{19}(t+1)^{29} \quad \text{and} \quad y = (t-1)^{31}t^{23}(t+1)^{17}.$$

Find the Newton polygon of its implicit equation $f(x,y) = 0$. How many terms do you expect the polynomial $f(x,y)$ to have?

(19) Let $\mathbf{v}_1, \mathbf{v}_2, \ldots, \mathbf{v}_m$ be vectors in $\mathbb{Z}^n$ that sum to zero: $\mathbf{v}_1 + \cdots + \mathbf{v}_m = 0$. Show that there exists an algebraic curve in $(\mathbb{C}^*)^n$ whose tropical curve in $\mathbb{R}^n$ consists of the rays spanned by $\mathbf{v}_1, \mathbf{v}_2, \ldots, \mathbf{v}_m$.

(20) Construct a rational parameterization (1.5.1) of a plane curve whose Newton polygon is an octagon. Give $\phi_1(t)$ and $\phi_2(t)$ explicitly.

(21) Let G_ξ be the group generated by the matrices A and X in (1.6.1) for $\xi = \frac{1}{4}(1 + \sqrt{33})$. Can you construct a finite presentation of G_ξ?

(22) Find a nonzero ideal I in $\mathbb{Z}[x_1^{\pm}, \ldots, x_n^{\pm}]$ with $\mathrm{trop}_{\mathbb{Z}}(I) = \mathbb{R}^n$.

(23) What can the integral tropical variety $\mathrm{trop}_{\mathbb{Z}}(I)$ look like for an ideal I generated by two linear forms in $\mathbb{Z}[x,y,z]$? List all possibilities.

(24) Given 14 general points in the plane $\mathbb{C}^2$, what is the number of rational curves of degree 5 that pass through these 14 points?

(25) The two curves in Figure 1.3.4 are simple. For each of them, compute the genus using the formula in (1.7.4).

(26) Consider a curve X in $(\mathbb{C}^*)^3$ cut out by two general polynomials of degree 2. What is the genus g and the number m of punctures of this Riemann surface? Describe its tropical compactification X^{trop}.

(27) The set of singular 3×3-matrices with nonzero complex entries is a hypersurface X in the torus $(\mathbb{C}^*)^{3\times 3}$. Describe its tropical compactification X^{trop}. How many irreducible components does the boundary $X^{\mathrm{trop}} \backslash X$ have? How do these components intersect?

(28) Prove the tropical Bézout Theorem 1.3.2.

(29) For which values of x are the following matrices tropically singular?

$$\begin{pmatrix} 1 & 2 & 3 \\ 2 & 4 & 6 \\ 3 & 6 & x \end{pmatrix}, \quad \begin{pmatrix} 1 & 2 & 3 & 4 \\ 2 & 4 & 6 & 8 \\ 3 & 6 & 9 & 12 \\ 4 & 8 & 12 & x \end{pmatrix}, \quad \begin{pmatrix} x & 2 & 3 \\ 2 & x & 6 \\ 3 & 6 & x \end{pmatrix}, \quad \begin{pmatrix} x & 2 & 3 & 4 \\ 2 & x & 6 & 8 \\ 3 & 6 & x & 12 \\ 4 & 8 & 12 & x \end{pmatrix}$$

(30) The variety $X \subset (\mathbb{C}^*)^5$ defined by the ideal in (1.8.4) is the complement of an arrangement of six lines in the projective plane $\mathbb{P}^2$. Draw those six lines, and describe X^{trop} in terms of your arrangement.

Chapter 2

Building Blocks

Tropical geometry is a marriage between algebraic and polyhedral geometry. In order to develop this properly, we need tools and building blocks from various parts of mathematics, such as abstract algebra, discrete mathematics, elementary algebraic geometry, and symbolic computation. The first three sections of this chapter review fields and valuations, algebraic varieties, and polyhedral geometry. In the last three sections, we begin our study of tropical geometry in earnest. We redefine Gröbner bases using valuations which leads us to Gröbner complexes and tropical bases. Unlike in Chapter 1, formal definitions and proofs will be given; our day at the beach is over.

2.1. Fields

Let K be a field. We denote by K^* the nonzero elements of K. A *valuation* on K is a function $\text{val}\colon K \to \mathbb{R} \cup \{\infty\}$ satisfying the following three axioms:

(1) $\text{val}(a) = \infty$ if and only if $a = 0$;

(2) $\text{val}(ab) = \text{val}(a) + \text{val}(b)$; and

(3) $\text{val}(a + b) \geq \min\{\text{val}(a), \text{val}(b)\}$ for all $a, b \in K$.

We often identify a valuation val with its restriction $K^* \to \mathbb{R}$. The image of val is an additive subgroup Γ_{val} of the real numbers $\mathbb{R}$, called the *value group* of (K, val). Every field K has a trivial valuation, defined by $\text{val}(a) = 0$ for all $a \in K^*$. If we are given a valuation val on K that is nontrivial, then we usually assume that the value group Γ_{val} contains 1. This is not a serious restriction because $(\lambda\cdot\text{val})\colon K \to \mathbb{R}\cup\{\infty\}$ is also a valuation for any $\lambda \in \mathbb{R}_{>0}$.

Lemma 2.1.1. *If* $\text{val}(a) \neq \text{val}(b)$, *then* $\text{val}(a + b) = \min(\text{val}(a), \text{val}(b))$.

Proof. Without loss of generality we may assume that $\mathrm{val}(b) > \mathrm{val}(a)$. Since $1^2 = 1$, we have $\mathrm{val}(1) = 0$, and so $(-1)^2 = 1$ implies $\mathrm{val}(-1) = 0$ as well. This implies $\mathrm{val}(-b) = \mathrm{val}(b)$ for all $b \in K$. The third axiom implies

$$\mathrm{val}(a) \geq \min(\mathrm{val}(a+b), \mathrm{val}(-b)) = \min(\mathrm{val}(a+b), \mathrm{val}(b)),$$

and therefore $\mathrm{val}(a) \geq \mathrm{val}(a+b)$. But we also have

$$\mathrm{val}(a+b) \geq \min(\mathrm{val}(a), \mathrm{val}(b)) = \mathrm{val}(a),$$

and hence $\mathrm{val}(a+b) = \mathrm{val}(a)$ as desired. □

Consider the set of all field elements with nonnegative valuation:

$$R = \{ c \in K : \mathrm{val}(c) \geq 0 \}.$$

The set R is a local ring. This means that it has a unique maximal ideal:

$$\mathfrak{m}_K = \{ c \in K : \mathrm{val}(c) > 0 \}.$$

The quotient ring $\Bbbk = R/\mathfrak{m}_K$ is a field, called the *residue field* of (K, val).

Example 2.1.2. One of the original motivations for the study of valuations is the *p-adic valuation* on the field $K = \mathbb{Q}$ of rational numbers. Here p is a prime number, and the valuation $\mathrm{val}\colon \mathbb{Q}^* \to \mathbb{R}$ is given by setting $\mathrm{val}_p(q) = k$, for $q = p^k a/b$, where $a, b \in \mathbb{Z}$ and p does not divide a or b. For example,

$$\mathrm{val}_2(4/7) = 2 \quad \text{and} \quad \mathrm{val}_2(3/16) = -4.$$

The local ring R is the localization of the ring of integers $\mathbb{Z}$ at the prime $\langle p \rangle$. Its elements are the rational numbers a/b where p does not divide b. The maximal ideal $\mathfrak{m}_K$ consists of rationals a/b where p divides a but not b. The residue field $\Bbbk$ is the finite field with p elements, denoted $\mathbb{Z}/\mathbb{Z}p = \mathbb{F}_p$. ◊

Our other main example of a field with valuation is the Puiseux series.

Example 2.1.3. Let K be the field of *Puiseux series* with coefficients in the complex numbers $\mathbb{C}$. The elements of this field are formal power series

$$c(t) = c_1 t^{a_1} + c_2 t^{a_2} + c_3 t^{a_3} + \cdots, \tag{2.1.1}$$

where the c_i are nonzero complex numbers for all i, and $a_1 < a_2 < a_3 < \cdots$ are rational numbers that have a common denominator. We use the notation $\mathbb{C}\{\{t\}\}$ for the field of Puiseux series over $\mathbb{C}$. We can write this as the union

$$\mathbb{C}\{\{t\}\} = \bigcup_{n \geq 1} \mathbb{C}((t^{1/n})),$$

where $\mathbb{C}((t^{1/n}))$ is the field of Laurent series in the formal variable $t^{1/n}$.

This field has a natural valuation $\mathrm{val}\colon \mathbb{C}\{\{t\}\}^* \to \mathbb{R}$ given by taking a nonzero element $c(t) \in \mathbb{C}\{\{t\}\}^*$ to the lowest exponent a_1 that appears in the series expansion of $c(t)$. The field of rational functions $\mathbb{C}(t)$ is a subfield of $\mathbb{C}\{\{t\}\}$ because every rational function $c(t)$ in one variable t has a unique expansion as a Laurent series in t. The valuation of a rational function $c(t)$ is a positive integer if $c(t)$ has a zero at $t = 0$. It is a negative integer if $c(t)$ has a pole at $t = 0$. Hence, $\text{val}(c(t))$ indicates the order of the zero or pole.

Here are three examples that illustrate the valuation on $\mathbb{C}\{\{t\}\}$ and the inclusion of $\mathbb{C}(t)$ into $\mathbb{C}\{\{t\}\}$:

$$\begin{aligned} c(t) &= \frac{4t^2 - 7t^3 + 9t^5}{6 + 11t^4} = \frac{2}{3}t^2 - \frac{7}{6}t^3 + \frac{3}{2}t^5 + \cdots \quad \text{has } \text{val}(c(t)) = 2; \\ \tilde{c}(t) &= \frac{14t + 3t^2}{7t^4 + 3t^7 + 8t^8} = 2t^{-3} + \frac{3}{7}t^{-2} + \cdots \quad \text{has } \text{val}(\tilde{c}(t)) = -3; \\ \pi &= 3.14159265358979323\,85\ldots \quad \text{has } \text{val}(\pi) = 0. \end{aligned}$$

We shall see in Theorem 2.1.5 that the field of Puiseux series is algebraically closed, so we also get an inclusion of $\overline{\mathbb{C}(t)}$ into $\mathbb{C}\{\{t\}\}$. Here is an illustration: Consider the two roots of the algebraic equation $x^2 - x + t = 0$. They are

$$\begin{aligned} x_1(t) &= \frac{1 + \sqrt{1-4t}}{2} = 1 - \sum_{k=1}^{\infty} \frac{1}{k+1}\binom{2k}{k} t^k \quad \text{with } \text{val}(x_1(t)) = 0, \\ x_2(t) &= \frac{1 - \sqrt{1-4t}}{2} = \sum_{k=1}^{\infty} \frac{1}{k+1}\binom{2k}{k} t^k \quad \text{with } \text{val}(x_2(t)) = 1. \end{aligned}$$

Similarly, every univariate polynomial with coefficients in $\mathbb{C}(t)$ has its roots in $\mathbb{C}\{\{t\}\}$. The method of Theorem 2.1.5 for computing such series solutions is available in computer algebra systems such as `maple` and `Mathematica`. ◊

Remark 2.1.4. We can replace $\mathbb{C}$ by another field $\Bbbk$ in Example 2.1.3 and construct the field $\Bbbk\{\{t\}\}$ of Puiseux series over $\Bbbk$. If $\Bbbk$ is algebraically closed of characteristic zero, then so is $\Bbbk\{\{t\}\}$. However, if $\Bbbk$ is algebraically closed of positive characteristic p, then the Puiseux series field $\Bbbk\{\{t\}\}$ would not be algebraically closed. Explicitly, if $\text{char}(\Bbbk) = p > 0$, then the Artin–Schreier polynomial $x^p - x - t^{-1}$ has no roots (see Remark 2.1.10 below for details).

Here now is the promised key property of the Puiseux series field:

Theorem 2.1.5. *The field $K = \Bbbk\{\{t\}\}$ of Puiseux series is algebraically closed when $\Bbbk$ is an algebraically closed field of characteristic zero.*

Proof. We need to show that, given a polynomial $F = \sum_{i=0}^{n} c_i x^i \in K[x]$, there is $y \in K$ with $F(y) = \sum_{i=0}^{n} c_i y^i = 0$. We shall describe an algorithm for constructing y as a Puiseux series by successively adding higher powers of t. We first note that we may assume that F has the following properties:

(1) $\text{val}(c_i) \geq 0$ for all i;

(2) there is some j with $\text{val}(c_j) = 0$;

(3) $c_0 \neq 0$; and

(4) $\operatorname{val}(c_0) > 0$.

We must justify that these assumptions can be made. Property (4) especially needs a justification. For (1) and (2), we note that if $\alpha = \min\{\operatorname{val}(c_i) : 0 \leq i \leq n\}$, then multiplying F by $t^{-\alpha}$ does not change the existence of a root of F. For (3) if $c_0 = 0$, then $y = 0$ is a root so there is nothing to prove.

To justify assumption (4), let us suppose that F satisfies the first three assumptions except $\operatorname{val}(c_0) = 0$. If $\operatorname{val}(c_n) > 0$, then we can form $G(x) = x^n F(1/x) = \sum_{i=0}^{n} c_{n-i} x^i$, which has the desired form, and if $G(y') = 0$ for $y' \in K^*$, then $F(1/y') = 0$. If $\operatorname{val}(c_0) = \operatorname{val}(c_n) = 0$, then consider $f := \overline{F} \in \Bbbk[x]$. This is the image of F modulo $\mathfrak{m}_K$. This is not constant since $\operatorname{val}(c_n) = 0$. Since $\Bbbk$ is algebraically closed, the polynomial f has a root $\lambda \in \Bbbk$. Then

$$\tilde{F}(x) \;:=\; F(x+\lambda) \;=\; \sum_{i=0}^{n} \left(\sum_{j=i}^{n} c_j \binom{j}{i} \lambda^{j-i} \right) x^i$$

has the constant term $\tilde{F}(0) = F(\lambda)$ with positive valuation and $\tilde{F}$ still satisfies the first three properties. If y' is a root of $\tilde{F}$, then $y' + \lambda$ is a root of F.

Set $F_0 = F$. We will construct a sequence of polynomials $F_l = \sum_{i=0}^{n} c_i^l x^i$. Each of the F_l is assumed to satisfy conditions (1) to (4) above, by the same reasoning as above. We define the *Newton diagram* to be the convex hull in $\mathbb{R}^2$ of the set

$$\{(i,j) \in \mathbb{N}^2 : \text{ there is } k \text{ with } k \leq i, \operatorname{val}(c_k^l) \leq j\}.$$

This is different from the Newton polygon in Definition 1.1.3. In fact, the Newton diagram is the Minkowski sum of the Newton polygon and the orthant $\mathbb{R}^2_{\geq 0}$.

The Newton diagram has an edge with negative slope connecting the vertex $(0, \operatorname{val}(c_0^l))$ to a vertex $(k_l, \operatorname{val}(c_{k_l}^l))$. Up to sign, that slope equals

$$w_l \;=\; \frac{\operatorname{val}(c_0^l) - \operatorname{val}(c_{k_l}^l)}{k_l}.$$

Let f_l be the image in $\Bbbk[x]$ of the polynomial $t^{-\operatorname{val}(c_0^l)} F_l(t^{w_l} x) \in K[x]$. Note that f_l has degree k_l and has nonzero constant term. Since $\Bbbk$ is algebraically closed, we can find a root λ_l of f_l. Let r_{l+1} be its multiplicity.

Then $f_l(x) = (x - \lambda_l)^{r_{l+1}} g_l(x)$, where $g_l(\lambda_l) \neq 0$. We define

$$F_{l+1}(x) = \sum_{j=0}^{n} c_j^{l+1} x^j := t^{-\operatorname{val}(c_0^l)} F_l\big(t^{w_l}(x + \lambda_l)\big).$$

The coefficients c_j^{l+1} of the new polynomial $F_{l+1}(x)$ are given by the formula

$$c_j^{l+1} = \sum_{i=j}^{n} c_i^l \, t^{iw_l - \operatorname{val}(c_0^l)} \binom{i}{j} \lambda_l^{i-j}. \tag{2.1.2}$$

The image of this Puiseux series in the residue field $\Bbbk$ equals

$$\overline{c_j^{l+1}} = \frac{1}{j!} \frac{\partial^j f_l}{\partial x^j}(\lambda_l).$$

For $0 \leq j < r_{l+1}$ this is zero, since λ_l is a root of f_l of multiplicity r_{l+1}. For $j = r_{l+1}$ this is nonzero. Thus $\operatorname{val}(c_j^{l+1}) > 0$ for $0 \leq l < r_{l+1}$, and $\operatorname{val}(c_j^{l+1}) = 0$ for $j = r_{l+1}$. Note that here we used the fact that $\operatorname{char}(\Bbbk) = 0$.

If $c_0^{l+1} = 0$, then $x = 0$ is a root of F_{l+1}, so $\lambda_l t^{w_l}$ is root of F_l. Further back substitutions reveal that $\sum_{j=0}^{l} \lambda_j t^{w_0 + \cdots + w_j}$ is a root of $F_0 = F$, and we are done. Thus, we may assume $c_0^{l+1} \neq 0$ for each l, so F_{l+1} satisfies conditions (1) to (4) above. This ensures that the construction can be continued.

The observation above on $\operatorname{val}(c_j^{l+1})$ implies $k_{l+1} \leq r_{l+1} \leq k_l$. Since n is finite, the value of k_l can only drop a finite number of times. Hence, there exist $k \in \{1, \ldots, n\}$ and $m \in \mathbb{N}$ such that $k_l = k$ for all $l \geq m$. This means that $r_l = k$ for all $l > m$, so $f_l = \mu_l (x - \lambda_l)^k$ for all $l > m$, and some $\mu_l \in \Bbbk$.

Let N_l be such that $c_j^l \in \Bbbk((t^{1/N_l}))$ for $0 \leq j \leq n$. By (2.1.2), we can take N_{l+1} to be the least common multiple of N_l and the denominator of w_l. We claim that $N_{l+1} = N_l$ for $l > m$. Indeed, we have $w_l = \operatorname{val}(c_0^l)/k$, so it suffices to show $\operatorname{val}(c_0^l) \in \frac{k}{N_l}\mathbb{Z}$ for $l > m$. Since f_l is a pure power, we have $\operatorname{val}(c_j^l) = (k - j) \operatorname{val}(c_{k-1}^l)$ for $0 \leq j \leq k$, and hence $\operatorname{val}(c_{k-1}^l) = 1/k \operatorname{val}(c_0^l)$ lies in $\frac{1}{N_l}\mathbb{Z}$. This ensures that $y_l = \sum_{j=0}^{l} \lambda_j t^{w_0 + \cdots + w_j}$ lies in $\Bbbk((t^{1/N_l}))$.

We have found $N \in \mathbb{N}$ such that $y_l \in \Bbbk((t^{1/N}))$ for all l. Hence the limit

$$y = \sum_{j \geq 0} \lambda_j t^{w_0 + \cdots + w_j} \qquad \text{lies in} \quad \Bbbk((t^{1/N})).$$

It remains to show that y is a root of F. To see this, consider $z_i = \sum_{j \geq i} \lambda_j t^{w_i + \cdots + w_j}$, and note that $y = y_{l-1} + t^{w_0 + \cdots + w_{l-1}} z_l$ for $l > 0$. We have

$$F_l(z_l) = t^{\operatorname{val}(c_0^l)} F_{l+1}(z_{l+1}).$$

Since $z_0 = y$, it follows that

$$\operatorname{val}(F(y)) = \sum_{j=0}^{l} \operatorname{val}(c_0^j) + \operatorname{val}(F_{l+1}(z_{l+1})) \geq \sum_{j=0}^{l} \operatorname{val}(c_0^j) \qquad \text{for all } l > 0.$$

Since $\operatorname{val}(c_0^j) \in \frac{1}{N}\mathbb{Z}$, we find $\operatorname{val}(F(y)) = \infty$, so $F(y) = 0$ as required. □

Remark 2.1.6. When char($\Bbbk$)=0, the Puiseux series field $\Bbbk\{\{t\}\}$ is the algebraic closure of the Laurent series field $\Bbbk((t))$. See [**Rib99**, 7.1.A(β), p. 186].

The fact that the field of Puiseux series is not algebraically closed when char($\Bbbk$) > 0 motivates the definition in Example 2.1.7. Recall that a group G is *divisible* if, for all $g \in G$ and all integers $n \geq 1$, there is $g' \in G$ with $ng' = g$.

Example 2.1.7. Fix an algebraically closed field $\Bbbk$ and a divisible group $G \subset \mathbb{R}$. The *Mal'cev–Neumann ring* $K = \Bbbk((G))$ of *generalized power series* is the set of formal sums $\alpha = \sum_{g \in G} \alpha_g t^g$, where $\alpha_g \in \Bbbk$ and t is a variable, with the property that $\operatorname{supp}(\alpha) := \{g \in G : \alpha_g \neq 0\}$ is a well-ordered set. If $\beta = \sum_{g\in G} \beta_g t^g$, then we set $\alpha + \beta = \sum_{g \in G}(\alpha_g + \beta_g)t^g$, and $\alpha\beta = \sum_{h \in G}(\sum_{g+g'=h} \alpha_g \beta_{g'})t^h$. Then $\operatorname{supp}(\alpha + \beta) \subseteq \operatorname{supp}(\alpha) \cup \operatorname{supp}(\beta)$ is well ordered and thus $\alpha + \beta$ is well defined. For $\alpha\beta$, let $\operatorname{supp}(\alpha) + \operatorname{supp}(\beta)$ denote $\{g + g' : g \in \operatorname{supp}(\alpha), g' \in \operatorname{supp}(\beta)\}$. This set is well ordered, and hence $\{(g, g') : g + g' = h\}$ is finite for all $h \in G$. Thus, multiplication is well defined. The same holds for division, so K is a field. For details see [**Pas85**, Theorem 13.2.11]. Moreover, it is known that the field K is algebraically closed. For a nonconstructive proof see [**Poo93**, Corollary 4]. ♢

Remark 2.1.8. One might be tempted to define the elements of a ring of generalized power series to be formal sums $\alpha = \sum_{g\in G} \alpha_g t^g$ with no restriction on $\operatorname{supp}(\alpha)$. However, with that definition, multiplication is not well defined. Without the well-ordering hypothesis, the set $\{(g, g') : g + g' = h\}$ summed over in the definition of the product of two series may be infinite.

The field of generalized power series is one of the most general fields with valuation that we need to consider. This is meant in the following sense.

Theorem 2.1.9 ([**Poo93**, Corollary 5])**.** *Fix a divisible group G. Let K be a valued field whose residue field $\Bbbk$ is algebraically closed and whose value group Γ_{val} equals G. If the valuation is trivial on the prime field ($\mathbb{F}_p$ or $\mathbb{Q}$) of K, then (K, val) is isomorphic to a subfield of $\Bbbk((G))$ with the induced valuation.*

Remark 2.1.10. Consider the case when $\Bbbk$ has characteristic $p > 0$. Then the Artin–Schreier polynomial $x^p - x - t^{-1}$ has roots

$$(t^{-1/p} + t^{-1/p^2} + t^{-1/p^3} + \cdots) + c,$$

where c runs over the prime field $\mathbb{F}_p$ of $\Bbbk$. These are well-defined elements of the ring of generalized power series, since $\{-1/p^i : i \geq 0\} \cup \{0\}$ is well ordered, but they are not Puiseux series. Since the Artin–Schreier polynomial of degree p has p such roots and the Puiseux series are a subfield of the generalized power series, we see that there are no Puiseux series roots. Hence, the Puiseux series field over $\Bbbk$ is not algebraically closed. See [**Ked01**] for a subfield of the field of generalized power series that contains the algebraic closure of the field of Laurent series in positive characteristic.

Example 2.1.11. Let $K = \overline{\mathbb{Q}(t)}$ be the algebraic closure of the field of rational functions in one variable with coefficients in $\mathbb{Q}$. Since $\mathbb{Q}(t) \subset \mathbb{C}((t))$, the field K is a subfield of $\mathbb{C}\{\{t\}\}$. An advantage of K over $\mathbb{C}\{\{t\}\}$ is that elements of K can be described in finite space as the roots of polynomials $g = \sum_{i=0}^{r} a_i x^i$ with coefficients $a_i \in \mathbb{Q}(t)$. This allows them to be represented in a computer. The valuation val: $K^* \to \mathbb{R}$ is inherited from $\mathbb{C}\{\{t\}\}$. The valuations of the roots of g can also be read from g as follows: Write $a_i = p_i/q_i$ for $1 \leq i \leq n$ where $p_i, q_i \in \mathbb{Q}[t]$. The valuation of $p = \sum_{j=0}^{s} b_j t^j \in \mathbb{Q}[t]$ is $\min\{j : b_j \neq 0\}$, and $\text{val}(a_i) = \text{val}(p_i) - \text{val}(q_i)$. Then the valuations of the roots α of g are the $w \in \mathbb{R}$ at which the graph of the function $x \mapsto \min\{\text{val}(a_i) + ix : 0 \leq i \leq r\}$ is not differentiable. There are at most r such values w. See Figure 1.1.1 and the surrounding discussion for related examples. The polynomial g is replaced by the associated tropical polynomial, and the valuations of the roots of g are the roots of that tropical polynomial, as in Section 1.1. See Example 2.1.16 for another example of this phenomenon and Section 3.1 for more underlying theory. ◊

Lemma 2.1.12. *Let K be algebraically closed with nontrivial valuation. Then the value group Γ_{val} is a divisible subgroup of $\mathbb{R}$ that is dense in $\mathbb{R}$.*

Proof. The fact that $\Gamma_{\text{val}} = \text{val}(K^*)$ is divisible follows from $\text{val}(a^{1/n}) = 1/n\,\text{val}(a)$. We assume for all valuations that $1 \in \Gamma_{\text{val}}$, so this means in addition that $\mathbb{Q} \subseteq \Gamma_{\text{val}}$, which implies that Γ_{val} is dense in $\mathbb{R}$. □

Example 2.1.13. In [**Mar10**] Thomas Markwig proposes using a subfield of $\Bbbk((\mathbb{R}))$ that contains the Puiseux series when $\text{char}(\Bbbk) = 0$. His field has the advantage that the valuation map $K^* \to \mathbb{R}$ is surjective. This is not the case for the Puiseux series, since the valuation of any series is rational. ◊

Example 2.1.14. Consider the p-adic valuation on $\mathbb{Q}$ described in Example 2.1.2. We use this valuation to construct the completion $\mathbb{Q}_p$ of $\mathbb{Q}$. Algebraically, this is the field of fractions of the completion $\mathbb{Z}_p$ of $\mathbb{Z}$ at the prime p. See [**Eis95**, Chapter 7] for details on completions.

More analytically, the field $\mathbb{Q}_p$ is the completion of $\mathbb{Q}$ with respect to the norm $|\cdot|_p$ induced by the p-adic valuation val_p. This norm is given by

$|a|_p = p^{-\operatorname{val}(a)}$. Every element $a \in \mathbb{Q}_p$ can be written in the form

$$a = \sum_{i=m}^{\infty} a_i p^i,$$

where $a_i \in \{0, \ldots, p-1\}$ and $m \in \mathbb{Z}$. The p-adic integers $\mathbb{Z}_p$ have the same representation but with $m \in \mathbb{N}$. The valuation val extends to $\mathbb{Q}_p$ by setting $\operatorname{val}(a) = \min\{i : a_i \neq 0\}$. This is consistent with the valuation on $\mathbb{Q}$ and the inclusion of $\mathbb{Q}$ into $\mathbb{Q}_p$; for example, $\operatorname{val}_2(6) = 1$, and $6 = 1 \cdot 2^1 + 1 \cdot 2^2$.

It is instructive to explore the topological properties of $\mathbb{Q}_p$. The ball with center 0 and radius 1 in this metric space equals $\mathbb{Z}_p$. The topology on $\mathbb{Z}_p$ is fractal in nature, and, in fact, $\mathbb{Z}_2$ is homeomorphic to the Cantor set. For an introduction to such topologies on valued fields and arithmetic applications, we recommend the lecture notes by Bosch [**Bos**] on rigid analytic geometry.

The field $\mathbb{Q}_p$ is not algebraically closed. For instance, $x^p - x - p^{-1}$ has no roots. Its algebraic closure $\overline{\mathbb{Q}}_p$ inherits the norm but it is not complete. The completion of $\overline{\mathbb{Q}}_p$ is the field $\mathbb{C}_p$, which is both complete and algebraically closed. Performing arithmetic in these fields is a challenge. ◊

We will frequently use the fact that there is a splitting of the surjection $K^* \twoheadrightarrow \Gamma_{\text{val}}$ from the multiplicative group of the field to the value group.

Lemma 2.1.15. *If K is algebraically closed, then the surjection $K^* \twoheadrightarrow \Gamma_{\text{val}}$ splits: there is a homomorphism $\psi \colon (\Gamma_{\text{val}}, +) \to (K^*, \cdot)$ with $\operatorname{val}(\psi(w)) = w$.*

Proof. Since K is algebraically closed, it contains the nth roots of its elements. Thus, for any $a \in K^*$, there is a group homomorphism $\psi \colon (\mathbb{Q}, +) \to (K^*, \cdot)$ with $\psi(1) = a$. Both $(K^*, \cdot)$ and $(\Gamma_{\text{val}}, +)$ are divisible abelian groups. Since Γ_{val} is an additive subgroup of $\mathbb{R}$, it is torsion-free, so Γ_{val} is a torsion-free divisible group, and thus isomorphic to a (possibly uncountable) direct sum of copies of $\mathbb{Q}$ (see [**Hun80**, Exercise 8, p. 198]). Given $w \in \Gamma_{\text{val}}$ in one of these summands of Γ_{val} isomorphic to $\mathbb{Q}$, for which w is taken to 1 by the isomorphism, and any $a \in K^*$ with $\operatorname{val}(a) = w$, there is a group homomorphism $\psi_w \colon \Gamma_{\text{val}} \to K^*$ taking w to a. By construction, this satisfies $\operatorname{val}(\psi_w(mw/n)) = mw/n$. The universal property of the direct sum implies the existence of a homomorphism $\psi : \Gamma_{\text{val}} \to K^*$ with $\operatorname{val}(\psi(w)) = w$. □

Throughout this book, we use the notation t^w to denote the element $\psi(w) \in K^*$. This is consistent with the canonical splitting for the Puiseux series field $\mathbb{C}\{\{t\}\}$. Here $\Gamma_{\text{val}} = \mathbb{Q}$, and the elements t^w are the powers of t.

Consider any field K with a valuation $\operatorname{val} \colon K \to \mathbb{R} \cup \{\infty\}$. The valuation induces a *norm* $|\cdot| \colon K \to \mathbb{R}$ by setting $|a| = \exp(-\operatorname{val}(a))$ for $a \neq 0$, and $|0| = 0$. Here "exp" can be the exponential function for any base. The standard norm axioms are satisfied: $|a| = 0$ if and only if $a = 0$,

$|ab| = |a||b|$, and $|a+b| \leq |a| + |b|$. The last condition can be strengthened to $|a+b| \leq \max\{|a|, |b|\}$. Norms satisfying this are called *non-archimedean.*

The norm on K allows the use of analytical and topological arguments. The field K is now a *metric space* with distance $|a-b|$ between two elements $a, b \in K$. A *ball* is the set of all elements whose distance to a fixed element is bounded by some real constant. Our metric space K has the following remarkable property: if two balls intersect, then one must be contained in the other. This suggests that K can be drawn as the leaves of a rooted tree. That is why pictures of trees are ubiquitous in arithmetic geometry.

In Theorem 2.1.9 we assumed that val is trivial on the prime field. That result does not apply to the fields K in Example 2.1.14, where the prime field is $\mathbb{Q}$ but with the p-adic valuation. There exists a generalization of the field $\Bbbk((G))$ of generalized power series which allows an extension of Theorem 2.1.9 to the case where val is the p-adic valuation on $\mathbb{Q}$. However, the arithmetic in such fields is very tricky. See [**Poo93**] for details.

Readers who wish to learn more about valued fields are referred to the book by Engler and Prestel [**EP05**]. The extension of valuations [**EP05**, Chapter 3] is a subtle issue that is important for geometric applications.

Example 2.1.16. What is the 2-adic valuation of the algebraic number

$$\alpha = \sqrt[3]{11} + \sqrt{17}\,?$$

This elementary question does not have a unique answer, since there are several ways to extend the 2-adic valuation from $\mathbb{Q}$ to the extension field $\mathbb{Q}(\alpha)$. To find all the possibilities, we first compute the minimal polynomial

$$\alpha^6 - 51 \cdot \alpha^4 - 2 \cdot 11 \cdot \alpha^3 + 867 \cdot \alpha^2 - 2 \cdot 561 \cdot \alpha - 2^3 \cdot 599.$$

From this we see that the valuation of α can be either 0, 1, or 2. In fact, these are the distinct roots, as in (1.1.1), of the corresponding tropical polynomial

$$0 \odot x^6 \oplus 0 \odot x^4 \oplus 1 \odot x^3 \oplus 0 \odot x^2 \oplus 1 \odot x \oplus 3.$$

Each coefficient of this tropical polynomial in x is the 2-adic valuation of the corresponding coefficient of the classical polynomial in α. This shows that computing with algebraic extensions of valued fields leads naturally to solving tropical polynomial equations in one variable. We saw this already in Example 2.1.11 and will return to the underlying theory in Section 3.1. ◇

We close this section with a remark about computational issues. It is impossible to enter a generalized power series or arbitrary Puiseux series into a computer, as it cannot be described by a finite amount of information. This suggests that the best pure characteristic zero field with which we can hope to compute is the algebraic closure $\overline{\mathbb{Q}(t)}$ of the ring of rational functions in

t with coefficients in $\mathbb{Q}$. Many of the examples in this book will be defined and computed over the field $\mathbb{Q}(t)$ of rational functions.

A typical computation one may wish to perform is finding a Gröbner basis of a homogeneous ideal in a polynomial ring, as in Section 2.4 below, or perhaps even a tropical basis of an ideal in a Laurent polynomial ring, as in Section 2.6 below. If $K = \mathbb{Q}(t)$, then this can be reduced to working over the field of constants $\Bbbk = \mathbb{Q}$. Namely, given an ideal $I \subset \mathbb{Q}(t)[x_1^{\pm 1}, \ldots, x_n^{+1}]$, we may instead do our computation for $I' = I \cap \mathbb{Q}[t^{\pm 1}, x_1^{\pm 1}, \ldots, x_n^{\pm 1}]$.

2.2. Algebraic Varieties

We now recall some concepts from the highly recommended undergraduate textbook on algebraic geometry by Cox, Little, and O'Shea [**CLO07**].

Definition 2.2.1. The n-dimensional *affine space* over a field K is

$$\mathbb{A}_K^n = \mathbb{A}^n = \{(a_1, a_2, \ldots, a_n) : a_i \in K\} = K^n.$$

The n-dimensional *projective space* over the field K is

$$\mathbb{P}_K^n = \mathbb{P}^n = (K^{n+1} \setminus \{\mathbf{0}\}) / \sim,$$

where $\mathbf{v} \sim \lambda \mathbf{v}$ for all $\lambda \neq 0$. The points of $\mathbb{P}^n$ are the equivalence classes of lines through the origin $\mathbf{0}$. We write $(v_0 : v_1 : \cdots : v_n)$ for the equivalence class of $\mathbf{v} = (v_0, v_1, \ldots, v_n) \in K^{n+1}$. The n-dimensional *algebraic torus* is

$$T_K^n = T^n = \{ (a_1, a_2, \ldots, a_n) : a_i \in K^* \}.$$

Definition 2.2.2. The *coordinate ring* of the affine space $\mathbb{A}^n$ is the polynomial ring $K[x_1, \ldots, x_n]$. The *homogeneous coordinate ring* of the projective space $\mathbb{P}^n$ is $K[x_0, x_1, \ldots, x_n]$, and the coordinate ring of the algebraic torus T^n is the Laurent polynomial ring $K[x_1^{\pm 1}, \ldots, x_n^{\pm 1}]$.

The *affine variety* defined by an ideal $I \subset K[x_1, \ldots, x_n]$ is

$$V(I) = \{\mathbf{a} \in \mathbb{A}^n : f(\mathbf{a}) = 0 \text{ for all } f \in I\}.$$

An ideal $I \subset K[x_0, x_1, \ldots, x_n]$ is *homogeneous* if it has a generating set consisting of homogeneous polynomials. The *projective variety* defined by a homogeneous ideal $I \subset K[x_0, x_1, \ldots, x_n]$ is

$$V(I) = \{\mathbf{v} \in \mathbb{P}^n : f(\mathbf{v}) = 0 \text{ for all } f \in I\}.$$

Any ideal I in $K[x_1^{\pm 1}, \ldots, x_n^{\pm 1}]$ defines a *very affine variety* in the torus:

$$V(I) = \{\mathbf{a} \in T^n : f(\mathbf{a}) = 0 \text{ for all } f \in I\}.$$

For any variety X we consider the ideal I_X of all polynomials (or homogeneous polynomials, or Laurent polynomials) that vanish on X. The *coordinate ring* $K[X]$ of a variety X is the quotient of the coordinate ring of the ambient space, namely $\mathbb{A}^n$, $\mathbb{P}^n$ or T^n, by the vanishing ideal I_X.

In tropical geometry, we are mostly concerned with Laurent polynomials and the very affine varieties they define. Frequently, our ground field will be $K = \mathbb{C}$, the complex numbers, or $K = \mathbb{C}\{\{t\}\}$, the Puiseux series. Very affine varieties over $\mathbb{C}$ are noncompact, as was discussed in Section 1.8.

The map that takes an ideal to its variety is not a bijection; for example, $V(\langle x \rangle) = V(\langle x^2 \rangle)$ in $\mathbb{A}^1$. Two ideals I and J satisfy $V(J) = V(I)$ if they have the same radical $\sqrt{J} = \sqrt{I}$. The converse holds when K is algebraically closed. Namely, assuming this hypothesis, *Hilbert's Nullstellensatz* states that $\sqrt{I} = I_X$, where $X = V(I)$ is the variety of I. For details, see any book on commutative algebra, such as [**Eis95**] or [**CLO07**].

We do not assume that our varieties are irreducible. A variety X is *irreducible* if it cannot be written as the union of two proper subvarieties. Every variety can be decomposed into a finite union of irreducible varieties. This decomposition can be computed algebraically (e.g., in `Macaulay2` [**M2**]) by means of the *primary decomposition* of the corresponding ideals. If X is an irreducible variety, then its vanishing ideal I_X is a prime ideal.

The simplest prime ideals are those generated by linear polynomials. The corresponding varieties are called *linear spaces*. An ideal is *principal* if it is generated by one polynomial, and in this case the variety is a *hypersurface*. Hypersurfaces are varieties of codimension 1. The *dimension* of a variety is its most basic invariant. The *codimension* is n minus the dimension. See [**CLO07**, Chapter 9] for the definition of dimension and how to compute it.

Linear algebra furnishes many interesting examples of varieties. For example, the set X of all $m \times n$-matrices of rank at most r is an irreducible variety. Its prime ideal I_X is generated by all $(r{+}1) \times (r{+}1)$-minors of an $m \times n$-matrix of variables. Such varieties are called *determinantal varieties*. Introductions to determinantal varieties, from two different perspectives, can be found in the textbooks by Harris [**Har95**] and by Miller and Sturmfels [**MS05**].

Example 2.2.3. Let $n = 8$. Fix any field K, and consider the affine space $\mathbb{A}^8 = \mathbb{A}^8_K$ whose points are pairs (A, B) of $2{\times}2$-matrices with entries in K:

$$A = \begin{pmatrix} a_{11} & a_{12} \\ a_{21} & a_{22} \end{pmatrix} \quad \text{and} \quad B = \begin{pmatrix} b_{11} & b_{12} \\ b_{21} & b_{22} \end{pmatrix}.$$

The *commuting variety* is defined by the four entries of the matrix equation $A \cdot B = B \cdot A$. These four polynomials generate a prime ideal of dimension 4. This means that the variety is irreducible of dimension 4.

For an example with different properties consider the five-dimensional variety defined by the matrix equation $A \cdot B = 0$. Its radical ideal equals

$$I \; = \; \langle\, a_{11}b_{11} + a_{12}b_{21},\, a_{11}b_{12} + a_{12}b_{22},\, a_{21}b_{11} + a_{22}b_{21},\, a_{21}b_{12} + a_{22}b_{22} \,\rangle.$$

This ideal I has the prime decomposition

$$(I + \langle a_{11}a_{22} - a_{12}a_{21}, b_{11}b_{22} - b_{12}b_{21}\rangle) \cap \langle b_{11}, b_{21}, b_{12}, b_{22}\rangle \cap \langle a_{11}, a_{12}, a_{21}, a_{22}\rangle.$$

Hence, the variety has three irreducible components. These live inside the affine space $\mathbb{A}^8$, and they have dimensions five, four, and four, respectively. The last two components correspond to one of A or B being the zero matrix.

Tropical geometers would study the variety $\{A \cdot B = 0\}$ not in affine space $\mathbb{A}^8$ but in the torus T^8. That variety in T^8 is irreducible because the components $\{A = 0\}$ and $\{B = 0\}$ disappear. In terms of algebra, the ideal I is a prime ideal in the Laurent polynomial ring $K[a_{11}^{\pm 1}, a_{12}^{\pm 1}, \ldots, b_{22}^{\pm 1}]$. $\diamond$

We place a topology on affine space $\mathbb{A}^n$ by taking the closed sets to be $\{V(I) : I \text{ is an ideal of } K[x_1, \ldots, x_n]\}$. This is the *Zariski topology.* To see that $\emptyset$ and $\mathbb{A}^n$ are Zariski closed, note that $\emptyset = V(\langle 1\rangle)$ and $\mathbb{A}^n = V(\{0\})$. It is an exercise to check that finite unions of closed sets and arbitrary intersections of closed sets are closed. We denote by $\overline{U}$ the closure in the Zariski topology of a set U. This is the smallest set of the form $V(I)$ that contains U. Similarly we can define the Zariski topology on $\mathbb{P}^n$ and T^n.

There are inclusions $T^n \xrightarrow{i} \mathbb{A}^n \xrightarrow{j} \mathbb{P}^n$, where the second map sends $\mathbf{x} \in \mathbb{A}^n$ to $(1 : \mathbf{x}) \in \mathbb{P}^n$. The *affine closure* of a variety $X \subset T^n$ is the Zariski closure $\overline{i(X)}$ of $i(X) \subset \mathbb{A}^n$. The *projective closure* of $X \subset \mathbb{A}^n$ is the Zariski closure $\overline{j(X)}$ of $j(X) \subset \mathbb{P}^n$. We now recall their algebraic descriptions.

Definition 2.2.4. The degree of a polynomial $f = \sum c_{\mathbf{u}} x^{\mathbf{u}}$ in $K[x_1, \ldots, x_n]$ is $W = \max\{|\mathbf{u}| : c_{\mathbf{u}} \neq 0\}$, where $|\mathbf{u}| = \sum_{i=1}^n u_i$. The homogenization $\tilde{f}$ of f is the homogeneous polynomial $\tilde{f} = \sum c_{\mathbf{u}} x_0^{W - |\mathbf{u}|} x^{\mathbf{u}} \in K[x_0, x_1, \ldots, x_n]$. The *homogenization* of an ideal I in $K[x_1, \ldots, x_n]$ is the ideal $I_{\text{proj}} = \langle \tilde{f} : f \in I\rangle$. We similarly define I_{proj} for a given Laurent ideal $I \subset K[x_1^{\pm 1}, \ldots, x_n^{\pm 1}]$.

Proposition 2.2.5. *Let $X = V(I)$ be a subvariety of the torus T^n with ideal $I \subset K[x_1^{\pm 1}, \ldots, x_n^{\pm 1}]$. Then $\overline{i(X)} = V(I_{\text{aff}})$, where $I_{\text{aff}} = I \cap K[x_1, \ldots, x_n]$. For an ideal $I \subset K[x_1, \ldots, x_n]$, the projective closure $\overline{j(X)}$ of $V(I)$ is the subvariety of projective space $\mathbb{P}^n$ defined by the homogeneous ideal I_{proj}.*

Proof. The sets $V(I_{\text{aff}})$ and $V(I_{\text{proj}})$ are Zariski closed in $\mathbb{A}^n$ and $\mathbb{P}^n$, respectively. They contain $i(X)$ and $j(X)$, so they contain $\overline{i(X)}$ and $\overline{j(X)}$. Conversely, suppose that $f \in K[x_1, \ldots, x_n]$ vanishes on $\overline{i(X)}$. Then $f(y) = 0$ for all $y \in X$, so $f \in I_X$ when regarded as a polynomial in $K[x_1^{\pm 1}, \ldots, x_n^{\pm 1}]$. Thus $f \in \sqrt{I}$, by the Nullstellensatz, and so f lies in $\sqrt{I_{\text{aff}}}$. This shows that $V(I_{\text{aff}}) \subseteq \overline{i(X)}$. Similarly, if a homogeneous polynomial $g \in K[x_0, \ldots, x_n]$ vanishes on $\overline{j(X)}$, then $g(1, y_1, \ldots, y_n) = 0$ for all $y = (y_1, \ldots, y_n) \in X$, and hence $g \in \sqrt{I_{\text{proj}}}$. This shows that $V(I_{\text{proj}}) \subseteq \overline{j(X)}$. $\square$

Example 2.2.6. Consider the very affine variety $X = V(I)$ in T^3 defined by

$$I \;=\; \langle \frac{1}{x_1}+\frac{1}{x_2}+\frac{1}{x_3}-1,\; \frac{1}{x_1}+\frac{2}{x_2}+\frac{3}{x_3} \rangle.$$

Its affine closure $\overline{i(X)} = V(I_{\text{aff}})$ in $\mathbb{A}^3$ is defined by the ideal

$$I_{\text{aff}} \;=\; I \cap K[x_1,x_2,x_3] \;=\; \langle\, x_2x_3+2x_2+x_3,\; 2x_1x_3+x_1-x_3 \,\rangle,$$

and its projective closure $\overline{j(X)} = V(I_{\text{proj}})$ in $\mathbb{P}^3$ is defined by the ideal

$$I_{\text{proj}} \;=\; \langle x_2x_3+2x_0x_2+x_0x_3,\; 2x_1x_3+x_0x_1-x_0x_3,\; 3x_1x_2-x_0x_1-2x_0x_2\rangle.$$

Such computations are based on *ideal quotients* as in [**CLO07**, §4.4]. ◊

A morphism $\phi : X \to Y$ of affine or very affine varieties is induced by a ring homomorphism $\phi^* : K[Y] \to K[X]$ between the respective coordinate rings. Note that the map ϕ^* takes the coordinate ring of Y to that of X. The transformation $X \mapsto K[X]$ is a contravariant functor. Computing the image of a morphism ϕ is known as *implicitization* (cf. Section 1.5).

For a morphism $\phi : T^n \to T^m$, we place the additional constraint that the map ϕ be a homomorphism of algebraic groups. This means that $\phi^* : K[x_1^{\pm 1},\ldots,x_m^{\pm 1}] \to K[y_1^{\pm 1},\ldots,y_n^{\pm 1}]$ is a monomial map, so $\phi^*(x_i)$ is a Laurent monomial in $y_1,\ldots,y_n$ for $1 \le i \le m$. Equivalently, the ring homomorphism ϕ^* is induced by a group homomorphism, which we also denote by ϕ^*, from $\mathbb{Z}^m$ to $\mathbb{Z}^n$. If $X = V(I)$ is a subvariety of T^n, then the Zariski closure of its image $\phi(X)$ in the torus T^m is the variety $V(\phi^{*-1}(I))$.

Recall that the group of automorphisms of the lattice $\mathbb{Z}^n$ is isomorphic to the group $\mathrm{GL}(n,\mathbb{Z})$ of invertible matrices with integer entries and determinant ± 1. We denote by $\mathbf{e}_1,\ldots,\mathbf{e}_n$ the standard basis for $\mathbb{Z}^n$.

Lemma 2.2.7. *Given any vector $\mathbf{v} \in \mathbb{Z}^n$ with the greatest common divisor of $|v_1|,\ldots,|v_n|$ equal to one, there is a matrix $U \in \mathrm{GL}(n,\mathbb{Z})$ with $U\mathbf{v} = \mathbf{e}_1$. Further, if L is a rank k subgroup of $\mathbb{Z}^n$ with $\mathbb{Z}^n/L$ torsion-free, then there is a matrix $U \in \mathrm{GL}(n,\mathbb{Z})$ with UL equal to the subgroup generated by $\mathbf{e}_1,\ldots,\mathbf{e}_k$.*

Proof. The first statement follows from the second. Indeed, if the greatest common divisor of $|v_1|,\ldots,|v_n|$ is one, then the group $\mathbb{Z}^n/\mathbb{Z}\mathbf{v}$ is torsion-free. Let A be a $k \times n$-matrix with rows an integer basis for the subgroup L. The condition that $\mathbb{Z}^n/L$ is torsion-free implies that the *Smith normal form* of A is the $k \times n$-matrix A' with first $k \times k$-block the identity matrix, and all other entries zero. The Smith normal form algorithm furnishes matrices $V \in \mathrm{GL}(k,\mathbb{Z})$ and $U' \in \mathrm{GL}(n,\mathbb{Z})$ that satisfy $A' = VAU'$. Multiplying on the left by an element of $\mathrm{GL}(k,\mathbb{Z})$ does not change the integer row span, so the integer row span of VA equals L. We now take $U = U'^T$. □

An automorphism of the torus T^n is an invertible map specified by n Laurent monomials in $x_1, \ldots, x_n$. Thus the automorphism group of T^n is isomorphic to $\mathrm{GL}(n, \mathbb{Z})$. Here the matrix entries are the exponents of the monomials. When we speak of a coordinate change in T^n, we mean the transformation given by such an invertible monomial map. These multiplicative changes of variables behave very differently from the more familiar linear changes of variables in affine space $\mathbb{A}^n$ or projective space $\mathbb{P}^n$. Automorphisms of T^n are essential for tropical geometry.

Example 2.2.8. The invertible integer matrix $U = \left(\begin{smallmatrix} 1 & -1 \\ 1 & -2 \end{smallmatrix}\right)$ represents the automorphism $(x,y) \mapsto (xy, x^{-1}y^{-2})$ of the torus T^2 and of its coordinate ring $\mathbb{C}[x^{\pm 1}, y^{\pm 1}]$. The image of the curve $X = \{(x,y) \in T^2 : f(x,y) = 0\}$ in Example 1.8.1 under the automorphism U is the curve defined by

$$(U \circ f)(x,y) \;=\; c_2 + c_5 x + c_1 y + c_3 xy + c_4 x^2 y^2.$$

Note how the linear map U moves the tropical curve $\mathrm{trop}(X)$. The compactifications $X^{\mathrm{hom}} = \overline{j(X)} \subset \mathbb{P}^2$ and $X^{\mathrm{bihom}} \subset \mathbb{P}^1 \times \mathbb{P}^1$ change under this automorphism. The tropical compactification X^{trop} remains the same. ◊

We close this section by introducing an important variety that we will use later in proofs and will study tropically in Section 4.3. The *Grassmannian* $G(r,m)$ is a fundamental parameter space in algebraic geometry. It is a smooth projective variety of dimension $r(m-r)$. Each of its points corresponds to an r-dimensional linear subspace of a fixed m-dimensional vector space V. The Grassmannian $G(r,m)$ embeds into $\mathbb{P}^{\binom{m}{r}-1}$ as follows.

Fix the vector space $V \simeq K^m$. Every r-dimensional subspace of V is the row-space of some $r \times m$-matrix of rank r. An issue with this representation is that different matrices can have the same row-space. If two $r \times m$-matrices A and B have the same row-space, then one can be obtained from the other by row operations, so there is an element $G \in \mathrm{GL}(r, K)$ with $A = GB$. We solve this ambiguity problem by mapping these matrices to the vector of length $\binom{m}{r}$ of their $r \times r$-minors. This *Plücker vector* has coordinates indexed by all subsets I of size r of $[m] = \{1, \ldots, m\}$. The coordinate indexed by I is the determinant of the $r \times r$-submatrix with columns indexed by I. If $A = GB$ for some $G \in \mathrm{GL}(r, K)$, then the Ith minor of A is $\det(G)$ times the Ith minor of B, so these represent the same point of $\mathbb{P}^{\binom{m}{r}-1}$. The subspace can be uniquely recovered from its Plücker vector.

Example 2.2.9. Let $U \subset \mathbb{C}^4$ be the row-space of the matrix

$$A = \begin{pmatrix} 1 & 1 & 1 & 1 \\ 0 & 1 & 2 & 3 \end{pmatrix}.$$

Note that U is also the row-space of the matrix

$$B = \begin{pmatrix} 3 & 2 & 1 & 0 \\ 0 & 1 & 2 & 3 \end{pmatrix}.$$

The 2×2-minors of a 2×4-matrix are indexed by the sets $\{1,2\}, \{1,3\}, \{1,4\}$, $\{2,3\}, \{2,4\}$, and $\{3,4\}$. The vector of 2×2-minors of A, listed in this order, is $(1,2,3,1,2,1)$, while the one for B is $(3,6,9,3,6,3)$. However, we have

$$(2.2.1) \qquad (1:2:3:1:2:1) \;=\; (3:6:9:3:6:3) \quad \text{in } \mathbb{P}^5.$$

It is instructive to recover the subspace U from this point in $\mathbb{P}^5$. ♢

The set of all such Plücker coordinate vectors forms a projective variety. We denote by $K[\mathbf{p}] = K[\,p_I : I \subset [m], |I| = r\,]$ the coordinate ring of $\mathbb{P}^{\binom{m}{r}-1}$. The *Plücker ideal* $I_{r,m}$ is the set of all polynomials in $K[\mathbf{p}]$ that vanish on the vectors of $r\times r$-minors for all $r\times m$-matrices. This is the homogeneous prime ideal of all polynomial relations among the $r \times r$-minors. It is generated by the Plücker relations, which are defined as follows.

Fix a subset $I \subset [m]$ of size $r-1$, and a subset $J \subset [m]$ of size $r+1$. For $j \in J$, the sign $\operatorname{sgn}(j; I, J)$ is $(-1)^\ell$, where ℓ is the number of elements $j' \in J$ with $j < j'$ plus the number of elements $i \in I$ with $i > j$. The *Plücker relation* $\mathcal{P}_{I,J}$ is the homogeneous quadric

$$\mathcal{P}_{I,J} \;=\; \sum_{j\in J} \operatorname{sgn}(j; I, J)\cdot p_{I\cup j}\cdot p_{J\backslash j},$$

where $p_{I\cup j} = 0$ if $j \in I$. Note that $\mathcal{P}_{I,J}$ is nonzero only if $|J\backslash I| \geq 3$. If $|J\backslash I| = 3$, then, after suitable reorderings and adjusting signs, we can write $I = I' \cup \{i\}$ and $J = I' \cup \{j,k,l\}$ with $i < j < k < l$, and this implies

$$\mathcal{P}_{I,J} \;=\; p_{I'ij}\cdot p_{I'kl} \;-\; p_{I'ik}\cdot p_{I'jl} \;+\; p_{I'il}\cdot p_{I'jk}.$$

Such Plücker relations are called *three-term Plücker relations.*

Proposition 2.2.10. *The Plücker relations generate the Plücker ideal*

$$I_{r,m} \;=\; \langle\, \mathcal{P}_{I,J} \;:\; I, J \subseteq [m],\ |I| = r-1,\ |J| = r+1 \,\rangle.$$

The Grassmannian $G(r,m)$ is the subvariety of $\mathbb{P}^{\binom{m}{r}-1}$ defined by this ideal.

See, for example, [**MS05**, Theorem 14.6] for a proof.

Example 2.2.11. Consider the case $r = 2, m = 4$. The six variables of $K[\mathbf{p}]$ are $p_{12}, p_{13}, p_{14}, p_{23}, p_{24}, p_{34}$. The Plücker ideal is principal:

$$I_{2,4} \;=\; \langle\, \mathcal{P}_{1,234} \,\rangle.$$

The generator is the three-term Plücker relation

$$\mathcal{P}_{1,234} \;=\; p_{12}p_{34} - p_{13}p_{24} + p_{14}p_{23}.$$

This quadric is equal, up to sign, to $\mathcal{P}_{2,134}, \mathcal{P}_{3,124}$, and $\mathcal{P}_{4,123}$. All other Plücker relations, such as $\mathcal{P}_{1,123}$, are zero. The Grassmannian $G(2,4) = V(I_{2,4})$ is a hypersurface in $\mathbb{P}^5$. Note that the point (2.2.1) lies in $G(2,4)$. ◊

Our next result relates the valuation on a field K with the Zariski topology on the torus T^n over K. It will be used in the proof of Proposition 3.1.5. Analogous statements hold for affine space $\mathbb{A}^n$ and projective space $\mathbb{P}^n$.

Lemma 2.2.12. *Let K be a valued field with a splitting $\Gamma_{\text{val}} \to K^*, w \mapsto t^w$, so that $\text{val}(t^w) = w$. Let $\alpha_1, \ldots, \alpha_n \in \Bbbk^*$ and $w_1, \ldots, w_n \in \Gamma_{\text{val}}$, and consider the set of all $\mathbf{y} = (y_1, \ldots, y_n)$ in T^n that satisfy $\text{val}(y_i) = w_i$ and $\overline{t^{-w_i} y_i} = \alpha_i$ for $i = 1, \ldots, n$. This set is dense in T^n with the Zariski topology.*

Proof. We will show that for any nonzero polynomial $h \in K[x_1^{\pm 1}, \ldots, x_n^{\pm 1}]$ there is a point $\mathbf{y}$ of the desired form with $h(\mathbf{y}) \neq 0$. For each i we fix an element z_i in the valuation ring R with $\overline{z_i} = \alpha_i$. Then $y_i = t^{w_i} z_i$ satisfies $\text{val}(y_i) = w_i$, as $\alpha_i \neq 0$ implies that $\text{val}(z_i) = 0$. Each coordinate y_i can be replaced by an infinite number of other choices in K^*; for example, $y_i + t^{w_i + j}$ also has the desired properties for all $j > 0$. We now show, by induction on n, that we can choose $\mathbf{y}$ of this form with $h(\mathbf{y}) \neq 0$. When $n = 1$, we choose y_1 from the infinite number of choices with $\text{val}(y_1) = w_1$ and $\overline{t^{-w_1} y_1} = \alpha_1$ to avoid the finitely many roots of h. When $n > 1$, write $h = \sum h_j x_n^j$ where $h_j \in K[x_1^{\pm 1}, \ldots, x_{n-1}^{\pm 1}]$. By induction there is $\mathbf{y}' = (y_1, \ldots, y_{n-1}) \in (K^*)^{n-1}$ with $\text{val}(y_i) = w_i$ and $\overline{t^{-w_i} y_i} = \alpha_i$ with $h_j(\mathbf{y}') \neq 0$ for all j. We then choose y_n with $\text{val}(y_n) = w_n$ and $\overline{t^{-w_n} y_n} = \alpha_n$ to avoid the finite number of roots of $h(y_1, \ldots, y_{n-1}, x_n) \in K[x_n^{\pm 1}]$. □

2.3. Polyhedral Geometry

We review here the notions from polyhedral geometry that are needed in this book. Polyhedral geometry is a rich and beautiful part of discrete mathematics. The reader unfamiliar with this area is encouraged to spend some time with the first few chapters of Ziegler's textbook [**Zie95**].

Definition 2.3.1. A set $X \subseteq \mathbb{R}^n$ is *convex* if, for all $\mathbf{u}, \mathbf{v} \in X$ and all $0 \leq \lambda \leq 1$, we have $\lambda \mathbf{u} + (1 - \lambda)\mathbf{v} \in X$. The *convex hull* $\text{conv}(U)$ of a set $U \subseteq \mathbb{R}^n$ is the smallest convex set containing U. If $U = \{\mathbf{u}_1, \ldots, \mathbf{u}_r\}$ is finite, then $\text{conv}(U) = \big\{ \sum_{i=1}^r \lambda_i \mathbf{u}_i \,:\, 0 \leq \lambda_i \leq 1, \sum_{i=1}^r \lambda_i = 1 \big\}$ is a *polytope*.

A *polyhedral cone* C in $\mathbb{R}^n$ is the positive hull of a finite subset of $\mathbb{R}^n$:

$$C \;=\; \text{pos}(\mathbf{v}_1, \ldots, \mathbf{v}_r) \;:=\; \Big\{ \sum_{i=1}^r \lambda_i \mathbf{v}_i \in \mathbb{R}^n : \lambda_i \geq 0 \;\text{ for all } i \Big\}.$$

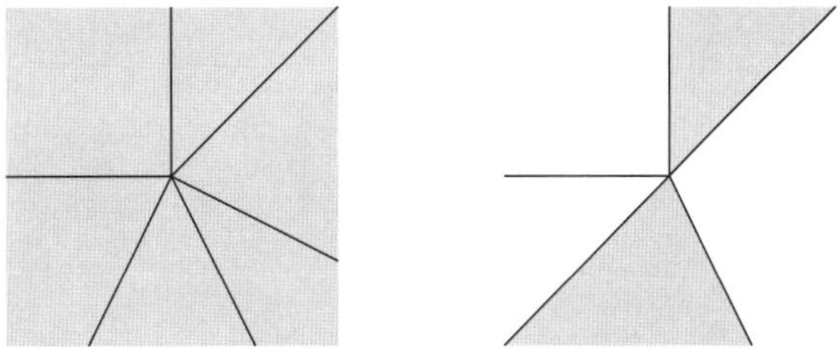

Figure 2.3.1. Polyhedral fans.

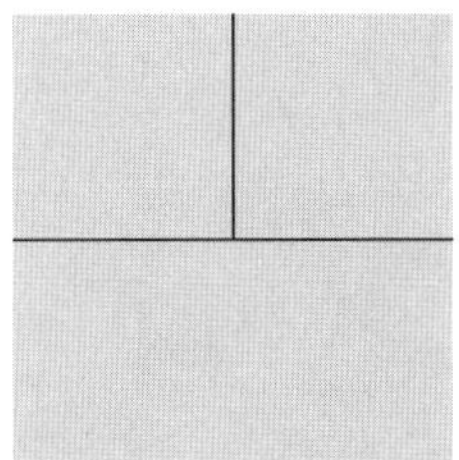

Figure 2.3.2. Not a polyhedral fan.

The cone is *simplicial* if the $\mathbf{v}_i$ are linearly independent. Every polyhedral cone has the alternate description as a set of the form

$$C = \{\mathbf{x} \in \mathbb{R}^n : A\mathbf{x} \le 0\},$$

where A is a $d \times n$-matrix. For a proof see [**Zie95**, Theorem 1.3].

A *face* of a cone is determined by a linear functional $\mathbf{w} \in (\mathbb{R}^n)^\vee$, via

$$\mathrm{face}_{\mathbf{w}}(C) = \{\mathbf{x} \in C : \mathbf{w} \cdot \mathbf{x} \le \mathbf{w} \cdot \mathbf{y} \text{ for all } \mathbf{y} \in C\}.$$

This has an alternate description as $\mathrm{face}_{\mathbf{w}}(C) = \{\mathbf{x} \in C : A'\mathbf{x} = 0\}$, where A' is a suitable $d' \times n$-submatrix of A derived from $\mathbf{w}$. A *(polyhedral) fan* is a collection of polyhedral cones satisfying two conditions: every face of a cone in the fan is in the fan, and the intersection of any two cones in the fan is a face of each. For an illustration of this definition see Figures 2.3.1 and 2.3.2. The fan is simplicial if every cone is simplicial. The notion of a fan is fundamental for toric varieties, as seen in Chapter 6.

A convex set is the intersection of a collection of half-spaces in $\mathbb{R}^n$. A *polyhedron* $P \subset \mathbb{R}^n$ is the intersection of finitely many closed half-spaces:

$$P = \{\mathbf{x} \in \mathbb{R}^n : A\mathbf{x} \le \mathbf{b}\},$$

where A is a $d \times n$-matrix, and $\mathbf{b} \in \mathbb{R}^d$. Polytopes are bounded polyhedra; see [**Zie95**, §1.1]. A *face* of a polyhedron is determined by a linear functional $\mathbf{w} \in (\mathbb{R}^n)^\vee$, via $\mathrm{face}_{\mathbf{w}}(P) = \{\mathbf{x} \in P : \mathbf{w} \cdot \mathbf{x} \le \mathbf{w} \cdot \mathbf{y} \text{ for all } \mathbf{y} \in P\}$.

A face of P that is not contained in any larger proper face is called *facet*. A *polyhedral complex* is a collection Σ of polyhedra satisfying two conditions:

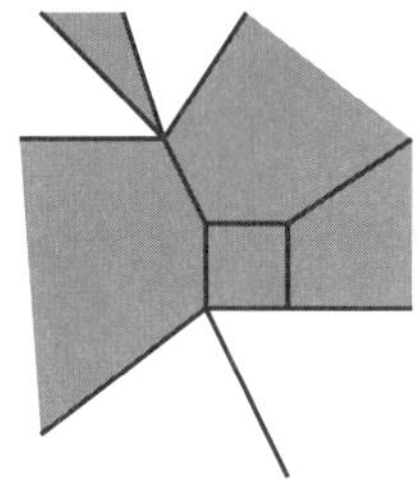

Figure 2.3.3. A polyhedral complex.

if P is in Σ, then so is any face of P, and if P and Q lie in Σ, then $P \cap Q$ is either empty or a face of both P and Q. An example is shown in Figure 2.3.3. The polyhedra in a polyhedral complex Σ are called the *cells* of Σ. Cells of Σ that are not faces of any larger cell are *facets* of the complex. Their facets are called *ridges* of the complex. The *support* $|\Sigma|$ of Σ is the set $\{\mathbf{x} \in \mathbb{R}^n : \mathbf{x} \in P \text{ for some } P \in \Sigma\}$. Note that polyhedral cones are special cases of polyhedra and fans are special cases of polyhedral complexes.

The *lineality space* of a polyhedron P is the largest linear subspace $V \subset \mathbb{R}^n$ with the property that $\mathbf{x} \in P, \mathbf{v} \in V$ implies $\mathbf{x} + \mathbf{v} \in P$. The lineality space of a polyhedral complex Σ is the intersection of all the lineality spaces of the polyhedra in the complex. The *affine span* of a polyhedron P is the smallest affine subspace containing P. This is the translate of a linear subspace of $\mathbb{R}^n$, which we call the *linear space parallel to* P. The *dimension* of P is the dimension of the linear space parallel to P.

A polyhedral complex Σ is *pure* of dimension d if every facet of Σ has dimension d. The *f-vector* $(f_0, f_1, \ldots, f_d)$ of a polyhedral complex Σ records the number f_i of cells of the complex of dimension i. The *relative interior* of P, denoted relint(P), is the interior of P inside its affine span. If $P = \{\mathbf{x} \in \mathbb{R}^n : A\mathbf{x} = \mathbf{b}, A'\mathbf{x} \leq \mathbf{b}'\}$, where each of the inequalities in $A'\mathbf{x} \leq \mathbf{b}'$ is strict for some $\mathbf{x} \in P$, then $\text{relint}(P) = \{\mathbf{x} \in \mathbb{R}^n : A\mathbf{x} = \mathbf{b}, A'\mathbf{x} < \mathbf{b}'\}$.

Definition 2.3.2. Let Γ be a subgroup of $(\mathbb{R}, +)$. A Γ-*rational polyhedron* is

$$P \;=\; \big\{\mathbf{x} \in \mathbb{R}^n \,:\, A\mathbf{x} \leq \mathbf{b}\big\},$$

where A is a $d \times n$-matrix with entries in $\mathbb{Q}$, and $\mathbf{b} \in \Gamma^d$. A polyhedral complex Σ is Γ-rational if every polyhedron in Σ is Γ-rational. We will be interested in the case where $\Gamma = \Gamma_{\text{val}}$ is the value group of a field K. If $\Gamma = \mathbb{Q}$, then we simply use the adjective *rational* instead of $\mathbb{Q}$-rational.

Definition 2.3.3. Let $P \subset \mathbb{R}^n$ be a polyhedron. The *normal fan* of P is the polyhedral fan $\mathcal{N}_P$ consisting of the cones

$$\mathcal{N}_P(F) \;=\; \text{cl}(\{\mathbf{w} \in (\mathbb{R}^n)^\vee : \text{face}_{\mathbf{w}}(P) = F\})$$

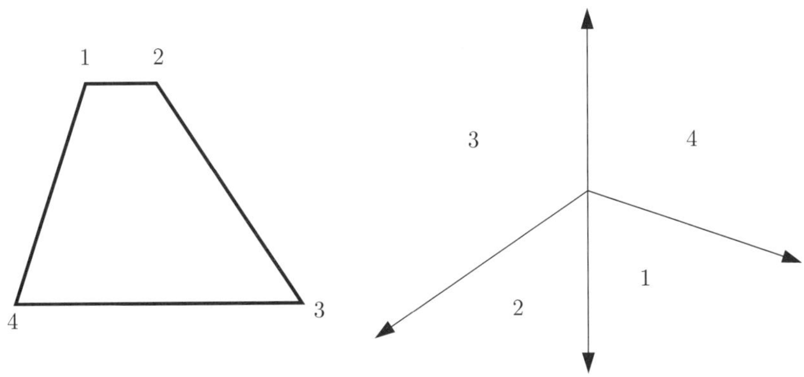

Figure 2.3.4. The normal fan of a polyhedron P.

as F varies over the faces of P. Here, $\mathrm{cl}(\cdot)$ denotes the closure in the Euclidean topology on $(\mathbb{R}^n)^\vee$, which is the vector space dual to $\mathbb{R}^n$. The fan $\mathcal{N}_P$ is also called the *inner normal fan* of P. Figure 2.3.4 shows the normal fan of a quadrangle P. This fan consists of nine cones, four of dimension 2, four of dimension 1, and one of dimension 0.

Definition 2.3.4. Let $S = K[x_1^{\pm 1}, \ldots, x_n^{\pm 1}]$ be the Laurent polynomial ring. Given $f = \sum_{\mathbf{u} \in \mathbb{Z}^n} c_{\mathbf{u}} \mathbf{x}^{\mathbf{u}} \in S$, the *Newton polytope* of f is the polytope

$$\mathrm{Newt}(f) \;=\; \mathrm{conv}(\mathbf{u} : c_{\mathbf{u}} \neq 0) \;\subset\; \mathbb{R}^n.$$

We call this a *Newton polygon* if it is two dimensional, as in Definition 1.1.3.

Example 2.3.5. Let $S = \mathbb{C}[x^{\pm 1}, y^{\pm 1}]$. Consider the polynomial

$$f \;=\; 7x + 8y - 3xy + 4x^2y - 17xy^2 + x^2y^2.$$

The Newton polygon of f is shown in Figure 2.3.5. Now, consider

$$g \;=\; x^{-1} - y^{-1} + 3x - 2y + xy.$$

The Newton polygon of the Laurent polynomial g is the translation of the Newton polygon of f by the vector $(-1, -1)$. $\Diamond$

Let Σ_1 and Σ_2 be two polyhedral complexes in $\mathbb{R}^n$. The *common refinement* of Σ_1 and Σ_2 is the polyhedral complex $\Sigma_1 \wedge \Sigma_2$ consisting of the polyhedra $\{P \cap Q : P \in \Sigma_1, Q \in \Sigma_2\}$. Note that $|\Sigma_1 \wedge \Sigma_2| = |\Sigma_1| \cap |\Sigma_2|$. We usually apply this operation when Σ_1 and Σ_2 have the same support.

The *Minkowski sum* of two subsets $A, B \subset \mathbb{R}^n$ is the set

$$A + B \;=\; \{\mathbf{a} + \mathbf{b} : \mathbf{a} \in A, \mathbf{b} \in B\} \;\subset\; \mathbb{R}^n. \tag{2.3.1}$$

If A and B are polyhedra in $\mathbb{R}^n$, then $A + B$ is also a polyhedron in $\mathbb{R}^n$. The same holds for polytopes, cones, and supports of polyhedral complexes. Here are two useful facts that relate Minkowski sums to other constructions.

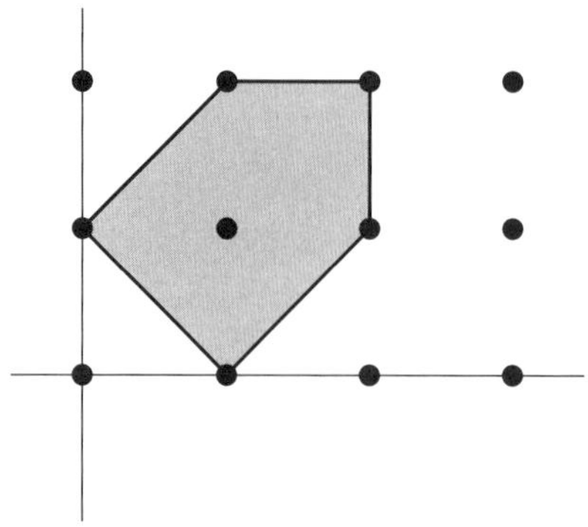

Figure 2.3.5. The Newton polygon in Example 2.3.5.

- If P and Q are polyhedra in $\mathbb{R}^n$, then the normal fan of their Minkowski sum is the common refinement of the two normal fans:

$$\mathcal{N}_{P+Q} \;=\; \mathcal{N}_P \wedge \mathcal{N}_Q. \tag{2.3.2}$$

- The Newton polytope of a product of two Laurent polynomials is the Minkowski sum of the two given Newton polytopes:

$$\mathrm{Newt}(f \cdot g) \;=\; \mathrm{Newt}(f) + \mathrm{Newt}(g). \tag{2.3.3}$$

Definition 2.3.6. Let Σ be a polyhedral complex in $\mathbb{R}^n$, and let σ be a cell in Σ. The *star* of σ in Σ is a fan in $\mathbb{R}^n$, denoted by $\mathrm{star}_\Sigma(\sigma)$. Its cones are indexed by those cells τ in Σ that contain σ as a face. The cone of $\mathrm{star}_\Sigma(\sigma)$ that is indexed by τ is the following subset of $\mathbb{R}^n$:

$$\bar{\tau} \;=\; \{\lambda(\mathbf{x}-\mathbf{y})\colon \lambda \geq 0, \mathbf{x} \in \tau, \mathbf{y} \in \sigma\}.$$

Example 2.3.7. The polyhedral complex Σ shown on the left of Figure 2.3.6 is a quadratic curve in the tropical plane, as seen in Section 1.3. The affine span of the vertex σ_1 in Σ is just the vertex itself. The star is shown on the right. For σ_2 the affine span is the entire y-axis, and this is also the star. $\diamondsuit$

A particularly interesting class of polyhedral complexes are the *regular subdivisions* of a polytope. Regular subdivisions were called *coherent subdivisions* in [**GKZ08**]. An excellent reference for all topics pertaining to triangulations and subdivisions is the book by De Loera et al. [**DRS10**].

Definition 2.3.8. Let $\mathbf{v}_1, \ldots, \mathbf{v}_r$ be an ordered list of vectors in $\mathbb{R}^{n+1}$. We fix a *weight vector* $\mathbf{w} = (w_1, \ldots, w_r) \in \mathbb{R}^r$. The *regular subdivision of* $\mathbf{v}_1, \ldots, \mathbf{v}_r$ *induced by* $\mathbf{w}$ is the polyhedral fan with support $\mathrm{pos}(\mathbf{v}_1, \ldots, \mathbf{v}_r)$ whose cones are $\mathrm{pos}(\mathbf{v}_i : i \in \sigma)$ for all subsets $\sigma \subseteq \{1, \ldots, r\}$ such that there exists $\mathbf{c} \in \mathbb{R}^{n+1}$ with $\mathbf{c} \cdot \mathbf{v}_i = w_i$ for $i \in \sigma$ and $\mathbf{c} \cdot \mathbf{v}_i < w_i$ for $i \notin \sigma$. When the fan is *simplicial*, the subdivision is called a *regular triangulation*.

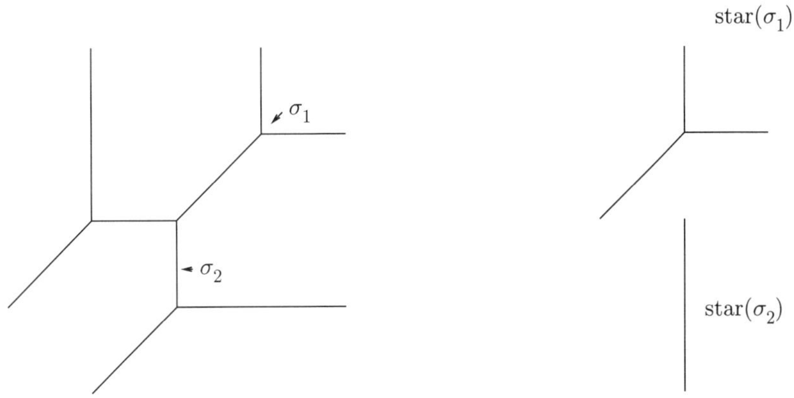

Figure 2.3.6. The star of a polyhedron in a polyhedral complex.

This construction is usually applied to vectors $\mathbf{v}_i = (\mathbf{u}_i, 1)$ representing a point configuration $\mathbf{u}_1, \ldots, \mathbf{u}_r$ in $\mathbb{R}^n$. In that case, the fan above corresponds to a subdivision of the polytope $P = \operatorname{conv}\{\mathbf{u}_i : 1 \leq i \leq r\}$ in $\mathbb{R}^n$. The regular subdivision of P induced by $\mathbf{w} = (w_1, \ldots, w_r) \in \mathbb{R}^r$ has the following geometric description. We form the polytope

$$P_{\mathbf{w}} = \operatorname{conv}\{(\mathbf{u}_i, w_i) : 1 \leq i \leq r\} \subset \mathbb{R}^{n+1}.$$

The *lower faces* of $P_{\mathbf{w}}$ are those with an inner normal vector $\mathbf{c} \in (\mathbb{R}^{n+1})^\vee$ with last coordinate positive. These lower faces project down to $P \subseteq \mathbb{R}^n$. They form a polyhedral complex whose support equals P. This is the *regular subdivision* of $\mathbf{u}_1, \ldots, \mathbf{u}_r$ induced by $\mathbf{w}$. When each polytope in the complex is a simplex, the subdivision is called a *regular triangulation* of P.

It can now be checked that the regular subdivision of $\mathbf{u}_1, \ldots, \mathbf{u}_r$ induced by $\mathbf{w}$ is the polyhedral complex obtained by intersecting the regular subdivision of the vectors $\{(\mathbf{u}_1, 1), \ldots, (\mathbf{u}_r, 1)\} \subset \mathbb{R}^{n+1}$ induced by $\mathbf{w}$ with the hyperplane obtained by setting the last coordinate equal to one. Indeed, if $(\mathbf{c}, 1)$ is an inner normal vector for a face $\operatorname{conv}\{(\mathbf{u}_i, w_i) : i \in \sigma\}$ of $P_{\mathbf{w}}$, then there is $c_0 \in \mathbb{R}$ with $(\mathbf{c}, 1) \cdot (\mathbf{u}_i, w_i) \geq c_0$ for all i, with equality exactly when $i \in \sigma$. Thus $(-\mathbf{c}, c_0) \cdot (\mathbf{u}_i, 1) \leq w_i$, with equality exactly when $i \in \sigma$.

Example 2.3.9. We present three low-dimensional examples to show the concepts of regular subdivisions and triangulations for cones and polytopes.

(1) Let $n = 1, r = 4$. The vectors $(0,1), (1,1), (2,1), (3,1)$ span the cone $\operatorname{pos}((0,1),(3,1))$ in $\mathbb{R}^2$. When $\mathbf{w} = (4,2,1,2)$, the regular triangulation has three cones: $\operatorname{pos}((0,1),(1,1)), \operatorname{pos}((1,1),(2,1))$, and $\operatorname{pos}((2,1),(3,1))$. The same is true for the weight vector $\mathbf{w} = (1,0,0,1)$. For $\mathbf{w} = (3,2,1,2)$ there are two cones: $\operatorname{pos}((0,1),(2,1))$

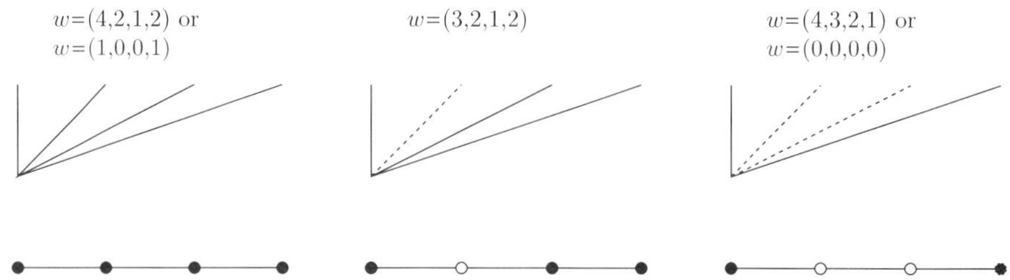

Figure 2.3.7. Some examples of regular subdivisions.

and pos((2, 1), (3, 1)). For $\mathbf{w} = (4, 3, 2, 1)$ there is only one cone: pos((0, 1), (3, 1)). The three subdivisions are shown in Figure 2.3.7.

(2) Consider the points $0, 1, 2, 3$ on the line $\mathbb{R}^1 = \mathbb{R}$. Their convex hull is the segment $[0, 3]$. When $\mathbf{w} = (4, 2, 1, 2)$, the regular triangulation consists of three line segments: $[0, 1]$, $[1, 2]$, and $[2, 3]$. The same is true for the weight vector $\mathbf{w} = (1, 0, 0, 1)$. For $\mathbf{w} = (3, 2, 1, 2)$ there are two line segments: $[0, 2]$ and $[2, 3]$, and for $\mathbf{w} = (4, 3, 2, 1)$ or $\mathbf{w} = (0, 0, 0, 0)$ there is only one line segment: $[0, 3]$. These are also shown in Figure 2.3.7. Note that this example is a slice of the previous one obtained by setting the last coordinate equal to one.

(3) Let $n = 2, r = 6$. Fix the points $(2, 0), (1, 1), (0, 2), (1, 0), (0, 1), (0, 0)$. For $\mathbf{w} = (1, 0, 1, 0, 0, 2)$, we get the triangulation with four triangles shown first in Figure 2.3.8. For $\mathbf{w} = (3, 0, 3, 1, 1, 0)$ the triangulation again has four triangles but is different from the first one; it is second in Figure 2.3.8. Finally, for $\mathbf{w} = (0, 0, 0, 0, 0, 0)$, the regular triangulation has only the one triangle shown at right in Figure 2.3.8. ◊

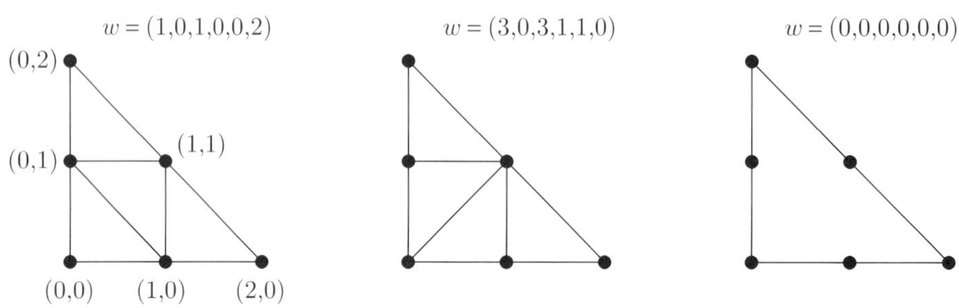

Figure 2.3.8. Some more examples of regular triangulations.

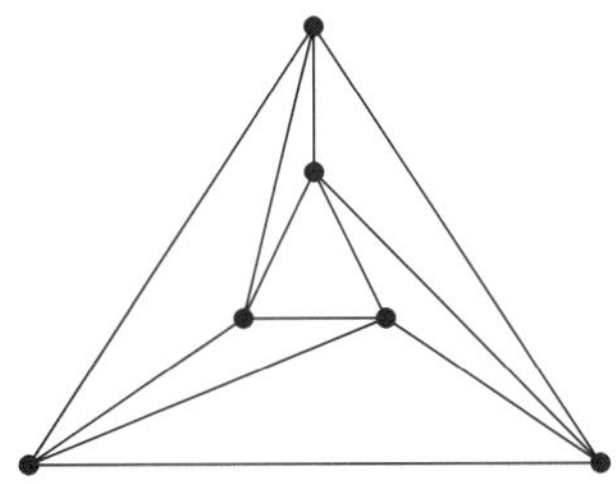

Figure 2.3.9. An example of a nonregular triangulation.

Remark 2.3.10. Not every subdivision of $\text{conv}(\mathbf{u}_1, \ldots, \mathbf{u}_r)$ is regular. The smallest nonregular example is shown in Figure 2.3.9. A subdivision is regular if and only if it is dual to a tropical curve, as defined in Section 1.3.

Computationally inclined readers may wonder what software is available for polytopes and polyhedra. An excellent general purpose platform is the software `polymake` [**GJ00**] due to Evgeny Gavrilov and Michael Joswig. For the specific study of polyhedral complexes and fans arising in tropical geometry, we also recommend Anders Jensen's package `Gfan` [**Jen**].

In this section we distinguished between a vector space $\mathbb{R}^n$ and its dual $(\mathbb{R}^n)^\vee$. In later sections we identify these two via the usual dot product.

2.4. Gröbner Bases

In this section we introduce Gröbner bases over a field K with a valuation. This is a generalization of the Gröbner basis theory familiar from [**CLO07**] and other standard references, such as [**Eis95**, Chapter 15]. The material in this section is not easy. Even those who are familiar with Gröbner bases will need to put some extra effort in now. For that reason, some readers may prefer to jump to Section 3.1 first and to return to this point later on.

In what follows, the field K not need be algebraically closed, but we will assume that it has a splitting $\phi\colon \Gamma_{\text{val}} \to K^*, w \mapsto t^w$. This hypothesis is needed for defining Gröbner bases. It holds if the valuation on K is trivial.

If $a \in K$ satisfies $\text{val}(a) \geq 0$, so a lies in the valuation ring R of K, we denote by $\overline{a}$ the image of a in the residue field $\Bbbk$. The splitting ϕ ensures that for every element $a \in K^*$ we get a nonzero element $\overline{t^{-\text{val}(a)}a} \in \Bbbk$. The resulting function $K^* \to \Bbbk^*$ is a homomorphism of multiplicative groups. For a polynomial f with coefficients in R, we write $\overline{f}$ for the polynomial obtained by replacing every coefficient a by $\overline{a}$.

Our first goal is to define Gröbner bases for a homogeneous ideal I in the polynomial ring $S = K[x_0, x_1, \ldots, x_n]$. As in standard Gröbner basis theory, this involves passing from I to an initial ideal. The difference is that now the valuations of the coefficients play a role in determining the initial ideal.

Let us start with a single polynomial $f = \sum_{\mathbf{u}\in\mathbb{N}^{n+1}} c_{\mathbf{u}} x^{\mathbf{u}}$ in S. The *tropicalization* of f is the piecewise linear function $\mathrm{trop}(f)\colon \mathbb{R}^{n+1} \to \mathbb{R}$ given by

$$(2.4.1) \qquad \mathrm{trop}(f)(\mathbf{w}) \;=\; \min\{\, \mathrm{val}(c_{\mathbf{u}}) + \mathbf{w}\cdot\mathbf{u} \;:\; \mathbf{u}\in\mathbb{N}^{n+1} \text{ and } c_{\mathbf{u}} \neq 0 \,\}.$$

Thus, $\mathrm{trop}(f)$ is the tropical polynomial induced by the classical polynomial f. Fix a *weight vector* $\mathbf{w} \in \mathbb{R}^{n+1}$, and let $W = \mathrm{trop}(f)(\mathbf{w}) = \min\{\mathrm{val}(c_{\mathbf{u}}) + \mathbf{w}\cdot\mathbf{u} : c_{\mathbf{u}} \neq 0\}$. The *initial form* of f with respect to $\mathbf{w}$ is

$$(2.4.2) \qquad \mathrm{in}_{\mathbf{w}}(f) \;=\; \sum_{\substack{\mathbf{u}\in\mathbb{N}^{n+1}:\\ \mathrm{val}(c_{\mathbf{u}})+\mathbf{w}\cdot\mathbf{u}=W}} \overline{c_{\mathbf{u}} t^{-\mathrm{val}(c_{\mathbf{u}})}}\, x^{\mathbf{u}} \;\in\; \Bbbk[x_0,\ldots,x_n].$$

When $\mathbf{w} \in \Gamma_{\mathrm{val}}^{n+1}$, the initial form can also be expressed as

$$\mathrm{in}_{\mathbf{w}}(f) \;=\; \overline{t^{-W} \sum_{\mathbf{u}\in\mathbb{N}^{n+1}} c_{\mathbf{u}} t^{\mathbf{w}\cdot\mathbf{u}} x^{\mathbf{u}}} \;=\; \overline{t^{-\mathrm{trop}(f)(\mathbf{w})} f(t^{w_0}x_0,\ldots,t^{w_n}x_n)}.$$

Example 2.4.1. Let $f = (t+t^2)x_0 + 2t^2x_1 + 3t^4x_2 \in \mathbb{C}\{\{t\}\}[x_0,x_1,x_2]$. If $\mathbf{w} = (0,0,0)$, then $W = 1$ and $\mathrm{in}_{\mathbf{w}}(f) = \overline{(1+t)x_0} = x_0$. If $\mathbf{w} = (4,2,0)$, then $W = 4$ and $\mathrm{in}_{\mathbf{w}}(f) = 2x_1 + 3x_2$. Note also $\mathrm{in}_{(2,1,0)}(f) = x_0 + 2x_1$. $\Diamond$

It is instructive to also consider the field $K = \mathbb{Q}$ with the p-adic valuation of Example 2.1.2. Here, if f is a polynomial with rational coefficients, then $\mathrm{in}_{\mathbf{w}}(f)$ is a polynomial with coefficients in the finite field $\mathbb{Z}/p\mathbb{Z}$. See Example 2.4.3 for an illustration with linear polynomials in the case $p = 2$.

If I is a homogeneous ideal in $K[x_0,\ldots,x_n]$, then its *initial ideal* is

$$\mathrm{in}_{\mathbf{w}}(I) \;=\; \langle \mathrm{in}_{\mathbf{w}}(f) : f \in I \rangle \;\subset\; \Bbbk[x_0,\ldots,x_n].$$

Note that $\mathrm{in}_{\mathbf{w}}(I)$ is an ideal in $\Bbbk[x_0,\ldots,x_n]$. A set $\mathcal{G} = \{g_1,\ldots,g_s\} \subset I$ is a *Gröbner basis* for I with respect to $\mathbf{w}$ if $\mathrm{in}_{\mathbf{w}}(I) = \langle\, \mathrm{in}_{\mathbf{w}}(g_1),\ldots,\mathrm{in}_{\mathbf{w}}(g_s)\rangle$.

Lemma 2.4.2. *Let $I \subset K[x_0,\ldots,x_n]$ be a homogeneous ideal. Fix $\mathbf{w} \in \mathbb{R}^{n+1}$. Then $\mathrm{in}_{\mathbf{w}}(I)$ is homogeneous, and we may choose a homogeneous Gröbner basis for I. Further, if $g\in\mathrm{in}_{\mathbf{w}}(I)$, then $g=\mathrm{in}_{\mathbf{w}}(f)$ for some $f\in I$.*

Proof. To see that $\mathrm{in}_{\mathbf{w}}(I)$ is homogeneous, consider $f = \sum_{i\geq 0} f_i \in S$ with each f_i homogeneous of degree i. The initial form $\mathrm{in}_{\mathbf{w}}(f)$ is the sum of initial forms of those f_i with $\mathrm{trop}(f)(\mathbf{w}) = \mathrm{trop}(f_i)(\mathbf{w})$. Since each homogeneous component f_i lives in I, the initial ideal $\mathrm{in}_{\mathbf{w}}(I)$ is generated by elements $\mathrm{in}_{\mathbf{w}}(f)$ with f homogeneous. The initial form of a homogeneous polynomial is homogeneous, so this means that $\mathrm{in}_{\mathbf{w}}(I)$ is homogeneous. As the polynomial ring is Noetherian, $\mathrm{in}_{\mathbf{w}}(I)$ is generated by a finite number of these $\mathrm{in}_{\mathbf{w}}(f)$, so the corresponding f form a homogeneous Gröbner basis for I. For the last claim, let $g = \sum a_{\mathbf{u}} x^{\mathbf{u}}\, \mathrm{in}_{\mathbf{w}}(f_{\mathbf{u}}) \in \mathrm{in}_{\mathbf{w}}(I)$, with $a_{\mathbf{u}} \in \Bbbk^*$

and $f_{\mathbf{u}} \in I$ for all $\mathbf{u}$. Then $g = \sum a_{\mathbf{u}} \operatorname{in}_{\mathbf{w}}(x^{\mathbf{u}} f_{\mathbf{u}})$. For each $a_{\mathbf{u}}$ choose a lift $c_{\mathbf{u}} \in R$ with $\operatorname{val}(c_{\mathbf{u}}) = 0$ and $\overline{c_{\mathbf{u}}} = a_{\mathbf{u}}$, and let $W_{\mathbf{u}} = \operatorname{trop}(f_{\mathbf{u}})(\mathbf{w}) + \mathbf{w} \cdot \mathbf{u}$. Let $f = \sum_{\mathbf{u}} c_{\mathbf{u}} t^{-W_{\mathbf{u}}} x^{\mathbf{u}} f_{\mathbf{u}}$. Then by construction $\operatorname{trop}(f)(\mathbf{w}) = 0$, and $\operatorname{in}_{\mathbf{w}}(f) = \sum_{\mathbf{u}} a_{\mathbf{u}} x^{\mathbf{u}} \operatorname{in}_{\mathbf{w}}(f) = g$. □

Example 2.4.3. Let $K = \mathbb{Q}$ with the 2-adic valuation, so $\Bbbk = \mathbb{Z}/2\mathbb{Z}$. Let $n = 3$ and consider the line in $\mathbb{P}^3_K$ defined by the ideal of linear forms

$$I \;=\; \langle\, x_0 + 2x_1 - 3x_2,\; 3x_1 - 4x_2 + 5x_3 \,\rangle.$$

If $\mathbf{w} = (0,0,0,0)$, then the two generators are a Gröbner basis and $\operatorname{in}_{\mathbf{w}}(I) = \langle x_0 + x_2,\; x_1 + x_3 \rangle$. This is an ideal in the polynomial ring with coefficients in $\Bbbk = \mathbb{Z}/2\mathbb{Z}$. If $\mathbf{w} = (1,0,0,1)$, then $\operatorname{in}_{\mathbf{w}}(I) = \langle x_1, x_2 \rangle$, and a Gröbner basis is $\{x_2 - 3x_0 + 10x_3, x_1 - 4x_0 + 15x_3\}$. One may ask, how many distinct initial ideals there are as $\mathbf{w}$ varies over $\mathbb{R}^4$? For an answer see Example 2.5.10. ◊

Remark 2.4.4. Our definition of Gröbner bases is restricted to polynomial ideals I that are homogeneous. With this restriction, every Gröbner basis $\mathcal{G}$ generates its ideal I. For a proof see [**CM13**]. The same definition of Gröbner bases makes sense also for nonhomogeneous polynomial ideals I, but these are generally not generated by their Gröbner bases. For instance, the singleton $\mathcal{G} = \{x - x^2\}$ is a Gröbner basis for the ideal $I = \langle x \rangle$ in the univariate polynomial ring $K[x]$ with $w = 1$, but $\mathcal{G}$ does not generate I. ◊

Remark 2.4.5. The material in this section contains the usual Gröbner basis theory using term orders [**CLO07**] as a special case. The latter arises when the field K has the trivial valuation. That situation is ubiquitous in this book. We refer to it as the case of *constant coefficients*. Here, if f is a polynomial and $\mathbf{w}$ is generic in $\mathbb{R}^{n+1}$, then $\operatorname{in}_{\mathbf{w}}(f)$ is the leading monomial of f with respect to the term order determined by $-\mathbf{w}$, as defined in [**Eis95**, §15.1]. For arbitrary $\mathbf{w}$, this is the leading form in the sense of [**Stu96**, §1].

We now iterate this construction, by taking initial forms of initial forms. In the outer iteration, the operator $\operatorname{in}_{\mathbf{v}}$ is applied to a polynomial f with coefficients in $\Bbbk$, where $\Bbbk$ has the trivial valuation. The same is done in Lemma 2.4.7 and Corollary 2.4.10 but with ideals I in place of polynomials f.

Lemma 2.4.6. *Fix $f \in K[x_0, \ldots, x_n]$ and $\mathbf{w}, \mathbf{v} \in \mathbb{R}^{n+1}$. There exists an $\epsilon > 0$ such that, for all ϵ' with $0 < \epsilon' < \epsilon$, we have*

$$\operatorname{in}_{\mathbf{v}}(\operatorname{in}_{\mathbf{w}}(f)) \;=\; \operatorname{in}_{\mathbf{w}+\epsilon'\mathbf{v}}(f). \tag{2.4.3}$$

Proof. Let $f = \sum_{\mathbf{u} \in \mathbb{N}^{n+1}} c_{\mathbf{u}} x^{\mathbf{u}}$. Then

$$\operatorname{in}_{\mathbf{w}}(f) \;=\; \sum_{\substack{\mathbf{u} \in \mathbb{N}^{n+1}: \\ \operatorname{val}(c_{\mathbf{u}}) + \mathbf{w}\cdot\mathbf{u} = W}} \overline{c_{\mathbf{u}} t^{\mathbf{w}\cdot\mathbf{u} - W}}\, x^{\mathbf{u}},$$

where $W = \mathrm{trop}(f)(\mathbf{w})$. Let $W' = \min(\mathbf{v} \cdot \mathbf{u} : \mathrm{val}(c_\mathbf{u}) + \mathbf{w} \cdot \mathbf{u} = W)$. Then

$$\mathrm{in}_\mathbf{v}(\mathrm{in}_\mathbf{w}(f)) \;=\; \sum_{\mathbf{v}\cdot\mathbf{u}=W'} \overline{c_\mathbf{u} t^{\mathbf{w}\cdot\mathbf{u}-W}} x^\mathbf{u}.$$

For all sufficiently small $\epsilon > 0$, we have

$$\mathrm{trop}(f)(\mathbf{w}+\epsilon\mathbf{v}) \;=\; \min(\mathrm{val}(c_\mathbf{u}) + \mathbf{w}\cdot\mathbf{u} + \epsilon\mathbf{v}\cdot\mathbf{u}) \;=\; W + \epsilon W', \quad \text{and}$$

$\{\mathbf{u} : \mathrm{val}(c_\mathbf{u}) + (\mathbf{w}+\epsilon'\mathbf{v})\cdot\mathbf{u} = W + \epsilon W'\} = \{\mathbf{u} : \mathrm{val}(c_\mathbf{u}) + \mathbf{w}\cdot\mathbf{u} = W, \mathbf{v}\cdot\mathbf{u} = W'\}$. This implies $\mathrm{in}_{\mathbf{w}+\epsilon'\mathbf{v}}(f) = \mathrm{in}_\mathbf{v}(\mathrm{in}_\mathbf{w}(f))$ for ϵ' with $0 < \epsilon' < \epsilon$. □

In Corollary 2.4.10 we shall see that (2.4.3) holds for any homogeneous ideal I in place of the polynomial f. The next lemma shows one containment.

Lemma 2.4.7. *Let I be a homogeneous ideal in $K[x_0, \ldots, x_n]$, and fix $\mathbf{w} \in \mathbb{R}^{n+1}$. There exists $\mathbf{v} \in \mathbb{R}^{n+1}$ and $\epsilon > 0$ such that $\mathrm{in}_\mathbf{v}(\mathrm{in}_\mathbf{w}(I))$ and $\mathrm{in}_{\mathbf{w}+\epsilon\mathbf{v}}(I)$ are monomial ideals, and $\mathrm{in}_\mathbf{v}(\mathrm{in}_\mathbf{w}(I)) \subseteq \mathrm{in}_{\mathbf{w}+\epsilon\mathbf{v}}(I)$ holds.*

Proof. Given $\mathbf{v} \in \mathbb{R}^{n+1}$, let $M_\mathbf{v}$ denote the ideal generated by all monomials in $\mathrm{in}_\mathbf{v}(\mathrm{in}_\mathbf{w}(I))$, and let $M_\mathbf{v}^\epsilon$ denote the ideal generated by all monomials in $\mathrm{in}_{\mathbf{w}+\epsilon\mathbf{v}}(I)$ for some $\epsilon > 0$. Choose $\mathbf{v} \in \mathbb{R}^{n+1}$ with $M_\mathbf{v}$ maximal, so there is no $\mathbf{v}' \in \mathbb{R}^{n+1}$ with $M_\mathbf{v} \subsetneq M_{\mathbf{v}'}$. This is possible since the polynomial ring is Noetherian. If $\mathrm{in}_\mathbf{v}(\mathrm{in}_\mathbf{w}(I))$ is not a monomial ideal, then there is $f \in I$ with none of the terms of $\mathrm{in}_\mathbf{v}(\mathrm{in}_\mathbf{w}(f))$ lying in $M_\mathbf{v}$. Choose $\mathbf{v}' \in \mathbb{R}^{n+1}$ with $\mathrm{in}_{\mathbf{v}'}(\mathrm{in}_\mathbf{v}(\mathrm{in}_\mathbf{w}(f)))$ a monomial; any $\mathbf{v}'$ for which the face of the Newton polytope of $\mathrm{in}_\mathbf{v}(\mathrm{in}_\mathbf{w}(f))$ minimizing $\mathbf{v}'$ is a vertex suffices. By Lemma 2.4.6 there is $\epsilon' > 0$ for which $\mathrm{in}_{\mathbf{v}+\epsilon'\mathbf{v}'}(\mathrm{in}_\mathbf{w}(f))$ is this monomial. For ϵ' sufficiently small, the ideal $\mathrm{in}_{\mathbf{v}+\epsilon'\mathbf{v}'}(\mathrm{in}_\mathbf{w}(I))$ contains each generator $x^\mathbf{u}$ of $M_\mathbf{v}$, as $x^\mathbf{u} = \mathrm{in}_\mathbf{v}(\mathrm{in}_\mathbf{w}(f))$ for some $f \in I$. This follows from applying Lemma 2.4.6 to $\mathrm{in}_\mathbf{w}(f)$. This contradicts the choice of $\mathbf{v}$, so we conclude that $\mathrm{in}_\mathbf{v}(\mathrm{in}_\mathbf{w}(I)) = M_\mathbf{v}$ for this choice of $\mathbf{v}$.

Let $M_\mathbf{v} = \langle x^{\mathbf{u}_1}, \ldots, x^{\mathbf{u}_s}\rangle$, and choose $f_1, \ldots, f_s \in I$ with $\mathrm{in}_\mathbf{v}(\mathrm{in}_\mathbf{w}(f_i)) = x^{\mathbf{u}_i}$. By Lemma 2.4.6, there is $\epsilon > 0$ with $\mathrm{in}_{\mathbf{w}+\epsilon\mathbf{v}}(f_i) = x^{\mathbf{u}_i}$ for all i. For this ϵ we have $\mathrm{in}_\mathbf{v}(\mathrm{in}_\mathbf{w}(I)) \subseteq \mathrm{in}_{\mathbf{w}+\epsilon\mathbf{v}}(I)$. We may assume that $\mathbf{v} \in \mathbb{R}^{n+1}$ has been chosen so that $M_\mathbf{v}^\epsilon$ is as large as possible, so there is no $\mathbf{v}'$ with $M_\mathbf{v} = M_{\mathbf{v}'}$ and $M_\mathbf{v}^\epsilon \subsetneq M_{\mathbf{v}'}^\epsilon$. Again, if $\mathrm{in}_{\mathbf{w}+\epsilon\mathbf{v}}(I)$ is not monomial, then there is $f \in I$ with no term of $\mathrm{in}_{\mathbf{w}+\epsilon\mathbf{v}}(f) \in M_\mathbf{v}^\epsilon$. We can choose $\mathbf{v}'$ as above so that $M_\mathbf{v}^\epsilon \subsetneq M_{\mathbf{v}+\epsilon'\mathbf{v}'}^\epsilon$ for small ϵ'. For sufficiently small ϵ' we have $M_{\mathbf{v}+\epsilon'\mathbf{v}'} = M_\mathbf{v}$. From this contradiction we conclude that $\mathrm{in}_{\mathbf{w}+\epsilon\mathbf{v}}(I)$ is also a monomial ideal. We then have the inclusion $\mathrm{in}_\mathbf{v}(\mathrm{in}_\mathbf{w}(I)) \subseteq \mathrm{in}_{\mathbf{w}+\epsilon\mathbf{v}}(I)$. □

In what follows we use the notations $S_K = K[x_0, \ldots, x_n]$ and $S_\Bbbk = \Bbbk[x_0, \ldots, x_n]$ for the polynomial rings that contain a given homogeneous ideal I and its various initial ideals $\mathrm{in}_\mathbf{w}(I)$. We measure the size of these ideals by their *Hilbert functions*. These are numerical functions $\mathbb{N} \to \mathbb{N}$,

$d \mapsto \dim(S_K/I)_d$. For $d \gg 0$, the Hilbert function agrees with a polynomial (called the *Hilbert polynomial*) whose degree is one less than the Krull dimension of the quotient of the polynomial ring modulo that ideal. Gröbner bases are used to compute invariants of I that are encoded in the Hilbert function, such as dimension, as the Hilbert function of an ideal and its initial ideal agree. We now extend this to our modified Gröbner theory.

Lemma 2.4.8. *Let $I \subseteq S_K$ be a homogeneous ideal, and let $\mathbf{w} \in \mathbb{R}^{n+1}$ be such that $\mathrm{in}_{\mathbf{w}}(I)_d$ is spanned over $\Bbbk$ by its monomials. Then the monomials $x^{\mathbf{u}}$ of degree d that are not in $\mathrm{in}_{\mathbf{w}}(I)$ form a K-basis for $(S_K/I)_d$.*

Proof. Let $\mathcal{B}_d$ be the set of monomials of degree d not contained in $\mathrm{in}_{\mathbf{w}}(I)$. We first show that the image of $\mathcal{B}_d$ in $(S_K/I)_d$ is linearly independent over K. This will imply $\dim_{\Bbbk} \mathrm{in}_{\mathbf{w}}(I)_d \geq \dim_K I_d$ because $|\mathcal{B}_d| = \binom{n+d}{n} - \dim_{\Bbbk} \mathrm{in}_{\mathbf{w}}(I)_d$. Indeed, if this set were linearly dependent there would exist $f = \sum c_{\mathbf{u}} x^{\mathbf{u}} \in I_d$, with $x^{\mathbf{u}} \notin \mathrm{in}_{\mathbf{w}}(I)$ whenever $c_{\mathbf{u}} \neq 0$. But then $\mathrm{in}_{\mathbf{w}}(f) \in \mathrm{in}_{\mathbf{w}}(I)_d$. Thus, every term of $\mathrm{in}_{\mathbf{w}}(f)$ is in $\mathrm{in}_{\mathbf{w}}(I)_d$, contradicting the construction of f.

For each monomial $x^{\mathbf{u}} \in \mathrm{in}_{\mathbf{w}}(I)_d$, choose $f_{\mathbf{u}} \in I_d$ with $\mathrm{in}_{\mathbf{w}}(f_{\mathbf{u}}) = x^{\mathbf{u}}$. This is possible by Lemma 2.4.2. We next note that $\{f_{\mathbf{u}} : x^{\mathbf{u}} \in \mathrm{in}_{\mathbf{w}}(I)_d\}$ is linearly independent in S_K. If it were not, there would be $a_{\mathbf{u}} \in K$ not all zero with $\sum a_{\mathbf{u}} f_{\mathbf{u}} = 0$. Write $f_{\mathbf{u}} = x^{\mathbf{u}} + \sum c_{\mathbf{u}\mathbf{v}} x^{\mathbf{v}}$. Let $\mathbf{u}'$ minimize $\mathrm{val}(a_{\mathbf{u}}) + \mathbf{w} \cdot \mathbf{u}$ for all $\mathbf{u} \in \mathbb{N}^{n+1}$ with $x^{\mathbf{u}} \in \mathrm{in}_{\mathbf{w}}(I)_d$. Then $a_{\mathbf{u}'} + \sum_{\mathbf{u} \neq \mathbf{u}'} a_{\mathbf{u}} c_{\mathbf{u}\mathbf{u}'} = 0$, so there is $\mathbf{u}'' \neq \mathbf{u}'$ with $\mathrm{val}(a_{\mathbf{u}''}) + \mathrm{val}(c_{\mathbf{u}''\mathbf{u}'}) \leq \mathrm{val}(a_{\mathbf{u}'})$. But then $\mathrm{val}(a_{\mathbf{u}''}) + \mathrm{val}(c_{\mathbf{u}''\mathbf{u}'}) + \mathbf{w} \cdot \mathbf{u}' \leq \mathrm{val}(a_{\mathbf{u}'}) + \mathbf{w} \cdot \mathbf{u}' \leq \mathrm{val}(a_{\mathbf{u}''}) + \mathbf{w} \cdot \mathbf{u}''$, which contradicts $\mathrm{in}_{\mathbf{w}}(f_{\mathbf{u}''}) = x^{\mathbf{u}''}$. This shows $\dim_K I_d \geq \dim_{\Bbbk} \mathrm{in}_{\mathbf{w}}(I)_d$. Thus $\dim_K(S_K/I)_d = \dim_{\Bbbk}(S_{\Bbbk}/\mathrm{in}_{\mathbf{w}}(I))_d$, and $\mathcal{B}_d$ is a K-basis for $(S_K/I)_d$. $\square$

Corollary 2.4.9. *For any $\mathbf{w} \in \mathbb{R}^{n+1}$ and any homogeneous ideal I in S_K, the Hilbert function of I agrees with that of its initial ideal $\mathrm{in}_{\mathbf{w}}(I) \subset S_{\Bbbk}$, i.e.,*

$$\dim_K(S_K/I)_d \;=\; \dim_{\Bbbk}(S_{\Bbbk}/\mathrm{in}_{\mathbf{w}}(I))_d \qquad \text{for all } d \geq 0.$$

Thus the Krull dimensions of the rings S_K/I and $S_{\Bbbk}/\mathrm{in}_{\mathbf{w}}(I)$ coincide.

Proof. By Lemma 2.4.7 there is $\mathbf{v} \in \mathbb{R}^{n+1}$ and $\epsilon > 0$ with $\mathrm{in}_{\mathbf{v}}(\mathrm{in}_{\mathbf{w}}(I)) \subseteq \mathrm{in}_{\mathbf{w}+\epsilon\mathbf{v}}(I)$, and both are monomial ideals. Let $x^{\mathbf{u}} \in \mathrm{in}_{\mathbf{w}+\epsilon\mathbf{v}}(I)_d \setminus \mathrm{in}_{\mathbf{v}}(\mathrm{in}_{\mathbf{w}}(I))_d$. By Lemma 2.4.8, monomials not in $\mathrm{in}_{\mathbf{w}+\epsilon\mathbf{v}}(I)_d$ span $(S/I)_d$. Thus there is a polynomial $f_{\mathbf{u}} = x^{\mathbf{u}} - f'_{\mathbf{u}} \in I_d$ with none of the monomials of $f'_{\mathbf{u}}$ lying in $\mathrm{in}_{\mathbf{w}+\epsilon\mathbf{v}}(I)_d$. But then $\mathrm{in}_{\mathbf{w}}(f_{\mathbf{u}})$ contains only monomials not in $\mathrm{in}_{\mathbf{v}}(\mathrm{in}_{\mathbf{w}}(I))$, so $\mathrm{in}_{\mathbf{v}}(\mathrm{in}_{\mathbf{w}}(f_{\mathbf{u}})) \notin \mathrm{in}_{\mathbf{v}}(\mathrm{in}_{\mathbf{w}}(I))_d$. From this contradiction we conclude that $\mathrm{in}_{\mathbf{w}+\epsilon\mathbf{v}}(I)_d = \mathrm{in}_{\mathbf{v}}(\mathrm{in}_{\mathbf{w}}(I))_d$. Lemma 2.4.8 applied to $\mathrm{in}_{\mathbf{w}}(I)$ gives

$$\dim_{\Bbbk}(S_{\Bbbk}/\mathrm{in}_{\mathbf{w}}(I))_d \;=\; \dim_{\Bbbk}(S_{\Bbbk}/\mathrm{in}_{\mathbf{v}}(\mathrm{in}_{\mathbf{w}}(I)))_d.$$

Applied to I, it gives $\dim_K(S_K/I)_d = \dim_{\Bbbk}(S_{\Bbbk}/\mathrm{in}_{\mathbf{w}+\epsilon\mathbf{v}}(I))_d$. Hence, for any $\mathbf{w} \in \mathbb{R}^{n+1}$, we have $\dim_K(S_K/I)_d = \dim_{\Bbbk}(S_{\Bbbk}/\mathrm{in}_{\mathbf{w}}(I))_d$ for all degrees d. $\square$

Corollary 2.4.10. *Let I be a homogeneous ideal in $K[x_0, \ldots, x_n]$. For any $\mathbf{w}, \mathbf{v} \in \mathbb{R}^{n+1}$ there exists $\epsilon > 0$ such that*

$$\mathrm{in}_{\mathbf{v}}(\mathrm{in}_{\mathbf{w}}(I)) = \mathrm{in}_{\mathbf{w}+\epsilon'\mathbf{v}}(I) \qquad \textit{for all } 0 < \epsilon' < \epsilon.$$

Proof. Let $\{g_1, \ldots, g_s\} \subset \Bbbk[x_0, \ldots, x_n]$ be a generating set for $\mathrm{in}_{\mathbf{v}}(\mathrm{in}_{\mathbf{w}}(I))$, with each generator g_i of the form $\mathrm{in}_{\mathbf{v}}(\mathrm{in}_{\mathbf{w}}(f_i))$ with $f_i \in I$. By Lemma 2.4.6, there exists $\epsilon > 0$ such that $g_i = \mathrm{in}_{\mathbf{v}}(\mathrm{in}_{\mathbf{w}}(f_i)) = \mathrm{in}_{\mathbf{w}+\epsilon'\mathbf{v}}(f_i)$ for $i = 1, \ldots, s$ whenever $0 < \epsilon' < \epsilon_i$. This implies that $\mathrm{in}_{\mathbf{v}}(\mathrm{in}_{\mathbf{w}}(I)) \subseteq \mathrm{in}_{\mathbf{w}+\epsilon'\mathbf{v}}(I)$. By Corollary 2.4.9, the two ideals $\mathrm{in}_{\mathbf{v}}(\mathrm{in}_{\mathbf{w}}(I))$ and $\mathrm{in}_{\mathbf{w}+\epsilon'\mathbf{v}}(I)$ have the same Hilbert function as I, so their containment cannot be strict. □

Example 2.4.11. The Hilbert function of the ideals in Example 2.4.3 equals

$$\dim_{\mathbb{Q}}(\mathbb{Q}[x_0, x_1, x_2, x_3]/I)_d = \dim_{\Bbbk}(\Bbbk[x_0, x_1, x_2, x_3]/\,\mathrm{in}_{\mathbf{w}}(I))_d = d + 1.$$

Here $\Bbbk = \mathbb{Z}/2\mathbb{Z}$ is the field with two elements. The Hilbert polynomial $d+1$ shows that the projective varieties $V(I)$ and $V(\mathrm{in}_{\mathbf{w}}(I))$ both have dimension 1 and degree 1. They are straight lines in $\mathbb{P}^3_{\mathbb{Q}}$ and in $\mathbb{P}^3_{\Bbbk}$, respectively. ◊

We finish this section with some results that will be useful later in the book. Corollary 2.4.9 implies that the varieties $V(I) \subset \mathbb{P}^n_K$ and $V(\mathrm{in}_{\mathbf{w}}(I)) \subset \mathbb{P}^n_{\Bbbk}$ always have the same dimension. In typical applications, the given variety $V(I)$ is irreducible but $V(\mathrm{in}_{\mathbf{w}}(I))$ can have many irreducible components.

The following lemma states that *every* irreducible component of $V(\mathrm{in}_{\mathbf{w}}(I))$ has the same dimension as $V(I)$. We shall phrase this in the algebraic language of primary decomposition. Recall that P is a *minimal associated prime* of an ideal I in S_K or $S_{\Bbbk}$ if $I \subseteq P$ and there is no prime ideal P' with $I \subseteq P' \subsetneq P$. For technical reasons we assume here that $\mathbf{w} \in \Gamma_{\mathrm{val}}^{n+1}$.

Lemma 2.4.12. *If $I \subset S_K$ is a homogeneous prime of dimension d and $\mathbf{w} \in \Gamma_{\mathrm{val}}^{n+1}$, then every minimal associated prime of $\mathrm{in}_{\mathbf{w}}(I)$ has dimension d.*

Proof. Since $\mathbf{w} \in \Gamma_{\mathrm{val}}^{n+1}$ we may perform the change of coordinates $\phi^*(x_i) = t^{w_i} x_i$ to reduce to the case $\mathbf{w} = \mathbf{0}$. This takes $g \in I$ to $g(t^{w_0}x_0, \ldots, t^{w_n}x_n) \in \phi^*(I)$, so $\mathrm{in}_{\mathbf{w}}(I) = \langle \overline{f} : f \in \phi^*(I) \rangle = \mathrm{in}_{\mathbf{0}}(\phi^*(I))$. Now, let $\mathcal{G} = \{g_1, \ldots, g_s\}$ be a Gröbner basis for I, with respect to $\mathbf{0}$. After multiplying by t^w for $w \gg 0$ we may assume that $g_i \in R[x_0, \ldots, x_n]$ and $\overline{g_i} \neq 0$ for $1 \leq i \leq s$.

We first pass to a Noetherian subring R' of R where I is defined, as dimension is better behaved over Noetherian rings. It has the following definition. Let $\mathbb{F}$ be the subring of R generated by 1. This is $\mathbb{F}_p$ if $\mathrm{char}(K) = p > 0$, and $\mathbb{Z}$ if $\mathrm{char}(K) = 0$. Let $\tilde{R}$ be the $\mathbb{F}$-algebra generated by the coefficients of the g_i. Let $\mathfrak{m}' = \mathfrak{m}_K \cap \tilde{R}$, and let R' be the localization $R' = \tilde{R}_{\mathfrak{m}'}$. Since $\tilde{R}$ is a finitely generated algebra over a Noetherian ring, and localization preserves Noetherianity, the ring R' is Noetherian. The fraction field K' of R' is a subfield of K, and $\Bbbk' = R'/\mathfrak{m}'$ is a subfield of $\Bbbk$.

Let $c = \dim(R')$. By the converse to the Principal Ideal Theorem [**Eis95**, Corollary 10.5] there are $a_1, \dots, a_c \in \mathfrak{m}'$ for which $\mathfrak{m}'$ is minimal over $\langle a_1, \dots, a_c \rangle \subset R'$. Since $\mathfrak{m}'$ is the maximal ideal of R', any proper prime ideal of R' containing $\langle a_1, \dots, a_c \rangle$ equals $\mathfrak{m}'$. Let $I' = I \cap R'[x_0, \dots, x_n]$, and let $I'' = I \cap K'[x_0, \dots, x_n]$. Since $I = I'' \otimes_{K'} K$, we have $\dim(K[x_0, \dots, x_n]/I) = \dim(K'[x_0, \dots, x_n]/I'') = d$. In addition, $\dim(R'[x_0, \dots, x_n]/I') = d + c$. This follows from [**Eis95**, Theorem 13.8] applied to the prime $Q = \langle x_0, \dots, x_n \rangle + \mathfrak{m}'$ of $R'[x_0, \dots, x_n]/I'$, since R' is universally catenary.

Let P be a prime ideal of $R'[x_0, \dots, x_n]$ minimal over $I' + \mathfrak{m}'$. Any proper prime containing $I' + \langle a_1, \dots, a_c \rangle$ must intersect R' in a proper prime containing $a_1, \dots, a_c$, so it contains $\mathfrak{m}'$. Thus P is minimal over $I' + \langle a_1, \dots, a_c \rangle$. By the Principal Ideal Theorem [**Eis95**, Theorem 10.2] applied to the domain $R'[x_0, \dots, x_n]/I'$, the codimension of P/I' is at most c, so the dimension of P is at least $d + c - c = d$. Since minimal primes of $(I' + \mathfrak{m}')/\mathfrak{m}'$ have the form $P/\mathfrak{m}'$ for minimal primes P of $I' + \mathfrak{m}'$, this shows that all minimal primes of $(I' + \mathfrak{m}')/\mathfrak{m}'$ have dimension at least d. We now show $(I' + \mathfrak{m}')/\mathfrak{m}' \otimes_{\Bbbk'} \Bbbk = \text{in}_{\mathbf{0}}(I)$. Each Gröbner basis element g_i lies in $R'[x_0, \dots, x_n]$ by construction. Its image $\overline{g_i}$ lies in $\Bbbk'[x_0, \dots, x_n]$, and equals $\text{in}_{\mathbf{0}}(g_i)$. This shows $\text{in}_{\mathbf{0}}(I) \subseteq (I' + \mathfrak{m}')/\mathfrak{m}' \otimes \Bbbk$. The other inclusion is immediate from the description $\text{in}_{\mathbf{0}}(I) = \langle \overline{f} : f \in I \rangle$. Since $\dim(\text{in}_{\mathbf{w}}(I)) = d$ by Corollary 2.4.9, every minimal prime of $\text{in}_{\mathbf{0}}(I)$ is thus d-dimensional. □

Remark 2.4.13. The ring $R = \{x \in K : \text{val}(x) \geq 0\}$ need not be Noetherian. For example, when $K = \mathbb{C}\{\{t\}\}$, the ideals $I_m = \{x \in R : \text{val}(x) > 1/m\}$ form an infinite ascending chain in R. This necessitated passing to the Noetherian subring R' of R in the proof of Lemma 2.4.12, as many of the fundamental theorems of dimension theory apply only to Noetherian rings.

A key geometric property of Gröbner bases is the existence of a flat family over $\mathbb{A}^1$ for which any general fiber over $t \neq 0$ is isomorphic to the scheme defined by I, and the special fiber over $t = 0$ is isomorphic to the scheme defined by $\text{in}_{\mathbf{w}}(I)$. See [**Eis95**, Chapter 15] for an exposition. This generalizes to our new setting, where K has a valuation, as we now illustrate.

If A and B are rings with B an A-module, then B is a flat A-module if the right exact functor $- \otimes_A B$ is exact; i.e., if $C \to D$ is an injective A-module homomorphism, then the induced map $C \otimes_A B \to D \otimes_A B$ is also injective. When A and B are coordinate rings of varieties X and Y, respectively, and the A-module structure on B is induced from a homomorphism $\phi^* \colon A \to B$, then the corresponding morphism $\phi \colon Y \to X$ is a *flat family*. The variety X

is the *base* of the family, and we say that Y is flat over X, or over A. The *fiber* of the family at a point $x \in X$ has coordinate ring equal to the tensor product $B \otimes_A A/P$, where P is the ideal of the point x. We refer to [**Eis95**, Chapter 6] for more on flatness and its geometric significance.

We will show in Lemma 2.4.14 that our Gröbner theory gives a flat family over the valuation ring R of K. We associate to R the scheme $\operatorname{Spec}(R)$. As a set, this has one point for each prime ideal of R. If the valuation is nontrivial, then R has exactly two prime ideals: the zero ideal and the maximal ideal $\mathfrak{m}_K$. Hence $\operatorname{Spec}(R)$ has only two points, namely the general point and the special point. The fibers over these points are the *general fiber* and the *special fiber*, respectively. They are isomorphic to the schemes defined by the ideals I and $\operatorname{in}_{\mathbf{w}}(I)$. If valuation on R is trivial, then $R = K = \Bbbk$, so $\operatorname{Spec}(R)$ has only one point; the general and special fibers coincide.

Fix $\mathbf{w} \in \Gamma_{\mathrm{val}}^{n+1}$, and let I_R be the ideal in $R[x_0, \ldots, x_n]$ defined by

$$I_R \;=\; \left\langle t^{-\operatorname{trop}(f)(\mathbf{w})} f(t^{w_0}x_0, \ldots, t^{w_n}x_n) : f \in I \right\rangle.$$

Lemma 2.4.14. *The quotient* $M = R[x_0, \ldots, x_n]/I_R$ *is a flat* R*-module, with* $M \otimes_R K \cong K[x_0, \ldots, x_n]/I$ *and* $M \otimes_R \Bbbk \cong \Bbbk[x_0, \ldots, x_n]/\operatorname{in}_{\mathbf{w}}(I)$. *Thus*

$$\operatorname{Spec}(M) \to \operatorname{Spec}(R)$$

is a flat family with general fiber isomorphic to $\operatorname{Spec}(K[x_0, \ldots, x_n]/I)$ *and special fiber isomorphic to* $\operatorname{Spec}(\Bbbk[x_0, \ldots, x_n]/\operatorname{in}_{\mathbf{w}}(I))$.

Proof. We first show that M is a flat R-module. By [**Eis95**, Proposition 6.1] this is the case if and only if $\operatorname{Tor}_1^R(R/J, M) = 0$ for all finitely generated ideals $J \subset R$. Since R is a valuation ring, if J is a finitely generated ideal, it is principal: if $J = \langle a_1, \ldots, a_r \rangle$ with $\operatorname{val}(a_i) \leq \operatorname{val}(a_{i+1})$ for $1 \leq i \leq r-1$, then $J = \langle a_1 \rangle$, as $a_i/a_1 \in R$. Now $\operatorname{Tor}_1^R(R/a, M) = (0 :_M a)$, so this equals zero if and only if a is a nonzerodivisor on M. To see this, first note that M is isomorphic as an R-algebra to $S^{\mathbf{w}}/I^{\mathbf{w}}$, where

$$S^{\mathbf{w}} = \{\textstyle\sum c_{\mathbf{u}} x^{\mathbf{u}} : \operatorname{val}(c_{\mathbf{u}}) + \mathbf{w} \cdot \mathbf{u} \geq 0\} \subset S_K = K[x_0, \ldots, x_n]$$

and $I^{\mathbf{w}} = I \cap S^{\mathbf{w}}$. The isomorphism sends x_i to $t^{-w_i}x_i$. The fact that M is torsion-free is then immediate; if $f = \sum c_{\mathbf{u}} x^{\mathbf{u}} \in S^{\mathbf{w}}$ with $af \in I^{\mathbf{w}}$, then $af \in I$, so $f \in I$, and thus $f \in I^{\mathbf{w}}$.

We now show the claim about the general and special fibers. For the general fiber, we again use the isomorphism $M \cong S^{\mathbf{w}}/I^{\mathbf{w}}$. Consider the homomorphism $\psi\colon S^{\mathbf{w}}/I^{\mathbf{w}} \otimes_R K \to S_K/I$ given by sending $f \otimes a$ to af. To see that this is well defined, note that if $f \in I^{\mathbf{w}}$, then $\psi(f \otimes 1) = f \in I$. To see that it is injective, suppose that $\psi(\sum \mu_i(f_i \otimes a_i)) = 0$, where $\mu_i \in \mathbb{Z}$, $f_i \in S^{\mathbf{w}}$, and $a_i \in K$ for each i. This means that $\sum \mu_i f_i a_i \in I$. Let $v = \min(\operatorname{val}(a_i))$, and note that $f_i \otimes a_i = f_i a_i/t^v \otimes t^v$, since $a_i/t^v \in R$ for all i. Thus $\sum \mu_i f_i \otimes a_i = \sum(\mu_i f_i a_i/t^v) \otimes t^v = (t^{-v} \sum \mu_i f_i a_i) \otimes t^v$. Since $\sum \mu_i f_i a_i \in I$, so is $t^{-v} \sum \mu_i f_i a_i$, and thus this latter sum is in $I^{\mathbf{w}}$. This shows that $\sum \mu_i f_i \otimes a_i = 0$, and thus ψ is injective. It remains to note that ψ is surjective. This follows from the facts that for $\mathbf{u} \in \mathbb{N}^{n+1}$ we have $cx^{\mathbf{u}} \in S^{\mathbf{w}}$ for any c with $\text{val}(c) > -\mathbf{w} \cdot \mathbf{u}$, and for such a c we have $\psi(cx^{\mathbf{u}} \otimes a/c) = ax^{\mathbf{u}}$ for any $a \in K$. Thus $S^{\mathbf{w}}/I^{\mathbf{w}} \otimes_R K \cong S_K/I$ as required.

Let $S_{\Bbbk} = \Bbbk[x_0, \dots, x_n]$ and consider the homomorphism

$$\psi\colon R[x_0, \dots, x_n]/I_R \otimes_R R/\mathfrak{m}_K \to S_{\Bbbk}/\operatorname{in}_{\mathbf{w}}(I)$$

given by $\psi(f \otimes c) = \overline{cf}$ for $f \in R[x_0, \dots, x_n]$ and $c \in R$. To see that ψ is well defined, note that if $f \in I$ and $f' = t^{-\operatorname{trop}(f)(\mathbf{w})} f(t^{w_0}x_0, \dots, t^{w_n}x_n) \in I_R$, then $\psi(f' \otimes 1) = \operatorname{in}_{\mathbf{w}}(f)$. The homomorphism ψ is surjective by construction. To see that it is injective, note that if $f \in R[x_0, \dots, x_n]$ with $\overline{f} \in \operatorname{in}_{\mathbf{w}}(I)$, then there is $g \in I$ with $\operatorname{in}_{\mathbf{w}}(g) = \overline{f}$. Set $g' = t^{-\operatorname{trop}(g)(\mathbf{w})} g(t^{w_0}x_0, \dots, t^{w_n}x_n)$. Then $\overline{g'} = \operatorname{in}_{\mathbf{w}}(g) = \overline{f}$, so $g' - f \in \mathfrak{m}_K R[x_0, \dots, x_n]$. Now suppose that $\sum m_i(f_i \otimes a_i) \in \ker(\psi)$ for $m_i \in \mathbb{Z}$, $f_i \in R[x_0, \dots, x_n]$, and $a_i \in R$. We have $\sum m_i(f_i \otimes a_i) = (\sum m_i a_i f_i) \otimes 1$, so $\overline{\sum m_i a_i f_i} = 0$. The previous observation then implies that $\sum m_i a_i f_i = h_1 + a h_2$, where $h_1 \in I_R$, $h_2 \in R[x_0, \dots, x_n]$ and $a \in \mathfrak{m}_K$. But then $(\sum m_i a_i f_i) \otimes 1 = (h_1 + a h_2) \otimes 1 = h_1 \otimes 1 + h_2 \otimes a = 0$, which shows that ψ is injective. We conclude that $M \otimes_R R/\mathfrak{m}_K \cong S_{\Bbbk}/\operatorname{in}_{\mathbf{w}}(I)$, which proves the claim for the special fiber. □

Remark 2.4.15. The assumption $\mathbf{w} \in \Gamma_{\text{val}}^{n+1}$ in Lemma 2.4.14 cannot be easily removed. For example, if the valuation on K is trivial, so that $\Gamma_{\text{val}} = \{0\}$, then the general and special fibers of any family over R coincide. This means that $K[x_0, \dots, x_n]/I$ equals $\Bbbk[x_0, \dots, x_n]/\operatorname{in}_{\mathbf{w}}(I)$, and hence $\operatorname{in}_{\mathbf{w}}(I) = I$. But this happens for all $\mathbf{w} \in \mathbb{R}^{n+1}$ only if I is a monomial ideal.

Remark 2.4.16. Our definition of initial ideal depends on the choice of a splitting $w \mapsto t^w$ of the valuation map. Under the assumption that K is algebraically closed, this always exists by Lemma 2.1.15. We note here that different choices do not lead to substantially different initial ideals. Indeed, suppose $\phi_1, \phi_2 : \Gamma_{\text{val}} \to K^*$ are two different splittings of val, so $\text{val} \circ \phi_1 = \text{val} \circ \phi_2 = \text{id} : \Gamma_{\text{val}} \to \Gamma_{\text{val}}$. Fix $\mathbf{w} \in \mathbb{R}^{n+1}$. We then have isomorphisms $\phi_i\colon K[x_0, \dots, x_n] \to K[x_0, \dots, x_n]$ given by $x_j \mapsto \phi_i(w_j)x_j$ for $i = 1, 2$. The composition $\psi = \phi_1 \circ \phi_2^{-1}\colon K[x_0, \dots, x_n] \to K[x_0, \dots, x_n]$ restricts to an automorphism $\psi\colon R[x_0, \dots, x_n] \to R[x_0, \dots, x_n]$, given by $\psi(x_j) = (\phi_1(w_j)/\phi_2(w_j)) \cdot x_j$. This maps the ideal I_R defined above using ϕ_2 to the ideal I_R defined using ϕ_1. The induced isomorphism $\overline{\psi}\colon \Bbbk[x_0, \dots, x_n] \to \Bbbk[x_0, \dots, x_n]$ maps the initial ideal defined using ϕ_2 to the one using ϕ_1. Thus any two initial ideals defined using different splittings are related by an automorphism of $\Bbbk[x_0, \dots, x_n]$, so all invariants of these ideals coincide.

2.5. Gröbner Complexes

The goal of this section is to construct a polyhedral complex from a given homogeneous ideal $I \subset K[x_0, x_1, \ldots, x_n]$. This will be the ambient space for the tropical variety of I. Our assumption on the field K, as in Section 2.4, is that there is a splitting $w \mapsto t^w$ of the valuation on K.

We begin by defining the polyhedra in our complex. For $\mathbf{w} \in \mathbb{R}^{n+1}$ set

$$C_I[\mathbf{w}] \;=\; \{\mathbf{w}' \in \mathbb{R}^{n+1} : \mathrm{in}_{\mathbf{w}'}(I) = \mathrm{in}_{\mathbf{w}}(I)\}.$$

Let $\overline{C_I[\mathbf{w}]}$ be the closure of $C_I[\mathbf{w}]$ in $\mathbb{R}^{n+1}$ in Euclidean topology.

Example 2.5.1. Let $n = 2$ and $K = \mathbb{Q}$ with the 2-adic valuation, and let I be the principal ideal generated by the homogeneous cubic polynomial

$$f \;=\; 2x_0^3 + 4x_1^3 + 2x_2^3 + x_0x_1x_2.$$

The initial ideal for $\mathbf{w} = (0,0,0)$ equals $\mathrm{in}_{\mathbf{w}}(I) = \langle x_0x_1x_2\rangle$. Note that

$$\overline{C_I[\mathbf{w}]} \;=\; \{(v_0, v_1, v_2) \in \mathbb{R}^3 : v_0 + v_1 + v_2 \leq \min(3v_0 + 1, 3v_1 + 2, 3v_2 + 1)\}.$$

The valuation is essential here because $x_0x_1x_2$ would not be an initial monomial of f in the usual Gröbner basis sense of [**CLO07**]. The polyhedron $\overline{C_I[\mathbf{w}]}$ is the product of a triangle with the line spanned by $\mathbf{1} = (1,1,1)$. In what follows we shall work in the quotient modulo that line. ◊

We denote by $\mathbf{1} = (1, 1, \ldots, 1)$ the all-one vector in $\mathbb{R}^{n+1}$.

Proposition 2.5.2. *The set $\overline{C_I[\mathbf{w}]}$ is a Γ_{val}-rational polyhedron whose lineality space contains the line $\mathbb{R}\mathbf{1}$. If $\mathrm{in}_{\mathbf{w}}(I)$ is not a monomial ideal, then there exists $\mathbf{w}' \in \Gamma_{\mathrm{val}}^{n+1}$ such that $\mathrm{in}_{\mathbf{w}'}(I)$ is a monomial ideal and $\overline{C_I[\mathbf{w}]}$ is a proper face of the polyhedron $\overline{C_I[\mathbf{w}']}$.*

Proof. By Lemma 2.4.7, there exists $\mathbf{v} \in \mathbb{R}^{n+1}$ with $\mathrm{in}_{\mathbf{v}}(\mathrm{in}_{\mathbf{w}}(I))$ a monomial ideal. By Corollary 2.4.10, we have $\mathrm{in}_{\mathbf{w}+\epsilon\mathbf{v}}(I) = \mathrm{in}_{\mathbf{v}}(\mathrm{in}_{\mathbf{w}}(I))$ for sufficiently small $\epsilon > 0$. Fix such an ϵ, and let $\mathbf{w}' = \mathbf{w}+\epsilon\mathbf{v}$. Let $\mathrm{in}_{\mathbf{w}'}(I) = \langle x^{\mathbf{u}_1}, \ldots, x^{\mathbf{u}_s}\rangle$. By Lemma 2.4.8, the monomials not in $\mathrm{in}_{\mathbf{w}'}(I)$ of degree $d = \deg(x^{\mathbf{u}_i})$ form a basis for $(S/I)_d$. Let g_i' be the expansion of $x^{\mathbf{u}_i}$ in this basis, so no monomial occurring in g_i' lies in $\mathrm{in}_{\mathbf{w}'}(I)$. We write $c_{i\mathbf{v}}$ for the coefficient of $x^{\mathbf{v}}$ in g_i'. The polynomial $g_i = x^{\mathbf{u}_i} - g_i'$ is in I. Since $\mathrm{in}_{\mathbf{v}}(\mathrm{in}_{\mathbf{w}}(g_i))$ must lie in $\mathrm{in}_{\mathbf{w}'}(I)$, we have $\mathrm{in}_{\mathbf{v}}(\mathrm{in}_{\mathbf{w}}(g_i)) = x^{\mathbf{u}_i}$, and thus $\mathrm{in}_{\mathbf{w}'}(g_i) = x^{\mathbf{u}_i}$. By construction, the polynomials $\{g_1, g_2, \ldots, g_s\}$ form a Gröbner basis for I with respect to $\mathbf{w}'$.

We claim that $\overline{C_I[\mathbf{w}']}$ has the following inequality description:

$$(2.5.1)\quad \overline{C_I[\mathbf{w}']} = \{\mathbf{z} \in \mathbb{R}^{n+1} : \mathbf{u}_i{\cdot}\mathbf{z} \leq \mathrm{val}(c_{i\mathbf{v}}) + \mathbf{v}{\cdot}\mathbf{z} \text{ for } 1 \leq i \leq s,\ \mathbf{v} \in \mathbb{N}^{n+1}\}.$$

This implies that $\overline{C_I[\mathbf{w}']}$ is a Γ_{val}-rational polyhedron.

We now prove (2.5.1). Suppose $\tilde{\mathbf{w}} \in C_I[\mathbf{w}']$, but one of the inequalities $\mathbf{u}_i \cdot \mathbf{z} \leq \mathrm{val}(c_{i\mathbf{v}}) + \mathbf{v} \cdot \mathbf{z}$ is violated for $\mathbf{z} = \tilde{\mathbf{w}}$. For that index i, we have $\mathrm{in}_{\tilde{\mathbf{w}}}(g_i) \neq x^{\mathbf{u}_i}$. Since $\mathrm{in}_{\mathbf{w}'}(I) = \mathrm{in}_{\tilde{\mathbf{w}}}(I)$ is a monomial ideal, every term of $\mathrm{in}_{\tilde{\mathbf{w}}}(g_i)$ lies in $\mathrm{in}_{\tilde{\mathbf{w}}}(I)$, so this would contradict the construction of the polynomials g_i. Thus $\overline{C_I[\mathbf{w}']}$ is contained in the right-hand side of (2.5.1).

For the reverse inclusion, we assume $\mathbf{u}_i \cdot \tilde{\mathbf{w}} < \mathrm{val}(c_{i\mathbf{v}}) + \mathbf{v} \cdot \tilde{\mathbf{w}}$ for all i. Then $\mathrm{in}_{\tilde{\mathbf{w}}}(g_i) = x^{\mathbf{u}_i}$ for all i, and hence $\mathrm{in}_{\mathbf{w}'}(I) \subseteq \mathrm{in}_{\tilde{\mathbf{w}}}(I)$. The two ideals have the same Hilbert function, so they are equal, and we conclude $\tilde{\mathbf{w}} \in C_I[\mathbf{w}']$.

The first paragraph of the proof shows $C_I[\mathbf{w}] \subset \overline{C_I[\mathbf{w}']}$. To see that $\overline{C_I[\mathbf{w}]}$ is a Γ_{val}-rational polyhedron, it suffices to show that it is a face of $\overline{C_I[\mathbf{w}']}$. Note that $\{\mathrm{in}_{\mathbf{w}}(g_1), \ldots, \mathrm{in}_{\mathbf{w}}(g_s)\}$ is a Gröbner basis for $\mathrm{in}_{\mathbf{w}}(I)$ with respect to $\mathbf{v}$. If $\widetilde{\mathbf{w}} \in \mathbb{R}^{n+1}$ satisfies $\mathrm{in}_{\widetilde{\mathbf{w}}}(I) = \mathrm{in}_{\mathbf{w}}(I)$, then $\mathrm{in}_{\widetilde{\mathbf{w}}}(g_i) = \mathrm{in}_{\mathbf{w}}(g_i)$ for all i. Otherwise, $\mathrm{in}_{\widetilde{\mathbf{w}}}(g_i)$ would still have $x^{\mathbf{u}_i}$ in its support, or $\mathrm{in}_{\mathbf{v}}(\mathrm{in}_{\widetilde{\mathbf{w}}}(I))$ would not be equal to the monomial ideal $\mathrm{in}_{\mathbf{w}'}(I)$. But then $\mathrm{in}_{\widetilde{\mathbf{w}}}(g_i) - \mathrm{in}_{\mathbf{w}}(g_i) \in \mathrm{in}_{\mathbf{w}}(I)$, and this polynomial does not contain any monomials from $\mathrm{in}_{\mathbf{w}'}(I)$, contradicting the fact that $\mathrm{in}_{\mathbf{v}}(\mathrm{in}_{\mathbf{w}}(I)) = \mathrm{in}_{\mathbf{w}'}(I)$. We conclude that $\overline{C_I[\mathbf{w}]}$ is the set of points $\mathbf{z}$ in the cone $\overline{C_I[\mathbf{w}']}$ that satisfy $\mathbf{u}_i \cdot \mathbf{z} = \mathrm{val}(c_{i\mathbf{v}}) + \mathbf{v} \cdot \mathbf{z}$ whenever $x^{\mathbf{v}}$ appears in $\mathrm{in}_{\mathbf{w}}(g_i)$. This shows that $\overline{C_I[\mathbf{w}]}$ is a face of $\overline{C_I[\mathbf{w}']}$.

Finally, for any homogeneous polynomial $f \in K[x_0, \ldots, x_n]$, we have $\mathrm{in}_{\mathbf{w}}(f) = \mathrm{in}_{\mathbf{w}+\lambda\mathbf{1}}(f)$ for all $\lambda \in \Gamma_{\mathrm{val}}$. Since all initial ideals of I are generated by homogeneous polynomials, by Lemma 2.4.2, this implies $\mathrm{in}_{\mathbf{w}}(I) = \mathrm{in}_{\mathbf{w}+\lambda\mathbf{1}}(I)$ for all $\lambda \in \Gamma_{\mathrm{val}}$. Therefore, $\overline{C_I[\mathbf{w}]} = \overline{C_I[\mathbf{w}]} + \mathbb{R}\mathbf{1}$. We conclude that the lineality space of the polyhedron $\overline{C_I[\mathbf{w}]}$ contains the line $\mathbb{R}\mathbf{1}$. □

Since the line $\mathbb{R}\mathbf{1}$ is in the lineality space of $\overline{C_I[\mathbf{w}]}$, we will from now on regard $\overline{C_I[\mathbf{w}]}$ as a polyhedron in the n-dimensional quotient space $\mathbb{R}^{n+1}/\mathbb{R}\mathbf{1}$. The key result of the section is the following theorem.

Theorem 2.5.3. *The polyhedra $\overline{C_I[\mathbf{w}]}$ as $\mathbf{w}$ varies over $\mathbb{R}^{n+1}$ form a Γ_{val}-rational polyhedral complex supported on the n-dimensional space $\mathbb{R}^{n+1}/\mathbb{R}\mathbf{1}$.*

We actually prove something stronger: the complex in Theorem 2.5.3 is a regular subdivision of the ambient space. This requires the following lemma.

Lemma 2.5.4. *Let I be a homogeneous ideal in $K[x_0, \ldots, x_n]$. There are only finitely many distinct monomial initial ideals $\mathrm{in}_{\mathbf{w}}(I)$ as $\mathbf{w}$ runs over $\mathbb{R}^{n+1}$.*

Proof. If this were not the case, by [**Mac01**, Theorem 1.1], there would be $\mathbf{w}_1, \mathbf{w}_2 \in \mathbb{R}^{n+1}$ with $\mathrm{in}_{\mathbf{w}_2}(I) \subsetneq \mathrm{in}_{\mathbf{w}_1}(I)$, where both are monomial ideals. Fix $x^{\mathbf{u}} \in \mathrm{in}_{\mathbf{w}_1}(I) \setminus \mathrm{in}_{\mathbf{w}_2}(I)$. By Corollary 2.4.9, the monomials not in $\mathrm{in}_{\mathbf{w}_1}(I)$ form a K-basis for S/I, so there is $f_{\mathbf{u}} \in I$ with $f_{\mathbf{u}} = x^{\mathbf{u}} + \sum c_{\mathbf{v}} x^{\mathbf{v}}$ where $c_{\mathbf{v}} \neq 0$ implies $x^{\mathbf{v}} \notin \mathrm{in}_{\mathbf{w}_1}(I)$. But then $\mathrm{in}_{\mathbf{w}_2}(f_{\mathbf{u}}) \in \mathrm{in}_{\mathbf{w}_1}(I)$. Since $\mathrm{in}_{\mathbf{w}_1}(I)$ is a monomial ideal, all terms of $\mathrm{in}_{\mathbf{w}_2}(f_{\mathbf{u}})$ lie in $\mathrm{in}_{\mathbf{w}_1}(I)$. However, all monomials of $\mathrm{in}_{\mathbf{w}_2}(f_{\mathbf{u}})$ appear in $f_{\mathbf{u}}$. This leads to a contradiction. We conclude that I has only finitely many monomial initial ideals. □

The following definition is important for the subsequent construction.

Definition 2.5.5. Given a tropical polynomial function $F : \mathbb{R}^{n+1} \to \mathbb{R}$, we write Σ_F for the coarsest polyhedral complex such that F is linear on each cell in Σ_F. The maximal cells of the polyhedral complex Σ_F have the form

$$\sigma \;=\; \{\mathbf{w} \in \mathbb{R}^{n+1} \;:\; F(\mathbf{w}) = a + \mathbf{w}\cdot\mathbf{u}\},$$

where $a \odot x^{\mathbf{u}}$ runs over monomials of F. We have $|\Sigma_F| = \mathbb{R}^{n+1}$. If the coefficients a lie in a subgroup $\Gamma \subset \mathbb{R}$, then the complex Σ_F is Γ-rational.

We shall also need the following lemma concerning linear algebra over K. For an $r \times s$ matrix A of rank r and a subset J of the column indices of size r, we denote by A^J the $r \times r$ submatrix of A with columns indexed by J. This lemma has no assumption on the field K other than that it has a valuation.

Lemma 2.5.6. *Let A be an $r \times s$ matrix of rank r with entries in a field K, and fix $\tilde{\mathbf{w}} \in \mathbb{R}^s$. There exists $U \in \mathrm{GL}(r, K)$ and an index set $J = \{l_1, \ldots, l_r\}$ such that $(UA)^J$ is the identity matrix and $\mathrm{val}((UA)_{ij}) + \tilde{w}_j \geq \tilde{w}_{l_i}$ for $j \notin J$.*

Proof. Choose $J' = \{l_1, \ldots, l_r\}$ for which $\mathrm{val}(\det(A^J)) + \sum_{j\in J} \tilde{w}_j$ is minimized. This minimum is not ∞ because A has rank r, so it has a nonzero $r \times r$-minor. This means that $\mathrm{val}(\det(A^{J'})) < \infty$, so $\det(A^{J'}) \neq 0$. Let $U = (A^{J'})^{-1}$. The matrix UA then has an identity matrix in the columns indexed by J'. To prove the lemma, it remains to show the inequality on valuations. To see this, note that, up to sign, we have $(UA)_{ij} = \det((UA)^{J_{ij}})$, where $J_{ij} = J' \backslash \{l_i\} \cup \{j\}$. This follows from the fact that the submatrix of UA with columns indexed by $J'\backslash\{l_i\}$ has only one nonzero entry in every column. The identity $\det(UA^J) = \det(U) \cdot \det(A^J)$ implies

$$\begin{aligned} \mathrm{val}((UA)_{ij}) &= \mathrm{val}(\det(U)) + \mathrm{val}(\det(A^{J_{ij}})) \\ &= -\,\mathrm{val}(\det(A^{J'})) + \mathrm{val}(\det(A^{J_{ij}})). \end{aligned}$$

Now $\mathrm{val}(\det(A^{J'})) + \sum_{l\in J'} \tilde{w}_l \leq \mathrm{val}(\det(A^{J_{ij}})) + \sum_{l \in J_{ij}} \tilde{w}_l$ by the choice of J'. Subtracting $\sum_{l\in J'\backslash\{l_i\}} \tilde{w}_l$ from both sides of the inequality, we get $\mathrm{val}(\det(A^{J'})) + \tilde{w}_{l_i} \leq \mathrm{val}(\det(A^{J_{ij}})) + \tilde{w}_j$, and so $\mathrm{val}((UA)_{ij}) + \tilde{w}_j \geq \tilde{w}_{l_i}$. □

For what follows we fix a homogeneous ideal I in $S = K[x_0, \ldots, x_n]$. Let $d \in \mathbb{N}$, and choose a K-basis $\{f_1, \ldots, f_r\}$ for I_d. Let A_d be the $(r \times \binom{n+d}{n})$-matrix that records the coefficients of the polynomials f_i. The columns of A_d are indexed by the set $\mathcal{M}_d$ of monomials in S_d. The entry $(A_d)_{i\mathbf{u}}$ is the coefficient of $x^{\mathbf{u}}$ in f_i. Each $J \subseteq \mathcal{M}_d$ with $|J| = r$ specifies an $r \times r$ minor $\det(A_d^J)$ of A_d. The vector with entries $\det(A_d^J)$ for all J is the vector of Plücker coordinates of the point I_d in the Grassmannian $G(r, S_d)$. In particular, this vector is independent of our choice $\{f_1, \ldots, f_r\}$ of basis.

By Lemma 2.5.4, there exists $D \in \mathbb{N}$ such that any initial monomial ideal $\mathrm{in}_{\mathbf{w}}(I)$ of I has generators of degree at most D. We define the polynomial

$$(2.5.2) \qquad g := \prod_{d=1}^{D} g_d, \quad \text{where } g_d := \sum_{\substack{J \subseteq \mathcal{M}_d \\ |J|=r}} \det(A_d^J) \prod_{\mathbf{u} \in J} x^{\mathbf{u}}.$$

We consider the associated piecewise-linear function $\mathrm{trop}(g) : \mathbb{R}^{n+1} \to \mathbb{R}$, as defined in (2.4.1), and let $\Sigma_{\mathrm{trop}(g)}$ be the complex in Definition 2.5.5.

Theorem 2.5.7. *Let $I \subseteq K[x_0, \ldots, x_n]$ and $g_d, g, \Sigma_{\mathrm{trop}(g)}$ as above. If $\mathbf{w} \in \mathbb{R}^{n+1}$ lies in the interior of a maximal cell σ of $\Sigma_{\mathrm{trop}(g)}$, then $\sigma = \overline{C_I[\mathbf{w}]}$.*

Proof. We need to show two things: first, if $\mathbf{w}' \in \mathbb{R}^{n+1}$ lies in the interior of σ, then $\mathrm{in}_{\mathbf{w}'}(I) = \mathrm{in}_{\mathbf{w}}(I)$; and second, if $\mathbf{w}'$ does not lie in the interior of σ, then $\mathrm{in}_{\mathbf{w}'}(I)$ is not equal to $\mathrm{in}_{\mathbf{w}}(I)$. Note that $\Sigma_{\mathrm{trop}(g)}$ is the common refinement of the polyhedral complexes $\Sigma_{\mathrm{trop}(g_d)}$ for $d \leq D$, where $\Sigma_{\mathrm{trop}(g_d)}$ is the coarsest polyhedral complex for which $\mathrm{trop}(g_d)$ is linear on each cell. Thus it suffices to restrict to a fixed degree $d \leq D$.

Let σ_d be the maximal cell of $\Sigma_{\mathrm{trop}(g_d)}$ containing σ. We will show that $\mathbf{w}' \in \mathbb{R}^{n+1}$ is in the interior of σ_d if and only if $\mathrm{in}_{\mathbf{w}'}(I)_d = \mathrm{in}_{\mathbf{w}}(I)_d$. This suffices because $\mathrm{in}_{\mathbf{w}'}(I) = \mathrm{in}_{\mathbf{w}}(I)$ if and only if $\mathrm{in}_{\mathbf{w}'}(I)_d = \mathrm{in}_{\mathbf{w}}(I)_d$ for $d \leq D$.

For the "only-if" direction, let $\mathbf{w}'$ be in the interior of σ_d. The minimum in $\mathrm{trop}(g_d)$ is achieved at the same term for $\mathbf{w}$ and for $\mathbf{w}'$. Since σ_d is a maximal cell, this minimum is achieved at only one term, indexed by $J \subset \mathcal{M}_d$. By Lemma 2.5.6 applied to A_d and the vector $\tilde{\mathbf{w}} \in \mathbb{R}^{\binom{n+d}{d}}$ with $\tilde{w}_{\mathbf{u}} = \mathbf{w} \cdot \mathbf{u}$, there is an $(r \times \binom{n+d}{n})$-matrix B with B^J an identity matrix and $\mathrm{val}(B_{\mathbf{uv}}) + \mathbf{w} \cdot \mathbf{v} > \mathbf{w} \cdot \mathbf{u}$ for all $x^{\mathbf{u}} \in J$, $x^{\mathbf{v}} \notin J$. The strict inequality comes because the minimum in $\mathrm{trop}(g_d)(\mathbf{w})$ is achieved only once. Each row of B gives a polynomial $\tilde{f}_{\mathbf{u}} = x^{\mathbf{u}} + \sum_{x^{\mathbf{v}} \notin J} B_{\mathbf{uv}} x^{\mathbf{v}}$ indexed by $x^{\mathbf{u}} \in J$. Then $\mathrm{in}_{\mathbf{w}}(\tilde{f}_{\mathbf{u}}) = x^{\mathbf{u}}$ and hence $x^{\mathbf{u}} \in \mathrm{in}_{\mathbf{w}}(I)_d$. Corollary 2.4.9 implies $\dim_{\Bbbk} \mathrm{in}_{\mathbf{w}}(I)_d = r$, so J consists of precisely the monomials in $\mathrm{in}_{\mathbf{w}}(I)_d$. Since $|J| = r = \dim_{\Bbbk} \mathrm{in}_{\mathbf{w}}(I)_d = \dim_{\Bbbk} \mathrm{in}_{\mathbf{w}'}(I)_d$, we have $\mathrm{in}_{\mathbf{w}}(I)_d = \mathrm{in}_{\mathbf{w}'}(I)_d$.

For the "if" direction, suppose that $\mathbf{w}'$ does not lie in the interior of σ_d. This means that there exists $J' \subset \mathcal{M}_d$ with

$$\mathrm{val}(A_d^{J'}) + \sum_{x^{\mathbf{u}} \in J'} \mathbf{w}' \cdot \mathbf{u} \leq \mathrm{val}(A_d^{J''}) + \sum_{x^{\mathbf{u}} \in J''} \mathbf{w}' \cdot \mathbf{u} \quad \text{for all } J'' \subset \mathcal{M}_d \backslash \{J'\}.$$

We may choose J' to index a vertex of the polytope

$$\text{(2.5.3)} \qquad \operatorname{conv}\left(\sum_{x^{\mathbf{u}} \in J''} \mathbf{u} \ : \ \operatorname{val}(A_d^{J''}) + \sum_{x^{\mathbf{u}} \in J''} \mathbf{w}' \cdot \mathbf{u} \text{ is minimal} \right).$$

Hence, there exists $\mathbf{v} \in \mathbb{R}^{n+1}$ with $\mathbf{v} \cdot \sum_{x^{\mathbf{u}} \in J'} \mathbf{u} < \mathbf{v} \cdot \sum_{x^{\mathbf{u}} \in J''} \mathbf{u}$ for all other J'' on the right-hand side of the last equation on page 77. For sufficiently small $\epsilon > 0$, we have

$$\operatorname{val}(A_d^{J'}) + \sum_{x^{\mathbf{u}} \in J'} (\mathbf{w}' + \epsilon \mathbf{v}) \cdot \mathbf{u} < \operatorname{val}(A_d^{J''}) + \sum_{x^{\mathbf{u}} \in J''} (\mathbf{w}' + \epsilon \mathbf{v}) \cdot \mathbf{u} \quad \text{for } J'' \subset \mathcal{M}_d \backslash \{J'\}.$$

So, the minimum in $\operatorname{trop}(g_d)(\mathbf{w}' + \epsilon \mathbf{v})$ is achieved uniquely. The proof of the "only-if" direction implies $\operatorname{in}_{\mathbf{w}'+\epsilon\mathbf{v}}(I)_d = \operatorname{span}\{x^{\mathbf{u}} : x^{\mathbf{u}} \in J'\}$. By Corollary 2.4.10, we have $\operatorname{in}_{\mathbf{w}'+\epsilon\mathbf{v}}(I) = \operatorname{in}_{\mathbf{v}}(\operatorname{in}_{\mathbf{w}'}(I))$. This means that $\operatorname{in}_{\mathbf{w}'}(I)_d$ is not the span of the monomials in J, and thus $\operatorname{in}_{\mathbf{w}'}(I)_d \neq \operatorname{in}_{\mathbf{w}}(I)_d$. □

At this point, Theorem 2.5.3 can be derived easily from Theorem 2.5.7.

Proof of Theorem 2.5.3. Theorem 2.5.7 states that all top-dimensional cells of the Γ_{val}-rational polyhedral complex $\Sigma_{\operatorname{trop}(g)}$ are of the form $\overline{C_I[\mathbf{w}]}$ for some $\mathbf{w} \in \mathbb{R}^{n+1}$ with $\operatorname{in}_{\mathbf{w}}(I)$ a monomial ideal. For such $\mathbf{w} \in \mathbb{R}^{n+1}$, Corollary 2.4.10 implies that $\operatorname{in}_{\mathbf{w}+\epsilon\mathbf{v}}(I) = \operatorname{in}_{\mathbf{w}}(I)$ for all $\mathbf{v} \in \mathbb{R}^{n+1}$ and small $\epsilon > 0$. Hence $\overline{C_I[\mathbf{w}]}$ is full dimensional, so it must be one of the top-dimensional cells of $\Sigma_{\operatorname{trop}(g)}$. For any $\mathbf{w}' \neq \mathbf{w}$ with $\operatorname{in}_{\mathbf{w}'}(I)$ monomial, the cells $C_I[\mathbf{w}]$ and $C_I[\mathbf{w}']$ are either disjoint or they coincide. Theorem 2.5.3 now follows from Proposition 2.5.2, namely, if $\operatorname{in}_{\mathbf{w}}(I)$ is not a monomial ideal, then $\overline{C_I[\mathbf{w}]}$ is a face of some $\overline{C_I[\mathbf{w}']}$ with $\operatorname{in}_{\mathbf{w}'}(I)$ a monomial ideal. □

Definition 2.5.8. The *Gröbner complex* $\Sigma(I)$ of a homogeneous ideal I in $K[x_0, x_1, \ldots, x_n]$ is the polyhedral complex constructed in Theorems 2.5.3 and 2.5.7. It consists of the polyhedra $\overline{C_I[\mathbf{w}]}$ as $\mathbf{w}$ ranges over $\mathbb{R}^{n+1}$.

The Gröbner complex $\Sigma(I)$ lives in the quotient space $\mathbb{R}^{n+1}/\mathbb{R}\mathbf{1}$. This space is obtained from $\mathbb{R}^{n+1}$ by identifying vectors that differ from each other by tropical scalar multiplication. For that reason, $\mathbb{R}^{n+1}/\mathbb{R}\mathbf{1}$ was denoted by $\mathbb{TP}^n$ in some early papers on tropical geometry. In this book, we retain the notation $\mathbb{R}^{n+1}/\mathbb{R}\mathbf{1}$, and we call this the *tropical projective torus.* The notation $\mathbb{TP}^n$ and the name *tropical projective space* are reserved for the natural compactification of $\mathbb{R}^{n+1}/\mathbb{R}\mathbf{1}$ obtained by allowing ∞ among the coordinates. This will be explained in Chapter 6. For a quick glance, see Figure 6.2.2. Points in $\mathbb{R}^{n+1}/\mathbb{R}\mathbf{1}$ can be uniquely represented by vectors of the form $(0, v_1, \ldots, v_n)$. This is the convention we use for drawing pictures.

In our construction, we realized the Gröbner complex as $\Sigma(I) = \Sigma_{\operatorname{trop}(g)}$, where g was the auxiliary polynomial (2.5.2) that represents the ideal I.

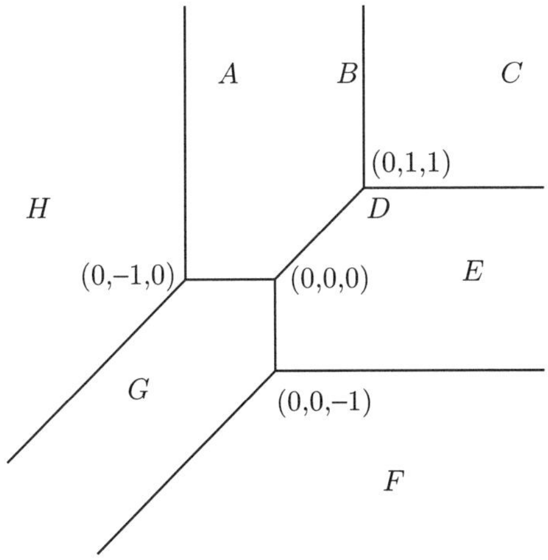

Figure 2.5.1. The Gröbner complex of a plane curve subdivides $\mathbb{R}^3/\mathbb{R}\mathbf{1}$.

Namely, $\Sigma(I)$ consists of the regions of linearity of the tropical polynomial trop(g), which is a piecewise linear function on $\mathbb{R}^{n+1}$. These regions are regarded as polyhedra in $\mathbb{R}^{n+1}/\mathbb{R}\mathbf{1}$. In the special case when $I = \langle f \rangle$ is a principal ideal, generated by a homogeneous polynomial f of degree d, we can take $D = d$. This gives $g_1 = \cdots = g_{d-1} = 1$ and $g = g_d = f$ in (2.5.2).

Example 2.5.9. Let $n{=}2$, let $K{=}\mathbb{C}\{\{t\}\}$, and let I be the ideal generated by

$$f \;=\; tx_1^2 + 2x_1x_2 + 3tx_2^2 + 4x_0x_1 + 5x_0x_2 + 6tx_0^2.$$

The Gröbner complex of I is the polyhedral complex in $\mathbb{R}^3/\mathbb{R}\mathbf{1}$ shown in Figure 2.5.1. It represents the regions of linearity of the map trop(f).

The ideal I has 19 distinct initial ideals, corresponding to the various cells of $\Sigma(I)$. There are six cells of dimension 2, nine cells of dimension 1, and four cells of dimension 0. Table 2.5.1 lists eight of the 19 initial ideals, namely, those corresponding to the labels in the diagram.

Table 2.5.1. Initial ideals.

Cell	Initial ideal	Cell	Initial Ideal
A	$\langle 4x_0x_1\rangle$	E	$\langle 5x_0x_2\rangle$
B	$\langle 4x_0x_1 + 6x_0^2\rangle$	F	$\langle 3x_2^2\rangle$
C	$\langle 6x_0^2\rangle$	G	$\langle 2x_1x_2\rangle$
D	$\langle 4x_0x_1 + 5x_0x_2 + 6x_0^2\rangle$	H	$\langle x_1^2\rangle$

The initial ideal $\text{in}_{\mathbf{w}}(I)$ contains a monomial if and only if the corresponding cell is full dimensional in the tropical projective torus $\mathbb{R}^3/\mathbb{R}\mathbf{1}$. ◊

Another special case that deserves particular attention is that of linear spaces. Suppose that I is generated by linear forms in $K[x_0, \ldots, x_n]$. Then we may take $D = 1$ in (2.5.2), and hence $g = g_1$. The polynomial g represents the vector of Plücker coordinates of the linear variety $V(I)$ in $\mathbb{P}^n$, and $\mathrm{trop}(g)$ represents the vector of tropicalized Plücker coordinates. The Gröbner complex $\Sigma(I)$ consists of the regions of linearity of the map $\mathrm{trop}(g)$. It is determined by the valuations of the Plücker coordinates of $V(I)$.

Example 2.5.10. Let $n = 3$, and consider the ideal of a general line in $\mathbb{P}^3$,

$$I = \langle a_1x_1 + a_2x_2 + a_3x_3 + a_4x_4 \, , \, b_1x_1 + b_2x_2 + b_3x_3 + b_4x_4 \rangle.$$

In the notation of the paragraph prior to (2.5.2), we have $d = 1, r = 2$, and

$$A_1 = \begin{pmatrix} a_1 & a_2 & a_3 & a_4 \\ b_1 & b_2 & b_3 & b_4 \end{pmatrix}.$$

The 2×2-minors p_{ij} of A_1 are scalars in K that satisfy the Plücker relation $p_{12}p_{34} - p_{13}p_{24} + p_{14}p_{23} = 0$ of Example 2.2.11. Up to relabeling of the variables, the following relation will hold for the scalars $p_{ij} = a_ib_j - a_jb_i$:

$$\text{(2.5.4)} \quad \mathrm{val}(p_{12}) + \mathrm{val}(p_{34}) = \mathrm{val}(p_{13}) + \mathrm{val}(p_{24}) \leq \mathrm{val}(p_{14}) + \mathrm{val}(p_{23}).$$

The inequality is strict for most choices of a_i and b_j. Let us now assume that this is the case. The polynomial g of (2.5.2) is the quadric

$$g = p_{12}x_1x_2 + p_{13}x_1x_3 + p_{14}x_1x_4 + p_{23}x_2x_3 + p_{24}x_2x_4 + p_{34}x_3x_4.$$

The Gröbner complex $\Sigma(I) = \Sigma_{\mathrm{trop}(g)}$ is a subdivision of the space $\mathbb{R}^4/\mathbb{R}\mathbf{1}$ into six three-dimensional regions. It has twelve unbounded two-dimensional walls, nine edges (one bounded and eight unbounded), and two vertices. ◇

The construction of the Gröbner complex allows us to define the concept of a *universal Gröbner basis* for a homogeneous ideal $I \subset K[x_0, \ldots, x_n]$. This is a finite subset $\mathcal{U}$ of I such that, for all $\mathbf{w} \in \mathbb{R}^{n+1}$, the set $\mathrm{in}_{\mathbf{w}}(\mathcal{U}) = \{\mathrm{in}_{\mathbf{w}}(f) : f \in \mathcal{U}\}$ generates the initial ideal $\mathrm{in}_{\mathbf{w}}(I)$ in $\Bbbk[x_0, \ldots, x_n]$.

Corollary 2.5.11. *Fix a field K with valuation. Every homogeneous ideal I in the polynomial ring $K[x_0, \ldots, x_n]$ has a finite universal Gröbner basis.*

Proof. The Gröbner complex $\Sigma(I)$ is finite. For each maximal cell σ, pick $\mathbf{w}$ in the interior of σ. The initial ideal $\mathrm{in}_{\mathbf{w}}(I)$ is a monomial ideal. For each generator $x^{\mathbf{u}}$ of $\mathrm{in}_{\mathbf{w}}(I)$ there is a polynomial $g_{\mathbf{u}} = x^{\mathbf{u}} - \sum c_{\mathbf{v}}x^{\mathbf{v}} \in I$ with $x^{\mathbf{v}} \notin \mathrm{in}_{\mathbf{w}}(I)$ whenever $c_{\mathbf{v}} \neq 0$. The set of all $g_{\mathbf{u}}$ as $x^{\mathbf{u}}$ varies over the minimal generators of $\mathrm{in}_{\mathbf{w}}(I)$ forms a Gröbner basis for I in $K[x_0, \ldots, x_n]$ with respect to any $\mathbf{w}' \in \sigma = \overline{C_I[\mathbf{w}]}$. For $\mathbf{w}' \in C_I[\mathbf{w}]$ this is immediate as we must have $\mathrm{in}_{\mathbf{w}'}(g_{\mathbf{u}}) = x^{\mathbf{u}}$. For $\mathbf{w}' \in \overline{C_I[\mathbf{w}]} \backslash C_I[\mathbf{w}]$, this can be derived from Corollary 2.4.10 since $\mathrm{in}_{\mathbf{w}}(I)$ is an initial ideal of $\mathrm{in}_{\mathbf{w}'}(I) \subset \Bbbk[x_0, \ldots, x_n]$ using the trivial valuation on $\Bbbk$. □

It follows that the Gröbner complex of a constant coefficient ideal I can be identified with the *Gröbner fan* of [**Stu96**, §2] up to the sign change.

Corollary 2.5.12. *Let I be a homogeneous ideal with constant coefficients. Then the negated Gröbner complex $-\Sigma(I)$ is equal to the Gröbner fan of I.*

In many of the geometric examples later in this book we will study a projective variety whose defining ideal I has coefficients in the field $\mathbb{Q}$ of rational numbers. Such an ideal I has a well-defined Gröbner fan. It arises as $-\Sigma(I)$ when $\mathbb{Q}$ has the trivial valuation. On the other hand, we can also consider the p-adic Gröbner complex of the same ideal I. The p-adic Gröbner complex $\Sigma(I)$ is generally not a fan.

2.6. Tropical Bases

In the last two sections we introduced Gröbner bases and the Gröbner complex for homogeneous ideals in a polynomial ring $K[x_0, x_1, \ldots, x_n]$. We now examine the case when the ambient ring is the Laurent polynomial ring $K[x_1^{\pm 1}, \ldots, x_n^{\pm 1}]$. There is no natural intrinsic notion of Gröbner bases for ideals in $K[x_1^{\pm 1}, \ldots, x_n^{\pm 1}]$. There is, however, a natural analogue to the notion of a universal Gröbner basis, namely, that of a tropical basis. This is our subject in this section, and it will be introduced formally in Definition 2.6.3.

We begin by defining initial ideals for ideals in $K[x_1^{\pm 1}, \ldots, x_n^{\pm 1}]$. As in the previous two sections, unless otherwise noted we assume that the valuation on the field K has a splitting $w \mapsto t^w$ with $\mathrm{val}(t^w) = w$. For $f \in K[x_1^{\pm 1}, \ldots, x_n^{\pm 1}]$ and $\mathbf{w} \in \mathbb{R}^n$, we define the *initial form* $\mathrm{in}_{\mathbf{w}}(f) \in \Bbbk[x_1^{\pm 1}, \ldots, x_n^{\pm 1}]$ by the same rule as in (2.4.2); namely, we set

$$\mathrm{in}_{\mathbf{w}}(f) \;=\; \sum_{\mathbf{u}:\mathrm{val}(c_{\mathbf{u}})+\mathbf{w}\cdot\mathbf{u}=W} \overline{t^{-\mathrm{val}(c_{\mathbf{u}})}\, c_{\mathbf{u}}} \cdot x^{\mathbf{u}}, \tag{2.6.1}$$

where $W = \mathrm{trop}(f)(\mathbf{w}) = \min\{\mathrm{val}(c_{\mathbf{u}}) + \mathbf{w}\cdot\mathbf{u}\}$.

Let I be any ideal in $K[x_1^{\pm 1}, \ldots, x_n^{\pm 1}]$. The *initial ideal* $\mathrm{in}_{\mathbf{w}}(I)$ is the ideal in $\Bbbk[x_1^{\pm 1}, \ldots, x_n^{\pm 1}]$ generated by the initial forms $\mathrm{in}_{\mathbf{w}}(f)$ for all $f \in I$. So far, this is the same as in the polynomial ring. But there is an important distinction that arises when we work with Laurent polynomials. For generic choices of $\mathbf{w} = (w_1, \ldots, w_n)$, the initial form $\mathrm{in}_{\mathbf{w}}(f)$ is a unit in $\Bbbk[x_1^{\pm 1}, \ldots, x_n^{\pm 1}]$, and the initial ideal $\mathrm{in}_{\mathbf{w}}(I)$ is equal to the whole ring. If this happens, then the initial ideal contains no information at all. Tropical geometry is concerned with the study of those special weight vectors $\mathbf{w} \in \mathbb{R}^n$ for which the initial ideal $\mathrm{in}_{\mathbf{w}}(I)$ is actually a proper ideal in $\Bbbk[x_1^{\pm 1}, \ldots, x_n^{\pm 1}]$.

In order to compute and study these initial ideals, we work with homogeneous polynomials as in Section 2.4. As in Definition 2.2.4, the homogenization I_{proj} is the ideal in $K[x_0, x_1, \ldots, x_n]$ generated by all polynomials

$$\tilde{f} \quad = \quad x_0^m \cdot f\left(\frac{x_1}{x_0}, \ldots, \frac{x_n}{x_0}\right),$$

where $f \in I$ and m is the smallest integer that clears the denominator.

The initial ideals $\text{in}_{\mathbf{w}}(I)$ of a *Laurent ideal* $I \subseteq K[x_1^{\pm 1}, \ldots, x_n^{\pm 1}]$ can be computed from the initial ideals of its homogenization I_{proj} as follows. The weight vectors for the homogeneous ideal I_{proj} naturally live in the quotient space $\mathbb{R}^{n+1}/\mathbb{R}\mathbf{1}$, and we identify this space with $\mathbb{R}^n$ via $\mathbf{w} \mapsto (0, \mathbf{w})$.

Proposition 2.6.1. *Let I be an ideal in $K[x_1^{\pm 1}, \ldots, x_n^{\pm 1}]$, and fix $\mathbf{w} \in \mathbb{R}^n$. Then $\text{in}_{\mathbf{w}}(I)$ is the image of $\text{in}_{(0,\mathbf{w})}(I_{\text{proj}})$ in $\Bbbk[x_1^{\pm 1}, \ldots, x_n^{\pm 1}]$ obtained by setting $x_0 = 1$. Every element of $\text{in}_{\mathbf{w}}(I)$ has the form $x^{\mathbf{u}} g$, where $x^{\mathbf{u}}$ is a Laurent monomial and $g = f(1, x_1, \ldots, x_n)$ for some $f \in \text{in}_{(0,\mathbf{w})}(I_{\text{proj}})$.*

Proof. Suppose $f = \sum c_{\mathbf{u}} x^{\mathbf{u}}$ is in $I \cap K[x_1, \ldots, x_n]$ and $\tilde{f} = \sum c_{\mathbf{u}} x^{\mathbf{u}} x_0^{j_{\mathbf{u}}}$ is its homogenization, where $j_{\mathbf{u}} = (\max_{c_{\mathbf{v}} \neq 0} |\mathbf{v}|) - |\mathbf{u}|$. We abbreviate

$$\begin{aligned} W \ := \ \text{trop}(f)(\mathbf{w}) \ &= \ \min(\text{val}(c_{\mathbf{u}}) + \mathbf{w} \cdot \mathbf{u}) \\ &= \ \min(\text{val}(c_{\mathbf{u}}) + (0, \mathbf{w}) \cdot (j_{\mathbf{u}}, \mathbf{u})) \ = \ \text{trop}(\tilde{f})((0, \mathbf{w})). \end{aligned}$$

Then $\text{in}_{(0,\mathbf{w})}(\tilde{f}) \; = \; \sum_{\text{val}(c_{\mathbf{u}}) + \mathbf{w} \cdot \mathbf{u} = W} \overline{c_{\mathbf{u}} t^{-\text{val}(c_{\mathbf{u}})}} x^{\mathbf{u}} x_0^{j_{\mathbf{u}}}$ in $\Bbbk[x_0, \ldots, x_n]$ and

$$\text{(2.6.2)} \qquad \text{in}_{(0,\mathbf{w})}(\tilde{f})|_{x_0=1} \quad = \sum_{\text{val}(c_{\mathbf{u}}) + \mathbf{w} \cdot \mathbf{u} = W} \overline{c_{\mathbf{u}} t^{-\text{val}(c_{\mathbf{u}})}} x^{\mathbf{u}} \quad = \quad \text{in}_{\mathbf{w}}(f).$$

By multiplying by monomials if necessary, we can choose polynomials $f_1, \ldots, f_s$ in $K[x_1, \ldots, x_n] \cap I$ such that $\text{in}_{\mathbf{w}}(I) = \langle \text{in}_{\mathbf{w}}(f_1), \ldots, \text{in}_{\mathbf{w}}(f_s) \rangle$. Since $\text{in}_{\mathbf{w}}(\tilde{f}_i)|_{x_0=1} = \text{in}_{\mathbf{w}}(f_i)$, we have $\text{in}_{\mathbf{w}}(I) \subseteq \text{in}_{(0,w)}(I_{\text{proj}})|_{x_0=1}$. For the reverse inclusion, note that if g is a homogeneous polynomial in I_{proj}, then $g = x_0^j \cdot \tilde{f}$ for some j, where $f(x) = g(1, x)$. By Lemma 2.4.2 we can choose a homogeneous Gröbner basis for I_{proj}. Hence (2.6.2) also implies the reverse inclusion.

The last sentence follows since each element of an ideal J in $\Bbbk[x_1^{\pm 1}, \ldots, x_n^{\pm 1}]$ is a Laurent monomial times an element of $J \cap \Bbbk[x_1, \ldots, x_n]$. □

Here are some basic facts about initial ideals of Laurent ideals.

Lemma 2.6.2. *Let I be an ideal in $K[x_1^{\pm 1}, \ldots, x_n^{\pm 1}]$, and fix $\mathbf{w} \in \mathbb{R}^n$.*

(1) *If $g \in \text{in}_{\mathbf{w}}(I)$, then $g = \text{in}_{\mathbf{w}}(h)$ for some $h \in I$.*

(2) *If $\text{in}_{\mathbf{v}}(\text{in}_{\mathbf{w}}(I)) = \text{in}_{\mathbf{w}}(I)$ for some $\mathbf{v} = (v_1, \ldots, v_n) \in \mathbb{R}^n$, then $\text{in}_{\mathbf{w}}(I)$ is homogeneous with respect to the grading given by $\deg(x_i) = v_i$.*

(3) *If $f, g \in K[x_1^{\pm 1}, \ldots, x_n^{\pm 1}]$, then $\text{in}_{\mathbf{w}}(fg) = \text{in}_{\mathbf{w}}(f)\, \text{in}_{\mathbf{w}}(g)$.*

Proof. For part (1) suppose $g \in \mathrm{in}_{\mathbf{w}}(I)$. By Proposition 2.6.1 we have $g = x^{\mathbf{u}} f(1, x_1, \ldots, x_n)$ for some $f \in \mathrm{in}_{(0,\mathbf{w})}(I_{\mathrm{proj}})$. By Lemma 2.4.2 there is $h \in I_{\mathrm{proj}}$ with $\mathrm{in}_{(0,\mathbf{w})}(h) = f$. Then $x^{\mathbf{u}} h \in I$ and $\mathrm{in}_{\mathbf{w}}(x^{\mathbf{u}} h) = g$, as required.

For part (2) suppose $\mathrm{in}_{\mathbf{v}}(\mathrm{in}_{\mathbf{w}}(I)) = \mathrm{in}_{\mathbf{w}}(I)$. Then $\mathrm{in}_{\mathbf{w}}(I)$ is generated by elements $\mathrm{in}_{\mathbf{v}}(g)$ where $g \in \mathrm{in}_{\mathbf{w}}(I)$. For any $g = \sum a_{\mathbf{u}} x^{\mathbf{u}} \in \Bbbk[x_1^{\pm 1}, \ldots, x_n^{\pm 1}]$, the initial form $\mathrm{in}_{\mathbf{v}}(g) = \sum_{\mathbf{v}\cdot\mathbf{u}=W} a_{\mathbf{u}} x^{\mathbf{u}}$ is $\mathbf{v}$-homogeneous of degree $W = \min_{a_{\mathbf{u}} \neq 0} \mathbf{v} \cdot \mathbf{u}$. Hence $\mathrm{in}_{\mathbf{w}}(I)$ has a $\mathbf{v}$-homogeneous generating set.

For part (3) consider $f = \sum c_{\mathbf{u}} x^{\mathbf{u}}$ and $g = \sum d_{\mathbf{u}} x^{\mathbf{u}}$. Then $fg = \sum_{\mathbf{v}} e_{\mathbf{v}} x^{\mathbf{v}}$ for $e_{\mathbf{v}} = \sum_{\mathbf{u}+\mathbf{u}'=\mathbf{v}} c_{\mathbf{u}} d_{\mathbf{u}'}$. Let $W_1 = \mathrm{trop}(f)(\mathbf{w})$ and $W_2 = \mathrm{trop}(g)(\mathbf{w})$. The definition (2.4.1) readily implies $\mathrm{trop}(fg)(\mathbf{w}) = W_1 + W_2$. We conclude

$$\begin{aligned}
\mathrm{in}_{\mathbf{w}}(fg) &= \sum_{\substack{\mathbf{v}:\mathrm{val}(e_{\mathbf{v}})+\mathbf{w}\cdot\mathbf{v}\\ =W_1+W_2}} \overline{e_{\mathbf{v}} t^{-\mathrm{val}(e_{\mathbf{v}})}} x^{\mathbf{v}} \\
&= \sum_{\substack{\mathbf{v}:\mathrm{val}(e_{\mathbf{v}})+\mathbf{w}\cdot\mathbf{v}\\ =W_1+W_2}} \sum_{\mathbf{u}+\mathbf{u}'=\mathbf{v}} \overline{c_{\mathbf{u}} d_{\mathbf{u}'} t^{-W_1-W_2+\mathbf{w}\cdot(\mathbf{u}+\mathbf{u}')}} x^{\mathbf{v}} \\
&= \Bigg(\sum_{\mathbf{u}:\mathrm{val}(c_{\mathbf{u}})+\mathbf{w}\cdot\mathbf{u}=W_1} \overline{c_{\mathbf{u}} t^{-\mathrm{val}(c_{\mathbf{u}})}} x^{\mathbf{u}} \Bigg) \Bigg(\sum_{\mathbf{u}':d_{\mathbf{u}'}+\mathbf{w}\cdot\mathbf{u}'=W_2} \overline{d_{\mathbf{u}'} t^{-\mathrm{val}(d_{\mathbf{u}'})}} x^{\mathbf{u}'} \Bigg) \\
&= \mathrm{in}_{\mathbf{w}}(f)\,\mathrm{in}_{\mathbf{w}}(g).
\end{aligned}$$

This completes the proof of all three parts. □

With the definition of initial forms in $K[x_1^{\pm 1}, \ldots, x_n^{\pm 1}]$ we can now define the notion of a tropical basis. For this we relax our assumption that the valuation of K has a splitting $w \mapsto t^w$. Now, K can be any valued field.

Definition 2.6.3. Let I be an ideal in $K[x_1^{\pm 1}, \ldots, x_n^{\pm 1}]$, where K is an arbitrary field with a valuation. A finite generating set $\mathcal{T}$ of I is a *tropical basis* if, for all vectors $\mathbf{w} \in \mathbb{R}^n$, there is a Laurent polynomial $f \in I$ for which the minimum in $\mathrm{trop}(f)(\mathbf{w})$ is achieved only once if and only if there is $g \in \mathcal{T}$ for which the minimum in $\mathrm{trop}(g)(\mathbf{w})$ is achieved only once.

If the valuation on K has a splitting so the notion of initial ideal and initial form are defined, this has the following reformulation. The set $\mathcal{T}$ is a tropical basis if, for all $\mathbf{w} \in \mathbb{R}^n$, the initial ideal $\mathrm{in}_{\mathbf{w}}(I)$ contains a unit if and only if the finite set $\mathrm{in}_{\mathbf{w}}(\mathcal{T}) = \{\mathrm{in}_{\mathbf{w}}(f) : f \in \mathcal{T}\}$ contains a unit.

Our first example of a tropical basis concerns principal ideals.

Example 2.6.4. If $f \in K[x_1^{\pm 1}, \ldots, x_n^{\pm 1}]$, then $\{f\}$ is a tropical basis for the ideal $I = \langle f \rangle$ it generates. Indeed, suppose that $\mathrm{in}_{\mathbf{w}}(I)$ contains a unit. Then there exists $g \in K[x_1^{\pm 1}, \ldots, x_n^{\pm 1}]$ such that $\mathrm{in}_{\mathbf{w}}(fg) = \mathrm{in}_{\mathbf{w}}(f) \cdot \mathrm{in}_{\mathbf{w}}(g)$ is a unit, and this implies that $\mathrm{in}_{\mathbf{w}}(f)$ is a unit. ◊

Our goal in this section is to show that every ideal $I \subset K[x_1^{\pm 1}, \dots, x_n^{\pm 1}]$ has a finite tropical basis. The proof will require passing to an extension field, so we begin by showing that this does not create serious problems.

A *valued field extension* L/K of a field K with a valuation val_K is a field extension L/K for which L has a valuation $\mathrm{val}_L \colon L \to \mathbb{R} \cup \{\infty\}$ with $\mathrm{val}_L|_K = \mathrm{val}_K$. Useful examples include the extension $\mathbb{C}\{\{t\}\}/\mathbb{C}$ of $\mathbb{C}$ by the Puiseux series, and the extension $K((\mathbb{R}))/K$ of an arbitrary field K with a trivial valuation by the field of generalized power series $K((\mathbb{R}))$.

Lemma 2.6.5. *Let L/K be a valued field extension with no hypotheses on L or K. Let $I \subset K[x_1^{\pm 1}, \dots, x_n^{\pm 1}]$ be an ideal, and let $I_L = IL[x_1^{\pm 1}, \dots, x_n^{\pm 1}]$ be its extension to an ideal in $L[x_1^{\pm 1}, \dots, x_n^{\pm 1}]$. If a tropical basis for I_L exists, then there is one for which all coefficients of all polynomials lie in K.*

Proof. Let $\mathcal{T}_L$ be the tropical basis for I_L. Our goal is to transform $\mathcal{T}_L$ into a tropical basis for I_L consisting of polynomials with coefficients in K.

Fix a polynomial $g \in \mathcal{T}_L$, and set

$$C_g := \{\mathbf{w} \in \mathbb{R}^n : \text{ the minimum in } \mathrm{trop}(g)(\mathbf{w}) \text{ is achieved only once}\}.$$

We shall construct a finite collection of $f \in I$ such that for all $\mathbf{w} \in C_g$ the minimum in $\mathrm{trop}(f)(\mathbf{w})$ is achieved only once for some f in this collection. Since $g \in I_L$, we can write $g = \sum_{i=1}^r h_i f_i$ for $f_i \in I$ and $h_i \in L[x_1^{\pm 1}, \dots, x_n^{\pm 1}]$. By writing each h_i as a sum of terms and absorbing the monomial into the polynomial f_i, we may assume that each h_i is a scalar $c_i \in L$. We may also assume that g cannot be written as a linear combination of a proper subset of $\{f_1, \dots, f_r\}$. Let $\mathcal{U} = \{x^{\mathbf{u}_1}, \dots, x^{\mathbf{u}_s}\}$ be the collection of monomials occurring in the f_i, and let A be the $r \times s$-matrix whose entry $A_{ij} \in K$ is the coefficient of $x^{\mathbf{u}_j}$ in f_i. Our assumptions imply that the matrix A has rank r, and the coefficient vector of g is an L-linear combination of the rows of A. Fix $\mathbf{w} \in C_g$. Define the vector $\mathbf{w}' \in \mathbb{R}^{|\mathcal{U}|}$ by $w'_j = \mathbf{w} \cdot \mathbf{u}_j$. Then by Lemma 2.5.6, after renumbering the entries of $\mathcal{U}$ if necessary, there is another $r \times s$-matrix B with entries in K with the same row space as A and the property that the first $r \times r$-submatrix of B is the identity matrix, and $\mathrm{val}(B_{ij}) + w'_j \geq w'_i$ for $j > r$. The rows of this matrix are the coefficient vectors of Laurent polynomials that are K-linear combinations of the f_i. These Laurent polynomials lie in I. The matrix B may depend on the specific choice of $\mathbf{w}$, but there are only finitely many choices for it.

Write $g = \sum_{j=1}^{s} a_j x^{\mathbf{u}_j}$. By assumption, $a_j = \sum_{i=1}^{r} c'_i B_{ij}$ for some $c'_i \in L$. The special form of B means that $a_i = c'_i$ for $1 \leq i \leq r$. We note that

$$\begin{aligned} \mathrm{val}(a_j) + \mathbf{w} \cdot \mathbf{u}_j &\geq \min_{1 \leq i \leq r} (\mathrm{val}(c'_i) + \mathrm{val}(B_{ij})) + \mathbf{w} \cdot \mathbf{u}_j \\ &\geq \min_{1 \leq i \leq r} (\mathrm{val}(c'_i) + \mathbf{w} \cdot \mathbf{u}_i) \\ &= \min_{1 \leq i \leq r} (\mathrm{val}(a_i) + \mathbf{w} \cdot \mathbf{u}_i). \end{aligned}$$

Hence the minimum in $\mathrm{trop}(g)(\mathbf{w})$ is achieved at a term involving $\mathbf{u}_{i'}$ for some $1 \leq i' \leq r$. We now claim that $\mathbf{w} \cdot \mathbf{u}_{i'} < \mathrm{val}(B_{i'j}) + \mathbf{w} \cdot \mathbf{u}_j$ for all $j \neq i'$. If not, then $\mathrm{val}(c'_{i'}) + \mathbf{w} \cdot \mathbf{u}_{i'} = \mathrm{val}(c'_{i'}) + \mathrm{val}(B_{i'j}) + \mathbf{w} \cdot \mathbf{u}_j$ for some $j \neq i'$. Fix $i \in \{1, \ldots, r\} \backslash \{i'\}$. Then

$$\begin{aligned} \mathrm{val}(c'_i) + \mathrm{val}(B_{ij}) + \mathbf{w} \cdot \mathbf{u}_j &\geq \mathrm{val}(c'_i) + \mathbf{w} \cdot \mathbf{u}_i \\ &= \mathrm{val}(a_i) + \mathbf{w} \cdot \mathbf{u}_i \\ &> \mathrm{val}(a_{i'}) + \mathbf{w} \cdot \mathbf{u}_{i'} \\ &= \mathrm{val}(c'_{i'}) + \mathbf{w} \cdot \mathbf{u}_{i'} \\ &= \mathrm{val}(c'_{i'}) + \mathrm{val}(B_{i'j}) + \mathbf{w} \cdot \mathbf{u}_j. \end{aligned}$$

The strict inequality $>$ holds because the minimum in $\mathrm{trop}(g)(\mathbf{w})$ is achieved uniquely at $\mathbf{u}_{i'}$. Thus the minimum in $\min_i(\mathrm{val}(c'_i) + \mathrm{val}(B_{ij}))$ is achieved uniquely at i', so $\mathrm{val}(a_j) = \mathrm{val}(\sum_i c'_i B_{ij}) = \mathrm{val}(c'_{i'}) + \mathrm{val}(B_{i'j})$. Hence, $\mathrm{val}(a_j) + \mathbf{w} \cdot \mathbf{u}_j = \mathrm{val}(c'_{i'}) + \mathrm{val}(B_{i'j}) + \mathbf{w} \cdot \mathbf{u}_j = \mathrm{val}(c'_{i'}) + \mathbf{w} \cdot \mathbf{u}_{i'} = \mathrm{val}(a_{i'}) + \mathbf{w} \cdot \mathbf{u}_{i'}$. This contradicts the fact that minimum in $\mathrm{trop}(g)(\mathbf{w})$ is unique.

Let $f = \sum_j B_{i'j} x^{\mathbf{u}_j} \in I$. In the previous paragraph we showed that $\mathbf{w} \cdot \mathbf{u}_{i'} < \mathrm{val}(B_{i'j}) + \mathbf{w} \cdot \mathbf{u}_j$ for $j \neq i'$. Since $B_{i'i'} = 1$, this means that the minimum in $\mathrm{trop}(f)(\mathbf{w})$ is achieved uniquely at the term involving $\mathbf{u}_{i'}$. As there were finitely many choices for the matrix B as $\mathbf{w}$ varied over C_g, we thus get a finite set of polynomials f with the property that for all $\mathbf{w} \in C_g$ there is some f in this set with the minimum in $\mathrm{trop}(f)(\mathbf{w})$ achieved uniquely at the term involving $x^{\mathbf{u}}$. Adding to our set a generating set for I, which also generates I_L, then gives a tropical basis for I_L with all coefficients in K. □

This allows us to prove the main theorem of this section.

Theorem 2.6.6. *Let K be an arbitrary valued field. Every ideal I in $K[x_1^{\pm 1}, \ldots, x_n^{\pm 1}]$ has a finite tropical basis.*

Proof. We first consider the case that the valuation on the field K has a splitting $w \mapsto t^w$, so we can apply the Gröbner theory of Sections 2.4 and 2.5. Consider the homogenization I_{proj} of I. Its Gröbner complex $\Sigma(I_{\mathrm{proj}})$ consists of finitely many polyhedra σ in $\mathbb{R}^{n+1}/\mathbb{R}\mathbf{1}$. For each σ, we select one representative vector $(0, \mathbf{w})$ in $\text{relint}(\sigma)$. The initial ideal $\text{in}_{\mathbf{w}}(I)$ depends only on σ, not on the choice of $\mathbf{w}$, by the definition of $\Sigma(I_{\text{proj}})$.

Fix any $\mathbf{w} \in \mathbb{R}^n$ with $\text{in}_{\mathbf{w}}(I) = \langle 1 \rangle$. There is a monomial $x^{\mathbf{u}} \in \text{in}_{(0,\mathbf{w})}(I_{\text{proj}})$. Choose $\mathbf{w}' = (0, \mathbf{w}) + \epsilon \mathbf{v}$ for $\mathbf{v} \in \mathbb{R}^{n+1}$ generic and $\epsilon > 0$ sufficiently small so that $\text{in}_{\mathbf{w}'}(I_{\text{proj}})$ is a monomial initial ideal of $\text{in}_{(0,\mathbf{w})}(I_{\text{proj}})$. Since $x^{\mathbf{u}} \in \text{in}_{\mathbf{w}'}(I_{\text{proj}})$, by Lemma 2.4.8, we can find $f = x^{\mathbf{u}} - g \in I_{\text{proj}}$ such that no monomial occurring in g lies in $\text{in}_{\mathbf{w}'}(I_{\text{proj}})$. For any $(0, \tilde{\mathbf{w}}) \in \text{relint}(\sigma)$ we have $\text{in}_{(0,\tilde{\mathbf{w}})}(f) = x^{\mathbf{u}}$, as $\text{in}_{(0,\tilde{\mathbf{w}})}(f) - x^{\mathbf{u}} \in \text{in}_{(0,\mathbf{w})}(I_{\text{proj}})$ and otherwise one of $\text{in}_{\mathbf{v}}(\text{in}_{(0,\tilde{\mathbf{w}})}(f) - x^{\mathbf{u}})$ and $\text{in}_{\mathbf{v}}(\text{in}_{(0,\tilde{\mathbf{w}})}(f))$ would not be in $\text{in}_{\mathbf{v}}(\text{in}_{(0,\mathbf{w})}(I_{\text{proj}})) = \text{in}_{\mathbf{w}'}(I_{\text{proj}})$. Set $f' = f|_{x_0=1}$. Then $\text{in}_{\tilde{\mathbf{w}}}(f')$ is a unit.

We define $\mathcal{T}$ by taking any finite generating set of I together with the Laurent polynomials f' constructed above. Then $\mathcal{T}$ also generates I. Consider an arbitrary weight vector $\mathbf{w} \in \mathbb{R}^n$. Then $(0, \mathbf{w}) \in \text{relint}(\sigma)$ for some cell σ in the Gröbner complex of I_{proj}. If $\text{in}_{\mathbf{w}}(I) = \langle 1 \rangle$, then $\text{in}_{\mathbf{w}}(f')$ is a unit for the corresponding polynomial $f' \in \mathcal{T}$. Hence the initial ideal $\text{in}_{\mathbf{w}}(I)$ equals $\langle 1 \rangle$ if and only if the finite set $\text{in}_{\mathbf{w}}(\mathcal{T})$ contains a unit.

When K is an arbitrary field, we can choose a valued field extension L/K for which the valuation on L has a splitting. For example, by Lemma 2.1.15, taking L to be the algebraic closure of K suffices. The above argument shows that there is a finite tropical basis $\mathcal{T}_L$ for $I_L = IL[x_1^{\pm 1}, \ldots, x_n^{\pm 1}]$. Lemma 2.6.5 then shows that there is a tropical basis $\mathcal{T}_L'$ for I_L with coefficients in K. Suppose that $f \in I$ and the minimum in $\text{trop}(f)(\mathbf{w})$ is achieved uniquely. Then since $f \in I_L$, there is $g \in \mathcal{T}_L'$ with the minimum in $\text{trop}(g)(\mathbf{w})$ achieved once. The set $\mathcal{T}_L'$ is thus also a tropical basis for I. □

In this section we used the finiteness of the Gröbner complex to prove the existence of finite tropical bases. An alternative proof, based on generic projections of tropical varieties, was given by Hept and Theobald [**HT09**]. Their approach shows that any generating set of I can be augmented to a tropical basis by a very small number of polynomials of high degree.

The concept of a tropical basis extends to ideals in a polynomial ring. If J is a homogeneous ideal in $K[x_0, \ldots, x_n]$, then a generating set $\mathcal{T}$ of J is a *tropical basis* if, for all $\mathbf{w} \in \mathbb{R}^{n+1}$, the ideal $\text{in}_{\mathbf{w}}(J)$ contains a monomial if and only if $\text{in}_{\mathbf{w}}(\mathcal{T})$ contains a monomial. This is different from the notion of a universal Gröbner basis for J, as the following example shows.

Example 2.6.7. Let $n = 2$, and consider the homogeneous polynomials

$$\begin{aligned} \mathcal{T} \;=\; \big\{ & x_0(x_1 + x_2 - x_0),\, x_1(x_0 + x_2 - x_1),\, x_2(x_0 + x_1 - x_2), \\ & x_0x_1(x_0 - x_1),\;\; x_0x_2(x_0 - x_2),\;\; x_1x_2(x_1 - x_2) \big\}. \end{aligned}$$

This set is a universal Gröbner basis for the polynomial ideal it generates:

$$I \;=\; \langle \mathcal{T} \rangle \;=\; \langle x_0 - x_1, x_2 \rangle \cap \langle x_0 - x_2, x_1 \rangle \cap \langle x_1 - x_2, x_0 \rangle.$$

However, the set $\mathcal{T}$ is not a tropical basis. For $\mathbf{w} = (1, 1, 1)$ the minimum in $\text{trop}(f)(\mathbf{w})$ is achieved at least twice for all $f \in \mathcal{T}$, but $x_0x_1x_2 \in I$. $\Diamond$

Tropical bases for specific polynomial ideals play an important role for combinatorial studies in tropical geometry. In this book, we shall encounter special generators that form a tropical basis for the ideals of linear spaces (in Proposition 4.1.6), Grassmannians (in Corollary 4.3.12), complete intersections (in Theorem 4.6.18), and determinantal varieties (in Theorem 5.3.25).

Example 2.6.8. Consider the linear forms $f = x_0 + x_1 + x_2 + x_3$ and $g = x_0 + t^3x_1 + t^7x_2 + tx_3$ over the Puiseux series field $K = \mathbb{C}\{\{t\}\}$. We claim that $\{f, g\}$ is a tropical basis for its ideal $J = \langle f, g\rangle$ in $K[x_0, x_1, x_2, x_3]$. This will follow from Theorem 4.6.18, but it can also be seen directly using the combinatorial analysis in Example 2.5.10. For the matrix A_1 derived from f and g, the inequality in (2.5.4) is strict, so the Gröbner complex $\Sigma(J)$ has $6 + 12 + 9 + 2 = 29$ cells. For vectors $\mathbf{w}$ in the interiors of the six maximal cells, either $\text{in}_{\mathbf{w}}(f)$ or $\text{in}_{\mathbf{w}}(g)$ is a monomial. For $\mathbf{w}$ in any of the lower-dimensional cells of $\Sigma(J)$, neither $\text{in}_{\mathbf{w}}(f)$ nor $\text{in}_{\mathbf{w}}(g)$ is a monomial. $\Diamond$

Our next goal is to show that the notion of a tropical basis is invariant under multiplicative coordinate changes in $K[x_1^{\pm1}, \ldots, x_n^{\pm1}]$. Along the way, we shall prove a general lemma that will be used in the proofs of Chapter 3.

Given a monomial map $\phi : T^n \to T^m$ with associated ring homomorphism $\phi^*\colon K[x_1^{\pm1}, \ldots, x_m^{\pm1}] \to K[z_1^{\pm1}, \ldots, z_n^{\pm1}]$, we also denote by ϕ^* the map $\mathbb{Z}^m \to \mathbb{Z}^n$ given by $\phi^*(\mathbf{e}_i) = \mathbf{u}$, where $\phi^*(x_i) = z^{\mathbf{u}}$. This gives an induced map, called the *tropicalization* of ϕ, by applying $\text{Hom}(-, \mathbb{Z})$ to ϕ^*:

$$\text{trop}(\phi)\colon \text{Hom}(\mathbb{Z}^n, \mathbb{Z}) \cong \mathbb{Z}^n \to \text{Hom}(\mathbb{Z}^m, \mathbb{Z}) \cong \mathbb{Z}^m.$$

Suppose the abelian group homomorphism ϕ^* is given by $\phi^*(x_i) = x^{\mathbf{a}_i}$ for $\mathbf{a}_i \in \mathbb{Z}^n$. Let A be the $n\times m$-matrix with ith column $\mathbf{a}_i$. Then $\text{trop}(\phi)\colon \mathbb{Z}^n \to \mathbb{Z}^m$ is the (classically) linear map given by the transpose A^T. We also denote by $\text{trop}(\phi)$ the vector space homomorphism $A^T : \mathbb{Z}^n \otimes \mathbb{R} \cong \mathbb{R}^n \to \mathbb{Z}^m \otimes \mathbb{R} \cong \mathbb{R}^m$ induced by tensoring with $\mathbb{R}$. Note that the image of the restriction of $\text{trop}(\phi)$ to Γ^n_{val} is contained in Γ^m_{val}. For any $\mathbf{y} = (y_1, \ldots, y_n) \in T^n$, we have

$$\begin{aligned} \text{val}(\phi(\mathbf{y})) &= (\text{val}(\mathbf{y}^{\mathbf{a}_1}), \ldots, \text{val}(\mathbf{y}^{\mathbf{a}_m})) \\ &= (\mathbf{a}_1 \cdot \text{val}(\mathbf{y}), \ldots, \mathbf{a}_m \cdot \text{val}(\mathbf{y})) \\ &= A^T\, \text{val}(\mathbf{y}) \;=\; \text{trop}(\phi)(\text{val}(\mathbf{y})). \end{aligned} \tag{2.6.3}$$

Example 2.6.9. Let $K = \mathbb{C}\{\{t\}\}$ and $\phi : T^3 \to T^2, (t_1, t_2, t_3) \mapsto (t_1t_2, t_2t_3)$. Then $\phi^* : K[x_1^{\pm1}, x_2^{\pm1}] \to K[z_1^{\pm1}, z_2^{\pm1}, z_3^{\pm1}]$ maps x_1 to z_1z_2, x_2 to z_2z_3, and

$$A = \begin{pmatrix} 1 & 0 \\ 1 & 1 \\ 0 & 1 \end{pmatrix}.$$

Consider the element $\mathbf{y} = (1+3t, t+t^5, 7) \in T^3$, with $\mathrm{val}(\mathbf{y}) = (0,1,0)$. Then $\phi(\mathbf{y}) = (t{+}3t^2{+}t^5{+}3t^6, 7t+7t^5)$, so $\mathrm{val}(\phi(\mathbf{y})) = (1,1) = A^T\,\mathrm{val}(\mathbf{y})$. ◊

We now reinstate our assumption that the valuation on K has a splitting, so that we can consider the effect of projections on initial ideals.

Lemma 2.6.10. *Let $\phi^*\colon K[x_1^{\pm 1},\dots,x_m^{\pm 1}] \to K[z_1^{\pm 1},\dots,z_n^{\pm 1}]$ be a monomial map. Let $I \subseteq K[z_1^{\pm 1},\dots,z_n^{\pm 1}]$ be an ideal, and let $I' = {\phi^*}^{-1}(I)$. Then*

$$\phi^*(\mathrm{in}_{\mathrm{trop}(\phi)(\mathbf{w})}(I')) \subseteq \mathrm{in}_{\mathbf{w}}(I) \qquad \textit{for all } \mathbf{w} \in \mathbb{R}^n.$$

Thus, in particular, if $\mathrm{in}_{\mathbf{w}}(I) \neq \langle 1\rangle$, then we also have $\mathrm{in}_{\mathrm{trop}(\phi)(\mathbf{w})}(I') \neq \langle 1\rangle$.

Proof. Let $\phi^*(x_i){=}z^{\mathbf{a}_i}$, where $\mathbf{a}_i \in \mathbb{Z}^n$. Then $\phi^*(x^{\mathbf{u}}) = z^{A\mathbf{u}}$, where $A = (\mathbf{a}_1,\dots,\mathbf{a}_m)$. Let $f = \sum c_{\mathbf{u}} x^{\mathbf{u}} \in I'$, so that $\phi^*(f) = \sum c_{\mathbf{u}} z^{A\mathbf{u}} \in I$. Then $W = \mathrm{trop}(f)(A^T\mathbf{w}) = \min_{c_{\mathbf{u}}\neq 0}(\mathrm{val}(c_{\mathbf{u}}) + \mathbf{w}\cdot A\mathbf{u}) = \mathrm{trop}(\phi^*(f))(\mathbf{w})$, and

$$\begin{aligned} \phi^*(\mathrm{in}_{\mathrm{trop}(\phi)(\mathbf{w})}(f)) &= \phi^*\Bigg(\sum_{\mathrm{val}(c_{\mathbf{u}})+\mathbf{w}\cdot A\mathbf{u}=W} \overline{t^{-\mathrm{val}(c_{\mathbf{u}})}c_{\mathbf{u}}}\cdot x^{\mathbf{u}}\Bigg) \\ &= \sum_{\mathrm{val}(c_{\mathbf{u}})+\mathbf{w}\cdot A\mathbf{u}=W} \overline{t^{-\mathrm{val}(c_{\mathbf{u}})}c_{\mathbf{u}}}\cdot x^{A\mathbf{u}} \;=\; \mathrm{in}_{\mathbf{w}}(\phi^*(f)). \end{aligned}$$

This implies $\phi^*(\mathrm{in}_{\mathrm{trop}(\phi)(\mathbf{w})}(I')) \subseteq \mathrm{in}_{\mathbf{w}}(I)$. This ideal contains $1 = \phi^*(1)$ if $\mathrm{in}_{\mathrm{trop}(\phi)(\mathbf{w})}(I') = \langle 1\rangle$. This proves the contrapositive of the last claim. □

Example 2.6.11. Fix ϕ as in Example 2.6.9. Consider the principal ideal $I = \langle z_1{+}z_3\rangle$ in $K[z_1^{\pm 1}, z_2^{\pm 1}, z_3^{\pm 1}]$. Then $I' = {\phi^*}^{-1}(I) = \langle x_1{+}x_2\rangle$. For $\mathbf{w} = (1,0,0)$, we have $\mathrm{in}_{\mathbf{w}}(I) = \langle z_3\rangle$ and $\mathrm{in}_{\mathrm{trop}(\phi)(\mathbf{w})}(I') = \mathrm{in}_{(1,0)}(I') = \langle x_2\rangle$. Here we have $\phi^*(\mathrm{in}_{\mathrm{trop}(\phi)(\mathbf{w})}(I')) = \langle z_2 z_3\rangle = \mathrm{in}_{\mathbf{w}}(I)$. ◊

Corollary 2.6.12. *Let ϕ^* be a monomial automorphism of $K[x_1^{\pm 1},\dots,x_n^{\pm 1}]$, let I be any ideal in this Laurent polynomial ring, and let $I' = {\phi^*}^{-1}(I)$. Then*

$$\mathrm{in}_{\mathbf{w}}(I) = \langle 1\rangle \quad \textit{if and only if} \quad \mathrm{in}_{\mathrm{trop}(\phi)(\mathbf{w})}(I') = \langle 1\rangle. \tag{2.6.4}$$

In this section we showed that every Laurent ideal I over a valued field possesses a finite tropical basis. In Chapter 3 we shall introduce the tropical variety of I, and we shall see that $\mathrm{in}_{\mathbf{w}}(I) \neq \langle 1\rangle$ is one of the characterizations for $\mathbf{w}$ that lie in that object. To cut out the tropical variety by tropical polynomials, it suffices to take a tropical basis of I. In that context, Theorem 2.6.6 can be regarded as a version of the Hilbert Basis Theorem for tropical geometry.

2.7. Exercises

(1) Show that the residue field of $\Bbbk\{\{t\}\}$ is isomorphic to $\Bbbk$.

(2) Let $K = \mathbb{Q}$ with the p-adic valuation. Show that the residue field of K is isomorphic to $\mathbb{Z}/p\mathbb{Z}$.

(3) The quotient ring $K = \mathbb{Q}[s]/\langle 3s^3+s^2+36s+162\rangle$ is a field. Describe all valuations on this field that extend the 3-adic valuation on $\mathbb{Q}$.

(4) Show that if K is an algebraically closed field with a valuation $\mathrm{val} : K^* \to \mathbb{R}$ and $\Bbbk = R/\mathfrak{m}$ its residue field, then $\Bbbk$ is algebraically closed. Give an example to show that if $\Bbbk$ is algebraically closed, then it does not automatically follow that K is algebraically closed.

(5) Apply the algorithm implicit in the proof of Theorem 2.1.5 to compute (the start of) a solution in $\mathbb{C}\{\{t\}\}$ to the equation $x^2+t+1 = 0$.

(6) In this exercise you will show that the splitting of Lemma 2.1.15 does not always exist if the field is not algebraically closed.

Let $\mathbb{F}$ be an arbitrary field, and let $K = \mathbb{F}(x_1, x_2, \dots)$ be the field of rational functions in countably many variables. This is the union of the rational function fields $\mathbb{F}(x_1, \dots, x_n)$ for all $n \geq 1$ so only finitely many variables appear in each rational function.

(a) Show that there is a valuation $\mathrm{val} : K^* \to \mathbb{R}$ with $\mathrm{val}(a) = 0$ for $a \in \mathbb{F}$ and $\mathrm{val}(x_j) = 1/j$.

(b) Show that for this valuation the value group Γ_{val} equals $\mathbb{Q}$.

(c) Suppose a splitting $\phi : \mathbb{Q} \to K^*$ exists. There exist $f_n, g_n \in \mathbb{F}[x_1, x_2, \dots]$ with $\phi(1/n) = f_n/g_n$ for $n \geq 1$. Derive a contradiction by comparing these polynomials for $n = 1$ and $n > 1$. *Hint*: The polynomial ring in finitely many variables is a UFD.

(7) In the setting of Definition 2.2.4, find an ideal I in $\mathbb{Q}[x_1^{\pm 1}, x_2^{\pm 1}, x_3^{\pm 1}]$ such that $I_{\mathrm{aff}} \subset \mathbb{Q}[x_1, x_2, x_3]$ requires more generators than I, and also $I_{\mathrm{proj}} \subset \mathbb{Q}[x_0, x_1, x_2, x_3]$ requires more generators than I_{aff}.

(8) List an explicit minimal set of generators for the Plücker ideal $I_{4,8}$.

(9) (a) Show that if $\phi\colon \mathbb{C}^* \to \mathbb{C}^*$ is a homomorphism of algebraic groups, then ϕ has the form $\phi(x) = x^n$ for some $n \in \mathbb{Z}$.

(b) Deduce that $\mathrm{Hom}_{\mathrm{alg}}(T^n, \mathbb{C}^*) \cong \mathbb{Z}^n$.

(c) Conclude that the group of automorphisms of T^n as an algebraic group is $\mathrm{GL}(n, \mathbb{Z})$.

(10) Pick two triangles P and Q that lie in nonparallel planes in $\mathbb{R}^3$. Draw their Minkowski sum $P + Q$, and verify the identity (2.3.2).

(11) Classify all regular triangulations of the three-dimensional cube.

(12) What is the maximal number of facets of any four-dimensional polytope with eight vertices? How many edges (= one-dimensional faces) are there in such a polytope?

(13) Show that for a polyhedron σ in a polyhedral complex Σ, the cone $\bar{\tau}$ of $\mathrm{star}_\Sigma(\sigma)$ defined in Definition 2.3.6 is

$$\bar{\tau} \;=\; \{\mathbf{v} \in \mathbb{R}^n : \exists \epsilon > 0 \text{ with } \mathbf{w} + \epsilon \mathbf{v} \in \tau\} \;+\; \mathrm{aff}(\sigma),$$

for any fixed $\mathbf{w} \in \mathrm{relint}(\sigma)$.

(14) Let $K = \mathbb{C}\{\{t\}\}$ and consider the homogeneous polynomial $f = (x_0 + t^{11}x_1 + t^{38}x_2)^{2000} + (x_0x_1 + t^{-9}x_0x_2 + t^{13}x_1x_2)^{1000}$. Determine $\mathrm{trop}(f)(\mathbf{w})$ for $\mathbf{w} = (3,4,5)$ and for $\mathbf{w} = (30,40,50)$.

(15) Let $K = \mathbb{Q}_7$, and let R be its valuation ring. Give an example of an ideal I in $R[x_0,x_1,x_2]$ such that $R[x_0,x_1,x_2]/I$ is not a flat R-module.

(16) Compute all initial ideals of $I = \langle 7x_0^2 + 8x_0x_1 - x_1^2 + x_0x_2 + 3x_2^2 \rangle \subseteq \mathbb{C}[x_0,x_1,x_2]$, and draw the Gröbner complex of I. Repeat for the ideal $I = \langle tx_1^2 + 3x_1x_2 - tx_2^2 + 5x_0x_1 - x_0x_2 + 2tx_0^2 \rangle \subseteq \mathbb{C}\{\{t\}\}[x_0,x_1,x_2]$.

(17) Draw the set $\{\mathbf{w} \in \mathbb{R}^2 : \mathrm{in}_\mathbf{w}(I) \neq \langle 1 \rangle\}$ for the principal ideal $I = \langle 7 + 8x_1 - x_1^2 + x_2 + 3x_2^2 \rangle$ in $\mathbb{C}[x_1^{\pm 1}, x_2^{\pm 1}]$. Repeat for the ideal

$$I \;=\; \big\langle tx_1^2 + 3x_1x_2 - tx_2^2 + 5x_1 - x_2 + 2t \big\rangle \;\subseteq\; \mathbb{C}\{\{t\}\}[x_1^{\pm 1}, x_2^{\pm 1}].$$

(18) Let I be the ideal (1.8.4) in Example 1.8.3. Determine the Gröbner fan, a universal Gröbner basis, and a tropical basis for I.

(19) One property of Gröbner bases as in [**CLO07**] is that the condition $\mathrm{in}_\mathbf{w}(I) = \langle \mathrm{in}_\mathbf{w}(g_1), \ldots, \mathrm{in}_\mathbf{w}(g_s) \rangle$ for $g_1, \ldots, g_s \in I$ implies $I = \langle g_1, \ldots, g_s \rangle$. Does this hold for ideals $I \subset K[x_1^{\pm 1}, \ldots, x_n^{\pm 1}]$? Can you formulate sufficient conditions under which it holds?

(20) Let I be an ideal in $K[x_0, \ldots, x_n]$. Let $m = \prod_{i=0}^n x_i$. Show that I contains a monomial if and only if the saturation $(I : m^\infty) := \{f : fm^k \in I \text{ for some } k > 0\}$ contains 1.

(21) The maximal ideal $\langle x_1 + x_2 + 3, x_1 + 5x_2 + 7 \rangle \subseteq \mathbb{C}[x_1^{\pm 1}, x_2^{\pm 1}]$ defines a point in the plane. Compute a tropical basis for this ideal.

(22) The set $\mathbb{Z}$ of integers with the 2-adic valuation is a metric space, by Example 1.8.4. Sketch a picture of this space. How about $\mathbb{Q}_2$?

(23) Let I be the homogeneous ideal in $\mathbb{Q}[x,y,z]$ generated by the set

$$\mathcal{G} \;=\; \{x + y + z, x^2y + xy^2, x^2z + xz^2, y^2z + yz^2\}.$$

Show that $\mathcal{G}$ is a *universal Gröbner basis*, that is, $\mathcal{G}$ is a Gröbner basis of I for all $\mathbf{w} \in \mathbb{R}^3$. Also, show that $\mathcal{G}$ is not a tropical basis.

(24) Draw the graph of F and the polyhedral complex Σ_F for the tropical polynomial functions $F = p$ and $F = p \odot p \odot p$ in Example 1.2.4. Work out a similar example in one dimension higher, and interpret your pictures in terms of integer linear programming.

(25) Fix two random quadrics in $K[x_0, x_1, x_2, x_3]$. Let I be the homogeneous ideal they generate. Compute the polynomial g in (2.5.2). Which D did you take and why? Describe $\mathrm{trop}(g)$ and $\Sigma_{\mathrm{trop}(g)}$.

(26) The Plücker ideal $I_{2,n}$ is minimally generated by the quadrics

$$\underline{p_{ik}p_{jl}} - p_{ij}p_{kl} - p_{il}p_{jk} \quad \text{for } 1 \leq i < j < k < l \leq n.$$

Find $\mathbf{w} \in \mathbb{Q}^{\binom{n}{2}}$ which selects the underlined initial monomials. Compute the cone $\overline{C_{I_{2,n}}[\mathbf{w}]}$. How many extreme rays does it have?

(27) Solve the equation $x^5 + tx^4 + t^3x^3 + t^6x^2 + t^{10}x = t^{15}$ in the Puiseux series field $\mathbb{C}\{\{t\}\}$. Also, solve the equation $x^5 + 2x^4 + 8x^3 + 64x^2 + 1024x = 32768$ in the field $\mathbb{Q}_2$. (*Hint*: See Exercise 1.9(3).)

(28) Let Σ_1 be the polyhedral complex consisting of all faces of the cube $[-1, 1]^3$, and let Σ_2 consist of all faces of the octahedron $\mathrm{conv}\{\pm\mathbf{e}_1, \pm\mathbf{e}_2, \pm\mathbf{e}_3\}$. Determine the common refinement $\Sigma_1 \wedge \Sigma_2$.

(29) Let $K = \mathbb{C}\{\{t\}\}$, and consider the matrix

$$A = \begin{pmatrix} t^2 & t^3 & t^5 & t^7 \\ t^{19} & t^{11} & t^{17} & t^{13} \end{pmatrix}.$$

For each of the six 2-element subsets J of $\{1, 2, 3, 4\}$, construct the transformation $U \in \mathrm{GL}(2, K)$ that is promised by Lemma 2.5.6.

Chapter 3

Tropical Varieties

We now introduce the main player of this book: the tropical variety. The two main results of this chapter are the Fundamental Theorem 3.2.3 and the Structure Theorem 3.3.5. The Fundamental Theorem gives several equivalent definitions of a tropical variety. We discuss this first for hypersurfaces and then for general varieties in Sections 3.1 and 3.2, respectively. The Structure Theorem strengthens the connection between tropical and polyhedral geometry. The main ideas are introduced in Section 3.3, with the proofs following in Sections 3.4 and 3.5. In Section 3.6 we develop the theory of stable intersections, which was previewed for tropical curves in Section 1.3.

We restrict our usage of the name *tropical variety* to mean the tropicalization of a classical variety over a field with a valuation. A more inclusive notion of tropical varieties allows for balanced polyhedral complexes that do not necessarily lift to a classical variety. In Chapter 4, we will see this distinction in the context of linear spaces. For now, we always start with Laurent polynomial ideals or, equivalently, with subvarieties of an algebraic torus.

3.1. Hypersurfaces

Let K be an arbitrary field with a possibly trivial valuation. We work in the ring $K[x_1^{\pm 1}, \ldots, x_n^{\pm 1}]$ of Laurent polynomials over K. Given a Laurent polynomial $f = \sum_{\mathbf{u} \in \mathbb{Z}^n} c_{\mathbf{u}} x^{\mathbf{u}}$, we define its *tropicalization* $\operatorname{trop}(f)$ as in (2.4.1). Namely, $\operatorname{trop}(f)$ is the real-valued function on $\mathbb{R}^n$ that is obtained by replacing each coefficient $c_{\mathbf{u}}$ by its valuation and by performing all additions and multiplications in the tropical semiring $(\mathbb{R}, \oplus, \odot)$. Explicitly,

$$\operatorname{trop}(f)(\mathbf{w}) = \min_{\mathbf{u}\in\mathbb{Z}^n} (\operatorname{val}(c_{\mathbf{u}}) + \sum_{i=1}^{n} u_i w_i) = \min_{\mathbf{u}\in\mathbb{Z}^n} (\operatorname{val}(c_{\mathbf{u}}) + \mathbf{u}\cdot\mathbf{w}).$$

The tropical polynomial $\operatorname{trop}(f)$ is a piecewise linear concave function $\mathbb{R}^n \to \mathbb{R}$. For an illustration of the graph of $\operatorname{trop}(f)$ when $n = 2$, see Figure 1.3.2.

The classical variety of the Laurent polynomial $f \in K[x_1^{\pm 1}, \dots, x_n^{\pm 1}]$ is a hypersurface in the algebraic torus T^n over the algebraic closure of K:

$$V(f) = \{\mathbf{y} \in T^n : f(\mathbf{y}) = 0\}.$$

We now define the tropical hypersurface associated with the same f.

Definition 3.1.1. The *tropical hypersurface* $\operatorname{trop}(V(f))$ is the set

$$\{\mathbf{w} \in \mathbb{R}^n : \text{ the minimum in } \operatorname{trop}(f)(\mathbf{w}) \text{ is achieved at least twice}\}.$$

This is the locus in $\mathbb{R}^n$ where the piecewise linear function $\operatorname{trop}(f)$ fails to be linear. When the valuation on K has a splitting $w \mapsto t^w$, this can be rephrased in terms of the initial forms we introduced in (2.6.1):

$$\operatorname{in}_{\mathbf{w}}(f) = \sum_{\substack{\mathbf{u}:\operatorname{val}(c_{\mathbf{u}})+\mathbf{w}\cdot\mathbf{u} \\ = \operatorname{trop}(f)(\mathbf{w})}} \overline{t^{-\operatorname{val}(c_{\mathbf{u}})} c_{\mathbf{u}}}\, x^{\mathbf{u}}. \tag{3.1.1}$$

The tropical hypersurface $\operatorname{trop}(V(f))$ is the set of weight vectors $\mathbf{w} \in \mathbb{R}^n$ for which the initial form $\operatorname{in}_{\mathbf{w}}(f)$ is not a unit in $\Bbbk[x_1^{\pm 1}, \dots, x_n^{\pm 1}]$. The equivalence of these two definitions is the easy direction in Theorem 3.1.3 below.

When F is a tropical polynomial, we write $V(F)$ for the set

$$\{\mathbf{w} \in \mathbb{R}^n : \text{ the minimum in } F(\mathbf{w}) \text{ is achieved at least twice}\}.$$

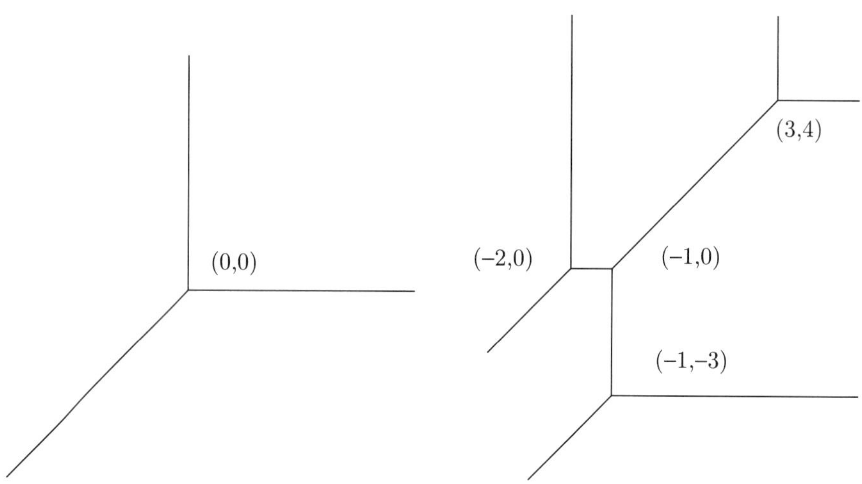

Figure 3.1.1. A tropical line and a tropical quadric.

With this notation, we have

$$\operatorname{trop}(V(f)) \;=\; V(\operatorname{trop}(f)).$$

Example 3.1.2. Let $K = \mathbb{C}\{\{t\}\}$ be the field of Puiseux series with complex coefficients. We examine bivariate Laurent polynomials $f \in K[x^{\pm 1}, y^{\pm 1}]$.

(1) Let $f = x + y + 1$. Then $\operatorname{trop}(f)(\mathbf{w}) = \min(w_1, w_2, 0)$, so

$$\operatorname{trop}(V(f))(\mathbf{w}) \;=\; \{w_1 = w_2 \leq 0\} \cup \{w_1 = 0 \leq w_2\} \cup \{w_2 = 0 \leq w_1\}.$$

This is the tropical line shown on the left in Figure 3.1.1.

(2) Let $f = t^2x^2 + xy + (t^2 + t^3)y^2 + (1 + t^3)x + t^{-1}y + t^3$. Then $\operatorname{trop}(f)(\mathbf{w}) = \min(2 + 2w_1, w_1 + w_2, 2 + 2w_2, w_1, -1 + w_2, 3)$, so $\operatorname{trop}(V(f))$ consists of the three line segments joining the pairs $\{(-1,0),(-2,0)\}$, $\{(-1,0),(-1,-3)\}$, $\{(-1,0),(3,4)\}$, and the six rays $\{(-2,0) + \lambda(0,1)\}$, $\{(-2,0) - \lambda(1,1)\}$, $\{(-1,-3) - \lambda(1,1)\}$, $\{(-1,-3) + \lambda(1,0)\}$, $\{(3,4) + \lambda(0,1)\}$, $\{(3,4) + \lambda(1,0)\}$. In these sets, λ runs over $\mathbb{R}_{\geq 0}$. This is shown on the right in Figure 3.1.1. ◊

The following theorem was stated in the early 1990s in an unpublished manuscript by Mikhail Kapranov. A proof appeared in [**EKL06**]. It establishes the link between classical hypersurfaces over a field K and tropical hypersurfaces in $\mathbb{R}^n$. In the next section, we present the more general Fundamental Theorem which works for varieties of arbitrary codimension. Kapranov's Theorem for hypersurfaces will serve as the base case for its proof.

We place the extra conditions here on the field K that it is algebraically closed and has a nontrivial valuation with a splitting. If K is an arbitrary field with a valuation, then we may pass to its algebraic closure $\overline{K}$ with an extension of the valuation. If the valuation on K is trivial, we further pass to the field of generalized power series $\overline{K}((\mathbb{R}))$. This yields a field satisfying these conditions. Note that passing to an extension field does not change the function $\operatorname{trop}(f)$, so does not alter the tropical hypersurface $\operatorname{trop}(V(f))$.

Theorem 3.1.3 (Kapranov's Theorem). *Let K be an algebraically closed field with a nontrivial valuation. Fix a Laurent polynomial $f = \sum_{\mathbf{u} \in \mathbb{Z}^n} c_{\mathbf{u}} x^{\mathbf{u}}$ in $K[x_1^{\pm 1}, \ldots, x_n^{\pm 1}]$. The following three sets coincide:*

(1) *the tropical hypersurface* $\operatorname{trop}(V(f))$ *in* $\mathbb{R}^n$;

(2) *the set* $\{\mathbf{w} \in \mathbb{R}^n : \operatorname{in}_{\mathbf{w}}(f)$ *is not a monomial* $\}$;

(3) *the closure in* $\mathbb{R}^n$ *of* $\{(\operatorname{val}(y_1), \ldots, \operatorname{val}(y_n)) : (y_1, \ldots, y_n) \in V(f)\}$.

Furthermore, if f is irreducible and $\mathbf{w}$ is any point in $\Gamma_{\mathrm{val}}^n \cap \operatorname{trop}(V(f))$, then the set $\{\mathbf{y} \in V(f) : \operatorname{val}(\mathbf{y}) = \mathbf{w}\}$ is Zariski dense in the hypersurface $V(f)$.

Example 3.1.4. Let $f = x - y + 1 \in K[x^{\pm 1}, y^{\pm 1}]$, where K is as above. Then $V(f) = \{(z, z+1) : z \in K, z \neq 0, -1\}$, and $\operatorname{trop}(V(f))$ is the tropical line in Figure 3.1.1. Note that $\mathrm{in}_{\mathbf{w}}(f)$ is a monomial unless either $\mathbf{w}$ is $(0,0)$, or $\mathbf{w}$ is a positive multiple of $(1,0),(0,1)$ or $(-1,-1)$. In the former case, $\mathrm{in}_{\mathbf{w}}(f)$ is $x-y+1$. In the latter cases, it is $-y+1, x+1$ or $x-y$. We have

$$\big(\mathrm{val}(z),\mathrm{val}(z+1)\big) = \begin{cases} (\mathrm{val}(z),0) & \text{if } \mathrm{val}(z)>0,\\ (\mathrm{val}(z),\mathrm{val}(z)) & \text{if } \mathrm{val}(z)<0,\\ (0,\mathrm{val}(z+1)) & \text{if } \mathrm{val}(z)=0,\ \mathrm{val}(z+1)>0,\\ (0,0) & \text{otherwise.}\end{cases}$$

As z runs over $K\backslash\{0,-1\}$, the above case distinction describes all points of Γ_{val}^2 that lie in the tropical line $\mathrm{trop}(V(f))$. Since K is algebraically closed, the value group Γ_{val} is dense in $\mathbb{R}$, so the closure of these points in Γ_{val}^2 is the entire tropical line in $\mathbb{R}^2$. This confirms Theorem 3.1.3 for this f. ◊

Proof of Theorem 3.1.3. Let $\mathbf{w}=(w_1,\ldots,w_n)\in\mathrm{trop}(V(f))$. By definition, the minimum in $W=\min_{\mathbf{u}:c_{\mathbf{u}}\neq 0}(\mathrm{val}(c_{\mathbf{u}})+\mathbf{u}\cdot\mathbf{w})=\mathrm{trop}(f)(\mathbf{w})$ is achieved at least twice. Therefore $\mathrm{in}_{\mathbf{w}}(f)$, as seen in (3.1.1), is not a monomial. Thus, set (1) is contained in set (2). Conversely, if $\mathrm{in}_{\mathbf{w}}(f)$ is not a monomial, then the minimum in W is achieved at least twice, so $\mathbf{w}\in\mathrm{trop}(V(f))$. This shows the other containment, and so the equality of sets (1) and (2).

We now prove that set (1) contains set (3). Since set (1) is closed, it is enough to consider points in set (3) of the form $\mathrm{val}(\mathbf{y}):=(\mathrm{val}(y_1),\ldots,\mathrm{val}(y_n))$ where $\mathbf{y}=(y_1,\ldots,y_n)\in(K^*)^n$ satisfies $f(\mathbf{y})=\sum_{\mathbf{u}\in\mathbb{Z}^n}c_{\mathbf{u}}\mathbf{y}^{\mathbf{u}}=0$. This means $\mathrm{val}(\sum_{\mathbf{u}\in\mathbb{Z}^n}c_{\mathbf{u}}\mathbf{y}^{\mathbf{u}})=\mathrm{val}(0)=\infty>\mathrm{val}(c_{\mathbf{u}'}\mathbf{y}^{\mathbf{u}'})$ for all $\mathbf{u}'$ with $c_{\mathbf{u}'}\neq 0$. Lemma 2.1.1 implies that the minimum of $\mathrm{val}(c_{\mathbf{u}'}\mathbf{y}^{\mathbf{u}'})=\mathrm{val}(c_{\mathbf{u}'})+\mathbf{u}'\cdot\mathrm{val}(\mathbf{y})$ for $\mathbf{u}'$ with $c_{\mathbf{u}'}\neq 0$ is achieved at least twice. Thus $\mathrm{val}(\mathbf{y})\in\mathrm{trop}(V(f))$.

It remains to be seen that set (3) contains set (1). This is the hard part of Kapranov's Theorem. It will be the content of Proposition 3.1.5. That proposition also shows that $\{\mathbf{y}\in V(f):\mathrm{val}(\mathbf{y})=\mathbf{w}\}$ is Zariski dense when f is irreducible, so it completes our proof. □

The next result, which finishes the proof of Theorem 3.1.3, states that every zero of an initial form lifts to a zero of the given polynomial.

Proposition 3.1.5. *Fix $f\in K[x_1^{\pm 1},\ldots,x_n^{\pm 1}]$, and let $\mathbf{w}\in\Gamma_{\mathrm{val}}^n$. Suppose $\mathrm{in}_{\mathbf{w}}(f)$ is not a monomial and $\alpha\in(\Bbbk^*)^n$ satisfies $\mathrm{in}_{\mathbf{w}}(f)(\alpha)=0$. There exists $\mathbf{y}\in(K^*)^n$ satisfying $f(\mathbf{y})=0$, $\mathrm{val}(\mathbf{y})=\mathbf{w}$, and $\overline{t^{-w_i}y_i}=\alpha_i$ for $1\leq i\leq n$. If f is irreducible, then the set of such $\mathbf{y}$ is Zariski dense in the hypersurface $V(f)$.*

Proof. We use induction on n. The base case is $n=1$. After multiplying by a unit, we may assume that $f=\sum_{i=0}^{s}c_ix^i=\prod_{j=1}^{s}(a_jx-b_j)$, where $c_0,c_s\neq 0$. Then $\mathrm{in}_w(f)=\prod_{j=1}^{s}\mathrm{in}_w(a_jx-b_j)$ by Lemma 2.6.2. Since

$\alpha \in \Bbbk^*$ and $\mathrm{in}_w(f)(\alpha) = 0$, the initial form $\mathrm{in}_w(f)$ is not a monomial, and $\mathrm{in}_w(a_j x - b_j)(\alpha) = 0$ for some j. This implies that $\mathrm{in}_w(a_j x - b_j)$ is not a monomial. Hence $\mathrm{val}(a_j) + w = \mathrm{val}(b_j)$, and $\alpha = \overline{t^{-w} b_j / a_j}$. Set $y = b_j/a_j \in K^*$. Then $f(y) = 0$, $\mathrm{val}(y) = w$, and $\overline{t^{-\mathrm{val}(y)} y} = \alpha$ as required.

We now assume $n > 1$ and that the proposition holds for smaller dimensions. We first reduce to the case where no two monomials appearing in f are divisible by the same power of x_n. This has the consequence that, when f is regarded as a polynomial in x_n with coefficients in $K[x_1^{\pm 1}, \ldots, x_{n-1}^{\pm 1}]$, the coefficients are all monomials of the form $cx^{\mathbf{u}}$ for $c \in K$ and $\mathbf{u} \in \mathbb{Z}^{n-1}$.

Consider the automorphism $\phi_l^* \colon K[x_1^{\pm 1}, \ldots, x_n^{\pm 1}]$ given by $\phi_l^*(x_j) = x_j x_n^{l^j}$ for $1 \leq j \leq n-1$, and $\phi_l^*(x_n) = x_n$, where $l \in \mathbb{N}$. For $\mathbf{u} \in \mathbb{Z}^{n-1}$, we have $\phi_l^*(x^{\mathbf{u}} x_n^i) = x^{\mathbf{u}} x_n^{i + \sum_{j=1}^{n-1} u_j l^j}$. For $l \gg 0$ each monomial in $\phi_l^*(f)$ is divisible by a different power of x_n as required. Suppose that $\mathbf{y} = (y_1, \ldots, y_n) \in T^n$ satisfies $\phi_l^*(f)(\mathbf{y}) = 0$, $\mathrm{val}(y_i) = w_i - l^i w_n$ and $\overline{t^{-w_i + l^i w_n} y_i} = \alpha_i \alpha_n^{-l^i}$ for $1 \leq i \leq n-1$, as well as $\mathrm{val}(y_n) = w_n$ and $\overline{t^{-w_n} y_n} = \alpha_n$. Define $\mathbf{y}' \in T^n$ by $y_i' = y_i y_n^{l^i}$ for $1 \leq i \leq n$ and $y_n' = y_n$. We then have $f(\mathbf{y}') = 0$, $\mathrm{val}(\mathbf{y}') = \mathbf{w}$, and $\overline{t^{-w_i} y_i'} = \alpha_i$. Hence it suffices to prove Proposition 3.1.5 for $\phi_l^*(f)$.

We now assume that f has the special form described above. Consider the set of all $(y_1, \ldots, y_{n-1})$ in T^{n-1} with $\mathrm{val}(y_i) = w_i$ and $\overline{t^{-w_i} y_i} = \alpha_i$ for $1 \leq i \leq n-1$. By Lemma 2.2.12, this set is Zariski dense in T^{n-1}. Moreover, for all such choices, $g(x_n) = f(y_1, \ldots, y_{n-1}, x_n)$ is not the zero polynomial.

Write $\mathbf{u}'$ for the projection of $\mathbf{u} \in \mathbb{Z}^n$ onto the first $n-1$ coordinates. Writing $g = \sum d_i x_n^i$, we have $d_i = \mathbf{c_u} \mathbf{y}^{\mathbf{u}'}$ for a unique $\mathbf{u} \in \mathbb{Z}^n$ that has $u_n = i$. Note that $\mathrm{val}(d_i) + w_n i = \mathrm{val}(c_{\mathbf{u}}) + \mathrm{val}(\mathbf{y}^{\mathbf{u}'}) + w_n i = \mathrm{val}(c_{\mathbf{u}}) + \mathbf{w}' \cdot \mathbf{u}' + w_n u_n = \mathrm{val}(c_{\mathbf{u}}) + \mathbf{w} \cdot \mathbf{u}$. Therefore $\mathrm{trop}(g)(w_n) = \mathrm{trop}(f)(\mathbf{w})$, and

$$\begin{aligned}
\mathrm{in}_{w_n}(g) &= \sum_{i:\mathrm{val}(d_i)+w_n i = \mathrm{trop}(g)(w_n)} \overline{t^{-\mathrm{val}(d_i)} d_i} x_n^i \\
&= \sum_{\mathbf{u}:\mathrm{val}(c_{\mathbf{u}} y^{\mathbf{u}'}) + w_n u_n = \mathrm{trop}(g)(w_n)} \overline{t^{-\mathrm{val}(c_{\mathbf{u}})} c_{\mathbf{u}} t^{-\mathbf{u}' \cdot \mathbf{w}'} y^{\mathbf{u}'}} x_n^{u_n} \\
&= \sum_{\mathbf{u}:\mathrm{val}(c_{\mathbf{u}}) + \mathbf{w} \cdot \mathbf{u} = \mathrm{trop}(f)(\mathbf{w})} \overline{t^{-\mathrm{val}(c_{\mathbf{u}})} c_{\mathbf{u}}} \cdot \alpha^{\mathbf{u}'} x_n^{u_n} \\
&= \mathrm{in}_{\mathbf{w}}(f)(\alpha_1, \ldots, \alpha_{n-1}, x_n).
\end{aligned}$$

Thus $\mathrm{in}_{w_n}(g)(\alpha_n) = 0$. By the $n = 1$ case there is $y_n \in K^*$ with $\mathrm{val}(y_n) = w_n$ and $\overline{t^{-w_n} y_n} = \alpha_n$ for which $g(y_n) = 0$, and thus $f(y_1, \ldots, y_{n-1}, y_n) = 0$. We conclude $\mathbf{y} = (y_1, \ldots, y_n)$ is the required point in the hypersurface $V(f)$.

We now show that if f is an irreducible polynomial, then the set $\mathcal{Y}$ of $\mathbf{y}$ with $\mathrm{val}(\mathbf{y}) = \mathbf{w}$ and $\overline{t^{-w_i} y_i} = \alpha_i$ for $1 \leq i \leq n$ is Zariski dense in $V(f)$. For any $(y_1,\ldots,y_{n-1}) \in T^n$, with $\text{val}(y_i) = w_i$ and $\overline{t^{-w_i}y_i} = \alpha_i$ for all i, we constructed a point $\mathbf{y} = (y_1,\ldots,y_{n-1},y_n) \in \mathcal{Y}$. The set of such $(y_1,\ldots,y_{n-1})$ is Zariski dense in T^{n-1} by Lemma 2.2.12. Hence the projection of $\mathcal{Y}$ onto the first $n-1$ coordinates is not contained in any hypersurface in T^{n-1}. Consider any $g \in K[x_1^{\pm 1},\ldots,x_n^{\pm 1}]$ with $g(\mathbf{y}) = 0$ for all $\mathbf{y} \in \mathcal{Y}$. Then $\langle f,g\rangle \cap K[x_1^{\pm 1},\ldots,x_{n-1}^{\pm 1}] = \{0\}$. Since f is irreducible, this implies that g is a multiple of f. We conclude that the set $\mathcal{Y}$ is Zariski dense in $V(f)$. □

In the rest of Section 3.1 we study the combinatorics of tropical hypersurfaces. This uses the notion of *regular subdivisions* from Section 2.3. By the *k-skeleton* of a polyhedral complex Σ, we mean the polyhedral complex consisting of all cells $\sigma \in \Sigma$ with $\dim(\sigma) \leq k$. The field K is again arbitrary.

Proposition 3.1.6. *Let $f \in K[x_1^{\pm 1},\ldots,x_n^{\pm 1}]$ be a Laurent polynomial. The tropical hypersurface* $\text{trop}(V(f))$ *is the support of a pure Γ_{val}-rational polyhedral complex of dimension $n-1$ in $\mathbb{R}^n$. It is the $(n-1)$-skeleton of the polyhedral complex dual to the regular subdivision of the Newton polytope of $f = \sum c_{\mathbf{u}}x^{\mathbf{u}}$ given by the weights* $\text{val}(c_{\mathbf{u}})$ *on the lattice points in* $\text{Newt}(f)$.

Proof. By definition, $\text{trop}(V(f))$ is the set of $\mathbf{w} \in \mathbb{R}^n$ for which the minimum in $\text{trop}(f)(\mathbf{w}) = \min_{\mathbf{u}}(\text{val}(c_{\mathbf{u}}) + \mathbf{w}\cdot\mathbf{u})$ is achieved at least twice. Let $P = \text{Newt}(f) = \text{conv}\{\mathbf{u} : c_{\mathbf{u}} \neq 0\} \subset \mathbb{R}^n$ be the Newton polytope of f, and define $P_{\text{val}} = \text{conv}\{(\mathbf{u}, \text{val}(c_{\mathbf{u}})) : c_{\mathbf{u}} \neq 0\} \subset \mathbb{R}^{n+1}$. Let $\pi : \mathbb{R}^{n+1} \to \mathbb{R}^n$ be the projection onto the first n coordinates. The regular subdivision of P induced by the weights $\text{val}(c_{\mathbf{u}})$, $c_{\mathbf{u}} \neq 0$, consists of the polytopes $\pi(F)$ as F varies over all lower faces of P_{val}. Being a *lower face* of P_{val} means that

$$F = \text{face}_{\mathbf{v}}(P_{\text{val}}) = \{\mathbf{x} \in P_{\text{val}} : \mathbf{v}\cdot\mathbf{x} \leq \mathbf{v}\cdot\mathbf{y} \text{ for all } \mathbf{y} \in P_{\text{val}}\}$$

for some $\mathbf{v} \in \mathbb{R}^{n+1}$ with last coordinate v_{n+1} positive. For such an F, let $\mathcal{N}(F) = \{\mathbf{v} : \text{face}_{\mathbf{v}}(P_{\text{val}}) = F\}$ be the normal cone. We denote by $\tilde{\pi}(\mathcal{N}(F))$ the restricted projection $\{\mathbf{w} \in \mathbb{R}^n : (\mathbf{w},1) \in \mathcal{N}(F)\}$. The collection of $\tilde{\pi}(\mathcal{N}(F))$ as F varies over all lower faces of P_{val} forms a polyhedral complex in $\mathbb{R}^n$ that is dual to the regular subdivision of P induced by the $\text{val}(c_{\mathbf{u}})$.

If $\mathbf{v} = (v_1,\ldots,v_n,1) \in \mathcal{N}(F)$, then $\text{in}_{\pi(\mathbf{v})}(f)$ is a sum of monomials with exponents in $\pi(F)$, and all vertices of the polytope $\pi(F)$ appear with nonzero coefficient. This means that $\mathbf{w} = (w_1,\ldots,w_n) \in \text{trop}(V(f))$ if and only if $\mathbf{w} \in \tilde{\pi}(\mathcal{N}(F))$ for some face F of P_{val} that has more than one vertex. So $\mathbf{w} \in \text{trop}(V(f))$ if and only if $F = \text{face}_{(\mathbf{w},1)}(P_{\text{val}})$ is not a vertex. This happens if and only if the face $\tilde{\pi}(\mathcal{N}(F))$ of the dual complex that contains $\mathbf{w}$ is not full dimensional. We conclude that $\text{trop}(V(f))$ is the $(n-1)$-skeleton of the dual complex. This is a pure Γ_{val}-rational polyhedral complex. □

Remark 3.1.7. The proof shows that the tropical hypersurface $\text{trop}(V(f))$ is precisely the $(n-1)$-skeleton of the complex $\Sigma_{\text{trop}(f)}$ in Definition 2.5.5.

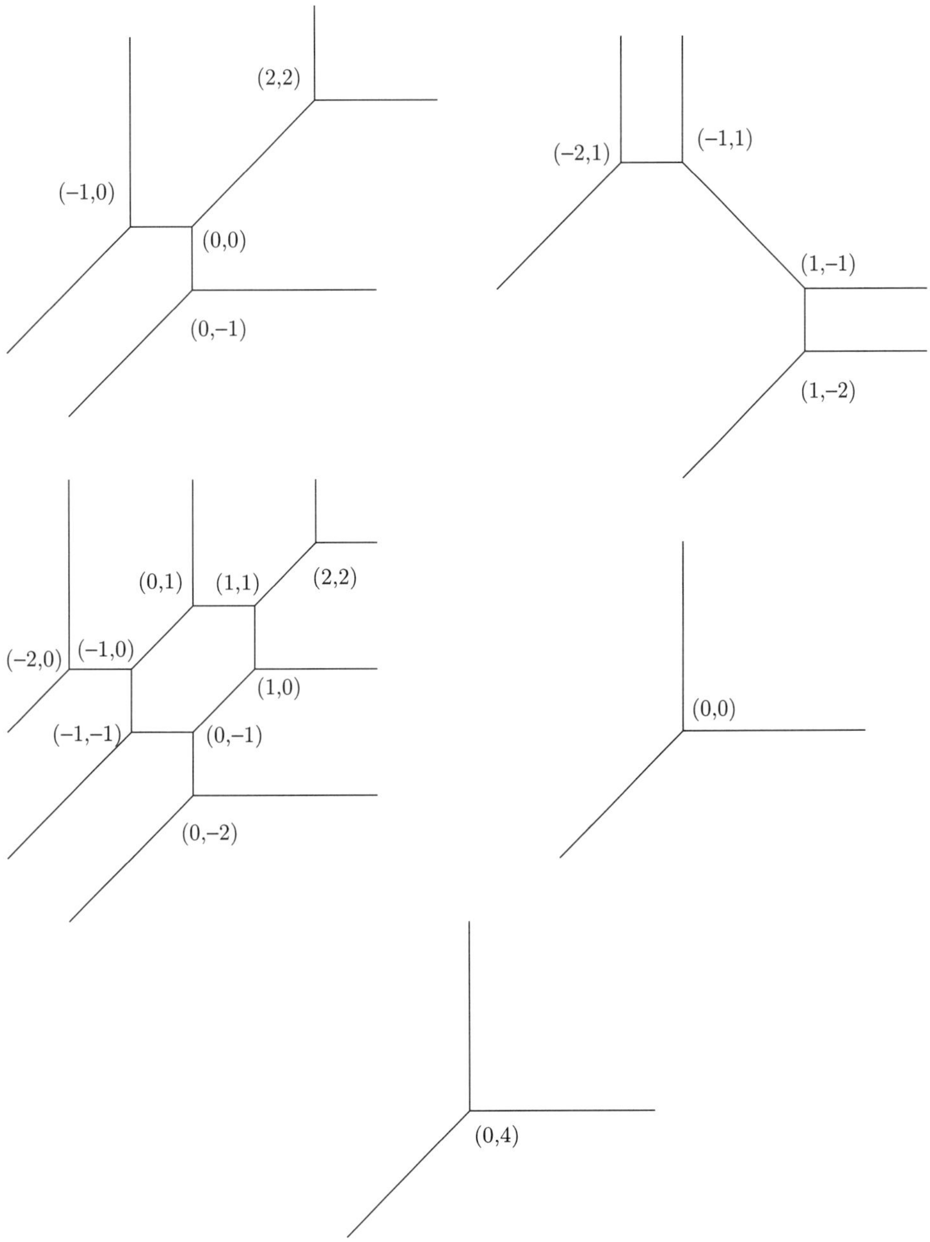

Figure 3.1.2. Five tropical curves from Example 3.1.8.

Example 3.1.8. Let $K = \mathbb{C}\{\{t\}\}$ and $n = 2$. Tropical hypersurfaces in $\mathbb{R}^2$ are tropical curves. The following five examples are depicted in Figure 3.1.2.

(1) If $f_1 = 3tx^2+5xy-7ty^2+8x-y+t^2$, then $\operatorname{trop}(V(f_1))$ is dual to the regular subdivision of $\mathcal{A}_2 = \{(2,0),(1,1),(0,2),(1,0),(0,1),(0,0)\}$ induced by $\mathbf{w} = (1,0,1,0,0,2)$. This subdivision is shown on the left in Figure 2.3.8. The curve $\operatorname{trop}(V(f_1))$ is shown in the upper left in Figure 3.1.2.

(2) Let $f_2 = 3t^3x^2 + 5xy - 7t^3y^2 + 8tx - ty + 1$. The tropical curve $\operatorname{trop}(V(f_2))$ is dual to the regular subdivision of $\mathcal{A}_2$ induced by $\mathbf{w} = (3,0,3,1,1,0)$. This subdivision is shown second in Figure 2.3.8, and $\operatorname{trop}(V(f_2))$ is second in Figure 3.1.2.

(3) Let $f_3 = 5t^3x^3+7tx^2y-8txy^2+9t^3y^3+8tx^2+5xy-ty^2+4tx+8ty+t^3$. The tropical curve $\operatorname{trop}(V(f_3))$ is dual to the regular subdivision of $\mathcal{A}_3=\{(3,0),(2,1),(1,2),(0,3),(2,0),(1,1),(0,2),(1,0),(0,1),(0,0)\}$ induced by $\mathbf{w} = (3,1,1,3,1,0,1,1,1,3)$. It consists of nine triangles. Note that $V(f_3)$ is an elliptic curve with nine points removed, and $\operatorname{trop}(V(f_3))$ has a cycle. See the second row of Figure 3.1.2.

(4) Let $f_4 = 5x^3+7x^2y+8xy^2+9y^3+8x^2+5xy-y^2+4x+8y+1$. The tropical cubic $\operatorname{trop}(V(f_4))$ is dual to the regular subdivision of $\mathcal{A}_3$ induced by $\mathbf{w} = (0,0,0,0,0,0,0,0,0,0)$. The subdivision consists of just the single triangle $\operatorname{conv}(\mathcal{A}_3)$. The picture of $\operatorname{trop}(V(f_4))$, on the right of the second row of Figure 3.1.2, looks like a tropical line. In Section 3.4 we will attach weights to tropical varieties. Those weights will distinguish our tropical cubic from a tropical line.

(5) Let $f_5 = (3t^3+5t^2)xy^{-1}+8t^2y^{-1}+4t^{-2}$. The curve $\operatorname{trop}(V(f_5))$ is dual to the regular triangulation of $\{(1,-1),(0,-1),(0,0)\}$ induced by $\mathbf{w} = (2,2,-2)$. This consists of a single triangle. The curve $\operatorname{trop}(V(f_5))$ is a tropical line, shifted so that the vertex is at $(0,4)$. This is shown at the bottom of Figure 3.1.2. ♢

It is instructive to also examine some tropical surfaces in $\mathbb{R}^3$.

Example 3.1.9. Let $K = \mathbb{Q}$, and fix the 2-adic valuation. The following polynomial defines a smooth surface in the three-dimensional torus T_K^3:

$$f = 12x^2 + 20y^2 + 8z^2 + 7xy + 22xz + 3yz + 5x + 9y + 6z + 4.$$

Its Newton polytope $P = \operatorname{Newt}(f)$ is the tetrahedron $\operatorname{conv}((2,0,0),(0,0,2),(0,0,2),(0,0,0))$. The 2-adic valuations of the coefficients of f define a regular triangulation of P into eight tetrahedra of volume $1/6$. That triangulation has 24 triangles, 25 edges, and ten vertices. It is a good exercise to verify these numbers. Eight triangles and one edge lie in the interior of P. The dual complex $\Sigma_{\operatorname{trop}(f)}$ is a subdivision of $\mathbb{R}^3$ into ten unbounded full-dimensional regions. Its 2-skeleton is the tropical quadric surface $\operatorname{trop}(V(f))$. That tropical surface consists of 25 two-dimensional polyhedra (24 unbounded

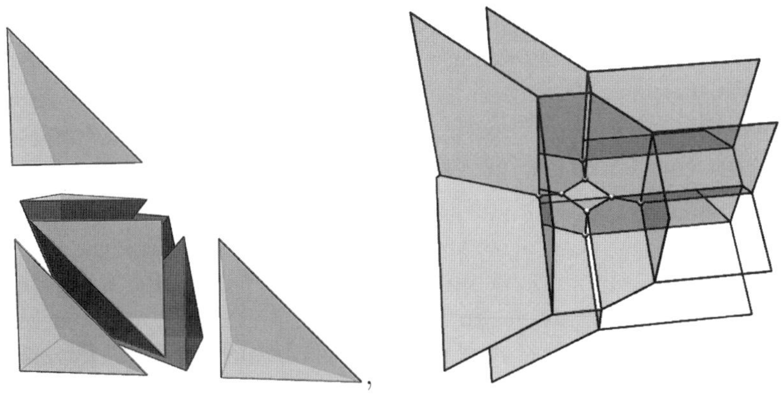

Figure 3.1.3. The regular triangulation and tropical surface of Example 3.1.9.

and one bounded square). It has eight vertices and 24 edges (16 unbounded and eight bounded). The regular triangulation of P is shown on the left of Figure 3.1.3, and the tropical surface is shown on the right. See Proposition 4.5.4 for the classification of all tropical surfaces of degree 2 in $\mathbb{R}^3$. ◊

An important special case of Proposition 3.1.6 arises when the valuations of the coefficients of f are all zero. In that case, the tropical hypersurface is a fan in $\mathbb{R}^n$. We saw an $n = 2$ instance of this in part (4) of Example 3.1.8.

Proposition 3.1.10. *Let $f \in K[x_1^{\pm 1}, \ldots, x_n^{\pm 1}]$ be a Laurent polynomial whose coefficients have valuation zero. The tropical hypersurface* $\mathrm{trop}(V(f))$ *is the support of an $(n-1)$-dimensional polyhedral fan in $\mathbb{R}^n$. That fan is the $(n-1)$-skeleton of the normal fan to the Newton polytope of f.*

Proof. Let $f = \sum c_{\mathbf{u}} x^{\mathbf{u}}$. If $\mathrm{val}(c_{\mathbf{u}}) = 0$ whenever $c_{\mathbf{u}} \neq 0$, then the regular subdivision of $\mathrm{Newt}(f)$ induced by the vector with coordinates $\mathrm{val}(c_{\mathbf{u}})$ is just the polytope $\mathrm{Newt}(f)$. The complex $\Sigma_{\mathrm{trop}(f)}$ of Definition 2.5.5 is the normal fan of $\mathrm{Newt}(f)$, so the claim follows from Proposition 3.1.6. □

Example 3.1.11. Let f denote the determinant of an $n \times n$-matrix (x_{ij}) whose entries are variables. We regard f as a polynomial of degree n with $n!$ terms in $K[x_{11}, x_{12}, \ldots, x_{nn}]$. Each coefficient is -1 or 1, so has valuation zero. The Newton polytope $P = \mathrm{Newt}(f)$ is the $(n-1)^2$-dimensional *Birkhoff polytope* of bistochastic matrices. The piecewise-linear function $\mathrm{trop}(f)$ is the tropical determinant from (1.2.6). The dual complex $\Sigma_{\mathrm{trop}(f)}$ is the normal fan of the Birkhoff polytope P. The polytope P has one vertex for each permutation of n. The normal fan thus divides $\mathbb{R}^{n^2}$ into $n!$ cones. The cones indexed by two permutations π and π' intersect in a common facet if and only if $\pi^{-1} \circ \pi'$ is a cycle. Checking this is a good exercise.

The tropical hypersurface $\text{trop}(V(f))$ is an (n^2-1)-dimensional fan with lineality space of dimension $2n-1$. The dimension of the lineality space comes from the calculation $n^2-(n-1)^2 = 2n-1$ for the codimension of the affine span of P. That fan has n^2 rays, one for each matrix entry, and its maximal cones are indexed by pairs (π, π') such that $\pi^{-1} \circ \pi'$ is a cycle.

If $n = 3$, then the Birkhoff polytope P is the cyclic 4-polytope with six vertices, whose f-vector is $(6, 15, 18, 9)$. The f-vector records the number of faces of P of each dimension; here P has 15 edges, 18 two-dimensional faces, and nine facets (three-dimensional faces). The tropical determinantal hypersurface $\text{trop}(V(f))$ is an eight-dimensional fan in $\mathbb{R}^9$. Modulo its five-dimensional lineality space, this fan has nine rays, 18 two-dimensional cones, and 15 maximal cones. It is the fan over a two-dimensional polyhedral complex with nine squares and six triangles, namely the 2-skeleton of the product of two triangles. ♢

3.2. The Fundamental Theorem

The goal of this section is to prove the Fundamental Theorem of Tropical Algebraic Geometry, which establishes a tight connection between classical varieties and tropical varieties. We must begin by defining the latter objects.

Definition 3.2.1. Let I be an ideal in $K[x_1^{\pm 1}, \ldots, x_n^{\pm 1}]$, and let $X = V(I)$ be its variety in the algebraic torus T^n. The *tropicalization* $\text{trop}(X)$ of the variety X is the intersection of all tropical hypersurfaces defined by elements of I:

$$\text{trop}(X) \; = \; \bigcap_{f \in I} \text{trop}(V(f)) \; \subseteq \; \mathbb{R}^n. \tag{3.2.1}$$

We shall see that the set $\text{trop}(X)$ depends only on the radical $\sqrt{I}$ of the ideal I. By a *tropical variety* in $\mathbb{R}^n$ we mean any subset of the form $\text{trop}(X)$ where X is a subvariety of the torus T^n over a field K with valuation.

In Definition 3.2.1, it does not suffice to take the intersection over the tropical hypersurfaces $\text{trop}(V(f))$ where f runs over a generating set of I. We usually have to pass to a larger set of Laurent polynomials in the ideal I. In other words, tropicalization of varieties does not commute with intersections. This fact is a salient feature of tropical geometry. A finite intersection of tropical hypersurfaces is known as a *tropical prevariety*.

Example 3.2.2. Let $n = 2$, $K = \mathbb{C}\{\{t\}\}$, and $I = \langle x + y + 1, \, x + 2y \rangle$. Then $X = V(I) = \{(-2, 1)\}$ and hence $\text{trop}(X) = \{(0, 0)\}$. However, the intersection of the two tropical lines given by the ideal generators equals

$$\text{trop}(V(x+y+1)) \cap \text{trop}(V(x+2y)) \; = \; \big\{(w_1, w_2) \in \mathbb{R}^2 : w_1 = w_2 \leq 0\big\}.$$

This half-ray is not a tropical variety. It is just a tropical prevariety. ♢

This example shows that a tropical variety $\text{trop}(X)$ is generally not the intersection of the tropical hypersurfaces corresponding to a given generating set of the ideal I of X. This brings us back to the notion of a tropical basis, as in Section 2.6. Definition 2.6.3 can be restated as follows: a finite generating set $\mathcal{T}$ for an ideal I in $K[x_1^{\pm 1}, \ldots, x_n^{\pm 1}]$ is a *tropical basis* for I if

$$\text{trop}(V(I)) \;=\; \bigcap_{f \in \mathcal{T}} \text{trop}(V(f)).$$

Theorem 2.6.6 states that every Laurent ideal has a finite tropical basis, and this implies that every tropical variety is a tropical prevariety.

In Example 3.2.2, the two given generators are not yet a tropical basis of the ideal I. However, we get a tropical basis if we add one more polynomial:

$$\mathcal{T} \;=\; \{x + y + 1,\, x + 2y,\, y - 1\}.$$

We now come to the main result of this section, which is the direct generalization of Theorem 3.1.3 from hypersurfaces to arbitrary varieties.

Theorem 3.2.3 (Fundamental Theorem of Tropical Algebraic Geometry). *Let K be an algebraically closed field with a nontrivial valuation, let I be an ideal in $K[x_1^{\pm 1}, \ldots, x_n^{\pm 1}]$, and let $X = V(I)$ be its variety in the algebraic torus $T^n \cong (K^*)^n$. Then the following three subsets of $\mathbb{R}^n$ coincide:*

(1) *the tropical variety* $\text{trop}(X)$ *as defined in equation* (3.2.1);

(2) *the set of all vectors* $\mathbf{w} \in \mathbb{R}^n$ *with* $\text{in}_{\mathbf{w}}(I) \neq \langle 1 \rangle$;

(3) *the closure of the set of coordinatewise valuations of points in* X,

$$\text{val}(X) = \{(\text{val}(y_1), \ldots, \text{val}(y_n)) : (y_1, \ldots, y_n) \in X\}.$$

Furthermore, if X is irreducible and $\mathbf{w}$ is any point in $\Gamma_{\text{val}}^n \cap \text{trop}(X)$, then the set $\{\mathbf{y} \in X : \text{val}(\mathbf{y}) = \mathbf{w}\}$ is Zariski dense in the classical variety X.

The rest of this section is devoted to proving Theorem 3.2.3. We first explain why the assumptions that K is algebraically closed and that the valuation is nontrivial are not serious restrictions for tropical geometry.

Fix a field extension L/K. If $Y \subset T_K^n$ is a variety defined by an ideal $I \subset K[x_1^{\pm 1}, \ldots, x_n^{\pm 1}]$, then the *extension of Y to T_L^n* is the subvariety Y_L of T_L^n defined by the ideal $I_L = IL[x_1^{\pm 1}, \ldots, x_n^{\pm 1}]$. Recall that L/K is a valued field extension if the valuations satisfy $\text{val}_L|_K = \text{val}_K$.

Theorem 3.2.4. *Let K be a field with a possibly trivial valuation, and let L/K be a valued field extension. Let $X \subset T_K^n$ be a subvariety of the torus T_K^n, and let X_L be its extension to T_L^n. Then*

$$\text{trop}(X_L) \;=\; \text{trop}(X) \;\subset\; \mathbb{R}^n.$$

Proof. Let $I \subset K[x_1^{\pm 1}, \dots, x_n^{\pm 1}]$ be the ideal of X so $I_L = IL[x_1^{\pm 1}, \dots, x_n^{\pm 1}]$ is the ideal of X_L. By definition we have $\mathrm{trop}(X_L) = \bigcap_{f \in I_L} \mathrm{trop}(V(f))$. Since every polynomial $f \in I$ is also a polynomial in I_L, we have $\mathrm{trop}(X_L) \subseteq \mathrm{trop}(X)$, so it suffices to show the other inclusion. By Theorem 2.6.6 the ideal I_L has a finite tropical basis, and by Lemma 2.6.5 there is a tropical basis $\mathcal{T}$ for I_L with all coefficients in K. Thus if $\mathbf{w} \notin \mathrm{trop}(X_L)$, there is $f \in \mathcal{T}$ with the minimum in $\mathrm{trop}(f)(\mathbf{w})$ achieved only once, so $\mathbf{w} \notin \mathrm{trop}(V(f))$. Since $f \in I$, we have $\mathbf{w} \notin \mathrm{trop}(X)$, which shows the other inclusion. □

Remark 3.2.5. Theorem 3.2.4 allows us to work over an extension field when this is necessary. In particular, if K has the trivial valuation, then we can take the field $L = K((\mathbb{R}))$ of generalized power series with coefficients in K (or the simpler field of Puiseux series if K has characteristic zero). We can thus assume that the given field has a nontrivial valuation. It also does not change the tropical variety to pass to the algebraic closure of the field K. This lets us assume that the value group Γ_{val} is dense in $\mathbb{R}$. We may also pass to an extension field L/K for which the valuation map $L^* \to \Gamma_{\mathrm{val}}$ splits; by Lemma 2.1.15 we can take $L = \overline{K}$, but smaller fields may also suffice. This allows us to use the Gröbner theory developed in Sections 2.4 and 2.5 when studying tropical varieties over an arbitrary field.

For the rest of this section we assume that K is an algebraically closed field with a nontrivial valuation that splits. We begin with a sequence of lemmas whose purpose is to prepare for the proof of Theorem 3.2.3.

Recall from commutative algebra that a *minimal associated prime* of an ideal I in $K[x_1^{\pm 1}, \dots, x_n^{\pm 1}]$ is a prime ideal $P \supset I$ for which there is no prime ideal Q with $P \supsetneq Q \supset I$. The variety $V(I)$ decomposes as $\bigcup_{P \text{ minimal}} V(P)$. See [**Eis95**, Chapter 3] or [**CLO07**, §4.7] for more details.

Lemma 3.2.6. *Let $X \subset T^n$ be an irreducible variety with prime ideal $I \subset K[x_1^{\pm}, \dots, x_n^{\pm 1}]$, and fix $\mathbf{w}$ with $\mathrm{in}_{\mathbf{w}}(I) \neq \langle 1 \rangle$. Then all minimal associated primes of the initial ideal $\mathrm{in}_{\mathbf{w}}(I)$ in $\Bbbk[x_1^{\pm 1}, \dots, x_n^{\pm 1}]$ have the same dimension as X.*

Proof. Let $d = \dim(X)$. The ideal $I_{\mathrm{proj}} \subseteq K[x_0, x_1, \dots, x_n]$, as in Definition 2.2.4, is prime of dimension $d+1$. Since we are assuming that K is algebraically closed with a nontrivial valuation, Γ_{val} is dense in $\mathbb{R}$. By Theorem 2.5.3, the Gröbner complex $\Sigma(I_{\mathrm{proj}})$ is Γ_{val}-rational, so the cell of $\Sigma(I_{\mathrm{proj}})$ containing $(0, \mathbf{w})$ contains a point $(0, \mathbf{w}') \in \Gamma_{\mathrm{val}}^{n+1}$. We may thus assume that $\mathbf{w} \in \Gamma_{\mathrm{val}}^n$. Hence, by Lemma 2.4.12, all minimal primes of $\mathrm{in}_{(0,\mathbf{w})}(I_{\mathrm{proj}})$ have dimension $d+1$. By the Principal Ideal Theorem, all minimal primes of $\mathrm{in}_{(0,\mathbf{w})}(I_{\mathrm{proj}}) + \langle x_0 - 1 \rangle$ have dimension at least d. Since $\mathrm{in}_{(0,\mathbf{w})}(I_{\mathrm{proj}})$ is homogeneous by Lemma 2.4.2, all minimal primes are homogeneous and contained in $\langle x_0, \dots, x_n \rangle$. None of them contains $x_0 - 1$.

Thus, the minimal primes of $\mathrm{in}_{(0,\mathbf{w})}(I_{\mathrm{proj}}) + \langle x_0 - 1\rangle$ have dimension exactly d. By Proposition 2.6.1 we have $\mathrm{in}_{\mathbf{w}}(I) = \mathrm{in}_{(0,\mathbf{w})}(I_{\mathrm{proj}})|_{x_0=1}$, viewed as an ideal in $\Bbbk[x_1^{\pm 1}, \dots, x_n^{\pm 1}]$. The minimal primes of $\mathrm{in}_{\mathbf{w}}(I)$ are the images in $\Bbbk[x_1^{\pm 1}, \dots, x_n^{\pm 1}]$ of those primes minimal over $\mathrm{in}_{(0,\mathbf{w})}(I_{\mathrm{proj}}) + \langle x_0 - 1\rangle$ that do not contain any monomial in $x_1, \dots, x_n$. These all have dimension d. □

The proof of Theorem 3.2.3 will proceed by projecting to the hypersurface case. The next result ensures that a sufficiently nice projection exists.

Proposition 3.2.7. *Fix a subvariety X in T^n and $m \geq \dim(X)$. There exists a morphism $\psi\colon T^n \to T^m$ whose image $\psi(X)$ is Zariski closed in T^m and satisfies $\dim(\psi(X)) = \dim(X)$. This map can be chosen so that the following hold.*

(1) *The kernel of the linear map $\mathrm{trop}(\psi)\colon \mathbb{R}^n \to \mathbb{R}^m$ intersects trivially with a fixed finite arrangement of m-dimensional subspaces in $\mathbb{R}^n$.*

(2) *When $n > m$, if we change coordinates so that ψ is the projection onto the first m coordinates, then the ideal I of X is generated by polynomials in $x_{m+1}, \dots, x_n$ whose coefficients are monomials in $x_1, \dots, x_m$.*

Proof. To prove this we derive a version of *Noether normalization* for the Laurent polynomial ring $K[x_1^{\pm 1}, \dots, x_n^{\pm 1}]$. We proceed by induction on $n - m$, the case $n = m$ being trivial. For $n > m$, the ideal I of X is not the zero ideal. For $l \in \mathbb{N}$, we define a monomial change of variables in T^n by

$$\phi_l^*(x_1) = x_1 x_n^l,\ \phi_l^*(x_2) = x_2 x_n^{l^2}, \dots,\ \phi_l^*(x_{n-1}) = x_{n-1} x_n^{l^{n-1}},\ \phi_l^*(x_n) = x_n.$$

For any f, choosing l sufficiently large, the transformed Laurent polynomial

$$g \;=\; \phi_l^*(f) \;=\; f\big(x_1 x_n^l, x_2 x_n^{l^2}, \;\dots,\; x_{n-1} x_n^{l^{n-1}}, x_n\big)$$

has the property that its monomials have distinct degrees in the variable x_n. Since ϕ^* is invertible, we may replace I by $\phi^*(I)$, and assume that I is generated by a set of polynomials with this property.

This suffices to show that the image of X under the monomial map

$$\pi \;:\; T^n \to T^{n-1},\ (x_1, x_2, \dots, x_n) \mapsto (x_1, \dots, x_{n-1})$$

is closed. By [**CLO07**, Theorem 3.2.2], the closure of $\pi(X)$ is the variety in T^{n-1} defined by $I \cap K[x_1^{\pm 1}, \dots, x_{n-1}^{\pm 1}]$. The difference $\overline{\pi(X)} \backslash \pi(X)$ is contained in the variety of the leading coefficients of the polynomials in a generating set of I when viewed as polynomials in x_n. As the leading coefficient of each generator is a monomial in $x_1, \dots, x_{n-1}$, the variety in T^{n-1} defined by these polynomials is empty. We conclude that $\overline{\pi(X)} = \pi(X)$.

To see that $\dim(X) = \dim(\pi(X))$, we note that the ideal I contains a polynomial that is monic when regarded as a polynomial in x_n. Hence $K[X]$

is generated by x_n as a $K[\pi(X)]$-algebra, and the field of fractions $K(X)$ is a finite extension of $K(\pi(X))$. This shows that their transcendence degrees agree, and thus $\dim(X) = \dim(\pi(X))$ (see [**CLO07**, Theorem 9.5.6]).

By induction on $n - m$, there is a morphism $\psi : T^{n-1} \to T^m$ with the desired properties. The claims on dimension, on the image being closed, and the second requirement on the form of the generators all follow.

We can choose our change of coordinates so that the kernel of $\text{trop}(\pi)$ avoids some subspaces, because in the original coordinates the kernel of $\text{trop}(\pi) : \mathbb{R}^n \to \mathbb{R}^{n-1}$ is the line spanned by $(1, l, l^2, \ldots, l^{n-1})$ in $\mathbb{R}^n$. For $l \gg 0$, this line intersects any fixed finite number of hyperplanes only in the origin. We obtain the general case using again induction on $n - m$. This shows that we can guarantee the map ψ to also satisfy property (1). $\square$

A key point of tropical geometry is that $\text{trop}(X)$ is the support of a Γ_{val}-rational polyhedral complex. One particular Γ_{val}-rational polyhedral structure on $\text{trop}(X)$ is derived from the Gröbner characterization in the Fundamental Theorem (part (2) of Theorem 3.2.3). In the following statement we identify $\mathbb{R}^n$ with $\mathbb{R}^{n+1}/\mathbb{R}\mathbf{1}$ via the map $\mathbf{w} \mapsto (0, \mathbf{w})$.

Proposition 3.2.8. *Let I be an ideal in $K[x_1^{\pm 1}, \ldots, x_n^{\pm 1}]$, and let $X = V(I)$ be its variety. Then $\{\mathbf{w} : \text{in}_{\mathbf{w}}(I) \neq \langle 1 \rangle\}$ is the support of a subcomplex of the Gröbner complex $\Sigma(I_{\text{proj}})$ and is thus the support of a Γ_{val}-rational polyhedral complex.*

Proof. The Gröbner complex $\Sigma(I_{\text{proj}})$ is a Γ_{val}-rational polyhedral complex in $\mathbb{R}^{n+1}/\mathbb{R}\mathbf{1}$ by Theorem 2.5.3. Let I_{proj} be as in Proposition 2.6.1. By Proposition 2.6.1 we have $\text{in}_{\mathbf{w}}(I) = \langle 1 \rangle$ if and only if $1 \in \text{in}_{(0,\mathbf{w})}(I_{\text{proj}})|_{x_0=1}$. This occurs if and only if there is an element in $\text{in}_{(0,\mathbf{w})}(I_{\text{proj}})$ that is a polynomial in x_0 times a monomial in $x_1, \ldots, x_n$, and thus if and only if there is a monomial in $\text{in}_{(0,\mathbf{w})}(I_{\text{proj}})$, since $\text{in}_{(0,\mathbf{w})}(I_{\text{proj}})$ is homogeneous by Lemma 2.4.2. So $\{\mathbf{w} \in \mathbb{R}^n : \text{in}_{\mathbf{w}}(I) \neq \langle 1 \rangle\}$ equals the set $\{\mathbf{w} \in \mathbb{R}^n : \text{in}_{(0,\mathbf{w})}(I_{\text{proj}})$ does not contain a monomial$\}$. This is a union of cells in the Gröbner complex $\Sigma(I_{\text{proj}})$. The set of $\mathbf{w} \in \mathbb{R}^n$ for which $\text{in}_{(0,\mathbf{w})}(I_{\text{proj}})$ contains a monomial is open by Corollary 2.4.10, so the complement is closed. Hence $\text{trop}(X)$ is the support of a subcomplex of the Gröbner complex. $\square$

The polyhedral complex structure defined by Proposition 3.2.8 depends on the choice of coordinates in T^n. The following example illustrates this.

Example 3.2.9. Let $K = \mathbb{C}$, let $n = 5$, and consider the ideal

$$I = \langle x_1 + x_2 + x_3 + x_4 + x_5,\ 3x_2 + 5x_3 + 7x_4 + 11x_5 \rangle \subset K[x_1^{\pm}, \ldots, x_5^{\pm}].$$

The generators are linear forms, so we can identify I with its homogenization I_{proj}. The tropical variety $\text{trop}(V(I_{\text{proj}}))$ is a three-dimensional fan with one-dimensional lineality space. It is a fan over the complete graph K_5 (cf. Example 4.2.13). That fan has ten maximal cones and five ridges. We consider the isomorphism $\phi : T^5 \to T^5$ defined by

$$\phi^* : x_1 \mapsto x_1,\ x_2 \mapsto x_2x_3,\ x_3 \mapsto x_3x_4,\ x_4 \mapsto x_4x_5,\ x_5 \mapsto x_5.$$

The transformed ideal $J = (\phi^*)^{-1}(I)$ has a finer Gröbner fan structure on its tropical variety $\mathrm{trop}(V(J_{\mathrm{proj}}))$. The support is still a fan over the complete graph K_5, but now two edges are subdivided, so the fan has 12 maximal cones. This can be verified using the software `Gfan` [**Jen**]. ◊

We now embark on the proof of the Fundamental Theorem 3.2.3. At this point we must treat the three sets described in Theorem 3.2.3 as distinct objects. We begin with proving a bound on the dimension of the polyhedral set $\{\mathbf{w} \in \mathbb{R}^n : \mathrm{in}_{\mathbf{w}}(I) \neq \langle 1 \rangle\}$, not yet knowing that it equals $\mathrm{trop}(X)$. That bound will be further improved to an equality in Theorem 3.3.8, whose proof in the next section will rely on Theorem 3.2.3.

Lemma 3.2.10. *Let X be a d-dimensional subvariety of T^n, with ideal $I \subset K[x_1^{\pm 1}, \ldots, x_n^{\pm 1}]$. Every cell in the Gröbner complex $\Sigma = \Sigma(I_{\mathrm{proj}})$ whose support lies in the set $\{\mathbf{w} \in \mathbb{R}^n : \mathrm{in}_{\mathbf{w}}(I) \neq \langle 1 \rangle\}$ has dimension at most d.*

Proof. Let $\mathbf{w} \in \Gamma_{\mathrm{val}}^n$ lie in the relative interior of a maximal cell $P \in \Sigma$. The affine span of P is $\mathbf{w} + L$, where L is a subspace of $\mathbb{R}^n$. By Lemma 2.2.7 and Corollary 2.6.12 we may assume that L is the span of $\mathbf{e}_1, \ldots, \mathbf{e}_k$ for some k. We need to show that $k = \dim(L) \leq d$. Since $\mathbf{w}$ lies in the relative interior of P, $\mathrm{in}_{\mathbf{w}+\epsilon\mathbf{v}}(I) \neq \langle 1 \rangle$ for all $\mathbf{v} \in \mathbb{Z}^n \cap L$ and ϵ sufficiently small. Lemma 2.4.6 and Proposition 2.6.1 imply $\mathrm{in}_{\mathbf{v}}(\mathrm{in}_{\mathbf{w}}(I)) = \mathrm{in}_{\mathbf{w}}(I)$ for all $\mathbf{v} \in L \cap \mathbb{Z}^n$. Choose a set $\mathcal{G}$ of generators for $\mathrm{in}_{\mathbf{w}}(I)$ so that no generator is the sum of two other polynomials in $\mathrm{in}_{\mathbf{w}}(I)$ having fewer monomials. Then $f \in \mathcal{G}$ implies that $\mathrm{in}_{\mathbf{v}}(f) = f$ for all $\mathbf{v} \in L$, as $\mathrm{in}_{\mathbf{v}}(f)$ is otherwise a polynomial in $\mathrm{in}_{\mathbf{w}}(I)$ having fewer monomials. In particular, we have $\mathrm{in}_{\mathbf{e}_i}(f) = f$ for $1 \leq i \leq k$, so $f = m\tilde{f}$, where m is a monomial, and $x_1, \ldots, x_k$ do not appear in $\tilde{f}$. Since monomials are units in $\Bbbk[x_1^{\pm 1}, \ldots, x_n^{\pm 1}]$, this means that $\mathrm{in}_{\mathbf{w}}(I)$ is generated by elements not containing $x_1, \ldots, x_k$. Hence $k \leq \dim(\mathrm{in}_{\mathbf{w}}(I)) \leq \dim(X) = d$ as required. □

We now use Theorem 3.1.3 to prove Theorem 3.2.3.

Proof of Theorem 3.2.3. The points in set (3) are $(\mathrm{val}(y_1), \ldots, \mathrm{val}(y_n))$ for $\mathbf{y} = (y_1, \ldots, y_n) \in X$. For any $f \in I$, these satisfy $f(\mathbf{y}) = 0$. By Theorem 3.1.3, $(\mathrm{val}(y_1), \ldots, \mathrm{val}(y_n))$ is in $\mathrm{trop}(V(f))$. Hence $(\mathrm{val}(y_1), \ldots, \mathrm{val}(y_n))$ lies in set (1). Since set (1) is closed by construction, set (1) contains set (3).

Next, let $\mathbf{w}$ lie in set (1). Then, for any $f = \sum c_{\mathbf{u}} x^{\mathbf{u}} \in I$, the minimum of $\{\mathrm{val}(c_{\mathbf{u}}) + \mathbf{u} \cdot \mathbf{w} : c_{\mathbf{u}} \neq 0\}$ is achieved twice. Thus $\mathrm{in}_{\mathbf{w}}(f)$ is not a monomial. By Lemma 2.6.2, we see that $\mathrm{in}_{\mathbf{w}}(I)$ is not equal to $\langle 1 \rangle$, so $\mathbf{w}$ lies in set (2).

It remains to prove that set (2) is contained in set (3). We first reduce to the case where I is prime. Since $\mathrm{in}_{\mathbf{w}}(f^r) = \mathrm{in}_{\mathbf{w}}(f)^r$ for all f, r by part (3) of Lemma 2.6.2, we have $\mathrm{in}_{\mathbf{w}}(I) = \langle 1 \rangle$ if and only if $\mathrm{in}_{\mathbf{w}}(\sqrt{I}) = \langle 1 \rangle$, so we may assume that I is radical. Thus we can write $I = \bigcap_{i=1}^{s} P_i$, where P_i is prime, and $V(P_1), \ldots, V(P_s)$ are the irreducible components of X. Note that if $\mathbf{w} \in \mathbb{R}^n$ has $\mathrm{in}_{\mathbf{w}}(I) \neq \langle 1 \rangle$, then there is $j \in \{1, 2, \ldots, s\}$ with $\mathrm{in}_{\mathbf{w}}(P_j) \neq \langle 1 \rangle$. Indeed, if not, by Lemma 2.6.2 there are $f_1, \ldots, f_s$ with $f_i \in P_i$ and $\mathrm{in}_{\mathbf{w}}(f_i) = 1$. Set $f = \prod_{i=1}^{s} f_i$. Then $\mathrm{in}_{\mathbf{w}}(f) = 1$ and $f \in I$, so $\mathrm{in}_{\mathbf{w}}(I) = \langle 1 \rangle$, contradicting our assumption.

We have shown that if $\mathbf{w}$ lies in set (2) for X, then $\mathbf{w}$ lies in set (2) for some irreducible component $V(P_j)$ of X. Thus to show that $\mathbf{w} = \mathrm{val}(\mathbf{y})$ for some $\mathbf{y} \in X$ it suffices to show that $\mathbf{w} = \mathrm{val}(\mathbf{y})$ for some $\mathbf{y} \in V(P_j)$. This remaining case is the content of Proposition 3.2.11 below. □

Proposition 3.2.11. *Let X be an irreducible subvariety of T^n, with prime ideal $I \subseteq K[x_1^{\pm 1}, \ldots, x_n^{\pm 1}]$. Fix $\mathbf{w} \in \Gamma_{\mathrm{val}}^n$ with $\mathrm{in}_{\mathbf{w}}(I) \neq \langle 1 \rangle$, and let $\alpha \in V(\mathrm{in}_{\mathbf{w}}(I)) \subset (\Bbbk^*)^n$. Then there exists a point $\mathbf{y} \in X$ with $\mathrm{val}(\mathbf{y}) = \mathbf{w}$ and $\overline{t^{-\mathbf{w}}\mathbf{y}} = \alpha$. The set of such $\mathbf{y}$ is dense in the Zariski topology on X.*

Proof. Let $d = \dim(X)$. The cases $n = 1$ and $n - d = 1$ follow from Proposition 3.1.5. So, we can assume $0 \leq d \leq n-2$. We shall use induction on n. By Lemma 3.2.10, the set $\{\mathbf{v} \in \mathbb{R}^n : \mathrm{in}_{\mathbf{v}}(I) \neq \langle 1 \rangle\}$ is the support of a polyhedral complex Σ, and every cell P in Σ has dimension at most $d = \dim(X)$. Let L_P denote the linear span of $P - \mathbf{w}$ in $\mathbb{R}^n$. Then $\dim(L_P) \leq d + 1 < n$, and $\mathbf{w} + L_P$ is the affine subspace spanned by P and $\mathbf{w}$.

Choose a monomial projection $\phi\colon T^n \to T^{n-1}$ so that the linear map $\mathrm{trop}(\phi) : \mathbb{R}^n \to \mathbb{R}^{n-1}$ satisfies $\ker(\mathrm{trop}(\phi)) \cap L_P = \{\mathbf{0}\}$ for all $P \in \Sigma$. This is possible by Proposition 3.2.7. We may also assume, after a change of coordinates, that ϕ maps onto the first $n-1$ coordinates, and the image $\phi(X)$ is closed in T^{n-1}. These assumptions ensure that we can uniquely recover $\mathbf{w}$ from its image under $\mathrm{trop}(\phi)$. Indeed, suppose some other vector $\mathbf{w}' \in \Gamma_{\mathrm{val}}^n$ satisfies $\mathrm{in}_{\mathbf{w}'}(I) \neq \langle 1 \rangle$ and $\mathrm{trop}(\phi)(\mathbf{w}') = \mathrm{trop}(\phi)(\mathbf{w})$. The first condition says that $\mathbf{w}' \in P$ for some cell $P \in \Sigma$. This implies $\mathbf{w}' \in \mathbf{w} + L_P$ and hence $\mathbf{w}' - \mathbf{w} \in L_P$. The kernel condition then gives $\mathbf{w} = \mathbf{w}'$.

Let $I' = \phi^{*-1}(I) = I \cap K[x_1^{\pm 1}, \ldots, x_{n-1}^{\pm 1}]$ and $X' = V(I')$. Since $\phi(X)$ is closed, we have $X' = \phi(X)$. By Lemma 2.6.10, $\mathrm{in}_{\mathrm{trop}(\phi)(\mathbf{w})}(I') \neq \langle 1 \rangle$. By induction, there is $\mathbf{y}' = (y_1, \ldots, y_{n-1}) \in X' \subset T^{n-1}$ with $\mathrm{val}(y_i) = w_i$, and $\overline{t^{-w_i}y_i} = \alpha_i$ for $1 \leq i \leq n-1$. Let $J = \langle f(y_1, \ldots, y_{n-1}, x_n) : f \in I \rangle \subseteq K[x_n^{\pm 1}]$. Since $K[x_n^{\pm 1}]$ is a principal ideal domain (PID), there exists a single polynomial $f \in I$ whose specialization generates the principal ideal J. By Proposition 3.2.7 we may assume that $f = x_n^l + f'$, where x_n^l does not divide any monomial in f'. So, the degree l is positive, hence $J \neq \langle 1 \rangle$, and we can find a point in $V(J)$.

By Proposition 3.2.7 we may also assume that all coefficients of f, regarded as a polynomial in x_n, are monomials, so $f = \sum_i c_i x^{\mathbf{u}_i} x_n^i$ for $\mathbf{u}_i \in \mathbb{Z}^{n-1}$. Let $g = f(y_1, \ldots, y_{n-1}, x_n) = \sum_i c_i y^{\mathbf{u}_i} x_n^i$. As in the proof of Proposition 3.2.11, $\operatorname{trop}(f)(\mathbf{w}) = \operatorname{trop}(g)(w_n)$, and thus $\operatorname{in}_{\mathbf{w}}(f)(\alpha_1, \ldots, \alpha_{n-1}, x_n) = \operatorname{in}_{w_n}(g)(x_n)$. This implies $\operatorname{in}_{w_n}(g)(\alpha_n) = 0$. As $\alpha_n \neq 0$, the polynomial $\operatorname{in}_{w_n}(g)$ is not a monomial. By the $n = 1$ case in Proposition 3.1.5, there is $y_n \in K^*$ with $g(y_n) = 0$, $\operatorname{val}(y_n) = w_n$ and $\overline{t^{-w_n} y_n} = \alpha_n$. We then have $\mathbf{y} = (y_1, \ldots, y_n) \in X$ with $\operatorname{val}(\mathbf{y}) = \mathbf{w}$, and $\overline{t^{w_i} y_i} = \alpha_i$ for all i, as required. This last step uses the "unique recovery" property from two paragraphs ago.

To finish the proof, we now argue that the set $\mathcal{Y}$ of all such $\mathbf{y}$ is Zariski dense in X. Suppose that there was a subvariety $X' \subsetneq X$ containing $\mathcal{Y}$. The projection $\phi(X')$ contains $\phi(\mathcal{Y})$, and so all $\mathbf{y}' \in \phi(X)$ with $\operatorname{val}(y_i') = w_i$ and $\overline{t^{-w_i} y_i} = \alpha_i$ for $1 \leq i \leq n-1$. By our choice of the map ϕ, we have $\dim(\phi(X)) = \dim(X) > 0$. This set is Zariski dense in $\phi(X)$, by induction, and thus $\phi(X') = \phi(X)$. This contradicts the fact that $\dim(\phi(X')) \leq \dim(X') < \dim(X)$. The last inequality uses that X is irreducible. We conclude that there is no such X', and so $\mathcal{Y}$ is Zariski dense. □

The observation that the set of preimages of $\mathbf{w}$ under the valuation map is Zariski dense is due to Payne; see [**Pay09b**], [**Pay12**].

Remark 3.2.12. It is a consequence of Theorems 3.2.3 and 3.2.4 that if L_1/K and L_2/K are two algebraically closed valued field extensions and $X \subset T_K^n$ is a variety, then the closures of $\operatorname{val}(X(L_1))$ and $\operatorname{val}(X(L_2))$ agree. Here $X(L_i)$ is the set of L_i-valued points of X; if I is the ideal of X in $K[x_1^{\pm 1}, \ldots, x_n^{\pm 1}]$ and $I_{L_i} = IL_i[x_1^{\pm 1}, \ldots, x_n^{\pm 1}]$, then $X(L_i) = V(I_{L_i}) \subset T_{L_i}^n$. To see the equality, note that $\operatorname{trop}(V(I_{L_1})) = \operatorname{trop}(V(I_{L_2}))$ by Theorem 3.2.4. By Theorem 3.2.3, we know that $\operatorname{trop}(V(I_{L_i}))$ is the closure of the valuations of points in $X(L_i)$ for each i, so these sets also agree.

At the end of Section 2.6, we introduced the tropicalization $\operatorname{trop}(\phi)$ of a monomial map ϕ. In this section we studied three equivalent characterizations of the tropicalization $\operatorname{trop}(X)$ of an algebraic variety X in a torus. The next corollary states that these two notions of tropicalization are compatible.

Corollary 3.2.13. *Let $\phi : T^n \to T^m$ be a monomial map. Consider any subvariety X of T^n and the Zariski closure $\overline{\phi(X)}$ of its image in T^m. Then*

$$\operatorname{trop}(\,\overline{\phi(X)}\,) = \operatorname{trop}(\phi)(\,\operatorname{trop}(X)\,). \tag{3.2.2}$$

Proof. If I is the ideal of X, then $I' = (\phi^*)^{-1}(I)$ is the ideal of $\overline{\phi(X)}$. By Lemma 2.6.10 if $\text{in}_{\mathbf{w}}(I) \neq \langle 1 \rangle$, then $\text{in}_{\text{trop}(\phi)(\mathbf{w})}(I') \neq \langle 1 \rangle$, which shows that $\text{trop}(\phi)(\text{trop}(X)) \subseteq \text{trop}(\overline{\phi(X)})$. For the converse we use the Fundamental Theorem. By part (3) of Theorem 3.2.3, $\text{trop}(\overline{\phi(X)})$ is the closure of the set $\{\text{val}(\mathbf{z}) \colon \mathbf{z} \in \overline{\phi(X)}\}$. Since $\text{trop}(\phi)(\text{trop}(X))$ is a closed subset of $\mathbb{R}^m$, it thus suffices to show that every $\mathbf{w} \in \Gamma_{\text{val}}^m \cap \text{trop}(\overline{\phi(X)})$ lies in $\text{trop}(\phi)(\text{trop}(X))$. By Theorem 3.2.3, the set of $\mathbf{z} \in \overline{\phi(X)}$ for which $\text{val}(z) = \mathbf{w}$ is Zariski dense in $\overline{\phi(X)}$, so there is $\mathbf{z} = \phi(\mathbf{y})$ for some $\mathbf{y} \in X$ with $\text{val}(\mathbf{z}) = \mathbf{w}$. Since $\text{val}(\phi(\mathbf{y})) = \text{trop}(\phi)(\text{val}(\mathbf{y}))$, this shows that $\mathbf{w} \in \text{trop}(\phi)(\text{trop}(X))$. □

Remark 3.2.14. Corollary 3.2.13 says that tropicalization commutes with morphisms of tori. This statement is *not* true if the morphism ϕ is replaced by a rational map of tori, with $\text{trop}(\phi)$ defined in each coordinate by the corresponding tropical polynomial as in (2.4.1).

For a simple example consider the map $t \mapsto (t, -1-t)$ from the affine line to the affine plane. This defines a rational map of tori $\phi \colon X = T^1 \dashrightarrow T^2$, which is undefined at $t = -1$. Its image is $\overline{\phi(X)} = V(x+y+1) \subset T^2$, and hence $\text{trop}(\overline{\phi(X)})$ is the standard tropical line seen in Figure 3.1.1. On the other hand, the tropicalization of ϕ is the piecewise-linear map

$$\text{trop}(\phi) : \text{trop}(X) = \mathbb{R} \to \mathbb{R}^2, \;\; w \mapsto (w, \min(w, 0)).$$

The image of this is the union of two of the three rays of the tropical line,

$$\text{trop}(\phi)(\,\text{trop}(X)\,) \;=\; \{(a,a) : a \leq 0\} \,\cup\, \{(a,0) : a \geq 0\}.$$

In this little example, the following inclusion holds and is strict:

$$\text{(3.2.3)} \qquad \text{trop}(\,\overline{\phi(X)}\,) \;\supset\; \text{trop}(\phi)(\,\text{trop}(X)\,).$$

The inclusion (3.2.3) holds for every rational map ϕ of tori, but the inclusion is generally strict unless ϕ is a monomial map. How to fill the gap is of central importance in *tropical implicitization*, which aims to compute the tropicalization of a rational variety directly from a parametric representation ϕ. In other words, one first computes $\text{trop}(V(I))$, and then one uses that balanced polyhedral complex in deriving generators for I. For more information see Theorems 5.5.1 and 6.5.16 as well as [**STY07, SY08**].

3.3. The Structure Theorem

We next explore the question of which polyhedral complexes are tropical varieties. The main result in this section is the Structure Theorem 3.3.5 which says that if X is an irreducible subvariety of T^n of dimension d, then $\text{trop}(X)$ is the support of a pure d-dimensional weighted balanced Γ_{val}-rational polyhedral complex that is connected through codimension 1.

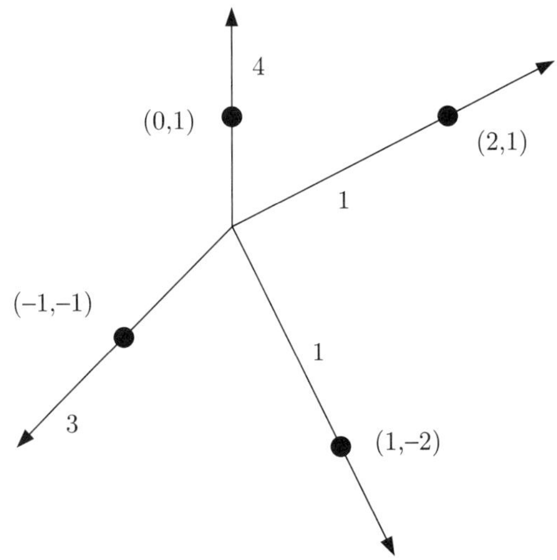

Figure 3.3.1. A balanced rational fan in $\mathbb{R}^2$.

We begin by defining these concepts, starting with the notion of a weighted balanced polyhedral complex. Let $\Sigma \subset \mathbb{R}^n$ be a one-dimensional rational fan with s rays. Let $\mathbf{v}_i$ be the first lattice point on the ith ray of Σ. We give Σ the structure of a *weighted fan* by assigning a positive integer weight $m_i \in \mathbb{N}$ to the ith ray of Σ. We say that the fan Σ is *balanced* if

$$\sum m_i \mathbf{v}_i = 0.$$

This is sometimes called the *zero-tension condition*: a tug-of-war game with ropes in the directions $\mathbf{v}_i$ and participants of strength m_i would have no winner. See Figure 3.3.1 for an example, where the weights are $1, 1, 3$, and 4. We now extend this concept to arbitrary weighted polyhedral complexes.

Definition 3.3.1. Let Σ be a rational fan in $\mathbb{R}^n$, pure of dimension d. Fix weights $m(\sigma) \in \mathbb{N}$ for all cones σ of dimension d. Given a cone $\tau \in \Sigma$ of dimension $d-1$, let L be the linear space parallel to τ. Thus L is a $(d-1)$-dimensional subspace of $\mathbb{R}^n$. Since τ is a rational cone, the abelian group $L_\mathbb{Z} = L \cap \mathbb{Z}^n$ is free of rank $d-1$, with $N(\tau) = \mathbb{Z}^n / L_\mathbb{Z} \cong \mathbb{Z}^{n-d+1}$. For each $\sigma \in \Sigma$ with $\tau \subsetneq \sigma$, the set $(\sigma + L)/L$ is a one-dimensional cone in $N(\tau) \otimes_\mathbb{Z} \mathbb{R}$. Let $\mathbf{v}_\sigma$ be the first lattice point on this ray. The fan Σ is *balanced at* τ if

$$\sum m(\sigma)\mathbf{v}_\sigma \;=\; 0. \tag{3.3.1}$$

The fan Σ is *balanced* if it is balanced at all $\tau \in \Sigma$ with $\dim(\tau) = d-1$. If Σ is a pure Γ_{val}-rational polyhedral complex of dimension d with weights $m(\sigma) \in \mathbb{N}$ on each d-dimensional cell in Σ, then for each $\tau \in \Sigma$ the fan $\mathrm{star}_\Sigma(\tau)$ inherits a weighting function m. The complex Σ is *balanced* if the fan $\text{star}_\Sigma(\tau)$ is balanced for all $\tau \in \Sigma$ with $\dim(\tau) = d - 1$. This condition is vacuous for zero-dimensional polyhedral complexes, so all are balanced.

We next explain the combinatorial meaning of the balancing condition for fans and complexes of codimension 1. Let P be a lattice polytope in $\mathbb{R}^n$ with normal fan $\mathcal{N}_P$, and let Σ denote the $(n-1)$-skeleton of $\mathcal{N}_P$. According to Proposition 3.1.10, the fan Σ is the tropical hypersurface $\text{trop}(V(f))$ of any constant-coefficient polynomial f with Newton polytope P. Equivalently, $\Sigma = V(F)$, where F is a tropical polynomial for which the coefficients are all 0 and the exponents of the monomials have convex hull P.

We can turn Σ into a weighted fan as follows. Each maximal cone $\sigma \in \Sigma$ is the inner normal cone of an edge $e(\sigma)$ of the polytope P. We define $m(\sigma)$ to be the *lattice length* of the edge $e(\sigma)$. This is one less than the number of lattice points in $e(\sigma)$. Proposition 3.3.2 below says that Σ is balanced.

For general tropical hypersurfaces we generalize from a lattice polytope P to regular subdivisions Δ of P. We can also define multiplicities from edge lengths in this case. Following Definition 2.3.8, Δ is given by a weight vector $\mathbf{c}$ with one entry $c_\mathbf{u}$ for each lattice point $\mathbf{u}$ in P. We construct from this a tropical polynomial $F = \min_{\mathbf{u} \in P}(c_\mathbf{u} + \mathbf{x} \cdot \mathbf{u})$. The subdivision Δ is dual to the polyhedral complex Σ_F. The tropical hypersurface $V(F)$ is the $(n-1)$-skeleton of Σ_F by Remark 3.1.7. Every facet σ of Σ corresponds to an edge $e(\sigma)$ of Δ, and we define $m(\sigma)$ to be the lattice length of $e(\sigma)$.

Proposition 3.3.2. *The $(n-1)$-dimensional polyhedral complex $V(F)$ given by a tropical polynomial F in n unknowns is balanced for the weights $m(\sigma)$ defined above.*

Proof. This statement is trivial for $n = 1$. If $n = \dim(P) = 2$, then $d = 1$ in Definition 3.3.1. We claim that $\text{star}_{V(F)}(\sigma)$ is balanced for all zero-dimensional cells σ. Such a cell is dual to a two-dimensional convex polygon Q in the regular subdivision Δ. The vectors $\mathbf{u}_\sigma$ in (3.3.1) are the primitive lattice vectors perpendicular to the edges of Q, and the vectors $m(\sigma)\mathbf{u}_\sigma$ are precisely the edges of Q rotated by 90 degrees. The equation (3.3.1) holds because the edge vectors of any convex polygon Q sum to zero. For $d \geq 3$ we reduce to the case $d = 2$ by working modulo L as in Definition 3.3.1. Here, L is the linear space parallel to σ. This is the lineality space of $\text{star}_{V(F)}(\sigma)$. Hence L is perpendicular to the polygon Q dual to σ in the regular subdivision Δ induced by F. Again, the edges of Q sum to zero. $\square$

Example 3.3.3. Let P be the Newton polytope of the *discriminant* of a univariate quartic $ax^4 + bx^3 + cx^2 + dx + e$. That discriminant equals

$$\underline{256a^3e^3} - 192a^2bde^2 - 128a^2c^2e^2 + 144a^2cd^2e + 144ab^2ce^2$$
$$- 80abc^2de - 6ab^2d^2e - \underline{27a^2d^4} + 18abcd^3 + \underline{16ac^4e}$$
$$- \underline{4ac^3d^2} - \underline{27b^4e^2} + 18b^3cde - \underline{4b^3d^3} - \underline{4b^2c^3e} + \underline{b^2c^2d^2}.$$

Its Newton polytope P is a three-dimensional cube that lives in $\mathbb{R}^5$. The eight vertices of P correspond to the underlined monomials. Here Σ is a fan with 12 cones σ of dimension 4, six cones τ of dimension 3, and one cone of dimension 2 (the lineality space). Eleven of the edges of P have lattice length 1, so $m(\sigma) = 1$ for these σ. However, the edge corresponding to $256a^3e^3 + 16ac^4e = 16ae(4ae + ic^2)(4ae - ic^2)$ has lattice length 2, so $m(\sigma) = 2$ for that maximal cone σ of Σ. To check that the fan Σ is balanced, we must examine the cones τ normal to the six square facets of P. The fan $\text{star}_\Sigma(\tau)$ is the normal fan of such a square, and (3.3.1) holds because the four edges of the square form a closed loop. The Newton polygon of the discriminant specialized with $b = 0$ is one of the two facets of P that contain the special edge above. For more information on such discriminants see Section 5.5. ◊

Every tropical polynomial F with coefficients in Γ_{val} has the form $F = \text{trop}(f)$ for some classical polynomial $f \in K[x_1^{\pm 1}, \ldots, x_n^{\pm 1}]$. Proposition 3.1.6 says that the tropical hypersurface $\text{trop}(V(f))$ is the balanced polyhedral complex Σ_F. We shall see in Lemma 3.4.6 that the combinatorial definition of multiplicity, where $m(\sigma)$ is the length of the edge in the subdivision Δ, is consistent with the general definition of multiplicities for tropical varieties. In Proposition 3.3.10 we prove a combinatorial converse to Proposition 3.3.2.

We next define what it means to be connected through codimension 1.

Definition 3.3.4. Let Σ be a pure d-dimensional polyhedral complex in $\mathbb{R}^n$. Then Σ is *connected through codimension* 1 if for any two d-dimensional cells $P, P' \in \Sigma$ there is a chain $P = P_1, P_2, \ldots, P_s = P'$ for which P_i and P_{i+1} share a common facet F_i for $1 \leq i \leq s-1$. Since the P_i are facets of Σ and the F_i are ridges, we call this a *facet-ridge path* connecting P and P'.

Every zero-dimensional polyhedral complex is connected through codimension 1. A pure one-dimensional polyhedral complex is connected through codimension 1 if and only if it is connected. An example of a connected two-dimensional polyhedral complex that is not connected through codimension 1 is shown in Figure 3.3.2.

This lets us state the second main theorem of this chapter. Its proof will straddle three sections.

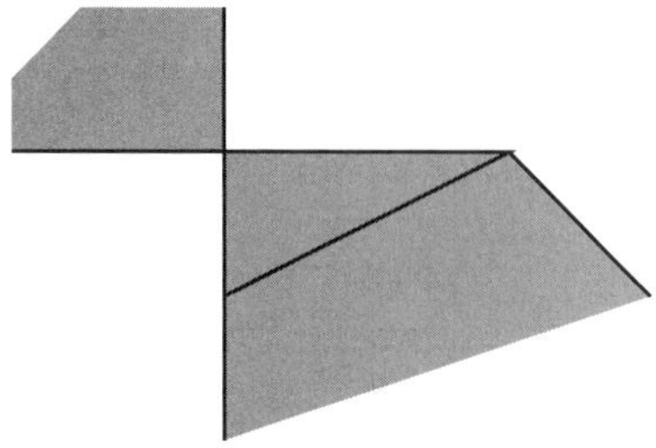

Figure 3.3.2. This complex is not connected through codimension 1.

Theorem 3.3.5 (Structure Theorem for Tropical Varieties). *Let X be an irreducible d-dimensional subvariety of T^n. Then* $\text{trop}(X)$ *is the support of a balanced weighted Γ_{val}-rational polyhedral complex pure of dimension d. Moreover, that polyhedral complex is connected through codimension* 1.

Proof. That $\text{trop}(X)$ is a pure Γ_{val}-rational d-dimensional polyhedral complex is Theorem 3.3.8. That it is balanced is Theorem 3.4.14. Theorem 3.5.1 shows that $\text{trop}(X)$ is connected through codimension 1. □

In the remainder of this section we prove the dimension part of the Structure Theorem. This will be stated separately in Theorem 3.3.8. Its proof will use the following lemma, which says that the star of any cell in a polyhedral complex with support $\text{trop}(X)$ is itself a tropical variety.

Lemma 3.3.6. *Let $X = V(I) \subset T^n_K$ where $I \subseteq K[x_1^{\pm}, \dots, x_n^{\pm 1}]$, and let Σ be a polyhedral complex with support* $\text{trop}(X) = \{\mathbf{w} \in \mathbb{R}^n : \text{in}_{\mathbf{w}}(I) \neq \langle 1 \rangle\} \subset \mathbb{R}^n$. *Fix $\mathbf{w} \in \Sigma$. If $\sigma \in \Sigma$ has $\mathbf{w}$ in its relative interior, then*

$$\text{star}_{\Sigma}(\sigma) = \big\{\mathbf{v} \in \mathbb{R}^n : \text{in}_{\mathbf{v}}(\text{in}_{\mathbf{w}}(I)) \neq \langle 1 \rangle\big\}.$$

Thus $\text{trop}(V(\text{in}_{\mathbf{w}}(I))) = \text{star}_{\text{trop}(X)}(\sigma)$.

Proof. We have

$$\begin{aligned} &\{\mathbf{v} \in \mathbb{R}^n : \text{in}_{\mathbf{v}}(\text{in}_{\mathbf{w}}(I)) \neq \langle 1 \rangle\} \\ &\quad = \{\mathbf{v} \in \mathbb{R}^n : \text{in}_{\mathbf{w}+\epsilon\mathbf{v}}(I) \neq \langle 1 \rangle \text{ for sufficiently small } \epsilon > 0\} \\ &\quad = \{\mathbf{v} \in \mathbb{R}^n : \mathbf{w} + \epsilon\mathbf{v} \in \Sigma \text{ for sufficiently small } \epsilon > 0\} \\ &\quad = \text{star}_{\Sigma}(\sigma), \end{aligned}$$

where the first equality follows from Lemma 2.4.6 and Proposition 2.6.1, and the third equality follows from Exercise 2.7(13). □

Example 3.3.7. Let $I = \langle tx^2 + x + y + xy + t \rangle$ in $\mathbb{C}\{\{t\}\}[x^{\pm 1}, y^{\pm 1}]$ and $X = V(I)$. The tropical curve $\text{trop}(X)$ is shown in Figure 3.3.3. The tropical curve of the initial ideal $\text{in}_{(1,1)}(I) = \langle x + y + 1 \rangle$ is the tropical line, with rays $(1,0)$, $(0,1)$, and $(-1,-1)$. This is the star of the vertex $(1,1)$. It is also the

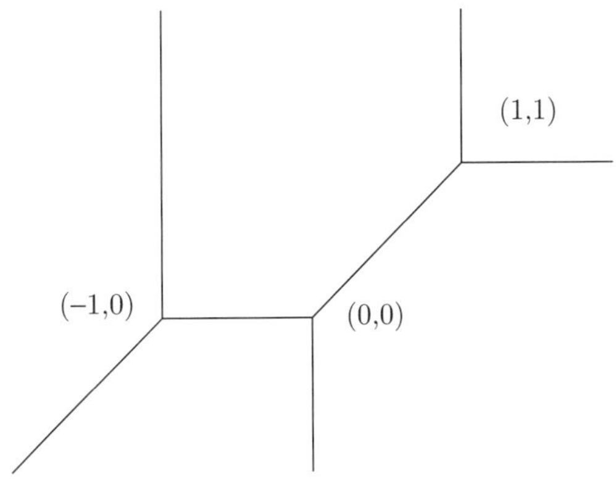

Figure 3.3.3. The tropical curve discussed in Example 3.3.7.

star of the vertex $(-1,0)$, since $\mathrm{in}_{(-1,0)}(I) = \langle x^2+x+xy\rangle = \langle x+1+y\rangle$. At the vertex $(0,0)$, the star has rays $(1,1)$, $(-1,0)$, and $(0,-1)$. This is the tropicalization of $V(\mathrm{in}_{(0,0)}(I)) = V(\langle x+y+xy\rangle)$. ◊

Theorem 3.3.8. *Let X be an irreducible subvariety of dimension d in the algebraic torus T^n over the field K. The tropical variety* $\mathrm{trop}(X)$ *is the support of a pure d-dimensional Γ_{val}-rational polyhedral complex in $\mathbb{R}^n$.*

Proof. Let I be the ideal of X. By part (2) of Theorem 3.2.3, the tropical variety $\mathrm{trop}(X)$ is the support of a Γ_{val}-rational polyhedral complex Σ. Lemma 3.2.10 shows that the dimension of each cell in Σ is at most d. It thus remains to show that each maximal cell in Σ has dimension at least d.

Let σ be a maximal cell in Σ, and fix $\mathbf{w} \in \mathrm{relint}(\sigma)$. Suppose that $\dim(\sigma) = k$. By Lemma 3.3.6, we have $\mathrm{trop}(V(\mathrm{in}_{\mathbf{w}}(I))) = |\,\mathrm{star}_\Sigma(\sigma)|$. This is the linear space parallel to σ, and thus a subspace L of $\mathbb{R}^n$ of dimension k. After a change of coordinates, we may assume that L is spanned by $\mathbf{e}_1,\ldots,\mathbf{e}_k$. Since $\mathrm{in}_{\mathbf{v}}(\mathrm{in}_{\mathbf{w}}(I)) = \mathrm{in}_{\mathbf{w}+\epsilon\mathbf{v}}(I) = \mathrm{in}_{\mathbf{w}}(I)$ for all $\mathbf{v} \in L$ and small $\epsilon > 0$, the ideal $\mathrm{in}_{\mathbf{w}}(I)$ is homogeneous with respect to the grading given by $\deg(x_i) = \mathbf{e}_i$ for $1 \le i \le k$ and $\deg(x_i) = 0$ for $i > k$. Hence, $\mathrm{in}_{\mathbf{w}}(I)$ is generated by Laurent polynomials which use only the variables $x_{k+1},\ldots,x_n$.

Let $J = \mathrm{in}_{\mathbf{w}}(I) \cap \Bbbk[x_{k+1}^{\pm 1},\ldots,x_n^{\pm 1}]$. We claim that $\mathrm{trop}(V(J)) = \{\mathbf{0}\}$. Indeed, let $\mathbf{v}' \in \mathrm{trop}(V(J))$ and $\mathbf{v} = (0,\mathbf{v}') \in \mathbb{R}^n$ with first k coordinates zero. If $\mathbf{v}' \neq \mathbf{0}$, then $\mathbf{v} \notin L$ and $\mathrm{in}_{\mathbf{v}}(\mathrm{in}_{\mathbf{w}}(I)) = \langle 1\rangle$, since σ is a maximal cell in Σ. Hence there is $f \in \mathrm{in}_{\mathbf{w}}(I)$ with $\mathrm{in}_{\mathbf{v}}(f) = 1$. We may choose f in J, as we can take it to be homogeneous in the $\mathbb{Z}^k$-grading discussed above. This shows that $\mathrm{in}_{\mathbf{v}'}(J) = \langle 1\rangle$, so $\mathrm{trop}(V(J)) \subseteq \{\mathbf{0}\}$. Since $1 \notin \mathrm{in}_{\mathbf{w}}(I)$, we have $\mathrm{in}_{\mathbf{0}}(J) = J \neq \langle 1\rangle$, and thus $\mathrm{trop}(V(J)) = \{\mathbf{0}\}$. Lemma 3.3.9 below then implies that $V(J)$ is finite, and so $\dim(\mathrm{in}_{\mathbf{w}}(I)) \le k$. From Lemma 3.2.6 we know that $\dim(\mathrm{in}_{\mathbf{w}}(I)) = d$, and hence $k = \dim(\sigma) \ge d$ as required. □

To complete the proof of Theorem 3.3.8, it now remains to show

Lemma 3.3.9. *Let X be a subvariety of T^n. If the tropical variety* $\mathrm{trop}(X)$ *is a finite set of points in $\mathbb{R}^n$, then X is a finite set of points in T^n.*

Proof. The proof is by induction on n. For $n = 1$, all nontrivial subvarieties are finite, and $\mathrm{trop}(T^1) = \mathbb{R}^1$. Suppose $n > 1$ and the lemma is true for all smaller n. If X is a hypersurface, then Proposition 3.1.6 implies that $\mathrm{trop}(X)$ is not finite. We thus assume $\dim(X) < n - 1$. Choose a map $\pi\colon T^n \to T^{n-1}$ with $Y := \overline{\pi(X)} = \pi(X)$ as guaranteed by Proposition 3.2.7. By changing coordinates, we may assume that π is the projection onto the first $n-1$ coordinates. By Corollary 3.2.13 we know that $\mathrm{trop}(Y)$ is a finite set of points in $\mathbb{R}^{n-1}$. By the induction hypothesis, the variety Y is finite: $Y = \{\mathbf{y}_1, \ldots, \mathbf{y}_r\} \subset T^{n-1}$. Since $\lambda \mathbf{e}_n \notin \mathrm{trop}(X)$ for $\lambda \gg 0$, the ideal I of X contains a polynomial of the form $1 + \sum_{i=1}^{s} f_i x_n^i$ with $f_i \in K[x_1^{\pm 1}, \ldots, x_{n-1}^{\pm 1}]$. Each $\mathbf{y}_i \in Y$ has at most s preimages $\mathbf{z} \in X$ with $\pi(\mathbf{z}) = \mathbf{y}_i$, so X is a finite set of points in T^n. □

According to the Structure Theorem 3.3.5, every tropical variety $\mathrm{trop}(X)$ is the support of a weighted balanced polyhedral complex Σ. One may wonder whether the converse is true: given such a complex Σ, can we always find a matching variety X with $\mathrm{trop}(X) = |\Sigma|$? We shall see in Chapter 4 that the answer is *no*, even in the context of linear spaces. That is why we distinguish between tropicalized linear spaces and tropical linear spaces. See Example 4.2.15 for a balanced fan Σ that is not $\mathrm{trop}(X)$ for any $X \subset T^n$. We close this section by showing that the answer is *yes* for hypersurfaces.

Proposition 3.3.10. *Let Σ be a weighted balanced Γ_{val}-rational polyhedral complex in $\mathbb{R}^n$ that is pure of dimension $n-1$. Then there exists a tropical polynomial F with coefficients in Γ_{val} such that $\Sigma = V(F)$. This ensures that $|\Sigma| = \mathrm{trop}(V(f))$ for some Laurent polynomial $f \in K[x_1^{\pm 1}, \ldots, x_n^{\pm 1}]$.*

Proof. We construct a tropical polynomial F in $u_1, \ldots, u_n$, with coefficients in Γ_{val}, such that $V(F) = \{\mathbf{w} \in \mathbb{R}^n : \text{the minimum in } F \text{ is achieved twice}\}$ equals Σ, and the weights in Σ are the edge lengths in the corresponding regular subdivision of the Newton polytope of F. Any Laurent polynomial f with $\mathrm{trop}(f) = F$ will satisfy the conclusion in the last sentence.

Fix an arbitrary generic basepoint $\mathbf{u}_0$ in $\mathbb{R}^n \backslash \Sigma$. For any facet σ of Σ let ℓ_σ be the unique primitive linear polynomial that vanishes on σ and satisfies $\ell_\sigma(\mathbf{u}_0) > 0$. Here *primitive* means that the coefficients of ℓ_σ are relatively prime integers. We also write $m(\sigma)$ for the multiplicity of σ in Σ. The linear forms ℓ_σ determine hyperplanes H_σ, which need not all be distinct. Let $\mathcal{A}$ be the hyperplane arrangement consisting of all these hyperplanes H_σ. By construction we have $\Sigma \subseteq \mathcal{A}$ and $\mathbf{u}_0 \notin \mathcal{A}$. We may refine the polyhedral complex structure on $\mathcal{A}$ so that Σ is a subcomplex of $\mathcal{A}$. For cells σ of $\mathcal{A}$ that are not contained in Σ, we set $m(\sigma) = 0$.

The complement $\mathbb{R}^n \backslash \mathcal{A}$ is the disjoint union of open convex polyhedra P. For each such polyhedron P we choose a path from $\mathbf{u}_0$ to P that crosses each hyperplane in $\mathcal{A}$ at most once and does so transversally. We define $\ell_P = \sum_{i=1}^n a_{P,i}x_i + b_P$ to be the sum of linear forms $m(\sigma)\ell_\sigma$, where σ is crossed by the path from $\mathbf{u}_0$ to P. The desired tropical polynomial is then

$$F(u) \; := \; \bigoplus_P b_P \odot u_1^{a_{P,1}} u_2^{a_{P,2}} \cdots u_n^{a_{P,n}},$$

where P ranges over all connected components of $\mathbb{R}^n \backslash \mathcal{A}$.

Since Σ is balanced, the definition of the linear form ℓ_P is independent of the choice of path from $\mathbf{u}_0$ to P. Indeed, any two such paths are connected by moves that cross codimension-2 faces τ of Σ. The balancing condition implies that $\sum_{\sigma \supset \tau} m(\sigma) \cdot \ell_\sigma = 0$ which ensures invariance of ℓ_P as τ is crossed. This means that the tropical polynomial $F(u)$ depends only on the choice of the basepoint $\mathbf{u}_0$. If $\mathbf{u}_0$ moves to a different component of $\mathbb{R}^n \backslash \Sigma$, then $F(u)$ is changed by tropical multiplication by a monomial, so the tropical hypersurface $V(F)$ remains unchanged.

By construction, the support of Σ is contained in $V(F)$ because F bends along each facet σ of Σ. This can be seen by choosing $\mathbf{u}_0$ just off σ. We need to show the reverse inclusion. Consider any region on which F is linear. That region corresponds to a vertex in the regular subdivision Δ that is dual to $V(F)$ in Proposition 3.1.6. By the remark above, we may assume that this vertex is the zero vector, and hence F is nonnegative. The region where F is zero lies in some connected component of $\mathbb{R}^n \backslash \Sigma$. By construction, every nonzero linear function ℓ_σ used in F is strictly positive on that connected component. Hence they are equal. Moreover, the linear function ℓ_σ is the corresponding edge direction, away from the vertex zero, in the regular subdivision Δ. The lattice length of that edge in Δ equals $m(\sigma)$. Hence Σ and $V(F)$ agree as weighted polyhedral complexes in $\mathbb{R}^n$. □

Remark 3.3.11. In this proof we described an algorithm for reconstructing a tropical polynomial F from the tropical hypersurface Σ it defines. This is interesting even in the constant coefficient case, when the input is a weighted balanced fan Σ of codimension 1, and the output is the corresponding Newton polytope P. In fact, that algorithm for computing P from Σ plays a central role in applications of tropical geometry, notably in implicitization [**STY07, SY08**]. Note that P is unique only up to translation.

3.4. Multiplicities and Balancing

In this section we define multiplicities that give a tropical variety the structure of a weighted balanced polyhedral complex. Another important result here is the Transverse Intersection Theorem (Theorem 3.4.12), which gives some control on the tropicalization of an intersection.

Given a subvariety $X \subset T^n$ with ideal $I \subset K[x_1^{\pm 1}, \dots, x_n^{\pm 1}]$, Proposition 3.2.8 implies that the tropical variety $\text{trop}(X)$ is the support of a polyhedral complex Σ. While $|\Sigma| = \text{trop}(X)$ is determined by I, the choice of Σ is not, as seen in Example 3.2.9. By Proposition 3.2.8 the polyhedral complex Σ can be chosen so that, for every $\sigma \in \Sigma$, we have $\text{in}_{\mathbf{w}}(I)$ constant for all $\mathbf{w} \in \text{relint}(\sigma)$. In what follows, we fix such a choice of Σ. Our aim is to define multiplicities on Σ that make it a weighted polyhedral complex.

We first recall some concepts from commutative algebra.

Definition 3.4.1. Let $S = \Bbbk[x_1^{\pm 1}, \dots, x_n^{\pm 1}]$. An ideal $Q \subset S$ is *primary* if $fg \in Q$ implies $f \in Q$ or $g^m \in Q$ for some $m > 0$. If Q is primary, then the radical of Q is a prime ideal P. Given an ideal $I \subset S$, we can write $I = \bigcap_{i=1}^{s} Q_i$ where each Q_i is primary with radical P_i, no P_i is repeated, and no Q_i can be removed from the intersection. This *primary decomposition* is not unique in general, but the set $\text{Ass}(I)$ of primes P_i that appear is determined by I. These are the *associated primes*. We are most interested in the *minimal (associated) primes* of I, i.e., those P_i that do not contain any other P_j. We denote this set by $\text{Ass}^{\min}(I) = \{P_1, \dots, P_t\}$. The primary ideal Q_i corresponding to a minimal prime P_i does not depend on the choice of a primary decomposition for I. For more information see [**Eis95**, Chapter 3], [**CLO07**, §4.7], or [**Stu02**, Chapter 5].

The *multiplicity* of a minimal prime $P_i \in \text{Ass}^{\min}(I)$ is the positive integer

$$\text{mult}(P_i, I) := \ell((S/Q_i)_{P_i}) = \ell(((I : P_i^{\infty})/I)_{P_i}). \tag{3.4.1}$$

Here $\ell(M)$ denotes the *length* of an S_{P_i}-module M. This is the length s of the longest chain of submodules $M = M_0 \supsetneq M_1 \supsetneq \cdots \supsetneq M_s$.

Example 3.4.2. Let $f = \alpha \prod_{i=1}^{r} (x - \lambda_i)^{m_i}$ with $\alpha, \lambda_i \in \Bbbk$ be a univariate polynomial in factored form. The set of associated primes of the ideal $\langle f \rangle$ is $\{\langle x - \lambda_i \rangle : 1 \leq i \leq r\}$, and the multiplicities are $\text{mult}(\langle x - \lambda_i \rangle, \langle f \rangle) = m_i$. ◊

Definition 3.4.3. Let I be a (not necessarily radical) ideal in $K[x_1^{\pm 1}, \dots, x_n^{\pm 1}]$. Let Σ be a polyhedral complex with support $|\Sigma| = \text{trop}(V(I))$ such that $\text{in}_{\mathbf{w}}(I)$ is constant for $\mathbf{w} \in \text{relint}(\sigma)$ for all $\sigma \in \Sigma$. For a top-dimensional cell $\sigma \in \Sigma$ the *multiplicity* $\text{mult}(\sigma)$ is defined by

$$\text{mult}(\sigma) = \sum_{P \in \text{Ass}^{\min}(\text{in}_{\mathbf{w}}(I))} \text{mult}(P, \text{in}_{\mathbf{w}}(I)) \qquad \text{for any } \mathbf{w} \in \text{relint}(\sigma).$$

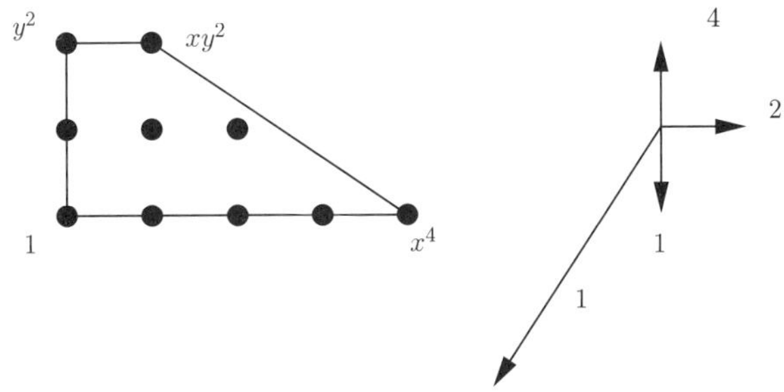

Figure 3.4.1. Multiplicities on the tropical curve in Example 3.4.5.

Remark 3.4.4. If $V(I)$ is irreducible of dimension d and σ is a maximal cell in Σ, then $\mathrm{in}_{\mathbf{w}}(I)$ is homogeneous with respect to a $\mathbb{Z}^d$-grading, so $V(\mathrm{in}_{\mathbf{w}}(I))$ has a d-dimensional torus action on it and is thus a union of d-dimensional torus orbits. The multiplicity $\mathrm{mult}(\sigma)$ is the number of such orbits, counted with multiplicity. See Lemma 3.4.7 below for an algebraic formulation.

Example 3.4.5. Let $f = xy^2 + 4y^2 + 3x^2y - xy + 8y + x^4 - 5x^2 + 4 \in \mathbb{C}[x^{\pm 1}, y^{\pm 1}]$. Then $\mathrm{trop}(V(f))$ consists of four rays perpendicular to the edges of the Newton polygon of f. The rays are generated by $\mathbf{u}_1 = (1, 0)$, $\mathbf{u}_2 = (0, 1)$, $\mathbf{u}_3 = (-2, -3)$, and $\mathbf{u}_4 = (0, -1)$; see Figure 3.4.1. The multiplicities on the rays of $\mathrm{trop}(V(f))$ are shown in Table 3.4.1.

Table 3.4.1

ray	$\mathrm{in}_{\mathbf{u}_i}(\langle f \rangle)$	$\mathrm{mult}(\mathrm{pos}(\mathbf{u}_i))$
$\mathbf{u}_1$	$\langle 4y^2 + 8y + 4 \rangle = \langle (y+1)^2 \rangle$	2
$\mathbf{u}_2$	$\langle x^4 - 5x^2 + 4 \rangle = \langle x-2 \rangle \cap \langle x-1 \rangle \cap \langle x+1 \rangle \cap \langle x+2 \rangle$	4
$\mathbf{u}_3$	$\langle xy^2 + x^4 \rangle = \langle y^2 + x^3 \rangle$	1
$\mathbf{u}_4$	$\langle xy^2 + 4y^2 \rangle = \langle x + 4 \rangle$	1

The variety of the initial ideal for $\mathbf{u}_1$ is one torus orbit with multiplicity two, the variety for $\mathbf{u}_2$ consists of four torus orbits, and the varieties for $\mathbf{u}_3$ and $\mathbf{u}_4$ are each a single torus orbit. The tropical curve $\mathrm{trop}(V(f))$ is balanced with these multiplicities because $2\mathbf{u}_1 + 4\mathbf{u}_2 + 1\mathbf{u}_3 + 1\mathbf{u}_4 = (0, 0)$. ◊

Example 3.4.5 is a special case of the following fact which holds for all tropical hypersurfaces and which was already addressed in Proposition 3.3.2.

Lemma 3.4.6. *Let* $f = \sum c_{\mathbf{u}} x^{\mathbf{u}} \in K[x_1^{\pm 1}, \ldots, x_n^{\pm 1}]$, *let* Δ *be the regular subdivision of* Newt(f) *induced by* $(\mathrm{val}(c_{\mathbf{u}}))$, *and let* Σ *be the polyhedral complex supported on* $\mathrm{trop}(V(f))$ *that is dual to* Δ. *The multiplicity of a maximal cell* σ *of* Σ *is the lattice length of the edge* $e(\sigma)$ *of* Δ *dual to* σ.

Proof. Pick $\mathbf{w}$ in the relative interior of σ. The initial ideal $\text{in}_{\mathbf{w}}(\langle f\rangle)$ is generated by $\text{in}_{\mathbf{w}}(f) = \sum_{\mathbf{u}\in e(\sigma)} \overline{t^{-\text{val}(c_{\mathbf{u}})}c_{\mathbf{u}}}x^{\mathbf{u}}$. The sum here is over those $\mathbf{u} \in e(\sigma)$ with $\text{val}(c_{\mathbf{u}}) + \mathbf{w}\cdot\mathbf{u} = \text{trop}(f)(\mathbf{w})$. Since $e(\sigma)$ is one dimensional, the vector $\mathbf{u}-\mathbf{u}'$ for $\mathbf{u}, \mathbf{u}' \in e(\sigma)$ is unique up to scaling, and there is a choice $\mathbf{v} = \mathbf{u} - \mathbf{u}'$ for which this has minimal length. The polynomial $\text{in}_{\mathbf{w}}(f)$ is then a monomial in $x_1, \ldots, x_n$ times a Laurent polynomial g in the variable $y = x^{\mathbf{v}}$. After multiplying f by a monomial, we may assume that $\text{in}_{\mathbf{w}}(f)$ is a (non-Laurent) polynomial in y with nonzero constant term. The degree of g in y is then the lattice length of the edge $e(\sigma)$. It follows from Example 3.4.2 that the multiplicity of σ is the lattice length of $e(\sigma)$, as required. □

Note that Lemma 3.4.6 and Proposition 3.3.2 together imply that tropical hypersurfaces are balanced with the multiplicities of Definition 3.4.3.

We now translate the geometric content of Remark 3.4.4 into a precise algebraic form. After a multiplicative change of variables, we may transport any cell in Σ to one with affine span parallel to the span of $\mathbf{e}_1, \ldots, \mathbf{e}_d$. The following lemma gives one method for computing the multiplicity of σ. This should be compared with the formula for multiplicity in Exercise 3.7(34).

Lemma 3.4.7. *Let $X \subset T^n$ be irreducible of dimension d with ideal $I \subset K[x_1^{\pm 1}, \ldots, x_n^{\pm 1}]$, and let Σ be a polyhedral complex on* $\text{trop}(X)$ *as above. Let σ be a maximal cell in Σ with affine span parallel to $\mathbf{e}_1, \ldots, \mathbf{e}_d$, and let $\mathbf{w} \in \text{relint}(\sigma) \cap \Gamma_{\text{val}}^n$. If $S' = \Bbbk[x_{d+1}^{\pm 1}, \ldots, x_n^{\pm 1}]$, then $\text{mult}(\sigma) = \dim_{\Bbbk}(S'/(\text{in}_{\mathbf{w}}(I) \cap S'))$.*

Proof. Since $\mathbf{w} \in \text{relint}(\sigma)$, by Corollary 2.4.10 and Proposition 2.6.1 we have $\text{in}_{\mathbf{w}+\epsilon\mathbf{e}_i}(I) = \text{in}_{\mathbf{w}}(I)$ for all sufficiently small $\epsilon > 0$ and $1 \le i \le d$. Thus, by part (2) of Lemma 2.6.2, the initial ideal $\text{in}_{\mathbf{w}}(I)$ is homogeneous with respect to the grading $\deg(x_i) = \mathbf{e}_i$ for $i \le d$ and $\deg(x_i) = 0$ for $i > d$. Hence $\text{in}_{\mathbf{w}}(I)$ has a generating set $\{f_1, \ldots, f_r\}$ not containing the variables $x_1, \ldots, x_d$. Let $\bigcap_{i=1}^s Q_i$ be a primary decomposition of $\text{in}_{\mathbf{w}}(I)$. Each Q_i is also generated by polynomials in $x_{d+1}, \ldots, x_n$, as they are also homogeneous with respect to the $\mathbb{Z}^d$-grading, so $\text{in}_{\mathbf{w}}(I) \cap S' = \bigcap_{i=1}^s (Q_i \cap S')$ is a primary decomposition of the zero-dimensional ideal $\text{in}_{\mathbf{w}}(I) \cap S'$, and $\text{mult}(P_i, Q_i) = \text{mult}(P_i \cap S', Q_i \cap S')$. This implies that each P_i is a minimal prime of $\text{in}_{\mathbf{w}}(I)$. Since $Q_i \cap S'$ is a zero-dimensional ideal in S', its multiplicity is its colength. Therefore,

$$\begin{aligned}\text{mult}(\sigma) &= \sum_{i=1}^{s} \text{mult}(P_i, Q_i)\\ &= \sum_{i=1}^{s} \dim_{\Bbbk} S'/(Q_i \cap S') = \dim_{\Bbbk} S'/(\text{in}_{\mathbf{w}}(I) \cap S').\end{aligned}$$

□

The goal of the rest of this section is to show that the multiplicities in Definition 3.4.3 force the polyhedral complex $\operatorname{trop}(X)$ to be balanced. We will reduce the proof of this to the case of constant coefficient curves. For this, we need following result about zero-dimensional ideals I. As before we write S_K and $S_{\Bbbk}$ for the Laurent polynomial rings in variables $x_1, \ldots, x_n$ with coefficients in K and $\Bbbk$, respectively. We denote by $\widetilde{S}_K$ and $\widetilde{S}_{\Bbbk}$ the corresponding polynomial rings with $n+1$ variables $x_0, x_1, \ldots, x_n$.

Proposition 3.4.8. *Let $I = \bigcap_{\mathbf{y}} Q_{\mathbf{y}} \subseteq K[x_1^{\pm 1}, \ldots, x_n^{\pm 1}]$, where each $Q_{\mathbf{y}}$ is primary to a maximal ideal $P_{\mathbf{y}} = \langle x_1 - y_1, \ldots, x_n - y_n \rangle$.*

(1) *Assume further that all $\mathbf{y} \in V(I) \subset T^n$ have the same tropicalization $\operatorname{val}(\mathbf{y}) = \mathbf{w}$, for some fixed $\mathbf{w} \in \Gamma_{\mathrm{val}}^n$. Then $\dim_{\Bbbk} S_{\Bbbk}/\operatorname{in}_{\mathbf{w}}(I) = \sum_{\mathbf{y}} \operatorname{mult}(P_{\mathbf{y}}, Q_{\mathbf{y}}) = \dim_K S_K/I$.*

(2) *Without that assumption, for $\mathbf{w} \in \operatorname{trop}(V(I))$, let*

$$I_{\mathbf{w}} = \bigcap_{\mathbf{y}:\operatorname{val}(\mathbf{y})=\mathbf{w}} Q_{\mathbf{y}}.$$

The multiplicity of the point $\mathbf{w}$ equals $\dim_K S_K/I_{\mathbf{w}}$.

Proof. The equation

$$\dim_K S_K/I \;=\; \sum_{\mathbf{y}\in V(I)} \operatorname{mult}(P_{\mathbf{y}}, Q_{\mathbf{y}})$$

holds for any zero-dimensional ideal $I = \bigcap_{\mathbf{y}} Q_{\mathbf{y}}$ where $P_{\mathbf{y}} = \operatorname{rad}(Q_{\mathbf{y}})$. The homogenization I_{proj} of such an ideal satisfies $\dim_K S_K/I = \dim_K(\widetilde{S}_K/I_{\mathrm{proj}})_d$ for any $d \gg 0$. These two facts also hold for ideals in $S_{\Bbbk}$. By Corollary 2.4.9 we have $\dim_K(\widetilde{S}_K/I_{\mathrm{proj}})_d = \dim_{\Bbbk}(\widetilde{S}_K/\operatorname{in}_{(0,\mathbf{w})}(I_{\mathrm{proj}}))_d$, so to show that $\dim_K S_K/I = \dim_{\Bbbk} S_{\Bbbk}/\operatorname{in}_{\mathbf{w}}(I)$ it suffices to show

$$(\operatorname{in}_{(0,\mathbf{w})}(I_{\mathrm{proj}}))_d = (\operatorname{in}_{\mathbf{w}}(I)_{\mathrm{proj}})_d \text{ for } d \gg 0.$$

The inclusion $\subseteq$ follows from Proposition 2.6.1, since $J \subseteq (J|_{x_0=1})_{\mathrm{proj}}$ for any homogeneous ideal $J \subset \Bbbk[x_0, \ldots, x_n]$. For the reverse inclusion, we first note that Proposition 2.6.1 also implies that $\operatorname{in}_{\mathbf{w}}(I)_{\mathrm{proj}} = (\operatorname{in}_{(0,\mathbf{w})}(I_{\mathrm{proj}}) : \prod_{i=0}^n x_i^\infty)$. Saturating by the irrelevant ideal $\langle x_0, \ldots, x_n \rangle$ does not change $(\operatorname{in}_{(0,\mathbf{w})}(I_{\mathrm{proj}}))_d$ for $d \gg 0$, and this saturation has only one-dimensional associated primes. These associated primes have the form $P_{\mathbf{y}'} = \langle y'_j x_i - y'_i x_j : 0 \leq i < j \leq n \rangle$ for some $\mathbf{y}' = (y'_0 : \cdots : y'_n) \in \mathbb{P}^n$. Write $(\operatorname{in}_{(0,\mathbf{w})}(I_{\mathrm{proj}}) : \langle x_0, \ldots, x_n \rangle^\infty) = \bigcap_{\mathbf{y}} Q_{\mathbf{y}'}$, where $Q_{\mathbf{y}'}$ is primary to $P_{\mathbf{y}'}$ Now

$$\Big(\bigcap Q_{\mathbf{y}'} : \prod x_i^\infty\Big) = \bigcap_{\mathbf{y}'} (Q_{\mathbf{y}'} : \prod x_i^\infty) = \bigcap_{\mathbf{y}' : x_i \notin P_{\mathbf{y}'}} Q_{\mathbf{y}'},$$

so it suffices to show that $x_i \notin P_{\mathbf{y}'}$ for all i and all $Q_{\mathbf{y}'}$.

Since each primary component $Q_{\mathbf{y}}$ of I is $P_{\mathbf{y}}$ primary, it contains $(x_i - y_i)^d$ for some $d \gg 0$. The product $\prod_{\mathbf{y}}(x_i - y_i x_0)^d$ is thus in I_{proj} for $d \gg 0$, and so since $\text{val}(\mathbf{y}) = \mathbf{w}$ for all $\mathbf{y}$, we have $\prod_{\mathbf{y}}(x_i - \tilde{y}_i x_0)^d \in \text{in}_{(0,\mathbf{w})}(I_{\text{proj}})$, where $\tilde{y}_i = \overline{t^{-w_i} y_i}$. This shows that $x_i \notin P_{\mathbf{y}'}$ for all $\mathbf{y}'$ and $0 \leq i \leq n$. Indeed, for each i the product $\prod_{\mathbf{y}}(x_i - \tilde{y}_i x_0)^d \in \text{in}_{(0,\mathbf{w})}(I_{\text{proj}})$, so for each $\mathbf{y}'$ there is there is $\tilde{y}_i$ with $x_i - \tilde{y}_i x_0 \in P_{\mathbf{y}'}$. If $x_i \in P_{\mathbf{y}'}$ for some i, then $x_j \in P_{\mathbf{y}'}$ for $0 \leq j \leq n$, since each $\tilde{y}_i$ is nonzero as $y_i \neq 0$. This contradicts the fact that $\mathbf{y}' \in \mathbb{P}^n$, so we conclude that the first claim holds.

For Proposition 3.4.8(2), we claim that $\text{in}_{\mathbf{w}}(I) = \text{in}_{\mathbf{w}}(I_{\mathbf{w}})$. The inclusion $\subseteq$ is immediate from $I \subseteq I_{\mathbf{w}}$. For the inclusion $\supseteq$, note that for any $\mathbf{y}$ with $\text{val}(\mathbf{y}) \neq \mathbf{w}$ we have $\mathbf{w} \notin \text{trop}(Q_{\mathbf{y}})$, so there is $f_{\mathbf{y}} \in Q_{\mathbf{y}}$ with $1 = \text{in}_{\mathbf{w}}(f)$. Given $f \in I_{\mathbf{w}}$, we then have $g = f \prod_{\text{val}(\mathbf{y}) \neq \mathbf{w}} f_{\mathbf{y}} \in I$ with $\text{in}_{\mathbf{w}}(g) = \text{in}_{\mathbf{w}}(f)$. This gives the other inclusion. The result now follows from the first part using the interpretation of the multiplicity of Lemma 3.4.7. □

The concept of transverse intersection is fundamental in algebraic, differential, and symplectic geometry. The same holds in tropical geometry.

Definition 3.4.9. Let Σ_1 and Σ_2 be two polyhedral complexes in $\mathbb{R}^n$, and let $\mathbf{w} \in \Sigma_1 \cap \Sigma_2$. The point $\mathbf{w}$ lies in the relative interior of a unique cell σ_i in Σ_i for $i = 1, 2$. The complexes Σ_1, Σ_2 *meet transversely* at $\mathbf{w} \in \Sigma_1 \cap \Sigma_2$ if the affine span of σ_i is $\mathbf{w} + L_i$ for $i = 1, 2$, and $L_1 + L_2 = \mathbb{R}^n$. Two tropical varieties $\text{trop}(X)$ and $\text{trop}(Y)$ *intersect transversely* at some $\mathbf{w} \in \text{trop}(X) \cap \text{trop}(Y)$ if there is *some* choice of polyhedral complexes Σ_1, Σ_2, with $\text{trop}(X) = |\Sigma_1|$ and $\text{trop}(Y) = |\Sigma_2|$, and these meet transversely at $\mathbf{w}$.

We next show, in Theorem 3.4.12, that if the tropicalizations of two varieties meet transversely at $\mathbf{w} \in \mathbb{R}^n$, then $\mathbf{w}$ lies in the tropicalization of the intersection of the varieties. This requires the following lemma.

Lemma 3.4.10. *Let I, J be homogeneous ideals in $K[x_0, \ldots, x_n, y_0, \ldots, y_m]$, and fix $\mathbf{w} \in \mathbb{R}^{n+m+2}$. If $\text{in}_{\mathbf{w}}(I)$ has a generating set only involving $x_0, \ldots, x_n$ and $\text{in}_{\mathbf{w}}(J)$ has a generating set only involving $y_0, \ldots, y_m$, then*

$$\text{in}_{\mathbf{w}}(I + J) \quad = \quad \text{in}_{\mathbf{w}}(I) + \text{in}_{\mathbf{w}}(J).$$

Proof. Suppose that this is not the case. Then there is some homogeneous polynomial $f + g$ in $I + J$ of degree d with $f \in I_d$, $g \in J_d$ and $\text{in}_{\mathbf{w}}(f + g) \notin \text{in}_{\mathbf{w}}(I) + \text{in}_{\mathbf{w}}(J)$. Fix a monomial term order $\prec$ on $\mathbb{k}[x_0, \ldots, y_m]$. We may further assume that $\text{in}_{\prec}(\text{in}_{\mathbf{w}}(f + g)) \notin \text{in}_{\prec}(\text{in}_{\mathbf{w}}(I) + \text{in}_{\mathbf{w}}(J))$. This implies

$$\text{(3.4.2)} \qquad \text{in}_{\prec}(\text{in}_{\mathbf{w}}(f + g)) \;\notin\; \text{in}_{\prec}(\text{in}_{\mathbf{w}}(I)) + \text{in}_{\prec}(\text{in}_{\mathbf{w}}(J)).$$

Let $x^{\mathbf{u}_1} y^{\mathbf{v}_1}$ and $x^{\mathbf{u}_2} y^{\mathbf{v}_2}$ be the monomials in $\text{in}_{\prec}(\text{in}_{\mathbf{w}}(f))$ and $\text{in}_{\prec}(\text{in}_{\mathbf{w}}(g))$, respectively, and let $\alpha_1, \alpha_2 \in K$ be their coefficients in f and g. From (3.4.2) we conclude that $x^{\mathbf{u}_1} y^{\mathbf{v}_1} = x^{\mathbf{u}_2} y^{\mathbf{v}_2}$, and $\text{val}(\alpha_1 + \alpha_2) > \text{val}(\alpha_1) = \text{val}(\alpha_2)$.

We may assume that this counterexample is maximal in the following sense: if $f' \in I_d, g' \in J_d$ is any other pair with $f + g = f' + g'$, then either $\mathrm{trop}(f')(\mathbf{w}) < \mathrm{trop}(f)(\mathbf{w})$ or $\mathrm{trop}(f')(\mathbf{w}) = \mathrm{trop}(f)(\mathbf{w})$ and $\mathrm{in}_\prec(\mathrm{in}_\mathbf{w}(f')) \succ \mathrm{in}_\prec(\mathrm{in}_\mathbf{w}(f))$. To see that such a maximal pair exists, note that if there were no such pair, we could find a sequence $f_i = f+h_i \in I, g_i = g - h_i \in J$ with $f_i + g_i = f + g$ for all i, and $\mathrm{trop}(f_i)(\mathbf{w})$ strictly increasing. The strict increase comes from the fact that there are only finitely many monomials of degree d, so we cannot have $\mathrm{trop}(f_{i+1})(\mathbf{w}) = \mathrm{trop}(f_i)(\mathbf{w})$ and $\mathrm{in}_\prec(\mathrm{in}_\mathbf{w}(f_{i+1})) \succ \mathrm{in}_\prec(\mathrm{in}_\mathbf{w}(f_i))$ an infinite number of times. By passing to a subsequence we may assume that the support of each f_i is the same.

Since $\mathrm{supp}(f + h_i) = \mathrm{supp}(f + h_{i+1})$, there are $\alpha, \beta \in K^*$ for which $\alpha(f + h_i) + \beta(f + h_{i+1}) = (\alpha + \beta)f + (\alpha h_i + \beta h_{i+1})$ has strictly smaller support. Since $f+h_i \neq f+h_{i+1}$, we may assume that one of the monomials removed from $\mathrm{supp}(f+h_i)$ in this manner has different coefficients in h_i and h_{i+1}, and thus $\alpha+\beta \neq 0$. Note that for any two polynomials p_1, p_2 we have $\mathrm{trop}(p_1 + p_2)(\mathbf{w}) \geq \min(\mathrm{trop}(p_1)(\mathbf{w}), \mathrm{trop}(p_2)(\mathbf{w}))$. Since $f_i - f = g - g_i \in I \cap J$ for all i, the resulting polynomial $h_i' = 1/(\alpha+\beta)(\alpha h_{i+1} + \beta h_i)$ is also in $I \cap J$, so $f_i' = f + h'$ lies in I and has $\mathrm{trop}(f_i')(\mathbf{w}) \geq \mathrm{trop}(f_i)(\mathbf{w})$ and $\mathrm{supp}(f_i') \subsetneq \mathrm{supp}(f_i)$. By passing to another subsequence, we may assume that the sequence $\mathrm{trop}(f+h_i')(\mathbf{w})$ is again increasing. Continuing to iterate this procedure would eventually yield the support of the new f_i being empty, which is impossible since $\mathrm{in}_\mathbf{w}(f_i+g_i) \notin \mathrm{in}_\mathbf{w}(J)$. This shows that the infinite increasing sequence does not exist, so we may assume that the pair f, g is maximal in the required sense.

Now $f \in I$ implies that $x^{\mathbf{u}_1}y^{\mathbf{v}_1} \in \mathrm{in}_\prec(\mathrm{in}_\mathbf{w}(I))$, so there is $f_1 \in I$ with $\mathrm{in}_\mathbf{w}(f_1) \in \Bbbk[x_0, \ldots, x_n]$, and $\mathrm{in}_\prec(\mathrm{in}_\mathbf{w}(f_1)) = x^{\mathbf{u}_3}$ dividing $x^{\mathbf{u}_1}$. We may assume that the coefficient of $x^{\mathbf{u}_3}$ in f_1 is one. We can thus write $f = \alpha_1 x^{\mathbf{u}_1-\mathbf{u}_3}y^{\mathbf{v}_1}f_1 + f_2$ where $\mathrm{trop}(f_2)(\mathbf{w}) \geq \mathrm{trop}(f)(\mathbf{w})$, and if equality holds, then $\mathrm{in}_\prec(\mathrm{in}_\mathbf{w}(f_2)) \prec \mathrm{in}_\prec(\mathrm{in}_\mathbf{w}(f))$. Similarly, $g = \alpha_2 x^{\mathbf{u}_1}y^{\mathbf{v}_1-\mathbf{v}_3}g_1 + g_2$ where $\mathrm{trop}(g_2)(\mathbf{w}) \geq \mathrm{trop}(g)(\mathbf{w})$, and if equality holds, then $\mathrm{in}_\prec(\mathrm{in}_\mathbf{w}(g_2)) \prec \mathrm{in}_\prec(\mathrm{in}_\mathbf{w}(g))$. Since $\mathrm{val}(\alpha_1 + \alpha_2) > \mathrm{val}(\alpha_1) = \mathrm{val}(\alpha_2)$, we can write $\alpha_2 = \alpha_1(-1+\beta)$, where $\mathrm{val}(\beta) > 0$. Then

$$\begin{aligned} f+g &= \alpha_1 x^{\mathbf{u}_1-\mathbf{u}_3}y^{\mathbf{v}_1}f_1 + f_2 + \alpha_2 x^{\mathbf{u}_1}y^{\mathbf{v}_1-\mathbf{v}_3}g_1 + g_2 \\ &= \alpha_1 x^{\mathbf{u}_1-\mathbf{u}_3}y^{\mathbf{v}_1-\mathbf{v}_3}(y^{\mathbf{v}_3}f_1 - x^{\mathbf{u}_3}g_1 + \beta x^{\mathbf{u}_3}g_1) + f_2 + g_2 \\ &= \alpha_1 x^{\mathbf{u}_1-\mathbf{u}_3}y^{\mathbf{v}_1-\mathbf{v}_3}(-(g_1-y^{\mathbf{v}_3})f_1 + (f_1-x^{\mathbf{u}_3})g_1 + \beta x^{\mathbf{u}_3}g_1) + f_2+g_2. \end{aligned}$$

Set

$$f' = \alpha_1 x^{\mathbf{u}_1-\mathbf{u}_3}y^{\mathbf{v}_1-\mathbf{v}_3}(-(g_1 - y^{\mathbf{v}_3})f_1) + f_2,$$

and

$$g' = \alpha_1 x^{\mathbf{u}_1-\mathbf{u}_3}y^{\mathbf{v}_1-\mathbf{v}_3}((f_1 - x^{\mathbf{u}_3})g_1 + \beta x^{\mathbf{u}_3}g_1) + g_2.$$

Then, by construction $f' \in I$, $g' \in J$, and $f' + g' = f + g$. In addition, either $\mathrm{trop}(f')(\mathbf{w}) > \mathrm{trop}(f)(\mathbf{w})$ or $\mathrm{in}_\prec(\mathrm{in}_\mathbf{w}(f')) \prec \mathrm{in}_\prec(\mathrm{in}_\mathbf{w}(f))$. This contradicts our choice of a maximal counterexample, so we conclude that none exists, and hence $\mathrm{in}_\mathbf{w}(I + J) = \mathrm{in}_\mathbf{w}(I) + \mathrm{in}_\mathbf{w}(J)$. □

Remark 3.4.11. Lemma 3.4.10 is a variant of Buchberger's second criterion for S-pairs. See, for example, [**CLO07**, §2.9] for details of the standard case. See [**CM13**] for more on this criterion for the Gröbner bases studied here.

We now use Lemma 3.4.10 to prove the Transverse Intersection Theorem. This states that when two tropical varieties meet transversely, their intersection equals the tropicalization of the intersections. This is a very useful tool for nontrivial computations. For some generalizations, see [**OP13**].

Theorem 3.4.12. *Let X and Y be subvarieties of T^n_K. If* $\mathrm{trop}(X)$ *and* $\mathrm{trop}(Y)$ *meet transversely at* $\mathbf{w} \in \Gamma^n_{\mathrm{val}}$, *then* $\mathbf{w} \in \mathrm{trop}(X \cap Y)$. *Therefore*

$$\mathrm{trop}(X \cap Y) \quad = \quad \mathrm{trop}(X) \cap \mathrm{trop}(Y)$$

if the polyhedral intersection on the right-hand side is transverse everywhere.

Proof. Let Σ_1 and Σ_2 be polyhedral complexes in $\mathbb{R}^n$ with $\mathrm{trop}(X) = |\Sigma_1|$ and $\mathrm{trop}(Y) = |\Sigma_2|$. Let $I, J \subset K[x_1^{\pm 1}, \ldots, x_n^{\pm 1}]$ be the ideals of X and Y. Let $\sigma_i \in \Sigma_i$ be the cell containing $\mathbf{w}$ in its relative interior for $i = 1, 2$, with the affine span of σ_i equal to $\mathbf{w} + L_i$. Our hypothesis says that $L_1 + L_2 = \mathbb{R}^n$.

We now reduce to the case that L_1 contains $\mathbf{e}_1, \ldots, \mathbf{e}_r, \mathbf{e}_{r+1}, \ldots, \mathbf{e}_s$, and L_2 contains $\mathbf{e}_1, \ldots, \mathbf{e}_r, \mathbf{e}_{s+1}, \ldots, \mathbf{e}_n$. By the assumption $L_1 + L_2 = \mathbb{R}^n$, there exists a basis $\mathbf{a}_1, \ldots, \mathbf{a}_n$ for $\mathbb{R}^n$ where $\mathbf{a}_1, \ldots, \mathbf{a}_r \in L_1 \cap L_2$, $\mathbf{a}_{r+1}, \ldots, \mathbf{a}_s \in L_1$, $\mathbf{a}_{s+1}, \ldots, \mathbf{a}_n \in L_2$ and all $\mathbf{a}_i \in \mathbb{Z}^n$. Write these as the rows of an $n \times n$-matrix A, with A_i the ith column of A. Let $\phi \colon T^n \to T^n$ be the morphism given by $\phi^*(x_i) = x^{A_i}$, so $\mathrm{trop}(\phi)$ is given by the matrix A^T. The morphism ϕ is finite but is not an isomorphism if $\det(A) \neq \pm 1$. Since A has full rank by construction, however, the linear map $\mathrm{trop}(\phi) \colon \mathbb{R}^n \to \mathbb{R}^n$ is an isomorphism. Let $I' = \phi^*(I)$, $X' = V(I')$, $J' = \phi^*(J)$, and $Y' = V(J')$. Then $\phi(X') = X$ and $\phi(Y') = Y$. By Corollary 3.2.13 we have

$$\mathrm{trop}(X) = \mathrm{trop}(\phi)(\mathrm{trop}(X')),$$
$$\mathrm{trop}(Y) = \mathrm{trop}(\phi)(\mathrm{trop}(Y')),$$
$$\mathrm{trop}(\phi)(\mathrm{trop}(X' \cap Y')) = \mathrm{trop}(X \cap Y).$$

By construction

$$\mathrm{trop}(\phi)(\mathrm{span}(\mathbf{e}_1, \ldots, \mathbf{e}_r)) \subseteq L_1 \cap L_2,$$
$$\mathrm{trop}(\phi)(\mathrm{span}(\mathbf{e}_{r+1}, \ldots, \mathbf{e}_s)) \subseteq L_1,$$
$$\mathrm{trop}(\phi)(\mathrm{span}(\mathbf{e}_{s+1}, \ldots, \mathbf{e}_n)) \subseteq L_2,$$

and $\mathrm{trop}(X')$ and $\mathrm{trop}(Y')$ intersect transversely at $\mathrm{trop}(\phi)^{-1}(\mathbf{w})$. It suffices to show that $\mathrm{trop}(\phi)^{-1}(\mathbf{w}) \in \mathrm{trop}(X' \cap Y')$. By replacing X and Y with X' and Y', we may thus assume that L_1 contains $\mathbf{e}_1, \ldots, \mathbf{e}_r, \mathbf{e}_{r+1}, \ldots, \mathbf{e}_s$ and L_2 contains $\mathbf{e}_1, \ldots, \mathbf{e}_r, \mathbf{e}_{s+1}, \ldots, \mathbf{e}_n$.

As in the proof of Theorem 3.3.8, $\mathrm{in}_{\mathbf{w}}(I)$ is homogeneous with respect to a $\mathbb{Z}^{\dim(L_1)}$-grading, and we can find polynomials $f_1, \ldots, f_l$ in $x_{s+1}, \ldots, x_n$ that generate $\mathrm{in}_{\mathbf{w}}(I)$. Similarly, there is a generating set $g_1, \ldots, g_m$ for $\mathrm{in}_{\mathbf{w}}(J)$ only using $x_{r+1}, \ldots, x_s$. Let $I_{\mathrm{proj}} \subseteq K[x_0, \ldots, x_{n+1}]$ be the ideal obtained by homogenizing $I \cap K[x_1, \ldots, x_n]$ using the variable x_{n+1}, and let J_{proj} be the ideal obtained by homogenizing $J \cap K[x_1, \ldots, x_n]$ using the variable x_0.

For $\overline{\mathbf{w}} = (0, \mathbf{w}, 0) \in \mathbb{R}^{n+2}$, the initial ideal $\mathrm{in}_{\overline{\mathbf{w}}}(I_{\mathrm{proj}})$ has a generating set only using $x_{s+1}, \ldots, x_{n+1}$, and $\mathrm{in}_{\overline{\mathbf{w}}}(J_{\mathrm{proj}})$ has a generating set only using $x_0, x_{r+1}, \ldots, x_s$. Thus by Lemma 3.4.10 we have $\mathrm{in}_{\overline{\mathbf{w}}}(I_{\mathrm{proj}} + J_{\mathrm{proj}}) = \mathrm{in}_{\overline{\mathbf{w}}}(I_{\mathrm{proj}}) + \mathrm{in}_{\overline{\mathbf{w}}}(J_{\mathrm{proj}})$. Furthermore, after setting $x_0 = x_{n+1} = 1$ as in Proposition 2.6.1, we obtain

$$\mathrm{in}_{\mathbf{w}}(I + J) = \mathrm{in}_{\mathbf{w}}(I) + \mathrm{in}_{\mathbf{w}}(J). \tag{3.4.3}$$

Since $\mathrm{in}_{\mathbf{w}}(I)$ and $\mathrm{in}_{\mathbf{w}}(J)$ are proper ideals, by the Nullstellensatz, there exist $\mathbf{y} = (y_{r+1}, \ldots, y_s) \in (\Bbbk^*)^{s-r}$ and $\mathbf{z} = (z_{s+1}, \ldots, z_n) \in (\Bbbk^*)^{n-s}$ with $f_i(\mathbf{y}) = g_j(\mathbf{z}) = 0$ for all i, j. Now, for any $(t_1, \ldots, t_r) \in (\Bbbk^*)^r$, the vector $(t_1, \ldots, t_r, y_{r+1}, \ldots, y_s, z_{s+1}, \ldots, z_n)$ lies in the variety $V(\mathrm{in}_{\mathbf{w}}(I)) \cap V(\mathrm{in}_{\mathbf{w}}(J)) = V(\mathrm{in}_{\mathbf{w}}(I) + \mathrm{in}_{\mathbf{w}}(J)) = V(\mathrm{in}_{\mathbf{w}}(I + J))$. We conclude that $\mathrm{in}_{\mathbf{w}}(I + J) \neq \langle 1 \rangle$, and hence $\mathbf{w} \in \mathrm{trop}(V(I + J)) = \mathrm{trop}(X \cap Y)$. □

If the two tropical varieties $\mathrm{trop}(X)$ and $\mathrm{trop}(Y)$ do not meet transversely at the point $\mathbf{w}$, then $\mathbf{w}$ may fail to lie in $\mathrm{trop}(X \cap Y)$. For instance, suppose X is a line and Y is a conic, both in the plane, and their tropicalizations intersect as in Figure 1.3.6. Then $\mathrm{trop}(X) \cap \mathrm{trop}(Y)$ contains the line segment $[A, B]$, while $\mathrm{trop}(X \cap Y) = \{A, B\}$ consists only of the two endpoints.

We next prove that the tropicalizations of constant coefficient curves are balanced. This is a key step in establishing the balancing property in Theorem 3.3.5. Our proof of Proposition 3.4.13 rests on basic commutative algebra and is self-contained. A shorter argument can be given if one uses intersection theory on toric varieties. This will be explained in Remark 6.7.8.

Proposition 3.4.13. *If C is a curve in $T^n_{\Bbbk} \cong (\Bbbk^*)^n$, then the one-dimensional fan* $\mathrm{trop}(C)$ *is balanced using the multiplicities of Definition* 3.4.3.

Proof. Let $\mathbf{u}_1, \ldots, \mathbf{u}_s$ be the first lattice points on the rays of $\mathrm{trop}(C)$, let $m_i = \mathrm{mult}(\mathrm{pos}(\mathbf{u}_i))$, and set $\mathbf{u} = \sum_{i=1}^{s} m_i \mathbf{u}_i$. We will show that if $\mathbf{v} \in \mathbb{Z}^n$ is primitive, in the sense that $\gcd(v_1, \ldots, v_n) = 1$, then $\mathbf{v} \cdot \mathbf{u} = 0$. This implies $\mathbf{v} \cdot \mathbf{u} = 0$ for all $\mathbf{v} \in \mathbb{Z}^n$, and so $\mathbf{u} = \mathbf{0}$. By Lemma 2.2.7 and Corollary 2.6.12, after a change of coordinates, it suffices to consider the case $\mathbf{v} = \mathbf{e}_1$.

Let $I \subset \Bbbk[x_1^{\pm 1}, \ldots, x_n^{\pm 1}]$ be the ideal of C. Let K' be the algebraic closure of $\Bbbk(t)$. Since $\Bbbk$ is algebraically closed, K' has residue field $\Bbbk$. We denote by I' the extension of I to $K'[x_1^{\pm 1}, \ldots, x_n^{\pm 1}]$, and write $C_{K'} \subset T^n_{K'}$ for the variety of I'. We have $\operatorname{trop}(C) = \operatorname{trop}(C_{K'})$ by Theorem 3.2.4.

Consider the ideal $J'_\alpha = I' + \langle x_1 - \alpha \rangle$ for $\alpha \in K'^*$. There exists $L \in \mathbb{N}$ and a finite subset $\mathcal{D} \subset K'^*$ such that $\dim_K(S_{K'}/J'_\alpha) = L$ for all $\alpha \in K'^* \backslash \mathcal{D}$. To see this, we apply classical Gröbner bases to the homogenization $(J'_\alpha)_{\text{proj}}$ of J'_α. The initial ideal is constant for all α outside a finite set $\mathcal{D}$. The number L equals the Hilbert polynomial of this initial ideal, which is constant because $x_1 - \alpha$ cannot be a zerodivisor on $S_{K'}/I'$ for infinitely many α.

Choose $\alpha_1, \alpha_2 \in K'^* \backslash \mathcal{D}$ with $\operatorname{val}(\alpha_1) = 1$ and $\operatorname{val}(\alpha_2) = -1$. Let $X^+ = V(I' + \langle x_1 - \alpha_1 \rangle) \subset T^n_{K'}$, and $X^- = V(I' + \langle x_1 - \alpha_2 \rangle) \subset T^n_{K'}$. The desired identity $u_1 = 0$ will be obtained by computing L tropically.

Set $\beta_1 = \overline{t^{-1}\alpha_1} \in \Bbbk^*$ and $\beta_2 = \overline{t^1\alpha_2} \in \Bbbk^*$. From (3.4.3) in the proof of Theorem 3.4.12, we conclude

$$\begin{array}{rcll} \operatorname{in}_{\mathbf{w}}(I' + \langle x_1 - \alpha_1 \rangle) & = & \operatorname{in}_{\mathbf{w}}(I') + \langle x_1 - \beta_1 \rangle \neq \langle 1 \rangle & \text{for } \mathbf{w} \in \operatorname{trop}(X^+), \\ \operatorname{in}_{\mathbf{w}}(I' + \langle x_1 - \alpha_2 \rangle) & = & \operatorname{in}_{\mathbf{w}}(I') + \langle x_1 - \beta_2 \rangle \neq \langle 1 \rangle & \text{for } \mathbf{w} \in \operatorname{trop}(X^-). \end{array}$$

We now focus on α_1. Let $H = \operatorname{trop}(V(x_1 - \alpha_1)) = \{\mathbf{w} \in \mathbb{R}^n : w_1 = 1\}$. We claim that $\operatorname{trop}(X^+) = \operatorname{trop}(C) \cap H$. Indeed, for any $\mathbf{w} \in \operatorname{trop}(C) \cap H$ the cone of $\operatorname{trop}(C)$ containing $\mathbf{w}$ in its relative interior is $\operatorname{pos}(\mathbf{w})$, so $\operatorname{trop}(C)$ intersects H transversely at $\mathbf{w}$. Since $\mathbf{w}$ was an arbitrary intersection point, the claim follows from Theorem 3.4.12. We now decompose $I' + \langle x_1 - \alpha_1 \rangle = \bigcap_{\mathbf{y}} Q_{\mathbf{y}}$, where $Q_{\mathbf{y}}$ is $P_{\mathbf{y}}$ primary for $\mathbf{y} \in T^n_{K'}$. The $\mathbf{y}$ appearing here are precisely the points of X^+. Let $X^+_{\mathbf{w}} = \{\mathbf{y} \in X^+ : \operatorname{val}(\mathbf{y}) = \mathbf{w}\}$. Note that for $\mathbf{w} \in \operatorname{trop}(X^+)$, we have $\operatorname{in}_{\mathbf{w}}(\bigcap_{\mathbf{y} \in X^+} Q_{\mathbf{y}}) = \operatorname{in}_{\mathbf{w}}(\bigcap_{\mathbf{y} \in X^+_{\mathbf{w}}} Q_{\mathbf{y}})$. The inclusion $\subseteq$ is automatic. For the other inclusion, note that for all $\mathbf{y} \in X^+ \setminus X^+_{\mathbf{w}}$, there is $f_{\mathbf{y}} \in Q_{\mathbf{y}}$ with $\operatorname{in}_{\mathbf{w}}(f_{\mathbf{y}}) = 1$. For any $g \in \bigcap_{\mathbf{y} \in X^+_{\mathbf{w}}} Q_{\mathbf{y}}$, we set $g' = g \prod_{\mathbf{y} \in X^+ \setminus X^+_{\mathbf{w}}} f_{\mathbf{y}}$ to get $\operatorname{in}_{\mathbf{w}}(g) = \operatorname{in}_{\mathbf{w}}(g')$. Combined with the first of the above equations, this gives $\operatorname{in}_{\mathbf{w}}(\bigcap_{\mathbf{y} \in X^+_{\mathbf{w}}} Q_{\mathbf{y}}) = \operatorname{in}_{\mathbf{w}}(I') + \langle x_1 - \beta_1 \rangle$.

At this point, Proposition 3.4.8 implies that

$$\begin{aligned} \dim_{K'} S_{K'} / \left(\bigcap_{\mathbf{y} \in X^+_{\mathbf{w}}} Q_{\mathbf{y}} \right) &= \sum_{\mathbf{y} \in X^+_{\mathbf{w}}} \operatorname{mult}(Q_{\mathbf{y}}, P_{\mathbf{y}}) \\ &= \dim_{\Bbbk}(S_{\Bbbk}/(\operatorname{in}_{\mathbf{w}}(I') + \langle x_1 - \beta_1 \rangle)). \end{aligned}$$

By summing these identities over all $\mathbf{w} \in \operatorname{trop}(X^+)$, we find

$$L = \sum_{\mathbf{y} \in X^+} \operatorname{mult}(Q_{\mathbf{y}}, P_{\mathbf{y}}) = \sum_{\mathbf{w} \in \operatorname{trop}(X^+)} \dim_{\Bbbk}(S_{\Bbbk}/(\operatorname{in}_{\mathbf{w}}(I') + \langle x_1 - \beta_1 \rangle)).$$

The same identities hold for X^- and β_2.

Let $\mathbf{u_w}$ be the first lattice point on the ray $\mathrm{pos}(\mathbf{w})$ of $\mathrm{trop}(C)$. Then $\lambda = (\mathbf{u_w})_1$ satisfies $\mathbf{u_w} = \lambda\mathbf{w}$ because $w_1 = 1$. We now claim that

$$(3.4.4)\qquad \lambda \cdot \mathrm{mult}(\mathrm{pos}(\mathbf{w})) \;=\; \dim_{\mathbb{k}}(S_{\mathbb{k}}/(\mathrm{in}_{\mathbf{w}}(I') + \langle x_1 - \beta_1\rangle)).$$

That claim implies

$$L \;=\; \sum_{\mathbf{w}\in \mathrm{trop}(X^+)} \mathrm{mult}(\mathrm{pos}(\mathbf{w}))\cdot(\mathbf{u_w})_1 \;=\; \sum_{i:(\mathbf{u}_i)_1>0} m_i\cdot(\mathbf{u}_i)_1,$$

and similarly $L = \sum_{i:(\mathbf{u}_i)_1<0} -m_i(\mathbf{u}_i)_1$. This completes the proof as follows:

$$u_1 \;=\; \sum_{i:(\mathbf{u}_i)_1>0} m_i(\mathbf{u}_i)_1 \;-\; \sum_{i:(\mathbf{u}_i)_1<0} m_i|(\mathbf{u}_i)_1| \;=\; L-L \;=\; 0.$$

Thus it remains to prove (3.4.4). To this end, we perform a change of coordinates that takes x_1 to $x^{\mathbf{u_w}}$, and thus $\mathbf{w}$ to $\lambda^{-1}\mathbf{e}_1$. Now, our claim (3.4.4) states $\lambda\cdot\mathrm{mult}(\mathrm{pos}(\mathbf{w})) = \dim_{\mathbb{k}}(S_{\mathbb{k}}/(\mathrm{in}_{\lambda^{-1}\mathbf{e}_1}(I') + \langle x^{\mathbf{u_w}} - \beta_1\rangle))$. The initial ideal $\mathrm{in}_{\lambda^{-1}\mathbf{e}_1}(I')$ has a generating set that does not contain x_1. Since $V(I')$ is a curve, by Corollary 2.4.9, the initial ideal is one dimensional, so for each $2 \le i \le n$ it contains a polynomial in $\mathbb{k}[x_i]$ with constant term one. After dividing by x_i, we obtain $x_i^{-1} - p_i' \in \mathrm{in}_{\lambda^{-1}\mathbf{e}_1}(I')$ for some $p_i' \in \mathbb{k}[x_i]$. Now $\langle x^{\mathbf{u_w}} - \beta_1\rangle = \langle x_1^{\lambda} - \beta_1 x^{\mathbf{u}'}\rangle$, where $u_1' = 0$ and $u_i' = -(\mathbf{u_w})_i$ for $2 \le i \le n$. This implies $\mathrm{in}_{\lambda^{-1}\mathbf{e}_1}(I') + \langle x^{\mathbf{u_w}} - \beta_1\rangle = \mathrm{in}_{\lambda^{-1}\mathbf{e}_1}(I') + \langle x_1^{\lambda} - f\rangle$ for some $f \in \mathbb{k}[x_2,\ldots,x_n]$. We next use the fact that $\dim_{\mathbb{k}} \mathbb{k}[x_1^{\pm1},\ldots,x_n^{\pm1}]/J = \dim_{\mathbb{k}} \mathbb{k}[x_1,\ldots,x_n]/J_{\mathrm{aff}}$ for any zero-dimensional Laurent ideal J. Fix the lexicographic term order $x_1 \succ x_2 \succ \cdots \succ x_n$ on $\mathbb{k}[x_1,\ldots,x_n]$. By Buchberger's criterion, the initial ideal of $(\mathrm{in}_{\lambda^{-1}\mathbf{e}_1}(I') + \langle x_1^{\lambda} - f\rangle)_{\mathrm{aff}}$ is generated by x_1^{λ} and the monomial generators of $\mathrm{in}_{\mathrm{lex}}((\mathrm{in}_{\lambda^{-1}\mathbf{e}_1}(I'))_{\mathrm{aff}})$. The right-hand side of (3.4.4) is λ times the $\mathbb{k}$-dimension of $\mathbb{k}[x_2^{\pm1},\ldots,x_n^{\pm1}]/\mathrm{in}_{\lambda^{-1}\mathbf{e}_1}(I')$. But, that last $\mathbb{k}$-dimension equals the multiplicity of $\mathrm{pos}(\mathbf{w})$ by Lemma 3.4.7. □

Last but not least, here is the main theorem in this section.

Theorem 3.4.14. *Let I be an ideal in $K[x_1^{\pm1},\ldots,x_n^{\pm1}]$ such that all irreducible components of $V(I)$ have the same dimension d. Fix a polyhedral complex Σ with support* $\mathrm{trop}(V(I))$ *such that* $\mathrm{in}_{\mathbf{w}}(I)$ *is constant for* $\mathbf{w}$ *in the relative interior of each cell in Σ. Then Σ is a weighted balanced polyhedral complex with the weight function* mult *of Definition* 3.4.3.

Proof. Write $\sqrt{I} = \bigcap P_i$ where each P_i is a d-dimensional prime ideal. By Theorem 3.2.3, the tropical variety $\mathrm{trop}(V(I))$ is the union $\bigcup \mathrm{trop}(V(P_i))$. By Theorem 3.3.8, $\mathrm{trop}(V(I))$ a pure d-dimensional polyhedral complex.

Fix a $(d-1)$-dimensional cell $\tau \in \Sigma$. Lemma 2.2.7 and Corollary 2.6.12 guarantee that, after a multiplicative change of coordinates, the affine span of τ is a translate of the span of $\mathbf{e}_1,\ldots,\mathbf{e}_{d-1}$. Fix $\mathbf{w} \in \mathrm{relint}(\tau)$. Part (2) of

Lemma 2.6.2 implies that $\mathrm{in}_{\mathbf{w}}(I)$ is homogeneous with respect to the $\mathbb{Z}^{d-1}$-grading given by $\deg(x_i) = \mathbf{e}_i$ for $1 \leq i \leq d-1$, and $\deg(x_i) = \mathbf{0}$ for $i \geq d$. This means that $\mathrm{in}_{\mathbf{w}}(I)$ has a generating set in which $x_1, \ldots, x_{d-1}$ do not appear.

Let $J = \mathrm{in}_{\mathbf{w}}(I) \cap \mathbb{k}[x_d^{\pm 1}, \ldots, x_n^{\pm 1}]$. By Lemma 3.3.6 the tropical variety of $V(\mathrm{in}_{\mathbf{w}}(I)) \subset T_{\mathbb{k}}^n$ is the star of τ in Σ, which has lineality space spanned by $\mathbf{e}_1, \ldots, \mathbf{e}_{d-1}$. Since $\mathrm{in}_{\mathbf{v}}(\mathrm{in}_{\mathbf{w}}(I)) \cap \mathbb{k}[x_d^{\pm 1}, \ldots, x_n^{\pm 1}] = \mathrm{in}_{\overline{\mathbf{v}}}(J)$, where $\overline{\mathbf{v}}$ is the projection of $\mathbf{v}$ onto the last $n-d+1$ coordinates, the fact that $\mathrm{trop}(V(I))$ is pure of dimension d implies that $\mathrm{trop}(V(J))$ is one dimensional.

Let $P_1, \ldots, P_r$ be the minimal associated primes of J. Then $V(J) = \bigcup_{i=1}^r V(P_i)$, so

$$\begin{aligned}\mathrm{trop}(V(J)) &= \mathrm{cl}(\mathrm{val}(\mathbf{y}) : \mathbf{y} \in V(J)) \\ &= \bigcup_{i=1}^{r} \mathrm{cl}(\mathrm{val}(\mathbf{y}) : \mathbf{y} \in V(P_i)) = \bigcup_{i=1}^{r} \mathrm{trop}(V(P_i)).\end{aligned}$$

By Theorem 3.3.8 we thus have $\dim(P_i) \leq 1$ and at least one index i satisfies $\dim(P_i) = 1$. Thus $\dim(V(J)) = 1$.

Suppose $\mathbf{v} \in \mathbb{Q}^n$ satisfies $\mathbf{w} + \epsilon \mathbf{v} \in \sigma$ for all sufficiently small $\epsilon > 0$, where σ is a maximal cell of Σ that has τ as a facet. The equality $\mathrm{in}_{\mathbf{v}}(\mathrm{in}_{\mathbf{w}}(I)) = \mathrm{in}_{\overline{\mathbf{v}}}(J)$ and Lemma 3.4.7 imply that the multiplicity of the cone $\mathrm{pos}(\overline{\mathbf{v}})$ in $\mathrm{trop}(V(J))$ equals the multiplicity of σ in $\mathrm{trop}(X)$. Thus, showing that Σ is balanced at τ is exactly the same as showing that $\mathrm{trop}(V(J))$ is balanced at $\mathbf{0}$. Thus proving the theorem for J suffices, so we may assume that X is a curve in $(\mathbb{k}^*)^n$. This is Proposition 3.4.13, so the result follows. □

Remark 3.4.15. In the statement of Theorem 3.4.14 we do not assume that I is radical. We have $\mathrm{trop}(V(I)) = \mathrm{trop}(V(\sqrt{I}))$, but the multiplicities might differ. If I is not radical, then the tropical variety together with its multiplicities records information about the affine scheme $X = \mathrm{Spec}(S/I)$.

3.5. Connectivity and Fans

The polyhedral complex underlying a tropical variety has a strong connectedness property, introduced in Definition 3.3.4.

Theorem 3.5.1. *Let X be an irreducible subvariety of T^n of dimension d. Then* $\mathrm{trop}(X)$ *is the support of a pure d-dimensional polyhedral complex that is connected through codimension* 1.

This result is important for the algorithmic computation of tropical varieties. Given a variety $X \subset T^n$, we can define a graph whose nodes are the d-dimensional cells of $\mathrm{trop}(X)$ and where two nodes are connected by an edge if the corresponding cells share a facet. Theorem 3.5.1 states that this graph is connected. To compute $\mathrm{trop}(X)$, one can start with one node and identify all neighbors using the initial ideal techniques of Section 2.5. This method is described in [**BJS⁺07**] and implemented in `Gfan` [**Jen**].

The proof of Theorem 3.5.1 is by induction on dimension d of X. The base case $d = 1$ is surprisingly nontrivial. It is proved in Proposition 6.6.22.

Proposition 3.5.2. *Let X be a one-dimensional irreducible subvariety of the torus T^n. Then* $\mathrm{trop}(X)$ *is connected.*

Let Δ be the standard tropical hyperplane $\mathrm{trop}(V(x_1+\cdots+x_n+1)) \subset \mathbb{R}^n$. The tropicalization of any hyperplane $H_{\mathbf{a}} = \{\mathbf{x} : a_1x_1 + \cdots + a_nx_n + a_0 = 0\} \subset T^n$ with all $a_i \neq 0$ equals $-\mathbf{v} + \Delta$, where $v_i = \mathrm{val}(a_i) - \mathrm{val}(a_0)$.

Proof of Theorem 3.5.1. Theorem 3.3.8 states that $\mathrm{trop}(X)$ is the support of a pure d-dimensional polyhedral complex Σ. We need to show that Σ is connected through codimension 1. The proof is by induction on $d = \dim(X)$. The base case $d = 1$ is Proposition 3.5.2. Indeed, a one-dimensional polyhedral complex Σ is a graph, and a graph is connected if and only if it is connected through codimension 1. Next, suppose that $d = \dim(X)$ satisfies $2 \leq d < n$, and that the theorem is true for all smaller dimensions. After a multiplicative change of coordinates in T^n, we may also assume that no facet σ of Σ lies in a tropical hyperplane $-\mathbf{v} + \Delta$.

Fix facets $\sigma, \sigma' \in \Sigma$. Pick $\mathbf{w} \in \mathrm{relint}(\sigma) \cap \Gamma_{\mathrm{val}}^n$ and $\mathbf{w}' \in \mathrm{relint}\, \sigma' \cap \Gamma_{\mathrm{val}}^n$. Choose $\mathbf{v} \in \Gamma_{\mathrm{val}}^n$ for which $-\mathbf{v} + \Delta$ contains $\mathbf{w}, \mathbf{w}'$. To see that this is possible, note that if $\mathbf{y}, \mathbf{y}' \in T^n$ with $\mathrm{val}(\mathbf{y}) = \mathbf{w}$, $\mathrm{val}(\mathbf{y}') = \mathbf{w}'$ and $H_{\mathbf{a}} = V(a_1x_1 + \cdots + a_nx_n + a_0)$ is any hyperplane passing through both $\mathbf{y}$ and $\mathbf{y}'$, then $\mathrm{trop}(H_{\mathbf{a}}) = -\mathbf{v} + \Delta$ is a tropical hyperplane passing through $\mathbf{w}$ and $\mathbf{w}'$. Since $\mathbf{w}, \mathbf{w}'$ lie in the relative interior of d-dimensional cells in Σ and these cells are not contained in $-\mathbf{v} + \Delta$, by replacing $\mathbf{w}, \mathbf{w}'$ with other points in the relative interior of their respective cells if necessary, we may assume that $\mathbf{w}$ and $\mathbf{w}'$ lie in top-dimensional cells of $-\mathbf{v} + \Delta$.

By Part 4 of Theorem 6.3 of [**Jou83**], the set U of $\mathbf{a} \in \mathbb{P}^n$ for which the intersection $X \cap H_{\mathbf{a}}$ is irreducible is Zariski open in $\mathbb{P}^n$. We write $U = \mathbb{P}^n \backslash V(f_1, \ldots, f_r)$ as the complement of a subvariety. By Lemma 2.2.12, there is $\mathbf{a} = (1 : a_1 : \cdots : a_n) \in U$ with $\mathrm{val}(a_i) = v_i$ for $i = 1, \ldots, n$.

The intersection $\mathrm{trop}(X) \cap \mathrm{trop}(H_{\mathbf{a}})$ inherits a polyhedral complex structure $\overline{\Sigma}$ from Σ and $-\mathbf{v} + \Delta$. Fix $\tilde{\mathbf{w}} \in \mathrm{trop}(X) \cap \mathrm{trop}(H_{\mathbf{a}})$ for which $\tilde{\mathbf{w}}$ lies in the relative interior of a top-dimensional cell σ of Σ and σ' of $-\mathbf{v} + \Delta$. By our assumption on $\mathrm{trop}(X)$, the cell σ does not lie in $-\mathbf{v} + \Delta$, so $\mathrm{trop}(X)$ and $\mathrm{trop}(H_{\mathbf{a}})$ intersect transversely at $\tilde{\mathbf{w}}$. Theorem 3.4.12 implies that $\mathrm{trop}(X) \cap \mathrm{trop}(H_{\mathbf{a}}) = \mathrm{trop}(X \cap H_{\mathbf{a}})$. By construction, $Y = X \cap H_{\mathbf{a}}$ is irreducible, so $\mathrm{trop}(Y)$ is connected through codimension 1 by induction.

If $\overline{\sigma}$ is a $(d-1)$-dimensional cell in $\overline{\Sigma}$, then $\overline{\sigma}$ is the intersection of $-\mathbf{v}+\Delta$ with a d-dimensional cell σ in Σ, by our assumption on $\operatorname{trop}(X)$. If $\overline{\sigma}$ and $\overline{\sigma}'$ are adjacent top-dimensional cells in $\overline{\Sigma}$, then either $\sigma = \sigma'$ or σ and σ' are adjacent in Σ. By construction, $\mathbf{w}$ and $\mathbf{w}'$ lie in the relative interiors of top-dimensional cells $\overline{\sigma}$ and $\overline{\sigma}'$ in $\overline{\Sigma}$, so there is a path $\overline{\sigma} = \overline{\sigma}_1, \overline{\sigma}_2, \ldots, \overline{\sigma}_r = \overline{\sigma}'$ in $\overline{\Sigma}$ connecting $\mathbf{w}$ to $\mathbf{w}'$. Lifting these and removing adjacent duplicates, we find that $\sigma_1, \ldots, \sigma_r$ is a path of adjacent top-dimensional cells in Σ connecting $\mathbf{w}$ to $\mathbf{w}'$. We conclude that $\operatorname{trop}(X)$ is connected through codimension 1. □

Remark 3.5.3. Theorem 3.5.1 is stronger than it may seem at first glance, as the property of being connected through codimension 1 in fact only depends on the underlying set. *Every* polyhedral complex Σ with support $|\Sigma| = \operatorname{trop}(X)$ is connected through codimension 1. To see this, it suffices to note that a polyhedral complex is connected through codimension 1 if and only if a refinement of it is connected through codimension 1. For the "if" direction, note that a path of adjacent top-dimensional cells in the refinement lifts to a path of adjacent or identical top-dimensional cells in the original complex. For the "only if" direction, it suffices to note that any subdivision of a cell is connected through codimension 1. Now let Σ' be an arbitrary polyhedral complex with support $\operatorname{trop}(X)$, and let Σ be a connected-through-codimension-1 polyhedral complex with support $\operatorname{trop}(X)$ whose existence is guaranteed by Theorem 3.5.1. Then the common refinement of Σ and Σ' is connected through codimension 1 since Σ is, and so Σ' is also connected through codimension 1.

When first entering the field of tropical geometry, a student might get the impression that every tropical variety $\operatorname{trop}(X)$ is the support of a *unique coarsest* polyhedral complex Σ. This would mean that Σ' refines Σ for any balanced complex Σ' with $|\Sigma'| = \operatorname{trop}(X)$. For instance, such a coarsest Σ exists when X is a hypersurface and also when $\dim(X) \leq 2$. However, it does not exist in general. The following is an explicit counterexample.

Example 3.5.4. Fix $K = \mathbb{C}$ with the trivial valuation. We present a three-dimensional variety $X \subset T^5$ for which there is no coarsest fan Σ in $\mathbb{R}^5$ with $|\Sigma| = \operatorname{trop}(X)$. Consider the torus T^3 with coordinates (x, y, z) and define X to be the closure of the image of the rational map

$$T^3 \dashrightarrow T^5,\ (x, y, z) \mapsto \big(x(1-x), x(1-y), x(1-z), y(1-z), z(1-z)\big).$$

The tropicalization of X is a three-dimensional fan in $\mathbb{R}^5$. It is constructed geometrically as follows. Start with three copies of the standard tropical line in $\mathbb{R}^2$. Consider their direct product. This is a three-dimensional fan in $\mathbb{R}^6 = \mathbb{R}^2 \times \mathbb{R}^2 \times \mathbb{R}^2$. It has nine rays and 27 maximal cones. Then $\operatorname{trop}(X)$

is the image of this fan under the classical linear map given by the matrix

$$A = (\mathbf{a}_1, \mathbf{a}_2, \ldots, \mathbf{a}_6) = \begin{pmatrix} 1 & 1 & 0 & 0 & 0 & 0 \\ 1 & 0 & 0 & 1 & 0 & 0 \\ 1 & 0 & 0 & 0 & 0 & 1 \\ 0 & 0 & 1 & 0 & 0 & 1 \\ 0 & 0 & 0 & 0 & 1 & 1 \end{pmatrix}.$$

The following two three-dimensional simplicial cones lie in trop(X):

$$\text{pos}\{\mathbf{a}_1, \mathbf{a}_3, \mathbf{a}_5\} \quad \text{and} \quad \text{pos}\{\mathbf{a}_2, \mathbf{a}_4, \mathbf{a}_6\}.$$

These two cones intersect in one ray. That ray is spanned by $\mathbf{1}=(1,1,1,1,1)^T$, and it lies in the relative interior of each of the two cones.

To see that there is no coarsest fan structure on trop(X), we note that trop(X) is the support of the cone over a two-dimensional polyhedral complex Π. That complex contains two triangles which meet in one point in their relative interiors. Any coarsest polyhedral subdivision of $|\Pi|$ would use that point as a 0-cell. Each triangle must be divided into three polygons that are either triangles or quadrilaterals. These coarsest subdivisions of a triangle are not unique. Hence no unique coarsest fan structure exists on trop(X). ◊

Our second topic in this section is the role of fans in tropical geometry. In Proposition 3.1.10 we saw that the tropicalization of a constant-coefficient hypersurface is a pure fan of codimension 1. We begin by generalizing this result to constant-coefficient varieties of arbitrary codimension.

Corollary 3.5.5. *Let $X \subset T^n$ be an irreducible d-dimensional variety where K is a field with the trivial valuation. Then the tropical variety* trop(X) *is the support of a balanced polyhedral fan of dimension d.*

Proof. We can choose the Gröbner fan structure given by Corollary 2.5.12. The statements about dimension and balancing follow from the Structure Theorem 3.3.5. Alternatively, we can choose a finite tropical basis $\mathcal{T}$ consisting of Laurent polynomials f whose coefficients have valuation zero. For each $f \in \mathcal{T}$, the tropical hypersurface trop($V(f)$) is the support of a fan, by Proposition 3.1.10. By taking the common refinement of these fans, we obtain a fan structure on the intersection $\text{trop}(X) = \bigcap_{f \in \mathcal{T}} \text{trop}(V(f))$. □

On the other hand, suppose $X \subset T^n$ is a d-dimensional variety over a field whose valuation is nontrivial. Then trop(X) is a polyhedral complex in $\mathbb{R}^n$ but usually not a fan. However, there are three different ways of associating fans to this complex. We summarize these below.

(1) By Lemma 3.3.6, $\mathrm{star}_{\mathrm{trop}(X)}(\sigma) = \mathrm{trop}(\mathrm{in}_{\mathbf{w}}(I))$ supports a fan for any cell σ of $\mathrm{trop}(X)$. Its dimension modulo the lineality space is $d - |\sigma|$. Every vertex of $\mathrm{trop}(X)$ determines a fan of dimension d.
(2) If $K = \Bbbk(t)$ and $X \subset T^n_K$, then we can construct the lift $X_t \subset T^{n+1}_{\Bbbk}$ by regarding t as a variable. The tropical variety $\mathrm{trop}(X_t)$ is a fan of dimension $d+1$ in $\mathbb{R}^{n+1}$ whose intersection with the affine hyperplane $w_t = 1$ is the tropical variety $\mathrm{trop}(X)$.
(3) By Theorem 3.5.6 below, the *recession fan* of $\mathrm{trop}(X)$ is the tropicalization of the same variety X, but with trivial valuation on K.

We need to explain what is meant by the recession fan. Fix a polyhedron

$$P = \{\mathbf{x} \in \mathbb{R}^n : A\mathbf{x} \leq \mathbf{b}\}.$$

The *recession cone* of P is

$$\mathrm{rec}(P) = \{\mathbf{x} \in \mathbb{R}^n : A\mathbf{x} \leq \mathbf{0}\}. \tag{3.5.1}$$

This is the unique cone satisfying $P = \mathrm{rec}(P) + Q$ for some polytope Q. In this decomposition, the polytope Q is not unique, but the recession cone is. Furthermore, $\mathrm{rec}(P)$ is the cone dual to the support of the normal fan $\mathcal{N}_P$.

If Σ is a polyhedral complex in $\mathbb{R}^n$, then its *recession fan* $\mathrm{rec}(\Sigma)$ is the union of all cones $\mathrm{rec}(P)$ where P runs over Σ. The set $\mathrm{rec}(\Sigma)$ is the support of a polyhedral fan. Burgos, Gil, and Sombra [**BGS11**] identify situations when the fan structure on Σ is not canonical. In particular, there is generally no unique coarsest fan structure on Σ. Our usage of the term "recession fan" simply means that such a fan structure exists, but it does not refer to any specific fan. With this understanding, the recession fan depends only on the support Σ, and we can write $\mathrm{rec}(|\Sigma|) = \mathrm{rec}(\Sigma)$.

Every field can be given a trivial valuation, for which $\mathrm{val}(a) = 0$ if $a \neq 0$. For $X \subset T^n$, we write $\mathrm{trop}(X_{\mathrm{triv}})$ for the tropicalization of X with respect to the trivial valuation. This may be different from the tropicalization $\mathrm{trop}(X)$ of X with respect to the original valuation on K, but there is a connection between the two.

Theorem 3.5.6. *Let X be a subvariety of T^n. Then the tropical variety* $\mathrm{trop}(X_{\mathrm{triv}})$ *is the recession fan of* $\mathrm{trop}(X)$:

$$\mathrm{trop}(X_{\mathrm{triv}}) = \mathrm{rec}(\mathrm{trop}(X)). \tag{3.5.2}$$

Proof. First suppose that $X = V(f)$ is a hypersurface with defining polynomial $f \in K[x_1^{\pm}, \ldots, x_n^{\pm 1}]$. Then $\mathrm{trop}(X)$ is the $(n-1)$-skeleton of the complex $\Sigma_{\mathrm{trop}(f)}$. Each unbounded cell in $\mathrm{trop}(X)$ corresponds to a face F of dimension ≥ 1 of the Newton polytope $\mathrm{Newt}(f)$, and its recession cone is the normal cone $\mathcal{N}(F)$. Moreover, every positive-dimensional face F occurs. Hence the right-hand side of (3.5.2) is the $(n-1)$-skeleton of the normal fan of Newt(f). By Proposition 3.1.10, this is also the left-hand side of (3.5.2).

For the general case, we use the identity

$$\operatorname{rec}(P \cap P') \;=\; \operatorname{rec}(P) \cap \operatorname{rec}(P'),$$

which holds for the recession cones of any two polyhedra P and P' in $\mathbb{R}^n$. As this extends to finite intersections of polyhedra in $\mathbb{R}^n$, we derive

$$\begin{aligned}\operatorname{trop}(X_{\text{triv}}) = \bigcap_{f \in \mathcal{T}} \operatorname{trop}(V(f)_{\text{triv}}) \;&=\; \bigcap_{f \in \mathcal{T}} \operatorname{rec}(\operatorname{trop}(V(f)))\\ = \operatorname{rec}\Big(\bigcap_{f \in \mathcal{T}} \operatorname{trop}(V(f))\Big) \;&=\; \operatorname{rec}\big(\operatorname{trop}(X)\big).\end{aligned}$$

Here, as before, the set $\mathcal{T}$ is a finite tropical basis for the variety X. □

3.6. Stable Intersection

In this section we introduce the notion of *stable intersection*. This was discussed for plane curves in Section 1.3. In general, it gives a combinatorial way to intersect any pair of weighted balanced polyhedral complexes so that the result is again a weighted balanced polyhedral complex. If the complexes are the tropicalizations of classical varieties, then the stable intersection represents their intersection after a generic multiplicative perturbation:

Theorem 3.6.1. *Let X_1, X_2 be subvarieties of the torus T^n, and let Σ_1, Σ_2 be weighted balanced Γ_{val}-rational polyhedral complexes whose supports are* trop(X_1) *and* trop(X_2)*, respectively. There exists a Zariski dense subset $U \subset T^n$, consisting of elements $\mathbf{t} = (t_1, \ldots, t_n)$ with* val$(\mathbf{t}) = \mathbf{0}$*, such that*

$$\operatorname{trop}(X_1 \cap \mathbf{t}X_2) \;=\; \Sigma_1 \cap_{st} \Sigma_2 \qquad \textit{for all } \mathbf{t} \in U. \tag{3.6.1}$$

Here $\Sigma_1 \cap_{st} \Sigma_2$ is the stable intersection of balanced polyhedral complexes. This purely combinatorial notion will be introduced in Definition 3.6.5. The set $\mathbf{t}X_2$ on the left-hand side is the translated variety $\{\mathbf{tx} : \mathbf{x} \in X_2\}$. A proof of Theorem 3.6.1 will be presented at the end of the section.

We begin by developing the formal theory of stable intersections. This requires some preliminary material. Let N denote the lattice $\mathbb{Z}^n$ of $\mathbb{R}^n$, and let N_σ be the sublattice of N generated by the lattice points in the linear space parallel to a Γ_{val}-rational polyhedron σ. The *index* $[N : N']$ of a sublattice $N' \subset N$ of the same rank is the order of the quotient group N/N'. A refinement Σ' of a weighted polyhedral complex Σ inherits a weighting from the complex Σ: if σ' is a maximal dimensional cell in Σ' with $\sigma' \subseteq \sigma$ for $\sigma \in \Sigma$, then we assign to σ' the weight of σ. We first note that if the complex Σ is balanced, then so is the refinement Σ'.

Lemma 3.6.2. *Let Σ be a pure weighted balanced Γ_{val}-rational polyhedral complex in $\mathbb{R}^n$, and let Σ' be a Γ_{val}-rational refinement of Σ. Then Σ' is balanced.*

Proof. For a codimension-1 cell τ' in Σ', let τ be the smallest cell in Σ containing τ'. If τ has codimension 1 in Σ, then balancing at τ' follows immediately from the balancing condition on Σ, since $\mathrm{star}_{\Sigma'}(\tau') = \mathrm{star}_{\Sigma}(\tau)$. If τ is top dimensional, then $\mathrm{star}_{\Sigma'}(\tau')$ has two cones that meet along the affine span of τ'. The generators $\mathbf{v}_1$ and $\mathbf{v}_2$ for the lattices of these two cones, modulo the lattice of τ', satisfy the balancing condition $\mathbf{v}_1 = -\mathbf{v}_2$. Since both cones come from the same cone τ of Σ, they have the same multiplicity m. Therefore, the weighted sum $m\mathbf{v}_1 + m\mathbf{v}_2$ equals zero, as required. □

We now consider how polyhedral complexes behave under projections. Let Σ be a pure weighted Γ_{val}-rational polyhedral complex in $\mathbb{R}^n$, and let $\phi : N \to N' \cong \mathbb{Z}^m$ be a homomorphism of lattices. We suppose that ϕ is given by an $m \times n$ integer matrix A. After refining Σ, we may assume that the projected polyhedra $\{\phi(\sigma) : \sigma \in \Sigma\}$ again form a polyhedral complex.

This image complex need not be pure. For example, consider a fan whose support is the union of the two planes $x_1 = x_2 = 0$ and $x_3 = x_4 = 0$ in $\mathbb{R}^4$. Let ϕ be the projection onto the first three coordinates. The image of the fan is the union of the plane $x_3 = 0$ and the line $x_1 = x_2 = 0$ in $\mathbb{R}^3$.

Let Σ' be the subcomplex of the image containing the projected polyhedra of maximum dimension and all their faces. Then Σ' inherits a weighting from Σ. Namely, we assign to a maximal cell σ' of Σ' the multiplicity

$$\mathrm{mult}(\sigma') \quad = \sum_{\sigma \in \Sigma, \phi(\sigma) = \sigma'} \mathrm{mult}(\sigma) \cdot [N'_{\sigma'} : \phi(N_\sigma)]. \tag{3.6.2}$$

If V is a matrix whose columns form a basis for N_σ, then the columns of the matrix AV form a basis for $\phi(N_\sigma)$. The lattice index $[N'_{\sigma'} : \phi(N_\sigma)]$ is the greatest common divisor of the maximal minors of the matrix AV. The sum of two sublattices of N is the smallest sublattice containing both.

Lemma 3.6.3. *If Σ is balanced, then the projection Σ' of Σ is balanced.*

Proof. Let τ' be a codimension-1 cell in Σ', and let $\tau_1, \ldots, \tau_r$ be the codimension-1 cells in Σ with $\phi(\tau_i) = \tau'$. There may be cells in Σ of both larger and smaller dimension that map to τ', but we consider only those of codimension 1. For each τ_i let $\sigma_{i1}, \ldots, \sigma_{il}$ be the maximal cells of Σ containing τ_i. For each i, j the quotient $N_{\sigma_{ij}}/N_{\tau_i}$ is isomorphic to $\mathbb{Z}$. Let $\mathbf{v}_{ij} \in N$ restrict to the generator for $N_{\sigma_{ij}}/N_{\tau_i}$ pointing in the direction of σ_{ij}. The balancing condition for Σ ensures that $\sum_j \mathrm{mult}(\sigma_{ij})\mathbf{v}_{ij}$ lies in N_{τ_i}.

Let $\sigma'_1, \ldots, \sigma'_s$ be the top-dimensional cells of Σ' containing τ'. Fix a vector $\mathbf{v}^k \in N'_{\sigma'_k} \subseteq N'$ whose image generates $N'_{\sigma'_k}/N'_{\tau'}$ and points in the direction of σ'_k. For each σ_{ij}, we have either $\phi(\sigma_{ij}) = \tau'$ or $\phi(\sigma_{ij}) = \sigma'_k$ for some $k = k(ij)$. In the former case we set $\mathbf{v}^{k(ij)} = \mathbf{0}$. In the latter case the projection of $\mathbf{v}_{ij}$ in $N'_{\sigma'_k}/N'_{\tau'}$ is a multiple of the corresponding $\mathbf{v}^{k(ij)}$ by the factor $[N'_{\sigma'_k} : N'_{\tau'} + \mathrm{span}_{\mathbb{Z}}(\phi(\mathbf{v}_{ij}))]$. Since

$$\begin{aligned}
[N'_{\sigma'_k} : \phi(N_{\sigma_{ij}})] &= [N'_{\sigma'_k} : \phi(N_\tau + \mathrm{span}_{\mathbb{Z}}(\mathbf{v}_{ij}))] \\
&= [N'_{\sigma'_k} : \phi(N_\tau) + \mathrm{span}_{\mathbb{Z}}(\phi(\mathbf{v}_{ij}))] \\
&= [N'_{\sigma'_k} : N'_{\tau'} + \mathrm{span}_{\mathbb{Z}}(\phi(\mathbf{v}_{ij}))][N'_{\tau'} : \phi(N_{\tau_i})],
\end{aligned}$$

this factor is $[N'_{\sigma'_{k(ij)}} : \phi(N_{\sigma_{ij}})]/[N'_{\tau'} : \phi(N_{\tau_i})]$.

Thus for each fixed index $i \in \{1, \ldots, r\}$ we have

$$\begin{aligned}
&\sum_j \mathrm{mult}(\sigma_{ij}) \cdot [N'_{\sigma'_{k(ij)}} : \phi(N_{\sigma_{ij}})] \cdot \mathbf{v}^{k(ij)} \\
&\quad = [N'_{\tau'} : \phi(N_{\tau_i})] \cdot \Big(\sum_j \mathrm{mult}(\sigma_{ij})[N'_{\sigma'} \colon N'_{\tau'} + \mathrm{span}_{\mathbb{Z}}(\phi(\mathbf{v}_{ij}))]\mathbf{v}^{k(ij)}\Big) \\
&\quad = [N'_{\tau'} : \phi(N_{\tau_i})] \cdot \phi\Big(\sum_j \mathrm{mult}(\sigma_{ij})\mathbf{v}_{ij}\Big) \\
&\quad = \mathbf{0} \ \in \ N'/N'_{\tau'}.
\end{aligned}$$

Summing this expression over all choices of τ'_i, we find

$$\sum_{ij} \mathrm{mult}(\sigma_{ij}) \cdot [N'_{\sigma'_{k(ij)}} : \phi(N_{\sigma_{ij}})] \cdot \mathbf{v}^{k(ij)} \quad = \quad \mathbf{0} \ \in \ N'/N'_{\tau'}.$$

The coefficient of $\mathbf{v}^k$ is $\sum_{i,j:k(ij)=k} \mathrm{mult}(\sigma_{ij})[N'_{\sigma'_k} : \phi(N'_{\sigma_{ij}})]$, which is the multiplicity we assigned to σ'_k in (3.6.2). This shows that Σ' is balanced. □

A key consequence of the balancing condition is the following technical lemma. This will be used to show that our definition of stable intersection is well defined. For two polyhedra σ_1, σ_2, recall from (2.3.1) that the Minkowski sum $\sigma_1 + \sigma_2$ is the polyhedron $\{\mathbf{a} + \mathbf{b} : \mathbf{a} \in \sigma_1, \mathbf{b} \in \sigma_2\}$.

Lemma 3.6.4. *Let Σ_1 and Σ_2 be pure weighted balanced Γ_{val}-rational polyhedral complexes in $\mathbb{R}^n$. Let $\sigma_1 \in \Sigma_1$, $\sigma_2 \in \Sigma_2$ be top-dimensional cells with $\dim(\sigma_1 + \sigma_2) = n$ and $\dim(\sigma_1 \cap \sigma_2) = \dim(\sigma_1) + \dim(\sigma_2) - n$. Choose refinements of Σ_1 and Σ_2 so that $\sigma_1 \cap \sigma_2$ is a cell in both complexes. For $\mathbf{v} \in \mathbb{R}^n$, consider the following sum over all maximal cones $\tau_1 \in \mathrm{star}_{\Sigma_1}(\sigma_1 \cap \sigma_2)$ and $\tau_2 \in \mathrm{star}_{\Sigma_2}(\sigma_1 \cap \sigma_2)$ with $\dim(\tau_1 + \tau_2) = n$ and $\tau_1 \cap (\mathbf{v} + \tau_2) \neq \emptyset$:*

$$\sum_{\tau_1, \tau_2} \mathrm{mult}(\tau_1)\,\mathrm{mult}(\tau_2)[N : N_{\tau_1} + N_{\tau_2}]. \tag{3.6.3}$$

This sum is constant for all vectors $\mathbf{v}$ in a dense open subset of $\mathbb{R}^n$.

Proof. Let $\tilde{\Sigma}_i = \mathrm{star}_{\Sigma_i}(\sigma_1 \cap \sigma_2)$ for $i = 1, 2$. Consider the product fan $\tilde{\Sigma}_1 \times \tilde{\Sigma}_2 \subseteq \mathbb{R}^{2n}$. This has cones $\tau_1 \times \tau_2$ for $\tau_1 \in \mathrm{star}_{\Sigma_1}(\sigma_1 \cap \sigma_2)$ and $\tau_2 \in \mathrm{star}_{\Sigma_2}(\sigma_1 \cap \sigma_2)$. This fan is balanced with the weight on $\tau_1 \times \tau_2$ given by $\mathrm{mult}(\tau_1)\,\mathrm{mult}(\tau_2)$. Balancing holds because each codimension-1 cone in this fan is the product of a maximal cone of one of the factors with a codimension-1 cone of the other. The balancing equation for this cone comes from the second factor.

Consider the projection $\pi : \mathbb{R}^{2n} \to \mathbb{R}^n$ given by $\pi(x, y) = x - y$. After refining $\widetilde{\Sigma}_1 \times \widetilde{\Sigma}_2$, this induces a map of fans: for each pair (τ_1, τ_2) with $\tau_i \in \Sigma_i$ for $i = 1, 2$, the Minkowski sum $\tau_1 + (-\tau_2)$ is a union of cones in the image. The condition $\dim(\sigma_1 + \sigma_2) = n$ means that the cones $\bar{\sigma}_1, \bar{\sigma}_2$ of the two stars corresponding to σ_1 and σ_2 satisfy $\dim(\bar{\sigma}_1 + (-\bar{\sigma}_2)) = n$. Let τ be a cone of the image fan. Each cone $\tau_1 \times \tau_2$ of the product fan that projects to τ contributes $\mathrm{mult}(\tau_1)\,\mathrm{mult}(\tau_2)[N : N_{\tau_1} + N_{\tau_2}]$ to the multiplicity of τ. The final multiplicity is obtained by adding up all these contributions. This image fan is balanced by Lemma 3.6.3.

Let V be the interior of a top-dimensional cone of the image fan. Now, $\mathbf{v}$ lies in the projection of a top-dimensional cone $\tau_1 \times \tau_2$ of $\widetilde{\Sigma}_1 \times \widetilde{\Sigma}_2$ if and only if $\mathbf{v} \in \tau_1 - \tau_2$, which occurs if and only if $\tau_1 \cap (\mathbf{v} + \tau_2) \neq \emptyset$. Thus, for $\mathbf{v} \in V$, the sum (3.6.3) is the multiplicity of the top-dimensional cone of the image fan that contains $\mathbf{v}$. Since that image is an n-dimensional balanced fan in $\mathbb{R}^n$, the multiplicity does not depend on the choice of cone (see Exercise 3.7(24)). The sum thus does not depend on the choice of $\mathbf{v}$, as long as $\mathbf{v}$ lies in the interior of a top-dimensional cone of the image fan. □

Definition 3.6.5. Let Σ_1 and Σ_2 be pure weighted balanced polyhedral complexes in $\mathbb{R}^n$. The *stable intersection* $\Sigma_1 \cap_{st} \Sigma_2$ is the polyhedral complex

$$\Sigma_1 \cap_{st} \Sigma_2 \quad = \bigcup_{\substack{\sigma_1 \in \Sigma_1, \sigma_2 \in \Sigma_2 \\ \dim(\sigma_1 + \sigma_2) = n}} \sigma_1 \cap \sigma_2. \tag{3.6.4}$$

The multiplicity of a top-dimensional cell $\sigma_1 \cap \sigma_2$ in $\Sigma_1 \cap_{st} \Sigma_2$ is

$$\mathrm{mult}_{\Sigma_1 \cap_{st} \Sigma_2}(\sigma_1 \cap \sigma_2) \;=\; \sum_{\tau_1, \tau_2} \mathrm{mult}_{\Sigma_1}(\tau_1)\,\mathrm{mult}_{\Sigma_2}(\tau_2)[N : N_{\tau_1} + N_{\tau_2}], \tag{3.6.5}$$

where the sum is over all $\tau_1 \in \mathrm{star}_{\Sigma_1}(\sigma_1 \cap \sigma_2), \tau_2 \in \mathrm{star}_{\Sigma_2}(\sigma_1 \cap \sigma_2)$ with $\tau_1 \cap (\mathbf{v} + \tau_2) \neq \emptyset$, for some fixed generic $\mathbf{v}$. This is independent of the choice of $\mathbf{v}$ by Lemma 3.6.4. In equation (3.6.4), the sum $\sigma_1 + \sigma_2$ is the Minkowski sum.

Note that the stable intersection of two polyhedral complexes is contained in their set-theoretic intersection. This containment can be strict.

We illustrate the concept of stable intersection with some examples.

Example 3.6.6. The standard tropical plane is the fan Σ with rays spanned by the vectors $\mathbf{e}_0 = (-1,-1,-1)$, $\mathbf{e}_1 = (1,0,0)$, $\mathbf{e}_2 = (0,1,0)$, and $\mathbf{e}_3 = (0,0,1)$. The two-dimensional cones C_{ij} of Σ are spanned by the pairs $\mathbf{e}_i, \mathbf{e}_j$. The multiplicity on each of the six cones C_{ij} is one. Two cones σ_1, σ_2 of Σ have $\dim(\sigma_1 + \sigma_2) = 3$ if and only if one is two dimensional and the other has dimension at least one and is not a ray of the first. For example, when $\sigma_1 = C_{12}$, then σ_2 can be any cone that contains $\mathbf{e}_0$ or $\mathbf{e}_3$. The intersection $\sigma_1 \cap \sigma_2$ in that case is either $\{\mathbf{0}\}$ or one of the rays of σ_1. The latter case occurs when $\sigma_2 = C_{ij}$ with $i \in \{1,2\}$ and $j \in \{0,3\}$. *The stable intersection $\Sigma \cap_{st} \Sigma$ is thus the one-skeleton of the fan Σ. The multiplicity of each ray is one.* To show this, it suffices to consider the case $C_{12} \cap C_{13}$. For $\mathbf{v} = (1,1,-1)$ the cones C_{12} and $\mathbf{v} + C_{13}$ intersect in the ray $(1,1,0) + \mathrm{pos}((1,0,0))$. The lattice $N_{C_{12}}$ is the span of $\{\mathbf{e}_1, \mathbf{e}_2\}$, while the lattice $N_{C_{13}}$ is the span of $\{\mathbf{e}_1, \mathbf{e}_3\}$, so $N_{C_{12}} + N_{C_{13}} = \mathbb{Z}^3 = N$. The lattice index $[N : N_{C_{12}} + N_{C_{13}}]$ equals one.

Next we consider the tropical curves shown in Figure 3.6.1, where we denote the solid curve by Σ_1 and the dotted curve by Σ_2. The stable intersection $\Sigma_1 \cap_{st} \Sigma_2$ consists of three points: $(-1,2)$ with multiplicity one, $(1,1)$ with multiplicity two, and $(3,-1)$ with multiplicity one. We verify the multiplicity of $\sigma = \{(1,1)\}$ using the formula (3.6.5). After refining Σ_1 and Σ_2 appropriately, $\mathrm{star}_{\Sigma_1}(\sigma)$ consists of three rays $\mathrm{pos}\{(1,0)\}$, $\mathrm{pos}\{(0,1)\}$, and $\mathrm{pos}\{(-1,-1)\}$. Likewise, $\mathrm{star}_{\Sigma_2}(\sigma)$ consists of two rays $\mathrm{pos}\{(1,-1)\}$ and $\mathrm{pos}\{(-1,1)\}$. For $\mathbf{v} = (1,1)$, the fan $\mathrm{star}_{\Sigma_1}(\sigma)$ intersects $\mathbf{v} + \mathrm{star}_{\Sigma_2}(\sigma)$ in two points, $(1,0)$ and $(0,1)$. The first of these comes from the rays $\tau_1 = \mathrm{pos}\{(1,0)\}$ and $\tau_2 = \mathrm{pos}\{(1,-1)\}$, while the second comes from the rays $\tau_1 = \mathrm{pos}\{(0,1)\}$ and $\tau_2 = \mathrm{pos}\{(-1,1)\}$. Since the weights of all cells in Σ_1 and Σ_2 are one, the multiplicity (3.6.5) is two:

$$(1)(1)[\mathbb{Z}^2 : \mathrm{span}_{\mathbb{Z}}((1,0),(1,-1))] + (1)(1)[\mathbb{Z}^2 : \mathrm{span}_{\mathbb{Z}}((0,1),(-1,1))] = 1+1.$$

For $\mathbf{v} = (-1,0)$, the only cones in the respective stars that intersect are $\mathrm{pos}\{(-1,-1)\}$ and $\mathrm{pos}\{(1,-1)\}$. The multiplicity (3.6.5) is now computed as $(1)(1)[\mathbb{Z}^2 : \mathrm{span}_{\mathbb{Z}}\{(-1,-1),(1,-1)\}] = 2$. Note that we get the same answer for the two different $\mathbf{v}$. ◊

The stable intersection of two pure weighted balanced polyhedral complexes is again pure and balanced. This requires the following three lemmas.

Lemma 3.6.7. *Let Σ_1 and Σ_2 be pure weighted balanced polyhedral complexes, and let σ be a cell of $\Sigma_1 \cap_{st} \Sigma_2$. We have the equality of weighted fans*

$$\mathrm{star}_{\Sigma_1 \cap_{st} \Sigma_2}(\sigma) = \mathrm{star}_{\Sigma_1}(\sigma) \cap_{st} \mathrm{star}_{\Sigma_2}(\sigma). \tag{3.6.6}$$

Proof. We first show the equality of sets. A vector $\mathbf{v}$ is in $\mathrm{star}_{\Sigma_1 \cap_{st} \Sigma_2}(\sigma)$ if and only if there is $\mathbf{w} \in \sigma$, a top-dimensional cell $\tau \in \Sigma_1 \cap_{st} \Sigma_2$, and an

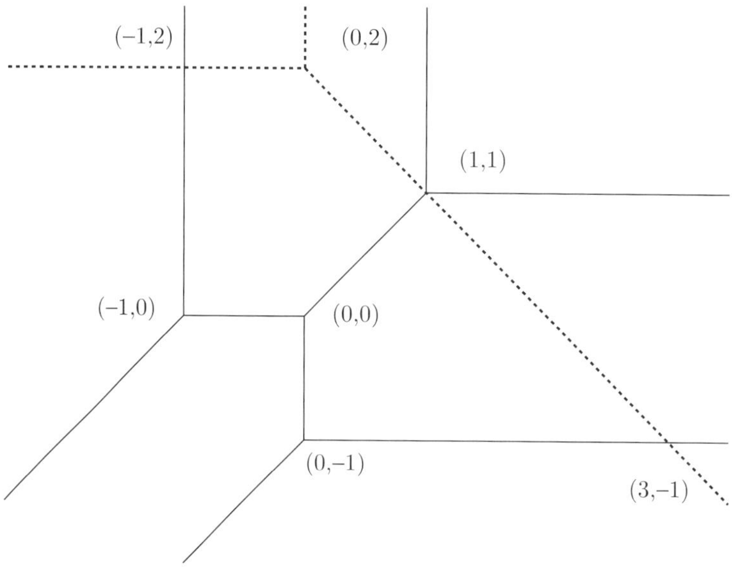

Figure 3.6.1. The stable intersection of two curves in Example 3.6.6.

$\epsilon > 0$ with $\mathbf{w} + \epsilon\mathbf{v} \in \tau$. By the definition of stable intersection, we can write $\tau = \tau_1 \cap \tau_2$ for $\tau_1 \in \Sigma_1$, $\tau_2 \in \Sigma_2$ with $\dim(\tau_1 + \tau_2) = n$. We have $\mathbf{w} + \epsilon\mathbf{v} \in \tau_i$ for $i = 1, 2$, so $\mathbf{v} \in \mathrm{star}_{\Sigma_1}(\sigma) \cap \mathrm{star}_{\Sigma_2}(\sigma)$. Then $\mathbf{v}$ is in the cone $\bar{\tau}_i$ of $\mathrm{star}_{\Sigma_i}(\sigma)$, which contains a translate of τ_i, for $i = 1, 2$, so $\dim(\bar{\tau}_1 + \bar{\tau}_2) = n$. Thus $\bar{\tau}_1 \cap \bar{\tau}_2 \in \mathrm{star}_{\Sigma_1}(\sigma) \cap_{st} \mathrm{star}_{\Sigma_2}(\sigma)$, and this shows "$\subseteq$" in (3.6.6).

For the reverse inclusion, it suffices to show that, for $\tau_1 \in \Sigma_1$, $\tau_2 \in \Sigma_2$, we have $\dim(\bar{\tau}_1 + \bar{\tau}_2) = n$ if and only if $\dim(\tau_1 + \tau_2) = n$. Since the linear space parallel to the sum of two polyhedra is the sum of the linear spaces parallel to the summands, it suffices to observe that the linear spaces parallel to $\bar{\tau}_i$ and to τ_i are equal. The linear space parallel to $\bar{\tau}_i$ is the span of $\mathbf{x} - \mathbf{y}$ with $\mathbf{x} \in \tau_i$ and $\mathbf{y} \in \sigma$, which is contained in the linear space parallel to τ_i. The opposite inclusion comes from the fact that $\bar{\tau}_i$ contains a translate of τ.

We now show that the multiplicities on the two fans in (3.6.6) agree. Let $\bar{\tau}$ be a top-dimensional cone in $\mathrm{star}_{\Sigma_1 \cap_{st} \Sigma_2}(\sigma)$. Its multiplicity is that of the corresponding cell τ in $\Sigma_1 \cap_{st} \Sigma_2$. This is the sum, over choices $\tau_1 \in \mathrm{star}_{\Sigma_1}(\sigma)$ and $\tau_2 \in \mathrm{star}_{\Sigma_2}(\sigma)$ with $\tau_1 \cap (\mathbf{v} + \tau_2) \neq \emptyset$ for fixed generic $\mathbf{v}$, of the quantities

$$\mathrm{mult}_{\Sigma_1}(\tau_1) \cdot \mathrm{mult}_{\Sigma_2}(\tau_2) \cdot [N : N_{\tau_1} + N_{\tau_2}].$$

The multiplicity on $\bar{\tau}$ in $\mathrm{star}_{\Sigma_1}(\sigma) \cap_{st} \mathrm{star}_{\Sigma_2}(\sigma)$ is the sum over all choices $\bar{\tau}_i \in \mathrm{star}_{\mathrm{star}_{\Sigma_i}(\sigma)}(\bar{\tau}) = \mathrm{star}_{\Sigma_i}(\tau)$ for $i = 1, 2$ with $\bar{\tau}_1 \cap (\mathbf{v} + \bar{\tau}_2) \neq \emptyset$ of

$$\mathrm{mult}_{\mathrm{star}_{\Sigma_1}(\sigma)}(\bar{\tau}_1) \cdot \mathrm{mult}_{\mathrm{star}_{\Sigma_2}}(\sigma)(\bar{\tau}_2) \cdot [N : N_{\bar{\tau}_1} + N_{\bar{\tau}_2}].$$

They are equal because $\mathrm{mult}_{\mathrm{star}_{\Sigma_i}(\sigma)}(\bar{\tau}_i) = \mathrm{mult}_{\Sigma_i}(\tau_i)$, and $N_{\tau_1} = N_{\bar{\tau}_2}$. $\square$

In this section we relax the notion of the lineality space of a polyhedral complex Σ to mean the largest subspace L for which if $\mathbf{x} \in |\Sigma|$ and $\mathbf{v} \in L$, then $\mathbf{x} + \mathbf{v} \in |\Sigma|$. This notion only depends on the support $|\Sigma|$ of the complex. For example, the fan in $\mathbb{R}$ consisting of the three cones pos(1), pos(−1), and $\{0\}$ has lineality space $\mathbb{R}$ for this definition.

Lemma 3.6.8. *Let Σ be a pure weighted balanced Γ_{val}-rational polyhedral complex in $\mathbb{R}^n$ of codimension d, and let H be the (classical) hyperplane $\{\mathbf{x} : x_1 = 0\}$. The stable intersection $\Sigma \cap_{st} H$ is either empty or a pure weighted balanced polyhedral complex of codimension $d+1$.*

Let l be the dimension of the intersection of the lineality space of Σ with H. If $d+l > n-1$, then the stable intersection is empty.

Proof. If $d+l > n-1$, then there is no pair $\tau \in \Sigma$ with $\dim(\tau+H) = n$, so the stable intersection is empty. Indeed, $\dim(\tau + H) = \dim(\tau) + \dim(H) - \dim(\text{aff}(\tau) \cap H) \leq (n-d) + (n-1) - l = 2n - (d+1+l)$ is less than n.

We now assume $d + l \leq n - 1$. Let τ be a top-dimensional cell of $\Sigma \cap_{st} H$, so $\tau = \sigma \cap H$, where σ is cell of Σ with $\dim(\sigma) = n - d$ and $\dim(\sigma + H) = n$. The image of H modulo the linear space parallel to σ has dimension d, so we can choose a d-dimensional subspace H' of H with $\dim(H' + \sigma) = n$. The fan $\text{star}_\Sigma(\tau)$ is balanced with the weights inherited from Σ. By Lemma 3.6.3, its image modulo H' is a balanced weighted $(n-d)$-dimensional polyhedral fan in $\mathbb{R}^n/H' \simeq \mathbb{R}^{n-d}$. Its support is all of $\mathbb{R}^{n-d}$, which means that $(\text{star}_\Sigma(\tau)+H')\cap H = (\text{star}_\Sigma(\tau)\cap H)+H'$ is all of H, and thus $\text{star}_\Sigma(\tau)\cap H$ has dimension at least $n-d-1$. Since $\dim(\sigma+H) = n$, we do not have $\sigma \subset H$, and so τ has dimension $n-d-1$. This shows that the stable intersection $\Sigma \cap_{st} H$ is pure of the expected codimension.

Let σ be a codimension-1 cell in $\Sigma \cap_{st} H$. To prove balancing at σ, we must show that the fan $\text{star}_{\Sigma \cap_{st} H}(\sigma)$ is balanced. By Lemma 3.6.7, it equals $\text{star}_\Sigma(\sigma) \cap_{st} \text{star}_H(\sigma)$. Since stable intersection commutes with projections, we can quotient by the linear space parallel to σ. This reduces balancing to the case where Σ is a two-dimensional fan, and hence $\Sigma \cap_{st} H$ is a one-dimensional fan. For generic small $\mathbf{v} \in \mathbb{R}^n$, the intersection $\Sigma \cap (\mathbf{v} + H)$ is transverse, so its relatively open one-dimensional cells lie in the relative interiors of two-dimensional cones of Σ. Therefore, the stable intersection $\Sigma \cap_{st} (\mathbf{v} + H)$ equals the actual intersection, and the multiplicity of a cone $\tau \cap (\mathbf{v} + H)$ is the lattice index $[N : N_\tau + N_H]$ times the multiplicity of τ. Each unbounded ray of $\Sigma \cap (\mathbf{v} + H)$ corresponds to a ray of $\Sigma \cap_{st} H$ plus the choice of a two-dimensional cone $\tau \in \Sigma$ with $\dim(\tau + H) = n$ and $\tau \cap (\mathbf{v} + H) \neq \emptyset$. The sum of the multiplicities of rays in $\Sigma \cap (\mathbf{v} + H)$ corresponding to a fixed ray σ of $\Sigma \cap_{st} H$ thus equals the multiplicity of σ.

We claim that it suffices to show that $\Sigma \cap (\mathbf{v} + H)$ is balanced. Indeed, when summing the left-hand side of the balancing equation (3.3.1) over all vertices of the intersection, each bounded edge contributes to two summands, with direction vectors $\pm\mathbf{u}_\sigma$ and the same multiplicity m_σ. These contributions cancel. Balancing implies that the sub-sum coming from each vertex adds to $\mathbf{0}$, so the entire sum is $\mathbf{0}$, and this equals the contribution coming from the unbounded rays, which is the equation (3.3.1) for $\Sigma \cap_{st} H$.

Let $\mathbf{u}$ be a vertex of $\Sigma \cap (\mathbf{v} + H)$. Let σ be the ray of Σ containing $\mathbf{u}$, and let $\tau_1, \ldots, \tau_s$ be the two-dimensional cones of Σ containing σ. Write $\mathbf{u}_i$ for the element of N_{τ_i} that projects to a generator of N_{τ_i}/N_σ, and m_i for the multiplicity of τ_i in Σ. Since Σ is balanced, we have $\sum_i m_i\mathbf{u}_i \in N_\sigma$.

We write $\mathbf{u}^i$ for the first lattice point of the ray of $\mathrm{star}_{\Sigma\cap(\mathbf{v}+H)}(\mathbf{u})$ corresponding to τ_i. The multiplicity of this cone in $\Sigma \cap_{st} (\mathbf{v} + H)$ is $m_i[N : N_{\tau_i} + N_H]$. We have

$$\mathbf{u}^i = [N_{\tau_i} : \mathbb{Z}\mathbf{u}^i + N_\sigma]\mathbf{u}_i + \mathbf{u}_{\sigma,i},$$

where $\mathbf{u}_{\sigma,i} \in N_\sigma$. By the second and third isomorphism theorems,

$$\begin{aligned} N/(N_H + N_{\tau_i}) &\cong \big(N/(N_H + N_\sigma)\big)/\big((N_H + N_{\tau_i})/(N_H + N_\sigma)\big) \\ &\cong \big(N/(N_H + N_\sigma)\big)/\big(N_{\tau_i}/((N_H \cap N_{\tau_i}) + N_\sigma)\big). \end{aligned}$$

Hence $[N : N_H + N_\sigma] = [N : N_H + N_{\tau_i}][N_{\tau_i} : \mathbb{Z}\mathbf{u}^i + N_\sigma]$. Thus,

$$\begin{aligned} &\sum_i m_i[N : N_{\tau_i} + N_H]\mathbf{u}^i \\ &= \sum_i m_i[N : N_{\tau_i} + N_H]([N_{\tau_i} : \mathbb{Z}\mathbf{u}^i + N_\sigma]\mathbf{u}_i + \mathbf{u}_{\sigma,i}) \\ &= \sum_i [N : N_H + N_\sigma]m_i\mathbf{u}_i + \sum_i m_i[N : N_{\tau_i} + N_H]\mathbf{u}_{\sigma,i} \\ &\in N_\sigma \cap N_H. \end{aligned}$$

Since $N_\sigma \cap N_H = \mathbf{0}$, we have that $\Sigma \cap_{st} (\mathbf{v} + H)$ is balanced as required. □

Lemma 3.6.9. *Let Σ be a pure weighted balanced Γ_{val}-rational polyhedral complex in $\mathbb{R}^n$. Let $H_1, \ldots, H_d$ be hyperplanes in $\mathbb{R}^n$ whose normal vectors are linearly independent. Write L for the linear space $\bigcap_{i=1}^d H_i$. We have*

$$\Sigma \cap_{st} L \;=\; ((\Sigma \cap_{st} H_1) \cap_{st} H_2) \cdots \cap_{st} H_d.$$

Proof. Without loss of generality we may assume that $H_i = \{\mathbf{x} \in \mathbb{R}^n : x_i = 0\}$. The proof is by induction on d. The base case $d = 1$ is a tautology. Let $L' = \{\mathbf{x} \in \mathbb{R}^n : x_1 = \cdots = x_{d-1} = 0\}$. By induction, $\Sigma \cap_{st} L' = ((\Sigma \cap_{st} H_1) \cap_{st} H_2) \cdots \cap_{st} H_{d-1}$, so we need to show $\Sigma \cap_{st} L = (\Sigma \cap_{st} L') \cap_{st} H_d$. Let σ be a maximal cell in $\Sigma \cap_{st} L$. Then there is a maximal cell $\tau \in \Sigma$ with $\sigma = \tau \cap L$ and $\dim(\tau + L) = n$, so the projection of $\mathrm{star}_\Sigma(\sigma) + L$ to $\mathbb{R}^n/L \cong \mathbb{R}^d$

is d-dimensional. As the projection is balanced by Lemma 3.6.3, it must be all of $\mathbb{R}^{n-d}$ (see Exercise 3.7(24)). This means that there is $\mathbf{x} \in \text{star}_\Sigma(\sigma) + L$ with $x_1 = \cdots = x_{d-1} = 0$ and $x_d \neq 0$. We may assume that $\mathbf{x} \in \text{star}_\Sigma(\sigma)$, as adding an element of L does not change the first d coordinates. Recall from Definition 2.3.6 that the cones of $\text{star}_\Sigma(\sigma)$ have the form $\bar{\tau}$ for $\tau \supset \sigma$. We must have $\mathbf{x} \in \bar{\tau}'$ for some cell τ' of Σ containing σ. The cell τ' has the property that $\bar{\tau}' + L$ has dimension n, and thus $\dim(\tau' + L) = n$. Since $\mathbf{x} \in \bar{\tau}'$, there is $\mathbf{x}' \in \tau'$ with $x'_1 = \cdots = x'_{d-1} = 0$ and $x'_d \neq 0$. Let $\sigma' = \tau \cap L'$. Since $L \subset L'$, we have $\dim(\tau + L') = n$, and so $\sigma' \in \Sigma \cap_{st} L'$. By construction, $\sigma = \sigma' \cap H_d$. Since $x'_d \neq 0$, we have $\sigma' \neq \sigma$, so $\dim(\sigma' + H_d) = n$. Thus $\sigma \in (\Sigma \cap_{st} L') \cap_{st} H_d$.

For the reverse inclusion, note that if σ is a maximal cell in $(\Sigma \cap_{st} L') \cap_{st} H_d$, then there is a maximal cell $\sigma' \in \Sigma \cap_{st} L'$ with $\sigma = \sigma' \cap H_d$ and $\dim(\sigma' + H_d) = n$. Furthermore, there is a maximal cell $\tau \in \Sigma$ with $\sigma' = \tau \cap L'$ and $\dim(\tau + L') = n$. We thus have $\tau \cap L = \tau \cap (L' \cap H_d) = \sigma$, and as before $\dim(\tau + L) = n$. This shows the equality as sets.

To see the equality of multiplicities, note that the multiplicities of L, L', and H_d are all one, and $N_{L'} + N_{H_d} = N_L$. This means that the multiplicity in both descriptions of a cell σ is the sum over all τ mentioned above with $\tau \cap (\mathbf{v} + L) \neq \emptyset$ for fixed generic $\mathbf{v}$ of the quantity $\text{mult}_\Sigma(\tau)[N : N_\tau + L]$. □

Theorem 3.6.10. *Let Σ_1 and Σ_2 be pure weighted balanced Γ_{val}-rational polyhedral complexes in $\mathbb{R}^n$ of codimensions d and e, respectively, and let l be the dimension of the intersection of their lineality spaces. Then the stable intersection $\Sigma_1 \cap_{st} \Sigma_2$ is either empty or it is a pure weighted balanced Γ_{val}-rational polyhedral complex of codimension $d + e$. If $d + e + l > n$, then the stable intersection is empty.*

Proof. Let Δ be the diagonal linear subspace $\{(\mathbf{w}, \mathbf{w}) : \mathbf{w} \in \mathbb{R}^n\} \subset \mathbb{R}^{2n}$. There is a natural identification of Δ with $N_\mathbb{R}$. We claim that

$$\Sigma_1 \cap_{st} \Sigma_2 \;=\; (\Sigma_1 \times \Sigma_2) \cap_{st} \Delta \;\subseteq\; \Delta \cong N_\mathbb{R}.$$

To see this, consider a cell $\tau_1 \times \tau_2$ of $\Sigma_1 \times \Sigma_2$. Let A_1 and A_2 be matrices whose columns form a basis for N_{τ_1} and N_{τ_2}, respectively. Then the matrix $A^{12} = \begin{pmatrix} A_1 & 0 & I \\ 0 & A_2 & I \end{pmatrix}$ has columns forming a basis for $N_{(\tau_1 \times \tau_2) + \Delta}$, so the Minkowski sum $(\tau_1 \times \tau_2) + \Delta$ has dimension $2n$ if and only if this matrix has rank $2n$. This is the case if and only if the matrix $A_{12} = \begin{pmatrix} A_1 & -A_2 \end{pmatrix}$ has rank n, which occurs if and only if $\dim(\tau_1 + \tau_2) = n$. This means that $(\tau_1 \times \tau_2) \cap \Delta \in (\Sigma_1 \times \Sigma_2) \cap_{st} \Delta$ if and only if $\tau_1 \cap \tau_2 \in \Sigma_1 \cap_{st} \Sigma_2$. Also, $(\tau_1 \times \tau_2) \cap ((\mathbf{0}, -\mathbf{v}) + \Delta) \neq \emptyset$ if and only if $\tau_1 \cap (\mathbf{v} + \tau_2) \neq \emptyset$. So, to compute the multiplicity of $\tau_1 \cap \tau_2$ or $(\tau_1 \times \tau_2) \cap \Delta$ we sum over the same pairs (τ_1, τ_2). To see that the multiplicity is the same in both cases, it suffices to observe that the index of $N_{(\tau_1\times\tau_2)+\Delta}$ in $N\oplus N$ is the greatest common divisor of the maximal minors of the matrix A^{12}, while the index $[N : N_{\tau_1} + N_{\tau_2}]$ is the greatest common divisor of the maximal minors of the matrix A_{12}. Since these coincide, the multiplicities coincide, and so the claim follows.

Change coordinates so that $\Delta = \{\mathbf{x} \in \mathbb{R}^{2n} : x_1 = \cdots = x_n = 0\}$. Write H_i for the hyperplane $\{\mathbf{x} \in \mathbb{R}^n : x_i = 0\}$, and $\Sigma = \Sigma_1 \times \Sigma_2$. By Lemma 3.6.9 we have $\Sigma \cap_{st} \Delta = ((\Sigma \cap_{st} H_1) \cap_{st} H_2) \cdots \cap_{st} H_n$. The result then follows from Lemma 3.6.8 since $\operatorname{codim}(\Sigma_1 \times \Sigma_2) = d+e$ and $\operatorname{codim}(\Delta) = n$ in $\mathbb{R}^{2n}$. Thus, if the stable intersection is nonempty, it has codimension $d+e+n$ in $\mathbb{R}^{2n}$, and so has codimension $d+e$ in $N_{\mathbb{R}} \cong \mathbb{R}^n$. □

Definition 3.6.11. Let Σ_1, Σ_2 be pure weighted balanced Γ_{val}-rational polyhedral complexes in $\mathbb{R}^n$ that meet transversely at a point $\mathbf{w}$ that lies in the relative interior of maximal cells $\sigma_1 \in \Sigma_1$ and $\sigma_2 \in \Sigma_2$. Here we use the notion of *meeting transversely* from Definition 3.4.9. The *tropical multiplicity* of the intersection at $\mathbf{w}$ is the product $\operatorname{mult}_{\Sigma_1}(\sigma_1)\operatorname{mult}_{\Sigma_2}(\sigma_2)[N : N_{\sigma_1}+N_{\sigma_2}]$.

If Σ_1 and Σ_2 intersect transversely at every point $\mathbf{w}$ of their intersection, then the stable intersection $\Sigma_1 \cap_{st} \Sigma_2$ equals the intersection $\Sigma_1 \cap \Sigma_2$, and the multiplicity of the stable intersection at $\mathbf{w}$ is the tropical multiplicity.

We now make the link to the construction for curves in Section 1.3. The stable intersection can be obtained by translating each Σ_i by a small amount so that the intersection is transverse, computing the intersection together with its tropical multiplicity, and then taking the limit as the translation becomes smaller and smaller. This definition is made precise as follows.

Recall that the *Hausdorff metric* on subsets of $\mathbb{R}^n$ is given by $d(A,B) = \max(\sup_{a\in A}\inf_{b\in B}\|a-b\|, \sup_{b\in B}\inf_{a\in A}\|a-b\|)$. This lets us speak about the limit of a sequence of subsets of $\mathbb{R}^n$. If the subsets are weighted polyhedral complexes Σ_i that converge to a polyhedral complex Σ, then the limit inherits a weighting in the following way. A top-dimensional cell σ of the limit complex Σ is the limit of top-dimensional cells σ_i of Σ_i if $\lim_{i\to\infty}\sigma_i = \sigma$. We consider the set of all such sequences of σ_i limiting to σ, where we identify cofinal sequences. If $\lim_{i\to\infty}\operatorname{mult}_{\Sigma_i}(\sigma_i)$ exists for all such sequences, then we define the multiplicity of σ to be the sum of all these limits.

We often apply these concepts to finite collections of weighted points. In this case the multiplicity of a limit point $\mathbf{u}$ is the sum of the multiplicities of all points that tend to $\mathbf{u}$. The following result, however, works in general.

Proposition 3.6.12. *Let Σ_1 and Σ_2 be weighted balanced polyhedral complexes that are pure of codimension d and e. For general $\mathbf{v} \in \mathbb{R}^n$, the limit*

$$\lim_{\epsilon\to 0} \Sigma_1 \cap (\epsilon\mathbf{v} + \Sigma_2)$$

exists and it equals $\Sigma_1 \cap_{st} \Sigma_2$ *as a weighted polyhedral complex. In particular, this intersection is independent of the choice of translate* $\mathbf{v}$.

Proof. We first give the condition for $\mathbf{v}$ to be generic. Consider any pair of cells $\tau_i \in \Sigma_1$ and $\tau_j \in \Sigma_2$ with nontrivial intersection. If $\dim(\tau_i + \tau_j) < n$, then there is a vector $\mathbf{u}_{ij}$ perpendicular to the affine spans of both τ_i and τ_j. For any $\mathbf{v}$ with $\mathbf{u}_{ij} \cdot \mathbf{v} \neq 0$, we have $\tau_i \cap (\mathbf{v} + \tau_j) = \emptyset$. Choose one such vector $\mathbf{u}_{ij}$ for each pair $\tau_i \in \Sigma_1$, $\tau_j \in \Sigma_2$ with $\tau_i \cap \tau_j \neq \emptyset$ and $\dim(\tau_i + \tau_j) < n$. Let V be the open set in $\mathbb{R}^n$ consisting of vectors $\mathbf{v}$ with $\mathbf{v} \cdot \mathbf{u}_{ij} \neq 0$ for all i, j.

Fix $\mathbf{v} \in V$. Suppose $\mathbf{w}$ lies in $\lim_{\epsilon \to 0} \Sigma_1 \cap (\epsilon \mathbf{v} + \Sigma_2)$. For all $\epsilon > 0$ there is $\mathbf{w}_\epsilon \in \Sigma_1 \cap (\epsilon' \mathbf{v} + \Sigma_2)$ with $\|\mathbf{w}_\epsilon - \mathbf{w}\| < \epsilon$. Here $\epsilon' < \epsilon$ depends on ϵ. Since Σ_1 is closed, we have $\mathbf{w} \in \Sigma_1$. Similarly, since $\mathbf{w}_\epsilon - \epsilon' \mathbf{v} \in \Sigma_2$, and Σ_2 is closed, we have $\mathbf{w} \in \Sigma_2$, so $\mathbf{w} \in \Sigma_1 \cap \Sigma_2$. Let σ be the smallest cell of Σ_1 containing $\mathbf{w}$. After refining if necessary, σ is also a cell of Σ_2. For sufficiently small ϵ, the point $\mathbf{w}_\epsilon$ lies in a cell τ_1 of Σ_1 containing σ. Similarly, for small ϵ, the point $\mathbf{w}_\epsilon - \epsilon' \mathbf{v}$ lies in a cell τ_2 of Σ_2 which must also have σ as a face. Since $\mathbf{v} \in V$, we have $\epsilon' \mathbf{v} \in V$. Since $\mathbf{w}_\epsilon \in \tau_1 \cap (\epsilon' \mathbf{v} + \tau_2)$, we must have $\dim(\tau_1 + \tau_2) = n$, which means that $\mathbf{w} \in \Sigma_1 \cap_{st} \Sigma_2$.

For the converse, let σ be a top-dimensional cell of $\Sigma_1 \cap_{st} \Sigma_2$, and let $\mathbf{w} \in \sigma$. There are $\tau_1 \in \Sigma_1$, $\tau_2 \in \Sigma_2$ with $\tau_1 \cap \tau_2 = \sigma$, $\dim(\tau_1 + \tau_2) = n$, and $\tau_1 \cap (\mathbf{v} + \tau_2) \neq \emptyset$. Choose $\mathbf{w}' \in \tau_1 \cap (\mathbf{v} + \tau_2)$. For any $0 < \epsilon' < 1$, we have $\mathbf{w}_{\epsilon'} = (1 - \epsilon')\mathbf{w} + \epsilon' \mathbf{w}' \in \tau_1 \cap (\epsilon' \mathbf{v} + \tau_2)$, since τ_1 and τ_2 are convex, and $\mathbf{w}' - \mathbf{v} \in \tau_2$. Given $\epsilon > 0$ we can choose $\epsilon' < \epsilon / \|\mathbf{w}' - \mathbf{w}\|$. Then $\|\mathbf{w}_{\epsilon'} - \mathbf{w}\| < \epsilon$. We conclude that $\mathbf{w} \in \lim_{\epsilon \to 0} \Sigma_1 \cap (\epsilon \mathbf{v} + \Sigma_2)$.

For the multiplicities, note that the intersection $\Sigma_1 \cap (\epsilon \mathbf{v} + \Sigma_2)$ is transverse for generic $\mathbf{v}$. A top-dimensional cell is the intersection of unique maximal cells $\tau_1 \in \Sigma_1$ and $\epsilon \mathbf{v} + \tau_2$ for $\tau_2 \in \Sigma_2$ with $\dim(\tau_1 + \tau_2) = n$. The multiplicity of such an intersection is $\mathrm{mult}_{\Sigma}(\tau_1)\, \mathrm{mult}_{\Sigma_2}(\tau_2)[N : N_{\tau_1} + N_{\tau_2}]$. Since the multiplicity of a top-dimensional cell σ in $\Sigma_1 \cap_{st} \Sigma_2$ is the sum of this quantity over all pairs τ_1, τ_2 with $\tau_1 \cap (\mathbf{v} + \tau_2) \neq \emptyset$, and such pairs are exactly those for which $\lim_{\epsilon \to 0} \tau_1 \cap (\epsilon \mathbf{v} + \tau_2) = \sigma$, this shows the equality. □

Example 3.6.13. Fix $K = \mathbb{Q}$ with the 2-adic valuation. Consider first $\Sigma_1 = \mathrm{trop}(V(4x^2 + xy + 12y^2 + y + 3))$ and $\Sigma_2 = \mathrm{trop}(V(4x + y + 4))$ in $\mathbb{R}^2$. This is shown in the first picture in Figure 3.6.2, with Σ_2 drawn with dotted lines. The vertical ray of Σ_1 has multiplicity 2. The second picture shows the intersections $\Sigma_1 \cap ((1, 0) + \Sigma_2)$ and $\Sigma_1 \cap ((0, 1) + \Sigma_2)$. Both intersection points have multiplicity 2. These are these cases $\epsilon = 1$ of the translations $\epsilon(1, 0) + \Sigma_2$ and $\epsilon(0, 1) + \Sigma_2$. As ϵ goes to zero, the intersection point in both cases approaches the point $(-1, 1)$. The multiplicity also does not change, so the stable intersection $\Sigma_1 \cap_{st} \Sigma_2$ is the point $(-1, 1)$ with multiplicity 2.

Consider next the tropical line $\Sigma_3 = \text{trop}(V(x+8y+1))$. This is shown in the third picture in Figure 3.6.2. The intersection $\Sigma_1 \cap \Sigma_3$ is not transverse. The translations $(1, 1/2) + \Sigma_3$ and $(-1/2, 0) + \Sigma_3$ are drawn in the last picture in Figure 3.6.2, and these give transverse intersections in two points. In both cases the tropical multiplicity is one at each point. As ϵ goes to zero, the limits of $\Sigma_1 \cap (\epsilon(1, 1/2) + \Sigma_3)$ and $\Sigma_1 \cap (\epsilon(-1/2, 0) + \Sigma_3)$ are both the two points $(0, 0)$ and $(0, -2)$. The limiting multiplicity is one in both cases, so the stable intersection $\Sigma_1 \cap_{st} \Sigma_2$ is these two points with multiplicity one.

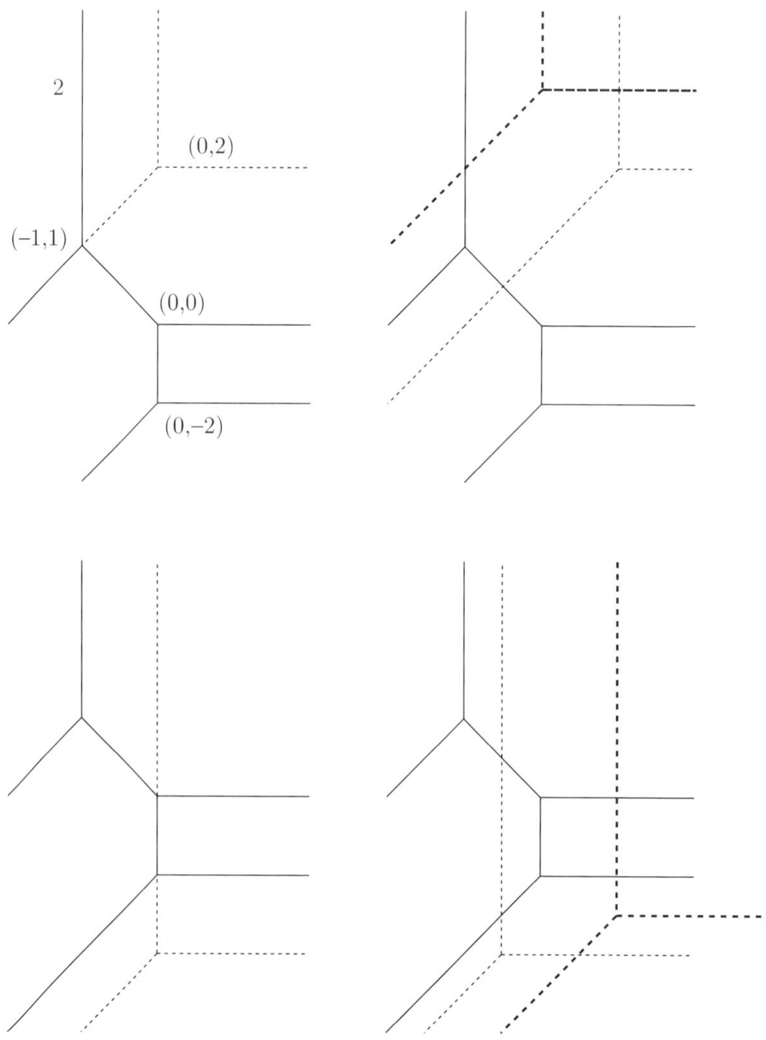

Figure 3.6.2. Stable intersections of lines and quadrics in the plane.

Both $\Sigma_1 \cap_{st} \Sigma_2$ and $\Sigma_1 \cap_{st} \Sigma_3$ are stable intersections of a quadric with a line. The intersections consist of two points, counted with multiplicity. This is a preview of the tropical complete intersections studied in Section 4.6. ◊

Remark 3.6.14. It can be shown that stable intersection is associative: if $\Sigma_1, \Sigma_2, \Sigma_3$ are weighted balanced Γ_{val}-rational polyhedral complexes, then

$$(\Sigma_1 \cap_{st} \Sigma_2) \cap_{st} \Sigma_3 \;=\; \Sigma_1 \cap_{st} (\Sigma_2 \cap_{st} \Sigma_3). \tag{3.6.7}$$

Thus stable intersection defines a multiplication on the set of weighted balanced Γ_{val}-rational polyhedral complexes, where complexes with the same support and weight function are identified. We can define an addition on this set by taking the unions of complexes (appropriately subdivided if necessary). If we also allow arbitrary real weights on maximal cells, then this makes this set into an $\mathbb{R}$-algebra. The subalgebra where all polyhedral complexes are fans appeared earlier in the work of McMullen as the *polytope algebra*. See [**JY13**] for details on the connection. Versions of this algebra have also arisen in the work of Allermann and Rau on tropical intersection theory [**AR10**] and Fulton and Sturmfels on toric intersection theory [**FS97**].

We now come to the derivation of Theorem 3.6.1 from the beginning of this section. We start with the following important special case.

Proposition 3.6.15. *Fix $X \subset T^n$. There is a finite set $B \subset \Bbbk$ for which for all $\alpha \in K$ with* $\text{val}(\alpha) = 0$ *and $\overline{\alpha} \notin B$ the hyperplane $H_\alpha = V(x_1 - \alpha)$ satisfies*

$$\text{trop}(X \cap H_\alpha) \;=\; \text{trop}(X) \cap_{st} \text{trop}(H_\alpha). \tag{3.6.8}$$

Proof. This proof is in two parts. We first show (3.6.8) set-theoretically and then check that the multiplicities coincide. Let I be the ideal of X in $K[x_1^{\pm 1}, \ldots, x_n^{\pm 1}]$. Fix $\mathbf{w} \in \mathbb{R}^n$ with $w_1 = 0$. We first claim that there is a finite set $B \subset \Bbbk$ for which if $\alpha \in K$ with $\text{val}(\alpha) = 0$ and $\overline{\alpha} \notin B$, then

$$\text{in}_{\mathbf{w}}(I + \langle x_1 - \alpha \rangle) \;=\; \text{in}_{\mathbf{w}}(I) + \langle x_1 - \overline{\alpha} \rangle. \tag{3.6.9}$$

To do this, we first consider the Gauss valuation on the field $K(s)$. This is given by setting $\text{val}(p) = \min(\text{val}(a_i))$ for a polynomial $p = \sum a_i s^i \in K[s]$, and then setting $\text{val}(p/q) = \text{val}(p) - \text{val}(q)$ for $p, q \in K[s]$. The valuation ring is $R(s)$, and the residue field is $\Bbbk(s)$. Let $I_s = IK(s)[x_1^{\pm 1}, \ldots, x_n^{\pm 1}]$. Let $J_s = (I_s)_{\text{proj}} + \langle x_1 - sx_0 \rangle \subset K(s)[x_0, \ldots, x_n]$. Note that $\text{in}_{(0,\mathbf{w})}(J_s) = \text{in}_{(0,\mathbf{w})}((I_s)_{\text{proj}}) + \langle x_1 - sx_0 \rangle$. The containment $\supseteq$ is immediate. By Corollary 2.4.9 the Hilbert function of $(I_s)_{\text{proj}}$ and $\text{in}_{(0,\mathbf{w})}((I_s)_{\text{proj}})$ agree. Since $x_1 - sx_0$ is a nonzerodivisor on both $(I_s)_{\text{proj}}$ and $\text{in}_{(0,\mathbf{w})}((I_s)_{\text{proj}})$, the Hilbert functions of J_s and $\text{in}_{(0,\mathbf{w})}((I_s)_{\text{proj}})) + \langle x_1 - sx_0 \rangle$ also agree, which implies the equality.

Let $\mathcal{G}$ be a Gröbner basis for J_s with respect to $(0, \mathbf{w})$. We may assume by the previous paragraph that $\mathcal{G}$ is the union of a Gröbner basis $\mathcal{G}'$ for $(I_s)_{\text{proj}}$ and the polynomial $x_1 - sx_0$. We may also assume, after multiplying by a common denominator, that all $g \in \mathcal{G}$ lie in $K[s][x_0, \ldots, x_n]$. Each $g \in \mathcal{G}'$ has the form $\sum h_i f_i$ for $f_i \in (I_s)_{\text{proj}} \cap K[x_0, \ldots, x_n]$ and $h_i \in K(s)[x_0, \ldots, x_n]$. Let $\mathcal{S} \subset K[s]$ be the set of all polynomials occurring in a numerator or denominator of a coefficient of some h_i. The set $\mathcal{S}$ only depends on the initial ideal $\text{in}_{(0,\mathbf{w})}((I_s)_{\text{proj}})$, and not on the particular choice of $\mathbf{w}$.

Fix $p = \sum a_i s^i \in K[s]$, so $p(\alpha) = \sum a_i \alpha^i \in K$ for $\alpha \in K$. There is a finite set $B_p \subset \Bbbk$ for which $\text{val}_K(p(\alpha)) = \text{val}_{K(s)}(p) = \min(\text{val}(a_i))$ for any $\alpha \in K$ with $\text{val}(\alpha) = 0$ and $\overline{\alpha} \notin B_p$. Let $B' = \bigcup_{p \in \mathcal{S}} B_p$. Then for any α with $\overline{\alpha} \notin B'$ we have $\text{in}_{(0,\mathbf{w})}(g)|_{s=\overline{\alpha}} = \text{in}_{(0,\mathbf{w})}(g|_{s=\alpha})$ for all $g \in \mathcal{G}$. The equality (3.6.9) then follows after setting $x_0 = 1$ and applying Proposition 2.6.1.

The classical hyperplane $H = \{\mathbf{w} : w_1 = 0\}$ is $\text{trop}(H_\alpha)$ for all $\alpha \in K$ with $\text{val}(\alpha) = 0$. Choose a polyhedral complex Σ with support $|\Sigma| = \text{trop}(X)$. We next show the following equivalence holds for α with $\text{val}(\alpha) = 0$ and $\overline{\alpha}$ outside a finite set B'':

$$\mathbf{w} \in \Sigma \cap_{st} H \text{ if and only if } \text{in}_{\mathbf{w}}(I) + \langle x_1 - \overline{\alpha} \rangle \neq \langle 1 \rangle.$$

For $\mathbf{w} \in \Sigma \cap H$ we have $\mathbf{w} \in \Sigma \cap_{st} H$ if and only if $\dim(\sigma + H) = n$ for some cone σ of Σ containing $\mathbf{w}$. This occurs if and only if $\text{star}_\Sigma(\sigma') \not\subseteq H$, where σ' is the cone of Σ containing $\mathbf{w}$ in its relative interior. Recall from Lemma 3.3.6 that $\text{star}_\Sigma(\sigma')$ has support $\text{trop}(V(\text{in}_{\mathbf{w}}(I)))$. Write $\pi : T^n \to K^*$ for the projection onto the first coordinate, and $Y = V(\text{in}_{\mathbf{w}}(I)) \subseteq T^n_{\Bbbk}$. By Corollary 3.2.13, the projection π satisfies $\text{trop}(\pi(Y)) = \pi(\text{trop}(Y))$, so $\text{trop}(Y) \subseteq H$ if and only if $\text{trop}(\pi(Y)) \subseteq \{0\}$, which by elimination theory [**CLO07**, Theorem 2, §3.2] is equivalent to the existence of a polynomial $f \in \text{in}_{\mathbf{w}}(I) \cap \Bbbk[x_1^{\pm 1}]$. Thus $\mathbf{w} \in \Sigma \cap_{st} H$ if and only if $\text{in}_{\mathbf{w}}(I) \cap \Bbbk[x_1^{\pm 1}] = \{0\}$.

If $f \in \text{in}_{\mathbf{w}}(I) \cap \Bbbk[x_1^{\pm 1}]$, then $f(\overline{\alpha}) \in \text{in}_{\mathbf{w}}(I) + \langle x_1 - \overline{\alpha} \rangle$. For a given nonzero polynomial $f \in \Bbbk[x_1^{\pm 1}]$, we have $f(\overline{\alpha}) \neq 0$ for all but finitely many $\overline{\alpha}$. Thus, if $\text{in}_{\mathbf{w}}(I) \cap \Bbbk[x_1^{\pm 1}] \neq \{0\}$, then $\text{in}_{\mathbf{w}}(I) + \langle x_1 - \overline{\alpha} \rangle = \langle 1 \rangle$ for all but finitely many $\overline{\alpha}$. Conversely, if $\text{in}_{\mathbf{w}}(I) \cap \Bbbk[x_1^{\pm 1}] = \{0\}$, the closure of $\pi(Y)$ is $\Bbbk^*$, so for all but finitely many $\overline{\alpha}$ there is $y_{\overline{\alpha}} \in (\Bbbk^*)^{n-1}$ with $(\overline{\alpha}, y_{\overline{\alpha}}) \in Y$. Since $(\overline{\alpha}, y_{\overline{\alpha}})$ also lies in $V(\text{in}_{\mathbf{w}}(I) + \langle x_1 - \overline{\alpha} \rangle)$, we conclude that $\text{in}_{\mathbf{w}}(I) + \langle x_1 - \overline{\alpha} \rangle \neq \langle 1 \rangle$. This gives a finite set $B'' \subset \Bbbk$ for which if $\overline{\alpha} \notin B''$ we have $\mathbf{w} \in \Sigma \cap_{st} H$ if and only if $\text{in}_{\mathbf{w}}(I) + \langle x_1 - \overline{\alpha} \rangle \neq \langle 1 \rangle$. Note that B'' only depends on $\text{in}_{\mathbf{w}}(I)$ and not on the particular choice of $\mathbf{w}$. Set $B_{\mathbf{w}} = B' \cup B''$. Since the Gröbner complex of $(I_s)_{\text{proj}}$ is finite, there are only a finite number of different choices for $\text{in}_{(0,\mathbf{w})}((I_s)_{\text{proj}})$, so there are only a finite number of $B_{\mathbf{w}}$ as $\mathbf{w}$ varies over $\Sigma \cap H$. Let B be the union of these finite sets. Thus if $\overline{\alpha} \notin B$, and $\mathbf{w} \in \Sigma \cap H$, we have $\mathbf{w} \in \Sigma \cap_{st} H$ if and only if $\mathrm{in}_{\mathbf{w}}(I + \langle x_1 - \alpha \rangle) \neq \langle 1 \rangle$, so if and only if $\mathbf{w} \in \mathrm{trop}(X \cap H_\alpha)$. This completes the first half of the proof.

For the second half of the proof we check that the multiplicities agree on the two sides of (3.6.8). Fix $\mathbf{w}$ in the relative interior of a maximal cell σ of $\mathrm{trop}(X) \cap_{st} \mathrm{trop}(H_\alpha)$. Since $\sigma \subset \{\mathbf{w} : w_1 = 0\}$, we may change coordinates while fixing w_1 so that the affine span of σ is $\mathrm{span}(\mathbf{e}_{n-d+2}, \ldots, \mathbf{e}_n)$, where $d = \dim(X)$. Part (2) of Lemma 2.6.2 implies that $\mathrm{in}_{\mathbf{w}}(I)$ is generated by polynomials in the variables $x_1, \ldots, x_{n-d+1}$. Let $J = \mathrm{in}_{\mathbf{w}}(I) \cap S_{n-d+1}$, where $S_{n-d+1} = \Bbbk[x_1^{\pm 1}, \ldots, x_{n-d+1}^{\pm 1}]$. The multiplicity of σ in $\mathrm{trop}(X \cap H_\alpha)$ equals

$$\text{(3.6.10)} \qquad \mathrm{mult}_{\mathrm{trop}(X \cap H_\alpha)}(\sigma) \;=\; \dim_{\Bbbk}(S_{n-d+1}/(J + \langle x_1 - \overline{\alpha} \rangle)),$$

by (3.6.9) and Lemma 3.4.7. We shall finish by showing that this is also the multiplicity of σ in $\Sigma \cap_{st} H$. We do this by computing the multiplicity of the stable intersection using the limit formulation of Proposition 3.6.12. By Lemma 3.6.7 we may pass to the star of σ and quotient by the linear space parallel to σ. This means that the stable intersection we need to consider is that of $V(\mathrm{in}_{\mathbf{w}}(J))$ and $\{\mathbf{w} \colon w_1 = 0\}$ in $\mathbb{R}^{n-d}$.

The dimension (3.6.10) equals $\dim_{K'} S_{K',n-d+1}/(J' + \langle x_1 - \overline{\alpha} \rangle)$ where $K' = \Bbbk((\mathbb{R}))$ is the field in Example 2.1.7, $S_{K',n-d+1} = K'[x_1^{\pm 1}, \ldots, x_{n-d+1}^{\pm 1}]$, and J' is the ideal in $S_{K',n-d+1}$ with the same generators as J. Indeed, this dimension can be computed using Buchberger's algorithm, which depends only on the field of definition of its input. Similar arguments show that $\dim_{K'} S_{K',n-d+1}/(J' + \langle x_1 - \beta \rangle)$ is a constant D for all but finitely many $\beta \in K'$. By Theorem 3.2.4 we have $\mathrm{trop}(V(J')) = \mathrm{trop}(V(J))$.

Choose $\alpha_\epsilon \in K'$ with $\mathrm{val}(\alpha_\epsilon) = \epsilon > 0$ that is generic in the sense above. Proposition 3.4.8 implies that $\dim_{K'}(S_{K',n-d+1}/J' + \langle x_1 - \alpha_\epsilon \rangle)$ is the sum of the multiplicities of the points in the finite set $\mathrm{trop}(V(J' + \langle x_1 - \alpha_\epsilon \rangle))$. Since the intersection of $\mathrm{trop}(V(J'))$ and $\mathrm{trop}(V(x_1 - \alpha_\epsilon)) = \{\mathbf{w} : w_1 = \epsilon\}$ is transverse at all points of their intersection, by Theorem 3.4.12 we have $\mathrm{trop}(V(J' + \langle x_1 - \alpha_\epsilon \rangle)) = \mathrm{trop}(V(J')) \cap \mathrm{trop}(V(x_1 - \alpha_\epsilon))$. Now $\mathrm{trop}(V(x_1 - \alpha_\epsilon)) = \epsilon \mathbf{v} + H$ for any generic $\mathbf{v}$ with $v_1 = 1$. By Proposition 3.6.12, the multiplicity of the origin in $\mathrm{trop}(V(J')) \cap_{st} H$ equals the limit as $\epsilon \to 0$ of the sum of the multiplicities of $\mathrm{trop}(V(J' + \langle x_1 - \alpha_\epsilon \rangle))$. For all but finitely many $\alpha \in K$, this is the dimension D of $S_{n-d+1}/(J + \langle x_1 - \overline{\alpha} \rangle)$ as required. □

Proof of Theorem 3.6.1. Write $x_1, \ldots, x_n, y_1, \ldots, y_n$ for coordinates on $\mathbb{R}^{2n}$. By Lemma 3.6.9, for any balanced weighted complex $\Sigma \in \mathbb{R}^{2n}$, $(\Sigma \cap_{st} \{\mathbf{w} : w_1{=}0\}) \cap_{st} \cdots \cap_{st} \{\mathbf{w} : w_n{=}0\} = \Sigma \cap_{st} \{\mathbf{w} : w_1{=}\cdots{=}w_n{=}0\}$. Using Proposition 3.6.15 and the change of coordinates $x_i \mapsto x_i/y_i$, $y_i \mapsto y_i$, this identity implies the following fact. For any variety $Z \subset T^{2n}$ there exists a dense set $U \subset T^n$ such that $\mathrm{val}(\alpha) = \mathbf{0}$ and

$$(\mathrm{trop}(Z) \cap_{st} \mathrm{trop}(V(x_1 - \alpha_1 y_1))) \cap_{st} \cdots \cap_{st} \mathrm{trop}(V(x_n - \alpha_n y_n)) \\ = \mathrm{trop}(Z \cap V(x_1 - \alpha_1 y_1, \ldots, x_n - \alpha_n y_n))$$

for all $\alpha \in U$. The fact that U is dense follows from Lemma 2.2.12.

Let I and J be the ideals for X_1 and X_2, respectively, in $K[x_1^{\pm 1}, \ldots, x_n^{\pm 1}]$. Given $\mathbf{t} = (t_1, \ldots, t_n) \in T^n$, write $J' = \mathbf{t}^{-1}J = \langle f(t_i^{-1}y_i) : f \in J \rangle$ for the ideal of $\mathbf{t}Y$. By the proof of Theorem 3.6.10, we have $\mathrm{trop}(X) \cap_{st} \mathrm{trop}(\mathbf{t}Y) \cong (\mathrm{trop}(X) \times \mathrm{trop}(\mathbf{t}Y)) \cap_{st} \Delta$, where $\Delta = \{(\mathbf{x}, \mathbf{y}) \in \mathbb{R}^{2n} : x_i = y_i \text{ for } 1 \leq i \leq n\}$ is the diagonal in $\mathbb{R}^{2n}$. So, the stable intersection we are interested in equals

$$(\mathrm{trop}(X) \times \mathrm{trop}(\mathbf{t}Y)) \cap_{st} \mathrm{trop}(V(x_i - y_i : 1 \leq i \leq n)). \tag{3.6.11}$$

The transformation $y_i' = t_i^{-1} y_i$ changes none of these tropical varieties when $\mathrm{val}(t_i) = 0$ for all i, so (3.6.11) equals

$$(\mathrm{trop}(X) \times \mathrm{trop}(Y)) \cap_{st} \mathrm{trop}((V(x_i - t_i y_i') : 1 \leq i \leq n)).$$

By the first paragraph, there is a dense set $U \subset T^n$ such that (3.6.11) equals $\mathrm{trop}((X \times \mathbf{t}Y) \cap V(x_i - y_i : 1 \leq i \leq n)) \simeq \mathrm{trop}(X \cap \mathbf{t}Y)$ for all $\mathbf{t} \in U$. □

We close this section with an application of Theorem 3.6.1. Further applications will be seen in Section 4.6. Recall that the *degree* of a projective variety $\overline{X} \subset \mathbb{P}^n$ of dimension d is the number of intersection points, counted with multiplicity, of $\overline{X}$ with a generic subspace of dimension $n-d$. Let L_{n-d} be the standard tropical linear space of dimension $n-d$ in $\mathbb{R}^{n+1}/\mathbb{R}\mathbf{1}$; this consists of all cones $\mathrm{pos}(\mathbf{e}_{i_1}, \ldots, \mathbf{e}_{i_{n-d}})$ where $0 \leq i_1 < \cdots < i_{n-d} \leq n$.

Corollary 3.6.16. *Let $\overline{X} \subseteq \mathbb{P}^n_K$ be an irreducible projective variety of dimension d, and let $X = \overline{X} \cap T^n$. Let K have the trivial valuation. The degree of $\overline{X}$ is the multiplicity of the origin in the stable intersection of* $\mathrm{trop}(X)$ *with the tropical linear space L_{n-d}:*

$$\deg(\overline{X}) \;=\; \mathrm{mult}_{\mathbf{0}}(\mathrm{trop}(X) \cap_{st} L_{n-d}).$$

Proof. We first show that there is an open set U_1 in the Grassmannian $G(n-d, n+1)$ parameterizing codimension-d subspaces of K^{n+1} for which if $L \in U_1$, with $L^\circ = (L \cap (K^*)^{n+1})/K^* \subset (K^*)^{n+1}/K^* \cong T^n$, then $\mathrm{trop}(L^\circ) = L_{n-d}$. Indeed, let U_1 be the open set consisting of those $L \in G(n-d, n+1)$ for which all Plücker coordinates are nonzero. Such an L has the property that for any subset $J = \{j_1, \ldots, j_d\} \subset \{0, 1, \ldots, n\}$ the ideal I_L of L has a generating set $\ell_1, \ldots, \ell_d$ with $\mathrm{supp}(\ell_i) \cap J = \{x_{j_i}\}$. Any $\mathbf{w} \notin L_{n-d}$ has $\min(w_0, \ldots, w_n)$ achieved at at most d indices, so choosing J to contain these indices, we see that $\mathrm{in}_{\mathbf{w}}(I_L)$ contains a monomial, and so $\mathbf{w} \notin \mathrm{trop}(L^\circ)$. This shows that $\operatorname{trop}(L^\circ) \subseteq L_{n-d}$. As L_{n-d} has the same dimension as L°, and no subfan can be balanced, we conclude that $\operatorname{trop}(L^\circ) = L_{n-d}$. (This is generalized in Example 4.2.13.) Since $\dim(\overline{X}\backslash X) < \dim(X)$, there is an open set $U_2 \subset G(n-d, n+1)$ for which if $L \in U_2$, then $L \cap \overline{X} = L \cap X$. There is also an open set $U_3 \subset G(n-d, n+1)$ for which if $L \in U_3$, then $\deg(\overline{X}) = |\overline{X} \cap L|$, where the latter is counted with multiplicity.

Fix $L \in U_1 \cap U_2 \cap U_3$. Let $U_4 \subset T^n$ be the open subset consisting of those $\mathbf{t} \in T^n$ for which $\mathbf{t}L \in U_1 \cap U_2 \cap U_3$. By Theorem 3.6.1 there is $\mathbf{t} \in U_4$ for which $\operatorname{trop}(X \cap \mathbf{t}L^\circ) = \operatorname{trop}(X) \cap_{st} \operatorname{trop}(L^\circ) = \operatorname{trop}(X) \cap_{st} L_{n-d}$. By Proposition 3.4.8 $\operatorname{trop}(X \cap \mathbf{t}L^\circ)$ is the origin with multiplicity equal to the number of points in $X \cap \mathbf{t}L^\circ$, counted with multiplicity. Since $\mathbf{t}L \in U_2 \cap U_3$, the multiplicity of the origin is thus the degree of $\overline{X}$, as required. □

3.7. Exercises

(1) Draw $\operatorname{trop}(V(f))$ for the following $f \in \mathbb{C}\{\{t\}\}[x^{\pm 1}, y^{\pm 1}]$:
  (a) $f = t^3x + (t + 3t^2 + 5t^4)y + t^{-2}$;
  (b) $f = (t^{-1} + 1)x + (t^2 - 3t^3)y + 5t^4$;
  (c) $f = t^3x^2 + xy + ty^2 + tx + y + 1$;
  (d) $f = 4t^4x^2 + (3t + t^3)xy + (5 + t)y^2 + 7x + (-1 + t^3)y + 4t$;
  (e) $f = tx^2 + 4xy - 7y^2 + 8$;
  (f) $f = t^6x^3 + x^2y + xy^2 + t^6y^3 + t^3x^2 + t^{-1}xy + t^3y^2 + tx + ty + 1$.

(2) By Example 3.1.11, the tropical hypersurface of the 3×3-determinant has 15 maximal cones. These come in two symmetry classes. Pick two representatives σ_1 and σ_2, and find matrices $\mathbf{w}_1$ and $\mathbf{w}_2$ in $\mathbb{Q}^{3\times 3}$ that satisfy $\mathbf{w}_i \in \operatorname{relint}(\sigma_i)$ for $i = 1, 2$. Next construct rank 2 matrices M_1 and M_2 in $\mathbb{C}\{\{t\}\}^{3\times 3}$ with $\operatorname{val}(M_i) = \mathbf{w}_i$ for $i = 1, 2$.

(3) The Pfaffian of a skew-symmetric 6×6-matrix is a polynomial of degree 3 in 15 variables. Compute its tropical hypersurface.

(4) Verify (as much as possible) the Fundamental Theorem 3.2.3 and the Structure Theorem 3.3.5 for the six curves in Exercise 3.7(1).

(5) Draw the recession fan for the six plane curves in Exercise 3.7(1).

(6) Let $K = \mathbb{Q}$ with the 3-adic valuation. Construct two explicit distinct quadratic polynomials $f, g \in K[x_1, x_2, x_3, x_4]$ which form a tropical basis for the Laurent polynomial ideal they generate.

(7) Using your f and g in Exercise 3.7(6), compute the elimination ideal $\langle f, g\rangle \cap K\left[x_1x_2^{-1}, x_2x_3^{-1}, x_3x_4^{-1}\right]$. Interpret your result geometrically.

(8) Let $Y = V(x_1 + x_2 + x_3 + x_4 + 1, x_2 - x_3 + x_4) \subseteq (\mathbb{C}^*)^4$. Compute $\mathrm{trop}(Y)$ and a polyhedral fan Σ with support $\mathrm{trop}(Y)$. Show that Σ is balanced if we put the weight one on each maximal cone.

(9) Give an example to show that the tropicalization of a hypersurface might be a fan even if some of the coefficients have nonzero valuation. What sort of converse can you give to Proposition 3.1.10?

(10) What is the largest multiplicity of any edge in the tropicalization of any plane curve of degree d? How about surfaces in 3-space?

(11) For f in Example 3.1.2(2) and the vertex $\mathbf{w} = (-1, 0)$ on the right in Figure 3.1.1, describe all points $\mathbf{y} \in V(f)$ with $\mathrm{val}(\mathbf{y}) = \mathbf{w}$. For this example, verify that the set of such $\mathbf{y}$ is Zariski dense in $V(f)$.

(12) Let I be the ideal in $\mathbb{C}[x_1^{\pm 1}, \ldots, x_4^{\pm 1}]$ generated by the five elements

$$\begin{aligned}
&(x_1 + x_3)^2(x_3 + x_4),\\
&(x_1 + x_2)(x_1 + x_4)^2,\\
&(x_1 + x_3)^2(x_1 + x_4),\\
&(x_1 + x_2)(x_1 + x_3)(x_1 + x_4),\\
&(x_1 + x_2)(x_1 + x_3)(x_3 + x_4)^2.
\end{aligned}$$

Find all associated primes of I and an explicit primary decomposition. Compute the tropical variety $\mathrm{trop}(V(I))$ with multiplicities.

(13) Let X and Y be subvarieties of T^n, and let $\Sigma = \mathrm{trop}(X) + \mathrm{trop}(Y)$ be the Minkowski sum of their tropicalizations in $\mathbb{R}^n$. Show that Σ is a tropical variety. Explain how to construct a subvariety $Z \subset T^n$ such that $\Sigma = \mathrm{trop}(Z)$.

(14) Compute generators for the ideal J_{proj} in Example 3.2.9. List the 12 maximal cones in the Gröbner fan structure on $\mathrm{trop}(V(J_{\mathrm{proj}}))$.

(15) True or false: The transverse intersection of two balanced polyhedral complexes in $\mathbb{R}^n$ is again a balanced polyhedral complex?

(16) Show that the k-skeleton of any n-dimensional polytope is connected through codimension 1. Get started with k=1 and n=3.

(17) Let $f(x, y)$ be the polynomial in Example 1.5.1. Compute the multiplicities of all rays in the one-dimensional fan $\mathrm{trop}(V(f))$.

(18) Describe a method for computing the multiplicity $\mathrm{mult}(P_i, I)$ defined in (3.4.1). Try it on some examples, e.g., using `Macaulay2`.

(19) Let P be the prime ideal generated by the 2×2-minors of a 3×3-matrix of unknowns. Compute $\mathrm{mult}(P, P^n)$ for $n = 1, 2, 3, \ldots$.

(20) Fix $\mathbf{w} = (1, 1)$ and the polynomials

$$\begin{aligned} f &= xy - tx - ty + t^2, \\ g &= x^2 - (t^2 + 2t)x + t^3 + t^2, \\ h_1 &= y^2 - (t^2 + t)y + t^3, \\ h_2 &= y^2 - (t^2 + 2t)y + t^3 + t^2. \end{aligned}$$

Which of the two ideals $I_1 = \langle f, g, h_1 \rangle$ and $I_2 = \langle f, g, h_2 \rangle$ in $\mathbb{C}\{\{t\}\}[x^{\pm 1}, y^{\pm 1}]$ satisfies the conclusion of Proposition 3.4.8?

(21) Consider the two tropical planes in $\mathbb{R}^3$ defined by

$$\begin{aligned} & a_1 \odot x \oplus a_2 \odot y \oplus a_3 \odot z \oplus a_4 \\ \text{and} \quad & b_1 \odot x \oplus b_2 \odot y \oplus b_3 \odot z \oplus b_4. \end{aligned}$$

Find necessary and sufficient conditions, in terms of $a_1, a_2, \ldots, b_4$, for these to meet transversally at every point in their intersection.

(22) Let $X \subset T^{12}$ be the variety of 3×4-matrices of rank at most 2. Determine a fan structure on $\operatorname{trop}(X)$. Verify that it is connected through codimension 1. Draw the graph on the maximal cones.

(23) Given two polyhedral complexes Σ and Σ' in $\mathbb{R}^n$, show that

$$\operatorname{rec}(\Sigma \cap \Sigma') = \operatorname{rec}(\Sigma) \cap \operatorname{rec}(\Sigma'),$$

and explain how to construct a fan structure on this set.

(24) Show that if Σ is an n-dimensional weighted balanced Γ_{val}-rational polyhedral complex in $\mathbb{R}^n$, then the support $|\Sigma|$ is all of $\mathbb{R}^n$ and the weight on each n-dimensional polyhedron is the same.

(25) Show that if L is a (classical) linear space contained in the lineality space of two weighted balanced Γ_{val}-rational polyhedral complexes $\Sigma_1, \Sigma_2 \subseteq \mathbb{R}^n$, then L is contained in the lineality space of the stable intersection $\Sigma_1 \cap_{st} \Sigma_2$, and $(\Sigma_1/L) \cap_{st} (\Sigma_2/L) = (\Sigma_1 \cap_{st} \Sigma_2)/L$.

(26) Let L be a sublattice of rank n in $\mathbb{Z}^n$ that is generated by the columns of an $n \times r$-matrix A. Show that the index $[\mathbb{Z}^n : L]$ is the greatest common divisor of the maximal nonzero minors of A.

(27) Let L_1 and L_2 be tropical linear spaces in $\mathbb{R}^n/\mathbb{R}\mathbf{1}$. Show that their stable intersection $L_1 \cap_{st} L_2$ is a tropical linear space. Express the Plücker coordinates of $L_1 \cap_{st} L_2$ in terms of those of L_1 and L_2.

(28) According to Example 3.1.11, the tropical 3×3-determinant $X = \operatorname{trop}(V(f))$ is an eight-dimensional fan in $\mathbb{R}^9$. Compute the fans $X \cap_{st} X$ and $X \cap_{st} X \cap_{st} X$. Realize these two fans as the tropicalizations of two explicit varieties in the torus of 3×3-matrices.

(29) The Grassmannian $\overline{X} = G(2, 5)$ is a variety of dimension 6 in $\mathbb{P}^9$. See Proposition 2.2.10. Use Corollary 3.6.16 to compute $\deg(\overline{X})$.

(30) Find two tropical surfaces in $\mathbb{R}^3$ whose stable intersection is empty. Show that your surfaces arise from projective surfaces in $\mathbb{P}^3$. Find an example where your two projective surfaces have degree 1000.

(31) Compute $\operatorname{trop}(X)$ for $X = V(\pi x^2 + ey^2 + \sqrt{2}, \zeta(3)xyz + 1) \subseteq T^3_{\mathbb{C}}$.

(32) Fix $K = \mathbb{Q}$ with the p-adic valuation where $p = 2$ or $p = 3$. The discriminant of Example 3.3.3 is a polynomial in $K[a, b, c, d, e]$ whose tropicalization F now has some nonzero coefficients. For both primes, compute the polyhedral complex Σ_F and the tropical variety $V(F)$. Find the weights, and explain why $V(F)$ is balanced.

(33) Let $X \subset T^5$ be the variety given by the parameterization in Example 3.5.4. Find the ideal of X and compute the tropicalization $\operatorname{trop}(X)$.

(34) Let I be a homogeneous ideal in $K[x_1, \ldots, x_n]$, and let $\mathbf{w}$ be in the relative interior of a maximal cell σ of $\operatorname{trop}(V(I))$. Let P be the toric ideal associated with the lattice $\{\mathbf{u} \in \mathbb{Z}^n : \operatorname{in}_{\mathbf{u}}(\operatorname{in}_{\mathbf{w}}(I)) = \operatorname{in}_{\mathbf{w}}(I)\}$, as in [**MS05, Stu96**]. Show that the multiplicity of σ can be computed in `Macaulay2` by the formula

$$\operatorname{mult}(\sigma) = \operatorname{degree}(\operatorname{in}_{\mathbf{w}}(I))/\operatorname{degree}(P).$$

Chapter 4

Tropical Rain Forest

Forests are made up of trees. This is also true for the tropical rain forest: trees and their parameter spaces are fundamental in tropical geometry. There is a lot of diversity in the forest we explore in this chapter. We begin with linear spaces, the simplest among classical varieties. Their tropical counterparts are similarly fundamental, and they are intimately connected to the study of hyperplane arrangements. On the combinatorial side this leads us to the theory of matroids. Tropicalized linear spaces are parameterized by the Grassmannian, and arbitrary linear spaces are parameterized by the Dressian. This mirrors the distinction between realizable and nonrealizable matroids. We focus on the Grassmannian $G(2, n)$, which parameterizes lines in the projective space $\mathbb{P}^{n-1}$, and we identify its tropicalization with the space of phylogenetic trees from computational biology. We then investigate surfaces in three-dimensional space, examining the tropical shadow of classical phenomena such as the rulings of a quadric surface. Finally, we study the tropicalization of a complete intersection. Bernstein's Theorem states that the expected number of solutions to a system of n Laurent polynomial equations in n unknowns equals the mixed volume of their Newton polytopes. We give a tropical proof of this result and explore what happens when the number of equations is less than the number of unknowns.

In this chapter, the theorems of Chapter 3 appear in practice. The focus is on investigating applications rather than on developing abstract theory.

4.1. Hyperplane Arrangements

Let $\mathcal{A} = \{H_i : 0 \leq i \leq n\}$ be an arrangement of $n + 1$ hyperplanes in $\mathbb{P}^d$. We are interested in its complement, $X = \mathbb{P}^d \setminus \cup\mathcal{A}$. In what follows we show that X is naturally a closed subvariety of the torus T^n, and it is cut out by a linear system of equations. This allows us to identify hyperplane arrangements and linear spaces. Our goal is to derive the tropicalization of X from the combinatorics of $\mathcal{A}$. Throughout this section we assume that all coefficients of the defining linear equations live in a subfield of K with trivial valuation. The general case, where the valuation matters, will be revisited in Section 4.4. The particular choice of field plays no role; only the characteristic matters. To gain a first intuition, it is best to think of $K = \mathbb{C}$.

Write $\mathbf{b}_i \in K^{d+1}$ for a normal vector of the hyperplane H_i, so $H_i = \{\mathbf{z} \in \mathbb{P}^d : \mathbf{b}_i \cdot \mathbf{z} = 0\}$. We assume that $\mathbf{b}_0, \ldots, \mathbf{b}_n$ span K^{d+1}. Geometrically, this means that the hyperplanes in $\mathcal{A}$ have no common intersection point.

We fix the torus $T^n = (K^*)^{n+1}/K^*$ in $\mathbb{P}^n$. The vectors $\mathbf{b}_i$ define a map

$$X \to T^n, \quad \mathbf{z} \mapsto (\mathbf{b}_0 \cdot \mathbf{z} : \mathbf{b}_1 \cdot \mathbf{z} : \cdots : \mathbf{b}_n \cdot \mathbf{z}). \tag{4.1.1}$$

This map is injective, since the $\mathbf{b}_i$ span K^{d+1}. The image is a closed subset of T^n which we now describe. Write B for the $(d+1) \times (n+1)$-matrix whose columns are the $\mathbf{b}_i$, and let $A = (a_{ij})$ be an $(n-d) \times (n+1)$-matrix whose rows are a basis for the kernel of B. Let I be the ideal in $K[x_0^{\pm 1}, \ldots, x_n^{\pm 1}]$ generated by the linear forms $f_i = \sum_{j=0}^{n} a_{ij} x_j$ for $1 \leq i \leq n-d$. Since I is homogeneous, its variety in $(K^*)^{n+1}$ is fixed by the diagonal action of K^*. Throughout this section we write $V(I)$ for the variety in $(K^*)^{n+1}/K^* = T^n$.

Proposition 4.1.1. *The map* (4.1.1) *defines an isomorphism between the arrangement complement* $X = \mathbb{P}^d \setminus \cup\mathcal{A}$ *and the subvariety* $V(I)$ *of* T^n.

Proof. The image of $X = \mathbb{P}^d \setminus \cup\mathcal{A}$ under the injective map (4.1.1) lies in $V(I)$ because the rows of B are in the kernel of A. Conversely, if $\mathbf{x} \in V(I)$, then $\mathbf{x}$ lies in the kernel of A, so $\mathbf{x} = B^T\mathbf{z}$ for a unique vector $\mathbf{z} \in K^{d+1}$. Since each coordinate of $\mathbf{x}$ is nonzero, we have $\mathbf{z} \notin \cup\mathcal{A}$, so $\mathbf{z} \in X$. This inverse map is given by a linear map, so it is a morphism as well. $\square$

Example 4.1.2. Let $\mathcal{A}$ be the arrangement in $\mathbb{P}^2$ consisting of the lines $H_0 = \{x_0{=}0\}$, $H_1 = \{x_1{=}0\}$, $H_2 = \{x_2{=}0\}$, $H_3 = \{x_0{=}x_1\}$, $H_4 = \{x_0{=}x_2\}$, and $H_5 = \{x_1{=}x_2\}$. See Figure 4.1.1. The matrix B is then

$$B = \begin{pmatrix} 1 & 0 & 0 & 1 & 1 & 0 \\ 0 & 1 & 0 & -1 & 0 & 1 \\ 0 & 0 & 1 & 0 & -1 & -1 \end{pmatrix},$$

and we can choose A to be the 3×6-matrix

$$A = \begin{pmatrix} 1 & -1 & 0 & -1 & 0 & 0 \\ 1 & 0 & -1 & 0 & -1 & 0 \\ 0 & 1 & -1 & 0 & 0 & -1 \end{pmatrix}.$$

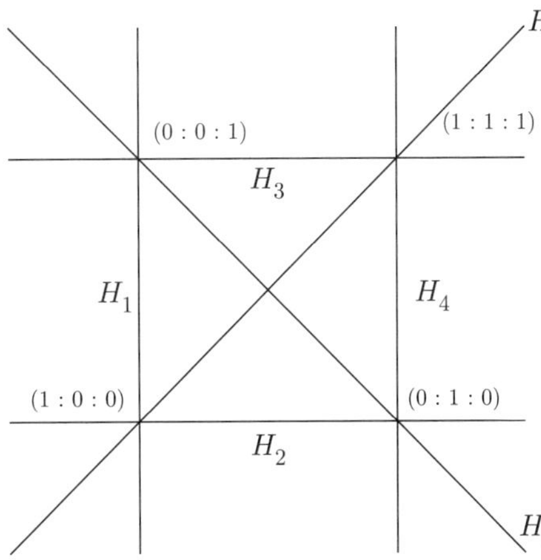

Figure 4.1.1. The line arrangement of Example 4.1.2.

The ideal defined by the matrix A equals

$$I \;=\; \langle x_0 - x_1 - x_3,\ x_0 - x_2 - x_4,\ x_1 - x_2 - x_5\rangle \;\subset\; K[x_0^{\pm 1},\ldots,x_5^{\pm 1}].$$

This linear ideal defines a plane in $\mathbb{P}^5$, and $V(I)$ is the intersection of that plane with the torus T^5. Proposition 4.1.1 identifies the linear variety $V(I)$ with the complement $\mathbb{P}^2 \backslash \cup \mathcal{A}$ of our arrangement of six lines in the plane. ◊

By reversing the construction in Proposition 4.1.1, we see that *any* ideal I generated by linear forms arises from some hyperplane arrangement. If the linear forms are not homogeneous, we can homogenize the ideal. We recover the tropical variety of X from its homogenization using Proposition 2.6.1.

Example 4.1.3. Consider the ideal

$$J = \langle x_1 + x_2 + x_3 + x_4 + 1, x_1 + 2x_2 + 3x_3\rangle \subset K[x_1^{\pm 1}, x_2^{\pm 1}, x_3^{\pm 1}, x_4^{\pm 1}],$$

which defines a two-dimensional subvariety X of T^4. The homogenization of J is the ideal

$$I = \langle x_0 + x_1 + x_2 + x_3 + x_4, x_1 + 2x_2 + 3x_3\rangle.$$

In this section, the ideal I lives in $K[x_0^{\pm 1}, x_1^{\pm 1}, x_2^{\pm 1}, x_3^{\pm 1}, x_4^{\pm 1}]$. The variety $X = V(I)$ is the complement of five lines in the plane $\mathbb{P}^2$. ◊

We now describe the tropical variety of $X = V(I)$. The *support* of a linear form $\ell = \sum a_i x_i \in I$ is $\mathrm{supp}(\ell) = \{i : a_i \neq 0\}$. A nonempty subset C of $\{0, 1, \ldots, n\}$ is a *circuit* of I if $C = \mathrm{supp}(\ell)$ for some nonzero linear form ℓ in the ideal I, and C is inclusion-minimal with this property. Equivalently, C is a minimal linearly dependent subset of the columns of the matrix B. In terms of the hyperplane arrangement $\mathcal{A}$, a set C is a circuit exactly when $\bigcap_{i\in C} H_i$ has codimension $|C| - 1$ and is equal to $\bigcap_{i\in C, i\neq j} H_i$ for all $j \in C$.

We record some facts about the circuits of a hyperplane arrangement.

Lemma 4.1.4. *Let I be the ideal in $K[x_0^{\pm 1}, \ldots, x_n^{\pm 1}]$ that is generated by the linear forms associated with a $(d+1) \times (n+1)$-matrix B of rank $d+1$.*

(1) *Up to scaling, every circuit $C \subseteq \{0, 1, \ldots, n\}$ uniquely determines the linear form ℓ_C that lies in I and satisfies $C = \mathrm{supp}(\ell_C)$.*

(2) *The set of linear forms ℓ_C in I that correspond to circuits C of I is the union of all reduced Gröbner bases for $I \cap K[x_0, \ldots, x_n]$.*

(3) *Each circuit C is determined by a spanning subset of $d+1$ columns of B plus one more column. Given such a set $L = \{i_1, \ldots, i_{d+2}\}$,*

$$\ell_C \quad = \quad \sum_{j=1}^{d+2} (-1)^{j-1} \det(B_{L \backslash i_j}) x_{i_j}, \tag{4.1.2}$$

where $B_{L \backslash i_j}$ is the square submatrix of B with column indices $L \backslash i_j$.

(4) *The ideal I has at most $\binom{n+1}{d+2}$ circuits. This bound is achieved for those matrices B whose $(d+1) \times (d+1)$-minors are all nonzero.*

Remark 4.1.5. Part (2) of Lemma 4.1.4 is essentially the statement that Buchberger's algorithm for computing Gröbner bases reduces to Gaussian elimination when the ideal in question is generated by linear forms. In part (3) the support C of the linear form ℓ_C is a (possibly proper) subset of L. The point is that C is uniquely determined by the $(d+2)$-element set L.

Proof. If ℓ_1, ℓ_2 are linear forms with the same support C and $i \in C$, then there is $\lambda \in K$ for which the coefficient of x_i in $\lambda \ell_1$ equals the coefficient of x_i in ℓ_2. The combination $\lambda \ell_1 - \ell_2$ has strictly smaller support. If C is a circuit, then this smaller support must be empty, which means that $\lambda \ell_1 = \ell_2$.

For part (2), let $\mathcal{G}$ be the union of all reduced Gröbner bases for $J := I \cap K[x_0, \ldots, x_n]$, with multiples removed. We first show that if $\ell := \sum_{i \in C} a_i x_i \in \mathcal{G}$, then C is a circuit of I. If not, there would be $\sum_{i \in C} b_i x_i \in I$ with some $b_j = 0$, and some $b_k \neq 0$. Suppose that ℓ is in the reduced Gröbner basis for the term order $\prec$ and $x_l = \mathrm{in}_\prec(\ell)$. This means that $a_l = 1$ and $x_i \notin \mathrm{in}_\prec(J)$ for $i \in C \backslash \{l\}$. The monomials not in $\mathrm{in}_\prec(J)$ form a basis for $K[x_0, \ldots, x_n]/J$, so there is no linear form in J with support in $C \backslash \{l\}$. This means that $b_l \neq 0$. But then $\ell - 1/b_l \sum_{i \in C} b_i x_i$ is a nonzero element of J and has support contained in C, which is a contradiction.

For the other inclusion, let C be a circuit of I, fix $j \in C$, and let ℓ be the corresponding linear form, scaled so that the coefficient of x_j in ℓ is 1. Choose a term order where x_j is the largest variable in C, and all x_i for $i \in C \backslash \{j\}$ are smaller than the remaining variables. The linear form ℓ

must appear in the reduced Gröbner basis; otherwise, it could be reduced by another element in the reduced Gröbner basis. If subtracting an element h of the reduced Gröbner basis cancels x_j, the resulting polynomial lies in I but has no term in the initial ideal of I, which is impossible. Thus h must cancel some x_i with $i \in C \setminus \{j\}$, so must have support in $C \backslash \{j\}$ since the variables in $C \backslash \{j\}$ are the smallest. Such a linear form cannot exist since C is a circuit. We conclude that ℓ is in $\mathcal{G}$.

For parts (3) and (4) it follows from the previous paragraphs that every circuit C is uniquely determined by giving an initial ideal $\text{in}_\prec(J)$ of J and a choice of generator x_j for this. Let $L' = \{i : x_i \notin \text{in}_\prec(J)\}$. The columns of B indexed by L' are a basis of K^{d+1}. Since there are $\binom{n+1}{d+2}$ subsets $L' \cup \{x_j\}$ of size $d+2$ in $\{0, 1, \ldots, n\}$, there are at most that many circuits.

The formula for ℓ_C in (4.1.2) is the determinant of the $(d+2) \times (d+2)$-matrix whose first row is $(x_{i_1}, \ldots, x_{i_{d+2}})$ and whose other rows are those of the submatrix B_L. That linear form is in J (and hence in I) because the determinant is zero if we replace that first row by any of the rows of B_L.

The bound on the number of circuits in part (4) is achieved exactly when every subset of $n - d$ variables spans an initial ideal for J, which happens exactly when all $(d+1) \times (d+1)$-minors of the matrix B are nonzero. □

Our first tropical result says that the circuits form a tropical basis for I.

Proposition 4.1.6. *Let $I \subseteq K[x_0^{\pm 1}, \ldots, x_n^{\pm 1}]$ be generated by linear forms where K has the trivial valuation, and consider the hyperplane arrangement complement $X = V(I)$. The set of linear polynomials ℓ_C in I whose supports are circuits is a tropical basis for I. Equivalently, a vector $\mathbf{w} \in \mathbb{R}^{n+1}/\mathbb{R}\mathbf{1}$ lies in* $\text{trop}(X)$ *if and only if for any circuit C of the ideal I the minimum of the coordinates w_i, as i ranges over C, is attained at least twice.*

Proof. Every circuit is the support of a linear form ℓ that lies in the ideal I. Hence the "only-if" direction is immediate from Definition 3.2.1 with $X = V(I)$. For the "if" direction suppose that $\mathbf{w} \in \mathbb{R}^{n+1}$ is not in $\text{trop}(X)$. Compute the reduced Gröbner basis of $J = I \cap K[x_0, \ldots, x_n]$ with respect to a term order that refines $\mathbf{w}$ in the usual sense of Gröbner bases (see [**Stu96**, Corollary 1.9]). By part (2) of Lemma 4.1.4 this consists of linear forms supported on circuits. The initial ideal $\text{in}_\mathbf{w}(J)$ is generated by the leading forms of these linear forms, which are themselves linear forms, so this initial ideal is prime. In addition these leading forms form a Gröbner basis for $\text{in}_\mathbf{w}(J)$. Our hypothesis states that $\mathbf{w} \notin \text{trop}(X)$, so $\text{in}_\mathbf{w}(J)$ contains a monomial. Since $\text{in}_\mathbf{w}(J)$ is prime, this implies that some variable x_i lies in $\text{in}_\mathbf{w}(J)$. Some element f of the reduced Gröbner basis has leading term x_i. In fact, the entire leading form must be x_i, as otherwise the remainder on division of x_i by $\mathrm{in}_{\mathbf{w}}(J)$ would not be zero. This means that the minimum of w_i for i in the corresponding circuit $C = \mathrm{supp}(f)$ is attained only once. □

Example 4.1.7. Let J be as in Example 4.1.3. The circuits are $\{1,2,3\}$, $\{0,2,3,4\}, \{0,1,3,4\}$, and $\{0,1,2,4\}$. These correspond to the linear polynomials $x_1+2x_2+3x_3, x_2+2x_3-x_4-1, x_1-x_3+2x_4+2$, and $2x_1+x_2+3x_4+3$. The circuits do not all have the same size here. By Proposition 4.1.6

$$\begin{aligned}\mathrm{trop}(X) = \mathrm{trop}(V(x_1+2x_2+3x_3)) &\cap \mathrm{trop}(V(x_2+2x_3-x_4-1)) \\ \cap\ \mathrm{trop}(V(x_1-x_3+2x_4+2)) &\cap \mathrm{trop}(V(2x_1+x_2+3x_4+3)).\end{aligned}$$

In fact, $\mathrm{trop}(X)$ is the intersection of the first three of these tropical hyperplanes. Hence the circuits are not always a minimal tropical basis. ◊

We now give a combinatorial description of the tropicalization $\mathrm{trop}(X)$ of a linear variety X in T^n. A key ingredient will be the *lattice of flats* of X.

Let $\mathcal{B} = \{\mathbf{b}_0, \ldots, \mathbf{b}_n\} \subset K^{d+1}$ be the columns of the matrix B. While $\mathcal{B}$ depends on the choice of the matrix B, it is determined up to the action of $\mathrm{GL}(d+1, K)$. A circuit in $I(X)$ is a minimal linear dependence among the vectors $\mathbf{b}_i$. The *lattice of flats* $\mathcal{L}(B)$ of the linear variety X is the set of subspaces (flats) of K^{d+1} that are spanned by subsets of $\mathcal{B}$. We make $\mathcal{L}(B)$ into a poset (partially ordered set) by setting $S_1 \preceq S_2$ if $S_1 \subseteq S_2$ for two subspaces S_1, S_2 of K^{d+1} spanned by subsets of $\mathcal{B}$. The poset $\mathcal{L}(B)$ is a lattice of rank $d+1$, so every maximal chain in $\mathcal{L}(B)$ has length $d+1$. See, for example, [**Sta12**, Chapter 3] for more on lattices.

Example 4.1.8. We continue Example 4.1.3. The matrices A and B are

$$A = \begin{pmatrix} 1 & 1 & 1 & 1 & 1 \\ 0 & 1 & 2 & 3 & 0 \end{pmatrix}, \qquad B = \begin{pmatrix} 0 & -2 & 1 & 0 & 1 \\ 1 & -2 & 1 & 0 & 0 \\ 0 & -1 & -1 & 1 & 1 \end{pmatrix}.$$

The set $\mathcal{B}$ consists of the vectors $\mathbf{b}_0 = (0,1,0)$, $\mathbf{b}_1 = (-2,-2,-1)$, $\mathbf{b}_2 = (1,1,-1)$, $\mathbf{b}_3 = (0,0,1)$, and $\mathbf{b}_4 = (1,0,1)$. The flats of X are the 15 subspaces of K^3 that are spanned by subsets of $\mathcal{B} = \{\mathbf{b}_0, \ldots, \mathbf{b}_4\}$. These are

(1) $\mathrm{span}(\emptyset) = \{\mathbf{0}\}$;

(2) $\mathrm{span}(\mathbf{b}_i)$ for $0 \le i \le 4$;

(3) $\mathrm{span}(\mathbf{b}_0, \mathbf{b}_1)$, $\mathrm{span}(\mathbf{b}_0, \mathbf{b}_2)$, $\mathrm{span}(\mathbf{b}_0, \mathbf{b}_3)$, $\mathrm{span}(\mathbf{b}_0, \mathbf{b}_4)$, $\mathrm{span}(\mathbf{b}_1, \mathbf{b}_2, \mathbf{b}_3)$, $\mathrm{span}(\mathbf{b}_1, \mathbf{b}_4)$, $\mathrm{span}(\mathbf{b}_2, \mathbf{b}_4)$, $\mathrm{span}(\mathbf{b}_3, \mathbf{b}_4)$;

(4) $\mathrm{span}(\mathcal{B}) = K^3$.

A Hasse diagram for the lattice of flats $\mathcal{L}(B)$ is shown in Figure 4.1.2. ◊

Associated to any poset is a simplicial complex, called the *order complex* of the poset. Its vertices are the elements of the poset, and its simplices are all proper chains, which are totally ordered subsets of the poset not using

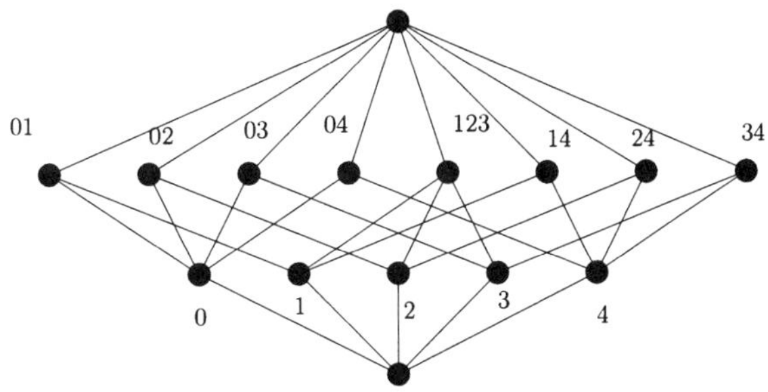

Figure 4.1.2. The lattice of flats for the linear space of Examples 4.1.3 and 4.1.8.

the bottom or top elements ($\{\mathbf{0}\}$ or K^{d+1} in our case). The order complex of the lattice of flats $\mathcal{L}(B)$ is pure of dimension $d-1$. There is a nice geometric realization of this simplicial complex as a fan, which we now describe.

Definition 4.1.9. Let $\mathbf{e}_i$ denote the ith standard basis vector for $\mathbb{R}^{n+1}$. For $\sigma \subset \{0, \ldots, n\}$, we set $\mathbf{e}_\sigma = \sum_{i \in \sigma} \mathbf{e}_i$. If V is a subspace of K^{d+1} spanned by some of the $\mathbf{b}_i$, we set $\sigma(V) = \{i : \mathbf{b}_i \in V\}$. We map the cone over the order complex of $\mathcal{L}(B)$ into $\mathbb{R}^{n+1}$ by sending a subspace V to $\text{pos}(\mathbf{e}_{\sigma(V)}) + \mathbb{R}\mathbf{1}$, and a simplex $\{V_1, \ldots, V_s\}$ to $\text{pos}(\mathbf{e}_{\sigma(V_i)} : 1 \leq i \leq s) + \mathbb{R}\mathbf{1}$, where $\mathbf{1} = (1, \ldots, 1) \in \mathbb{R}^{n+1}$. This gives a fan in $\mathbb{R}^{n+1}$ with $\mathbf{1}$ contained in the lineality space. We write $\Delta(\mathcal{B})$ for the image of this fan in $\mathbb{R}^{n+1}/\mathbb{R}\mathbf{1}$. The fact that this is a fan will be proved in a more general setting in Theorem 4.2.6.

Example 4.1.10. We continue Example 4.1.3. The fan $\Delta(\mathcal{B}) \subseteq \mathbb{R}^5/\mathbb{R}\mathbf{1} \cong \mathbb{R}^4$ has 13 rays, corresponding to the five rays spanned by the $\mathbf{b}_i$ and the eight planes spanned by them. There is a two-dimensional cone for each of the 17 inclusions of a ray into a plane. The intersection of this fan with the 3-sphere gives a graph, illustrated in Figure 4.1.3. This graph is the order complex of $\mathcal{L}(B)$, which is the one-dimensional simplicial complex given by the 17 edges connecting the middle two layers in Figure 4.1.2. ◊

We next show that the tropical variety $\text{trop}(V(I))$ is equal to the support $|\Delta(\mathcal{B})|$ of the fan $\Delta(\mathcal{B})$ associated with the order complex.

Theorem 4.1.11. *Let I be a homogeneous linear ideal in $K[x_0^{\pm 1}, \ldots, x_n^{\pm 1}]$. The tropical variety of $X = V(I) \cap T^n$ equals the support of the fan $\Delta(\mathcal{B})$.*

Proof. A vector $\mathbf{v}$ lies in the relative interior of the cone of $\Delta(\mathcal{B})$ indexed by the chain $F_1 \subsetneq \cdots \subsetneq F_r$ in $\mathcal{L}(B)$ if and only if the following holds for all k: we have $v_i = v_j$ when $i, j \in F_k \backslash F_{k-1}$ and $v_i > v_j$ if $i \in F_k$ and $j \notin F_k$.

We first show that $\text{trop}(X) \subseteq |\Delta(\mathcal{B})|$. Suppose $\mathbf{v} \notin |\Delta(\mathcal{B})|$. Let $V^j = \{\mathbf{b}_i : v_i \geq v_j\}$. Let $l = \min\{j :$ there exists $\mathbf{b}_k \in \text{span}(V^j) \backslash V^j\}$. If no such l existed, then the subspaces $\text{span}(V^j)$ would be flats of $\mathcal{B}$, forming a chain

$\emptyset \subsetneq V^{j_1} \subsetneq \cdots \subsetneq V^{j_s} \subsetneq V^{j_{s+1}} = \mathcal{B}$ in the lattice $\mathcal{L}(B)$. However, this would imply that $\mathbf{v}$ is in the corresponding cone of $\Delta(\mathcal{B})$, by the observation at the start of the proof. Let $F = \mathrm{span}(V^l)$. Pick $\mathbf{b}_k \in F \backslash V^l$. Then $v_k < v_l$ by the definition of V^l. Since V^l spans F, we can write $\mathbf{b}_k = \sum_{i \in V^l} \lambda_i \mathbf{b}_i$ with $\lambda_i \in K$. This means $\mathbf{e}_k - \sum \lambda_i \mathbf{e}_i \in \ker(B)$. Thus $f = x_k - \sum_{i \in V^l} \lambda_i x_i$ is in I. Now $\mathrm{in}_{\mathbf{v}}(f) = x_k$, so $\mathrm{in}_{\mathbf{v}}(I) = \langle 1 \rangle$, and hence $\mathbf{v} \notin \mathrm{trop}(X)$.

We next prove $|\Delta(\mathcal{B})| \subseteq \mathrm{trop}(X)$. By Proposition 4.1.6 it suffices to show that $|\Delta(\mathcal{B})| \subseteq \mathrm{trop}(V(\ell_C))$ for every circuit C of I. Fix a chain of flats $V_1 \subsetneq \cdots \subsetneq V_{d+1} = K^{d+1}$, where $\dim(V_i) = i$. A vector $\mathbf{w}$ in the relative interior of the cone corresponding to that chain has $w_i \neq w_j$ if $\mathbf{b}_i \in V_k$ and $\mathbf{b}_j \notin V_k$ for some k. Let C be a circuit of I with linear form $\ell_C = \sum_{i \in C} a_i x_i$. Let $k = \min\{j : \mathbf{b}_i \in V_j \text{ for all } i \in C\}$, and let $\mathcal{F} = \{i : i \in C, \mathbf{b}_i \notin V_{k-1}\}$. If $|\mathcal{F}| = 1$ with $j \in C$ satisfying $\mathbf{b}_j \in V_k \backslash V_{k-1}$, then the equality $\sum_{i \in C} a_i \mathbf{b}_i = 0$ implies that $\mathbf{b}_j$ is a linear combination of elements of V_{k-1}, and thus is itself in V_{k-1}, which is a contradiction. Hence $|\mathcal{F}| \geq 2$. For $i, j \in \mathcal{F}$ we have $w_i = w_j \leq w_l$ for all $l \in C$. Thus $\mathrm{in}_{\mathbf{w}}(\ell_C)$ is not a monomial, so $\mathbf{w} \in \mathrm{trop}(V(\ell_C))$. Since $\mathrm{trop}(V(\ell_C))$ is closed, this shows that $|\Delta(\mathcal{B})| \subseteq \mathrm{trop}(V(\ell_C))$, and so $|\Delta(\mathcal{B})| \subseteq \mathrm{trop}(X)$. □

Example 4.1.12. Let $\mathcal{A}$, B, and I be as in Example 4.1.2. The lattice of flats has six elements $0, 1, 2, 3, 4, 5$ at the lowest level, and the seven elements $05, 14, 23, 013, 024, 125, 345$ at the next level. The fan $\Delta(\mathcal{B})$ is two-dimensional and lives in $\mathbb{R}^6/\mathbb{R}\mathbf{1}$. It has 13 rays and 18 two-dimensional cones. Combinatorially, it is a graph with 13 vertices and 18 edges. This is the *Petersen graph*, with three edges subdivided, as in Figure 4.1.4. ♢

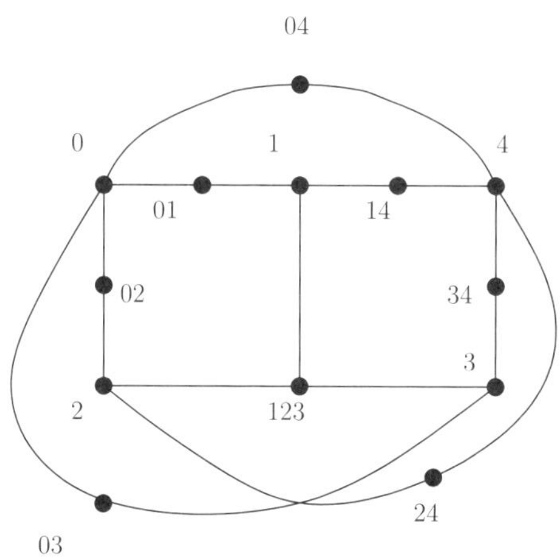

Figure 4.1.3. The graph of Example 4.1.10.

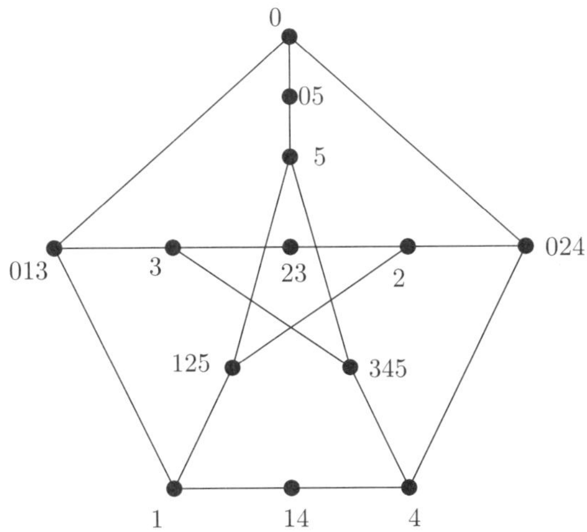

Figure 4.1.4. The fan over the Petersen graph is a tropicalized linear space.

When I is a linear ideal, every initial ideal $\text{in}_{\mathbf{w}}(I) \subset \Bbbk[x_0^{\pm 1}, \ldots, x_n^{\pm 1}]$ is generated by linear forms, so it is prime, and all the multiplicities are one. Note that the dimension of $\Delta(\mathcal{B})$ is d by construction, so $\dim(\text{trop}(X)) = \dim(X)$ as expected. We leave it as an exercise to verify the balancing condition and the rest of the conditions guaranteed by Theorem 3.3.5. In the more general settings of matroids, this is Exercise 4.7(13). See [**Rin13**] for algorithms and software to compute $\text{trop}(X)$ from the matrix B.

The title of this section emphasizes the point that $X = V(I)$ is the complement of a hyperplane arrangement. In later sections, we shall refer to X simply as a *linear subspace* of T^n and to $\text{trop}(X)$ as the corresponding *tropicalized linear space*. The fan structure $\Delta(\mathcal{B})$ on $\text{trop}(X)$ defines a *tropical compactification* of the open variety X. We saw a first glimpse of such compactifications in Section 1.8, and we will introduce them formally in Chapter 6. Thus, the material here can be read as a recipe for finding a good compactification of the complement of a hyperplane arrangement.

4.2. Matroids

Matroid theory is a branch of discrete mathematics that abstracts linear algebra. It aims to characterize the combinatorial structure of dependence relations among vectors in a linear space over a field K. In this section we will see that the constructions of the previous section are special cases of constructions for general matroids. This is the first hint of the importance of matroids in tropical geometry. In matroid theory, one distinguishes between *matroids* and *realizable matroids*, and our extension here will be the distinction between *tropical linear spaces* and *tropicalized linear spaces*.

Definition 4.2.1. A *tropicalized linear space* over K is a tropical variety of the form $\text{trop}(X)$ where X is a linear space in $T_K^n \cong (K^*)^{n+1}/K^*$. By this we mean that X is cut out by homogeneous linear forms in $K[x_0^{\pm 1}, \ldots, x_n^{\pm 1}]$.

In this section we restrict ourselves to the constant coefficient case, so we assume that K is a field with trivial valuation. Our aim is to explain the distinction between tropicalized and tropical linear spaces. The same distinction appears and is important when we study the extension to arbitrary fields K, where the valuation is nontrivial. This will be done in Section 4.4.

We are now prepared to define matroids. Fix an arbitrary finite set E. In the set-up of Section 4.1, we have $E = \{0, 1, 2, \ldots, n\}$. This is the *ground set* of the matroid M. There are many different but equivalent axiom systems for matroids. One of them is the following axiom system for circuits.

Definition 4.2.2. A *matroid* is a pair $M = (E, \mathcal{C})$ where E is a finite set and $\mathcal{C}$ is a collection of nonempty subsets of E, called *circuits* of M, such that

(C1) No proper subset of a circuit is a circuit.

(C2) If C_1, C_2 are distinct circuits and $e \in C_1 \cap C_2$, then $(C_1 \cup C_2)\backslash\{e\}$ contains a circuit.

Let $X \subset T^n$ be a linear subspace and consider its circuits as in Lemma 4.1.4. The set $\mathcal{C}$ of circuits of the ideal I of X satisfies (C1) and (C2). Indeed, if ℓ_1 and ℓ_2 are linear forms in I with respective supports C_1 and C_2, then a suitable linear combination of ℓ_1 and ℓ_2 has zero coordinate in position e but remains nonzero. This implies that some circuit in I has its support contained in $(C_1 \cup C_2)\backslash\{e\}$. A matroid M that arises in this manner from a linear subspace X is said to be *realizable* over the field K. We shall see that nonrealizable matroids exist.

Matroids provide a convenient language for linear algebra. Here are some basic definitions. An *independent set* of M is a subset of E that contains no circuit. A *basis* of M is a maximal independent set. All bases of M have the same cardinality. That number is called the *rank* of M. A *flat* of a matroid M is a set F such that $|C\backslash F| \neq 1$ for any circuit C. The poset of all flats, ordered by inclusion, is the *geometric lattice* of M. Each of these objects comes with its own axiom system for matroids. For example:

Definition 4.2.3. A *matroid* is a pair $M = (E, \rho)$ where E is a finite set and ρ is a function $2^E \to \mathbb{N}$, called the *rank function* of M, which satisfies the following.

(R1) $\rho(A) \leq |A|$ for all subsets A of E.

(R2) If A and B are subset of E with $A \subseteq B$, then $\rho(A) \leq \rho(B)$.

(R3) $\rho(A \cup B) + \rho(A \cap B) \leq \rho(A) + \rho(B)$ for any two subsets A, B of E.

The *rank* of the matroid M is defined to be the rank of E, and we write $\rho(M) := \rho(E)$. Starting with the axiom system (R1)–(R3), the other descriptions of matroids are derived as follows. A subset A of E is independent if $\rho(A) = |A|$ and is dependent otherwise. As before, a basis is a maximal independent set, and a circuit is a minimal dependent set. A flat is a subset $A \subseteq E$ such that $\rho(A) < \rho(A \cup \{e\})$ for all $e \in E\backslash A$. We can also characterize matroids via their bases, using the *basis exchange axiom.*

Definition 4.2.4. A *matroid* is a pair $M = (E, \mathcal{B})$, where E is a finite set and $\mathcal{B}$ is a collection of subsets of E, called the *bases* of M, that satisfies the following property: whenever σ and σ' are bases and $i \in \sigma\backslash\sigma'$, then there exists an element $j \in \sigma'\backslash\sigma$ such that $(\sigma\backslash\{i\}) \cup \{j\}$ is a basis as well.

This axiom implies the following stronger property (see [**Oxl11**, Exercise 11, page 22]): the element $j \in \sigma'\backslash\sigma$ can be chosen so that $(\sigma'\backslash\{j\}) \cup \{i\}$ is also a basis. Full proofs of the equivalence of Definitions 4.2.2, 4.2.3, and 4.2.4 plus the other axiom systems for matroids can be found in any book on matroids, such as [**Oxl11**], [**Wel76**], or [**Whi86, Whi87, Whi92**].

In Proposition 4.1.6 we saw that the circuits of X are a tropical basis. We now turn this result into a definition. This will associate a tropical linear space $\text{trop}(M)$ with any given matroid M, realizable or not.

Definition 4.2.5. Let M be a matroid on a finite set E, which we identify with $\{0, 1, 2, \ldots, n\}$. The *tropical linear space* $\text{trop}(M)$ is the set of vectors $\mathbf{w} = (w_0, w_1, \ldots, w_n) \in \mathbb{R}^{n+1}$ such that, for any circuit C of M, the minimum of the numbers w_i is attained at least twice as i ranges over C. If $\mathbf{w} \in \text{trop}(M)$, then $\mathbf{w} + \lambda\mathbf{1} \in \text{trop}(M)$ for any $\lambda \in \mathbb{R}$ ("$\text{trop}(M)$ is invariant under tropical scalar multiplication"), so we regard it as a subset of the quotient space $\mathbb{R}^{n+1}/\mathbb{R}\mathbf{1}$. Thus, by a *tropical linear space* we mean a subset of $\mathbb{R}^{n+1}/\mathbb{R}\mathbf{1}$ of the form $\text{trop}(M)$, where M is a matroid on $E = \{0, 1, 2, \ldots, n\}$.

Definition 4.2.5 will be further extended in Definition 4.4.3 to generalize the notion of tropicalized linear spaces to fields with valuations. Note, however, that our definition of $\text{trop}(M)$ does not involve the choice of a field.

We next describe a fan structure on the tropical linear space $\text{trop}(M)$ that is natural from a combinatorial perspective. This generalizes the construction in Definition 4.1.9 of the simplicial fan $\Delta(\mathcal{B})$ from the lattice $\mathcal{L}(B)$.

A flat F of the matroid M is represented by its incidence vector $\mathbf{e}_F = \sum_{i \in F} \mathbf{e}_i$. We regard $\mathbf{e}_F$ as an element in $\mathbb{R}^{n+1}/\mathbb{R}\mathbf{1}$. For any chain of flats $\emptyset \subset F_1 \subset \cdots \subset F_r \subset E$, where every inclusion is proper, we consider the polyhedral cone spanned by their incidence vectors:

$$\sigma = \text{pos}(\mathbf{e}_{F_1}, \ldots, \mathbf{e}_{F_r}) + \mathbb{R}\mathbf{1} = \{\lambda_0\mathbf{1} + \lambda_1\mathbf{e}_{F_1} + \cdots + \lambda_r\mathbf{e}_{F_r} \,:\, \lambda_1, \ldots, \lambda_r \geq 0\}.$$

Since $\mathbf{1}, \mathbf{e}_{F_1}, \mathbf{e}_{F_2}, \ldots, \mathbf{e}_{F_r}$ are linearly independent, σ is an r-dimensional simplicial cone in $\mathbb{R}^{n+1}/\mathbb{R}\mathbf{1}$, so it is the cone over an $(r-1)$-dimensional simplex. The following result generalizes Theorem 4.1.11 from realizable matroids to arbitrary matroids, and it also establishes the fan property.

Theorem 4.2.6. *Let M be a matroid on $E = \{0, 1, \ldots, n\}$. The collection of cones* $\text{pos}(\mathbf{e}_{F_1}, \ldots, \mathbf{e}_{F_r}) + \mathbb{R}\mathbf{1}$, *where $\emptyset \subset F_1 \subset \cdots \subset F_r \subset E$ runs over all chains of flats of M, forms a pure simplicial fan of dimension $\rho(M) - 1$ in $\mathbb{R}^{n+1}/\mathbb{R}\mathbf{1}$. The support of this fan equals the tropical linear space* $\text{trop}(M)$.

Proof. We first show that $\sigma \subset \text{trop}(M)$ for any $\sigma := \text{pos}(\mathbf{e}_{F_1}, \ldots, \mathbf{e}_{F_r})$ coming from a chain of flats with $\emptyset \subsetneq F_i \subsetneq F_{i+1} \subsetneq E$ for all i. Let $\mathbf{w} = \lambda_1 \mathbf{e}_{F_1} + \cdots + \lambda_r \mathbf{e}_{F_r}$ where $\lambda_1, \ldots, \lambda_r \geq 0$. Set $F_{r+1} = E$. Consider any circuit C of M, and let i be the largest index such that $(F_i \cap C) \backslash F_{i-1}$ is nonempty. We claim that this set has at least two elements. If not, then it is a singleton, and $C \subseteq F_i$. But then $|C \backslash F_{i-1}| = 1$, which contradicts the definition of flat of a matroid. Hence $(F_i \cap C) \backslash F_{i-1}$ has cardinality at least two. For $j \in C \backslash F_{i-1}$, we have $w_j = \lambda_i + \cdots + \lambda_r$. This is zero if $i = r+1$, while for all other $j \in C$ we have $w_j = \sum_{l=k}^{r} \lambda_l$, where $k < i$. Thus the minimum $\min_{i \in C} w_i$ is attained at those $j \in C$ with $j \notin F_{i-1}$, so is attained at least twice. Since this holds for any circuit C, we conclude $\mathbf{w} \in \text{trop}(M)$.

We next show that every $\mathbf{w} \in \text{trop}(M)$ lies in the relative interior of a unique cone σ as above. By adding a scalar multiple of $\mathbf{1}$, we obtain a nonnegative representative $\mathbf{w} \in \mathbb{R}^{n+1}$ whose support is a proper subset of E. Then there exists a unique chain $F_1 \subset F_2 \subset \cdots \subset F_k$ of proper nonempty subsets of E such that $\mathbf{w}$ lies in the relative interior of $\text{pos}(\mathbf{e}_{F_1}, \mathbf{e}_{F_2}, \ldots, \mathbf{e}_{F_k})$. The F_i are defined by the criterion that the function $j \mapsto w_j$ is constant on $F_i \backslash F_{i-1}$ and its value strictly decreases as i increases.

We claim that each F_i is a flat. Suppose that F_i were not a flat. By the definition of flats in terms of circuits, there would exist a circuit C such that $C \backslash F_i = \{e\}$ is a singleton. Then $w_e = \min\{w_i : i \in C\}$, and that minimum is uniquely attained. This is a contradiction to our hypothesis that $\mathbf{w}$ lies in the tropical linear space $\text{trop}(M)$. We conclude that the cones σ indexed by all chains of proper nonempty flats form a simplicial fan in $\mathbb{R}^{n+1}/\mathbb{R}\mathbf{1}$.

Each chain of flats of the matroid M can be extended to a maximal chain, and each maximal chain of flats involves precisely $\rho(M) - 1$ proper flats. Hence the fan is a pure fan of dimension $\rho(M) - 1$, as desired. □

We have shown that $\text{trop}(M)$ has the structure of a fan over a simplicial complex Δ_M of dimension $\rho(M) - 2$. We will sometimes identify $\text{trop}(M)$ with Δ_M. The simplicial complex Δ_M is the order complex of the geometric lattice of M. The order complex Δ_M has excellent combinatorial and topological properties. For instance, Δ_M is *shellable*, and hence its homology is free abelian and concentrated in the top dimension. The rank of that top homology group is denoted $\mu(M)$ and is known as the *Möbius number* of the matroid. It coincides with the Euler characteristic of Δ_M, so $\mu(M)$ is the absolute value of the alternating sum of the number of flats of rank i in M. For more information on these topics see Björner's Chapter 7 in [**Whi92**].

There is another fan structure on the tropical linear space $\text{trop}(M)$, which is much coarser than the one given by the order complex. That fan structure is known as the *Bergman fan*. We shall give a purely combinatorial description below. When M is realizable by a classical linear space $X = V(I)$, then the Bergman fan on $\text{trop}(M)$ comes from the Gröbner fan of I as in Corollary 2.5.12. We ask for a proof of this in Exercise 4.7(7).

Definition 4.2.7. For any $\mathbf{w} \in \mathbb{R}^{n+1}$, we define the *initial matroid* $M_{\mathbf{w}}$ as follows. The ground set is $E = \{0, 1, \ldots, n\}$, just as for M. The circuits of $M_{\mathbf{w}}$ are the sets $\{\, j \in C \,:\, w_j = \min_{i\in C}(w_i)\,\}$, where C runs over all circuits of M, but we only take sets that are minimal with respect to inclusion.

The reader is asked in Exercise 4.7(19) to check directly that $M_{\mathbf{w}}$ is again a matroid by showing that this set of circuits obeys axioms (C1) and (C2) of Definition 4.2.2. This also follows from Proposition 4.2.10 below.

Example 4.2.8. Let M be the uniform matroid (Example 4.2.13) of rank 3 on $E = \{0, 1, 2, 3, 4\}$. The bases of M are the ten subsets of E of size three. The circuits of M are the five subsets of E of size four. Let $\mathbf{w} = (0, 0, 0, 1, 1)$. Then $M_{\mathbf{w}}$ is the rank 3 matroid on E whose circuits and bases are

$$\mathcal{C} = \{\{0,1\}, \{0,2\}, \{1,2\}\} \quad \text{and} \quad \mathcal{B} = \{\{0,3,4\}, \{1,3,4\}, \{2,3,4\}\}.$$

While $C = \{0, 1, 2, 3\}$ is a circuit of M, the set $\{0, 1, 2\} = \{j \in C : w_j = \min_{i\in C}(w_i)\}$ is not minimal with respect to inclusion. An important point to note is that minimum is used to define the circuits of $M_{\mathbf{w}}$, while the bases of $M_{\mathbf{w}}$ are those of maximum weight $w_0 + w_1 + w_2 + w_3 + w_4 = 2$. In what follows we give a polyhedral interpretation of this. ♢

Definition 4.2.9. Let M be a matroid on $E = \{0, 1, \ldots, n\}$. The *matroid polytope* P_M is the convex hull in $\mathbb{R}^{n+1}$ of the indicator vectors of all bases:

$$P_M \;=\; \text{conv}\{\mathbf{e}_B : B \text{ is a basis of } M\} \;\subset\; \mathbb{R}^{n+1}.$$

For instance, in Example 4.2.8, the matroid polytope P_M is four-dimensional, and $P_{M_{\mathbf{w}}}$ is the two-dimensional face of P_M at which the linear form given by $\mathbf{w}$ is maximized. Here P_M is a hypersimplex and the face is triangle.

Matroid polytopes give a geometric representation of matroids which will be important for our study of tropical linear spaces in Section 4.4. The following proposition characterizes the faces of the matroid polytope P_M. Here, the *outer normal fan* of a polytope is the negative of its normal fan.

Proposition 4.2.10. *For any* $\mathbf{w} \in \mathbb{R}^{n+1}$*, the matroid polytope of* $M_{\mathbf{w}}$ *is the face of the matroid polytope* P_M *at which* $\mathbf{w}$ *is maximized. Thus* $M_{\mathbf{w}}$ *is constant on the relative interior of cones in the outer normal fan of* P_M*.*

Proof. The *weight* of a basis B of the given matroid M is the quantity $\sum_{i \in B} w_i$. The face of P_M maximizing $\mathbf{w}$ is the convex hull of those vectors $\mathbf{e}_B$ for which the basis B has maximal weight. We claim that these are precisely the bases of $M_{\mathbf{w}}$. Since each circuit of $M_{\mathbf{w}}$ is a subset of a circuit of M, each independent set of $M_{\mathbf{w}}$ is also independent in M. In particular, $\mathrm{rank}(M_{\mathbf{w}}) \leq \mathrm{rank}(M)$. Our argument will also show that equality holds.

Let W be the maximal weight of any basis in M. Fix a basis B_1 of M that has weight less than W. Choose a basis B_2 of weight W with $|B_2 \backslash B_1|$ as small as possible. Fix $i \in B_1 \backslash B_2$ with $w_i = \max_{l \in B_1 \backslash B_2} w_l$. By the stronger form of the basis exchange axiom given after Definition 4.2.4, there is $j \in B_2$ for which $B_1 \backslash \{i\} \cup \{j\}$ and $B_3 = B_2 \backslash \{j\} \cup \{i\}$ are both bases of M. Since $|B_3 \backslash B_1| < |B_2 \backslash B_1|$, the basis B_3 has weight less than W, so $w_j > w_i$. The set $B_1 \cup \{j\}$ is not independent, since bases are maximal independent sets. Hence $B_1 \cup \{j\}$ contains some circuit C of M. Since B_1 is a basis, we must have $j \in C$. The inequality $w_j > w_i = \max_{l \in B_1 \backslash B_2} w_l$ implies that $\{m \in C : w_m = \min_{j \in C} w_j\} \subseteq B_1$. This means that B_1 is not a basis of $M_{\mathbf{w}}$.

For the other inclusion, let B be a basis of M that has maximal weight W. We must show that B is independent in $M_{\mathbf{w}}$. Suppose otherwise. Then B contains some circuit $\{i \in C : w_i = \min_{j \in C} w_j\}$ of $M_{\mathbf{w}}$. We may assume that C is a circuit of M such that $|C \backslash B|$ is minimal with this property.

We claim that $|C \backslash B| = 1$. Pick $r \in C \backslash B$ with $w_r > \min_{j \in C} w_j$. Such an r exists since $C \not\subset B$, as B is a basis of M. The set $B \cup \{r\}$ contains a circuit C' of M, which must in turn contain r. If $|C \backslash B| > 1$, then $C' \neq C$, so by axiom (C2) there is a circuit $C'' \subset (C \cup C') \backslash \{r\}$. But then $C'' \backslash B \subsetneq C \backslash B$, contradicting the minimality of C. Hence $C \backslash B = \{r\}$.

Pick $i \in C \cap B$ with w_i minimal. The set $B' = B \backslash \{i\} \cup \{r\}$ is again a basis for M. Indeed, if not there would be a circuit C' contained in B', which must contain r, and axiom (C2) applied to C and C' would imply the existence of a circuit contained in the basis B. The weight of the basis B' is greater than the weight of B. This is a contradiction to the choice of B. We thus conclude that bases B of M of maximal weight are bases of $M_{\mathbf{w}}$. □

Proposition 4.2.10 implies that the tropical linear space $\mathrm{trop}(M)$ arises as a subfan of the outer normal fan of the matroid polytope P_M.

Corollary 4.2.11. *The tropical linear space of a matroid* M *is the union of those cones of the outer normal fan of* P_M *for which* $M_{\mathbf{w}}$ *has no loops:*

$$\mathrm{trop}(M) \;=\; \bigl\{ \mathbf{w} \in \mathbb{R}^{n+1}/\mathbb{R}\mathbf{1} \,:\, \textit{the matroid } M_{\mathbf{w}} \textit{ has no loops} \bigr\}.$$

Here, a *loop* of a matroid is a circuit $C = \{e\}$ of size one.

Proof. A vector $\mathbf{w}$ lies in $\text{trop}(M)$ if and only if the minimum $\min_{i\in C} w_i$ is achieved at least twice for all circuits of M. This occurs if and only if all circuits of $M_{\mathbf{w}}$ have size at least two, which means that $M_{\mathbf{w}}$ has no loops. □

The subfan described in Proposition 4.2.10 is the fan structure on $\text{trop}(M)$ specified by the distinct initial matroids $M_{\mathbf{w}}$. This fan is called the *Bergman fan* of the matroid M. See [**AK06, FS05**] for details. We note that the Bergman fan is the coarsest possible fan structure on $\text{trop}(M)$. This follows from its representation as a subfan in the normal fan of the polytope P_M. Write $\rho(M) = d + 1$. The d-dimensional Bergman fan is the fan over a $(d-1)$-dimensional subcomplex in the boundary of the polytope dual to P_M. This polyhedral complex is the *Bergman complex* of M. It is triangulated by the order complex Δ_M.

The following characterization of the polytopes P_M due to Gel′fand et al. [**GGMS87**] can be used as yet another axiom system to define matroids. It will be useful when we revisit tropical linear spaces in Section 4.4.

Theorem 4.2.12. *A polytope P with vertices in $\{0,1\}^{n+1}$ is a matroid polytope if and only if every edge of P is parallel to $\mathbf{e}_i - \mathbf{e}_j$ for some i, j.*

Proof. The "only-if" direction follows from Proposition 4.2.10. The point is that every edge of P_M is itself a matroid polytope. Such an edge is the convex hull of two vertices, $\mathbf{e}_B$ and $\mathbf{e}_{B'}$, and the basis exchange axiom (Definition 4.2.4) implies that B and B' differ in precisely one element.

For the "if" direction, let P be any polytope with vertices in $\{0,1\}^{n+1}$ such that each edge is a translate of some $\mathbf{e}_i - \mathbf{e}_j$. Let $\mathcal{B}$ be the collection of subsets σ of $E = \{0, \ldots, n\}$ such that $\mathbf{e}_\sigma$ is a vertex of P. We must verify the basis exchange axiom: given any two distinct vertices $\mathbf{e}_\sigma$ and $\mathbf{e}_{\sigma'}$ of P, we must identify a vertex of the form $\mathbf{e}_{(\sigma\backslash\{i\})\cup\{j\}}$ with $j \in \sigma'$ for any $i \in \sigma$. Our hypothesis ensures that σ and σ' have the same cardinality r, so $\sigma'\backslash\sigma \neq \emptyset$. Define a linear functional $\phi : \mathbb{R}^{n+1} \to \mathbb{R}$ by $\phi(\mathbf{x}) = r\sum_{j\in\sigma'\backslash\sigma} x_j + \sum_{j\in\sigma} x_j$. The polytope $Q = \{\mathbf{x} \in P : \phi(\mathbf{x}) \geq r\}$ contains both vertices $\mathbf{e}_\sigma$ and $\mathbf{e}_{\sigma'}$. Note that for any $\mathbf{w}$, if $\text{face}_{\mathbf{w}}(P) \subseteq Q$, then $\text{face}_{\mathbf{w}}(P) = \text{face}_{\mathbf{w}}(Q)$. Thus if $\mathbf{v}, \mathbf{v}'$ are two vertices of P contained in Q that are connected by an edge in P, then they are also connected by an edge in Q. Let $\mathbf{v}$ be a vertex of the face of P maximizing $\phi(\mathbf{x})$, which is contained in Q by construction. There are paths from both $\mathbf{e}_\sigma$ and $\mathbf{e}_{\sigma'}$ to $\mathbf{v}$ along edges of P for which ϕ increases, so these are also edges of Q. This follows, for example, from the simplex algorithm for linear programming. This means that there is a path from $\mathbf{e}_\sigma$ to $\mathbf{e}_{\sigma'}$ along edges of P that lie in Q. Suppose the first step of this path goes to $\mathbf{e}_{(\sigma\backslash\{l\})\cup\{k\}}$. If $k \notin \sigma'$, then $\phi(\mathbf{e}_{(\sigma\backslash\{l\})\cup\{k\}}) = r - 1$, so $\mathbf{e}_{(\sigma\backslash\{l\})\cup\{k\}}$ does not lie in Q. Hence we must have $k \in \sigma'$. This completes the proof. □

We finish the section with some examples of linear spaces $\mathrm{trop}(M)$.

Example 4.2.13 (Uniform matroids)**.** Suppose that all maximal minors of the matrices A and B in Section 4.1 are nonzero. This holds for generic subspaces X. The corresponding matroid is the *uniform matroid* $M = U_{d+1,n+1}$, whose bases are all subsets of $\{0, 1, \ldots, n\}$ of size $d + 1$. The circuits of $U_{d+1,n+1}$ are all subsets of size $d + 2$. The tropical linear space $\mathrm{trop}(U_{d+1,n+1})$ is the union of all orthants spanned by any d of the unit vectors $\mathbf{e}_0, \ldots, \mathbf{e}_n$ in $\mathbb{R}^{n+1}/\mathbb{R}\mathbf{1}$. This is the Bergman fan on $\mathrm{trop}(M)$. In the finer fan structure of Theorem 4.2.6, each orthant is subdivided. The new ray in the relative interior of a cone $\mathrm{pos}(\mathbf{e}_i : i \in \tau)$ is generated by $\sum_{i\in\tau} \mathbf{e}_i$. Here $\tau \subset \{0, \ldots, n\}$ has size at most d. The Bergman complex is the $(d - 1)$-skeleton of the n-simplex, while the order complex of M is the barycentric subdivision of the Bergman complex. The matroid polytope of $U_{d+1,n+1}$ is the *hypersimplex* $\Delta_{d+1,n+1}$, which is the convex hull of all vectors $\sum_{i\in\tau} \mathbf{e}_i \in \mathbb{R}^{n+1}$ as τ ranges over subsets of $\{0, \ldots, n\}$ of size $d + 1$. ◊

Example 4.2.14 (Graphic matroids)**.** Let G be a connected graph with d vertices and $n+1$ edges. We associate to G a *graphic matroid* M_G as follows. The ground set E of M_G is the set of edges of G. The circuits of M_G are the edges appearing in a circuit of G. This is a closed path in G that does not revisit vertices. An independent set of M_G is a collection of edges of G that do not contain any circuits, so the corresponding subgraph of G is a forest. A basis of M_G is thus the edges in a spanning tree of G, so the rank $\rho(M_G)$ of M_G is $d - 1$. This is one fewer than the number of vertices of G.

This matroid is realizable for any graph G. Choose an (arbitrary) orientation on each edge of G. The associated $(d-2)$-dimensional linear space $X \subseteq T^n$ has the parametric representation $x_{ij} = t_i - t_j$ for all directed edges (i, j). The set $\mathcal{B}$ of Section 4.1 consists of the vectors $\mathbf{b}_{ij} = \mathbf{e}_i - \mathbf{e}_j$, so the matrix B is the vertex-edge incidence matrix of G. Note that while this matrix B has d rows, it has rank $d - 1$, so $\mathrm{trop}(X)$ is a $(d - 2)$-dimensional fan in $\mathbb{R}^{n+1}/\mathbb{R}\mathbf{1}$. The circuits of this linear space are precisely the circuits of M_G, and these do not depend on the choice of orientation of the edges.

An important special case is when G is the complete graph K_d. The tropical variety $\mathrm{trop}(M_{K_d})$ is a fan in $\mathbb{R}^{\binom{d}{2}}/\mathbb{R}\mathbf{1}$. The smallest circuits have size three, and these form a tropical basis for $\mathrm{trop}(M_{K_d})$. This follows from the fact that every circuit in K_d of size $l > 3$ can be split, using a chord, into a circuit of length 3 and one of length $l - 1$. The condition that the minimum of the set $\{w_{ij}, w_{ik}, w_{jk}\}$ is achieved at least twice translates into the requirement that $w_{ij} \geq \min\{w_{ik}, w_{jk}\}$ for all i, j, k (including permutations

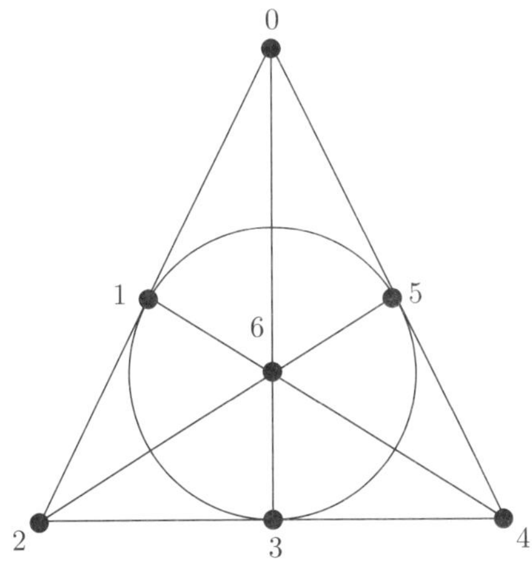

Figure 4.2.1. The lines of the Fano plane.

of the i, j, k). Up to a global sign change, these are precisely the *ultrametrics* on a set with d elements. The cones of the Bergman fan of M_{K_d} are indexed by rooted trees with d labeled leaves, and these correspond to unrooted trees with $d+1$ labeled leaves. This identifies the combinatorics of $\text{trop}(M_{K_d})$ with that of the tropical Grassmannian $\text{trop}(G(2, d+1))$ studied in the next section. See Lemma 4.3.9 for the precise connection. ◊

Example 4.2.15 (The Fano plane)**.** The Fano matroid M is defined by the projective plane $\mathbb{P}^2$ over the field $\mathbb{F}_2$. It has $\rho(M) = 3$ and $E = \{0, 1, \ldots, 6\}$. One realization takes the vectors $\mathbf{b}_i \in \mathcal{B}$ to be the seven nonzero vectors in $\mathbb{F}_2^3$, or equivalently the points of $\mathbb{P}^2_{\mathbb{F}_2}$. This matroid M has 14 circuits, seven of size three and seven of size four. The 3-element circuits of M are labeled

$$012\,,\ 036\,,\ 045\,,\ 135\,,\ 146\,,\ 234\,,\ 256. \tag{4.2.1}$$

The 4-element circuits are the complements of these. See Figure 4.2.1.

The simplicial complex Δ_M is one dimensional: it is a bipartite graph with 14 vertices and 21 edges. The vertices are the points $0, 1, \ldots, 6$ and the seven triples in (4.2.1). There is an edge from i to each triple that contains it. This matroid can be realized over a field K only if the characteristic of K equals 2. Thus, if $\text{char}(K) \neq 2$, then the tropical linear space $\text{trop}(M)$ is not a tropicalized linear space. There are many other tropical linear spaces that are not tropicalized linear spaces over any field. See Exercise 4.7(14). ◊

The tropical linear spaces introduced in this section are fundamental objects that can be used to prove purely combinatorial results in matroid theory. One example is the log-concavity theorem in [**HK12b**].

4.3. Grassmannians

Moduli spaces are fundamental objects in algebraic geometry. These spaces parameterize families of varieties. Each point in a moduli space corresponds to a different algebraic variety in the family of interest. The study of moduli spaces is also an important research direction in tropical algebraic geometry.

In this section we study a basic case, namely, the family of r-dimensional subspaces of the vector space K^m. This family is parameterized by the Grassmannian $G(r,m)$, which is a smooth projective variety of dimension $r(m-r)$. An r-dimensional linear subspace of K^m defines an $(r-1)$-dimensional subspace of $\mathbb{P}^{m-1}_K$. The Grassmannian $G(r,m)$ thus also parameterizes $(r-1)$-dimensional subspaces of $\mathbb{P}^{m-1}$. This is sometimes denoted by $\mathbb{G}(r-1,m-1)$, but we stick to the notation $G(r,m)$ in this book. Note that we made the shift $r=d+1$ and $m=n+1$ from Sections 4.1 and 4.2.

In Section 2.2 we realized the Grassmannian $G(r,m)$ as a subvariety of $\mathbb{P}^{\binom{m}{r}-1}$. Elements of $\mathbb{P}^{\binom{m}{r}-1}$ are represented by vectors $\mathbf{p}$ in $K^{\binom{m}{r}}$ whose coordinates p_I are indexed by subsets I of $[m]=\{1,2,\ldots,m\}$ with $|I|=r$. The Grassmannian $G(r,m)$ is the variety defined by the prime ideal

$$I_{r,m} \;=\; \big\langle\, \mathcal{P}_{I,J} \;:\; I,J\subseteq[m],\ |I|=r-1,\ |J|=r+1 \,\big\rangle \;\subset\; K[p_I],$$

whose generators are the *quadratic Plücker relations*

$$\mathcal{P}_{I,J} \;=\; \sum_{j\in J} \operatorname{sgn}(j;I,J)\cdot p_{I\cup j}\cdot p_{J\backslash j}.$$

Here, $|J\backslash I|\geq 3$, and $\operatorname{sgn}(j;I,J)$ equals $(-1)^\ell$, where ℓ is the number of elements $j'\in J$ with $j<j'$ plus the number of elements $i\in I$ with $i>j$.

As always, we focus on the open variety $G^0(r,m)=G(r,m)\cap T^{\binom{m}{r}-1}$ that is obtained by removing the coordinate hyperplanes in $\mathbb{P}^{\binom{m}{r}-1}$. The torus $T^{\binom{m}{r}-1}$ is the set of points $\mathbf{p}$ in $\mathbb{P}^{\binom{m}{r}-1}$ with nonzero coordinates p_I.

We shall study the tropicalization of $G^0(r,m)=G(r,m)\cap T^{\binom{m}{r}-1}$. Since

$$\dim(G^0(r,m)) \;=\; \dim(G(r,m)) \;=\; r(m-r),$$

Theorem 3.3.5 implies that the *tropical Grassmannian* $\operatorname{trop}(G^0(r,m))$ is a pure $r(m-r)$-dimensional rational polyhedral fan in $\mathbb{R}^{\binom{m}{r}-1}\cong\mathbb{R}^{\binom{m}{r}}/\mathbb{R}\mathbf{1}$.

The Plücker ideal $I_{r,m}$ is homogeneous with respect to the $\mathbb{Z}^m$-grading $\deg(p_I)=\sum_{i\in I}\mathbf{e}_i\in\mathbb{Z}^m$. Hence the lift of $\operatorname{trop}(G^0(r,m))$ to $\mathbb{R}^{\binom{m}{r}}$ has an m-dimensional lineality space L, namely the image of the linear map $\mathbb{R}^m\to\mathbb{R}^{\binom{m}{r}}$, $(u_1,\ldots,u_m)\mapsto\big(\sum_{i\in I}u_i\big)_{I\in\binom{[m]}{r}}$. That image equals

$$L \;=\; \operatorname{span}\Big(\sum_{I:i\in I}\mathbf{e}_I : 1\leq i\leq m\Big) \;\subseteq\; \mathbb{R}^{\binom{m}{r}}.$$

This gives an $(m-1)$-dimensional lineality space for $\operatorname{trop}(G^0(r,m))$, since $\mathbf{1} \in L$. Geometrically, this lineality space comes from the torus action on $G(r,m)$ induced from the $(m-1)$-dimensional torus action on $\mathbb{P}^{m-1}$, where we view $G(r,m)$ as parameterizing $(r-1)$-planes in $\mathbb{P}^{m-1}$.

Example 4.3.1. In Example 2.2.11 we saw that the Grassmannian $G(2,4)$ is the hypersurface in $\mathbb{P}^5$ defined by the equation $p_{12}p_{34}-p_{13}p_{24}+p_{14}p_{23}$. The tropical Grassmannian $\operatorname{trop}(G^0(2,4))$ is the tropical hypersurface in $\mathbb{R}^6/\mathbb{R}\mathbf{1}$ defined by this polynomial. The lineality space of this hypersurface is

$$L = \operatorname{span}(\mathbf{e}_{12}+\mathbf{e}_{13}+\mathbf{e}_{14},\ \mathbf{e}_{12}+\mathbf{e}_{23}+\mathbf{e}_{24},\ \mathbf{e}_{13}+\mathbf{e}_{23}+\mathbf{e}_{34},\ \mathbf{e}_{14}+\mathbf{e}_{24}+\mathbf{e}_{34}).$$

The Grassmannian $\operatorname{trop}(G^0(2,4))$ has three maximal cones:

$$L+\operatorname{pos}(\mathbf{e}_{12}+\mathbf{e}_{34}), \qquad L+\operatorname{pos}(\mathbf{e}_{13}+\mathbf{e}_{24}), \qquad \text{and } L+\operatorname{pos}(\mathbf{e}_{23}+\mathbf{e}_{14}).$$

We can identify $\mathbb{R}^6/L$ with $\mathbb{R}^2$ by sending $\mathbf{e}_{12}, \mathbf{e}_{34}$ to $(1,0)$, $\mathbf{e}_{13}, \mathbf{e}_{24}$ to $(0,1)$, and $\mathbf{e}_{14}, \mathbf{e}_{23}$ to $(-1,-1)$. The image of $\operatorname{trop}(G^0(2,4))$ in $\mathbb{R}^2$ is the standard tropical line of Figure 3.1.1. ♢

Example 4.3.2. The tropical Grassmannian $\operatorname{trop}(G^0(2,5))$ is a fan of dimension 6 in $\mathbb{R}^{10}/\mathbb{R}\mathbf{1}$. Its image modulo the lineality space $L \simeq \mathbb{R}^4$ is two-dimensional in $\mathbb{R}^5 \simeq \mathbb{R}^{10}/L$. That fan has 10 rays and 15 two-dimensional cones. The symmetric group S_5 acts naturally on these. Combinatorially, $\operatorname{trop}(G^0(2,5))$ is the Petersen graph, shown in Figures 4.1.4 and 4.3.2. This coincidence is not an accident, as we shall see in Lemma 4.3.9. ♢

In this section we first focus on the case $r=2$. The tropical Grassmannian $\operatorname{trop}(G^0(2,m))$ has an important connection to evolutionary biology [**PS05**, §4]. We then highlight some of the phenomena that make the cases $r>2$ more difficult, and we finish in Theorem 4.3.17 by explaining how the role of the Grassmannian as a moduli space extends to the tropical world.

A *phylogenetic tree* is a tree with m labeled leaves and no vertices of degree 2. These arise in biology, where the labels represent different taxa (e.g., species or DNA sequences), and the tree structure records their evolutionary history. This connection is discussed in detail in [**PS05**, §3.5]. See Figure 4.3.1 for phylogenetic trees with four and seven leaves, respectively. The m edges adjacent to the leaves of a tree τ are the *pendant edges* of τ.

Definition 4.3.3. A *tree distance* is a vector $\mathbf{d}=(d_{ij}) \in \mathbb{R}^{\binom{m}{2}}$ constructed as follows. Let τ be a phylogenetic tree. Assign a length $\ell_e \in \mathbb{R}$ to each edge e of τ. Between any two leaves i and j of τ there is a unique path in τ. We define the distance d_{ij} between leaves i and j as the sum of all edge lengths ℓ_e along this path. The resulting vector $\mathbf{d}=(d_{ij}) \in \mathbb{R}^{\binom{m}{2}}$ is called a *tree distance*. When all the ℓ_e are nonnegative, a tree distance specifies a finite metric on the set $[m]=\{1,\dots,m\}$. Finite metric spaces that arise from such a metric tree τ are called *tree metrics*.

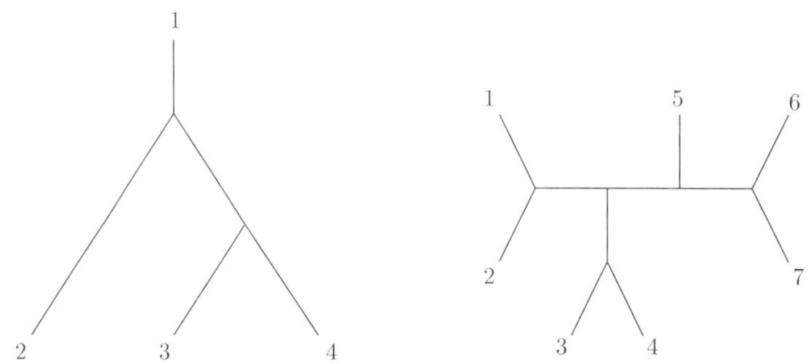

Figure 4.3.1. Some phylogenetic trees.

Definition 4.3.4. Let Δ denote the set of all tree distances in $\mathbb{R}^{\binom{m}{2}}$. This set is known as the *space of phylogenetic trees*.

Adding the vector $\lambda \sum_{j:j\neq i} \mathbf{e}_{ij}$ to a tree distance $\mathbf{d}$ corresponds to adding the constant λ to the length of the ith pendant edge of the tree τ. This shows that $\mathbf{d} + \lambda \sum_{j:j\neq i} \mathbf{e}_{ij} \in \Delta$ for all $\mathbf{d} \in \Delta$ and all $1 \leq i \leq m$. We conclude that the subspace L is contained in the lineality space of Δ:

$$\Delta + L = \Delta. \tag{4.3.1}$$

We thus view Δ as a subset of $\mathbb{R}^{\binom{m}{2}}/L$. Our aim is to prove

Theorem 4.3.5. *Up to sign, the tropicalization of the open Grassmannian $G^0(2,m)$ coincides with the space of phylogenetic trees. In symbols,*

$$\operatorname{trop}(G^0(2,m)) \;=\; -\Delta. \tag{4.3.2}$$

Our proof of Theorem 4.3.5 proceeds in three steps. We first prove the inclusion $\subseteq$ in (4.3.2). Thereafter, we give two derivations of the inclusion $\supseteq$. These will illustrate the two views on tropical varieties unified in the Fundamental Theorem 3.2.3.

We begin with the *four-point condition* which characterizes membership in tree space Δ. Its proof uses the notion of a *quartet*, which for any tree τ is the smallest subtree containing four leaves i, j, k, l. This subtree contains exactly one internal (nonpendant) edge. We say that i is adjacent to j if the unique path from i to j in this subtree does not use the internal edge. We denote the quartet by $(ij; kl)$ if i is adjacent to j and k is adjacent to l.

Lemma 4.3.6 (Four-point condition). *A metric $\mathbf{d}$ on the set $[m]$ is a tree metric if and only if for any four elements $u, v, x, y \in [m]$ the maximum of the three numbers $d_{uv}+d_{xy}$, $d_{ux}+d_{vy}$, and $d_{uy}+d_{vx}$ is attained at least twice.*

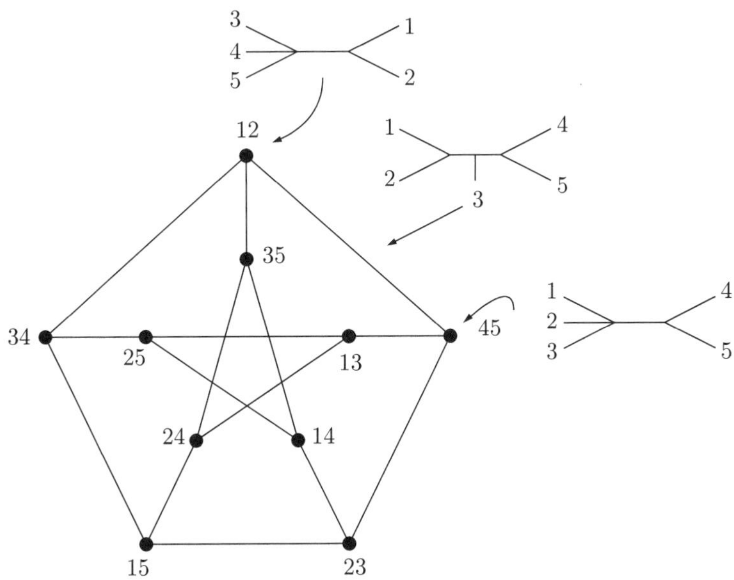

Figure 4.3.2. The space of phylogenetic trees for $m = 5$.

Proof. The proof is borrowed from [**PS05**, §2.4]. Suppose $\mathbf{d}$ equals the metric $\mathbf{d}_\tau$ defined by a tree τ on $[m]$. Then for any quartet $(uv; xy)$ of τ,

$$d_{uv} + d_{xy} \;\leq\; d_{ux} + d_{vy} \;=\; d_{uy} + d_{vx}. \tag{4.3.3}$$

Hence the "only if" direction in Lemma 4.3.6 holds.

We prove the "if" direction by induction on the number m of leaves. The result holds trivially for $m = 3$. Suppose that $m \geq 4$ and Lemma 4.3.6 holds for all metric spaces with fewer than m elements. Let $\mathbf{d}$ be a metric on $[m] = \{1, 2, \ldots, m\}$ which satisfies the four-point condition.

Choose a triple i, j, k that maximizes $d_{ik} + d_{jk} - d_{ij}$. By the induction hypothesis there is a tree τ' on $[m]\backslash i$ that realizes $\mathbf{d}$ restricted to $[m]\backslash i$. Let λ be the length of the edge e of τ' adjacent to j. We subdivide e by attaching the leaf i next to the leaf j. The edge adjacent to i is assigned length $\lambda_i = (d_{ij} + d_{ik} - d_{jk})/2$, the edge adjacent to j is assigned length $\lambda_j = (d_{ij} + d_{jk} - d_{ik})/2$, and the remaining part of e is assigned length $\lambda - \lambda_j$. We claim that the resulting tree τ has nonnegative edge weights, and it satisfies $\mathbf{d} = \mathbf{d}_\tau$. By construction, $\mathbf{d}$ and $\mathbf{d}_\tau$ agree on all pairs $x, y \in [m]\backslash i$.

Let l be any leaf of τ' other than i, j, k. Our choice of the triple i, j, k implies that $d_{ik} + d_{jk} - d_{ij} \geq d_{kl} + d_{ik} - d_{il}$ and $d_{ik} + d_{jk} - d_{ij} \geq d_{kl} + d_{jk} - d_{jl}$. The four-point condition in our hypothesis then gives

$$d_{ij} + d_{kl} \;\leq\; d_{ik} + d_{jl} \;=\; d_{il} + d_{jk}.$$

Since d is a metric, we have $\lambda_i \geq 0$ and $\lambda_j \geq 0$. To see that $\lambda - \lambda_j$ is nonnegative, we fix a leaf $l \neq j, k$ of τ'. Then $\lambda = (d_{jk} + d_{jl} - d_{kl})/2$, so

$$\lambda - \lambda_j \;=\; (d_{ik} + d_{jl} - d_{ij} - d_{kl})/2 \;\geq\; 0.$$

Thus our tree τ has nonnegative edge weights. We have $(d_\tau)_{ij} = \lambda_i + \lambda_j = d_{ij}$ and $(d_\tau)_{il} \;=\; (d_\tau)_{jl} - \lambda_j + \lambda_i \;=\; d_{jl} + d_{ik} - d_{jk} \;=\; d_{il}$ for $l \neq i$. □

Remark 4.3.7. Lemma 4.3.6 implies that if $\mathbf{d}$ is a tree distance, then $\mathbf{d}+\lambda\mathbf{1}$ is a tree metric for $\lambda \gg 0$. Indeed, any tree distance satisfies the four-point condition, and this is preserved by adding multiples of $\mathbf{1}$. For $\lambda \gg 0$, the vector $\mathbf{d} + \lambda\mathbf{1}$ has all entries positive, and satisfies the triangle inequality, so is a metric. Lemma 4.3.6 then implies that $\mathbf{d} + \lambda\mathbf{1}$ is a tree metric.

Note that $\mathbf{1} = 1/(m-1)(\sum_{i=1}^{m}\sum_{j\neq i}\mathbf{e}_{ij})$, so from the distances ℓ_e giving the tree metric $\mathbf{d} + \lambda\mathbf{1}$ we can recover distances ℓ'_e giving $\mathbf{d}$ by subtracting $\lambda/(m-1)$ from each pendant edge. We may thus assume that the distances ℓ_e defining a tree distance are nonnegative except on the pendant edges.

The proof above furnishes an algorithm whose input is a metric $\mathbf{d}$ satisfying the four-point condition and whose output is the unique metric tree τ with $\mathbf{d}_\tau = \mathbf{d}$. Another method for this is the *neighbor-joining algorithm* [**PS05**, Algorithm 2.41] from computational biology. The lengths of the interior edges of τ can be expressed as linear functions of $\mathbf{d}$. For small values of m, it is instructive to derive the formulas for the edge lengths explicitly.

Example 4.3.8. Let $m = 4$, and let $\mathbf{d} \in \mathbb{R}^6$ be a metric on the set $\{1, 2, 3, 4\}$ that satisfies the four-point condition. Then, after relabeling, we have

$$d_{13} + d_{24} \;=\; d_{23} + d_{14} \;\geq\; d_{12} + d_{34}.$$

The corresponding tree τ is depicted in Figure 4.3.3. Its edge lengths are

$$\begin{aligned} \alpha &= (d_{13} + d_{12} - d_{23})/2,\\ \beta &= (d_{12} + d_{23} - d_{13})/2,\\ \gamma &= (d_{14} + d_{23} - d_{12} - d_{34})/2,\\ \delta &= (d_{13} + d_{34} - d_{14})/2,\\ \epsilon &= (d_{14} + d_{34} - d_{13})/2. \end{aligned} \tag{4.3.4}$$

Indeed, this is the unique solution to the system of linear equations

$$\begin{array}{ccc} \alpha + \beta = d_{12}, & \alpha + \gamma + \delta = d_{13}, & \alpha + \gamma + \epsilon = d_{14},\\ \beta + \gamma + \delta = d_{23}, & \beta + \gamma + \epsilon = d_{24}, & \delta + \epsilon = d_{34}, \end{array}$$

modulo the hypothesis $d_{13}+d_{24} \;=\; d_{23}+d_{14}$. We invite the reader to perform the analogous calculation for the tree on the right in Figure 4.3.1. What are the expressions for the 11 edge lengths in terms of the 21 distances d_{ij}? ◊

The four-point condition suffices to prove one inclusion.

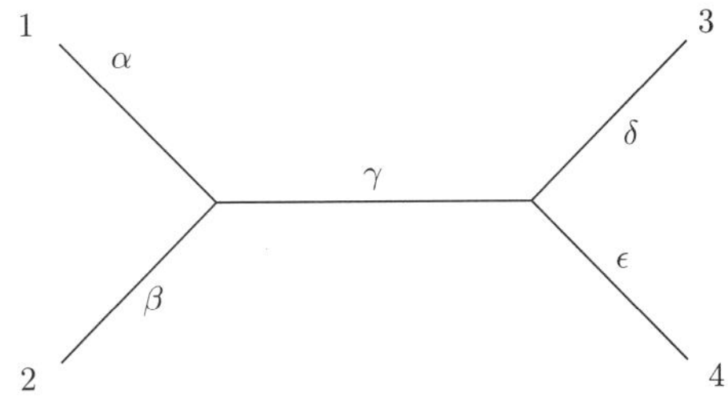

Figure 4.3.3. A trivalent tree with four leaves has five edges.

Proof of $\subseteq$ in Theorem 4.3.5. Fix a point $\mathbf{u}$ in $\text{trop}(G^0(2,m))$. Set $\mathbf{d} = -\mathbf{u}$. After adding the positive vector $\lambda\mathbf{1} \in L$ for $\lambda \gg 0$, we may assume that $\mathbf{d}$ is a metric. By the Fundamental Theorem 3.2.3, there is a classical point $\mathbf{p} \in G^0(2,m)$ with $\mathbf{u} = \text{val}(\mathbf{p})$. Its coordinates satisfy $u_{ij} = \text{val}(p_{ij}) = -d_{ij}$. Consider any quadruple $i,j,k,l \in [m]$. The Plücker relation $p_{ij}p_{kl} - p_{ik}p_{jl} + p_{il}p_{jk} = 0$ shows that $\min(u_{ij}+u_{kl}, u_{ik}+u_{jl}, u_{il}+u_{jk})$ is attained at least twice, so $\max(d_{ij}+d_{kl}, d_{ik}+d_{jl}, d_{il}+d_{jk})$ is attained at least twice. Lemma 4.3.6 implies $\mathbf{d} \in \Delta$. Hence $\mathbf{u} \in -\Delta$. □

An *ultrametric* on $[m] = \{1,\ldots,m\}$ is a metric $\mathbf{d}$ such that the maximum of d_{ij}, d_{ik}, d_{jk} is attained at least twice for any $i,j,k \in [m]$. Every ultrametric satisfies the four-point condition and is hence a tree metric. In fact, for an ultrametric, the corresponding phylogenetic tree is rooted and all leaves have the same distance to the root. Such trees are also known as *equidistant trees*. The matroid M_{K_m} associated to the complete graph K_m in Example 4.2.14 has ground set $E = \{\{i,j\} : 1 \leq i < j \leq m\}$ and a tropical basis given by the set of triangles in K_m. A vector $\mathbf{d} \in \mathbb{R}^{\binom{m}{2}-1}$ lies in $\text{trop}(M_{K_m})$ if and only if $\min(d_{ij}, d_{ik}, d_{jk})$ is attained at least twice for all $1 \leq i < j < k \leq m$. This shows that if $\mathbf{d} \in \mathbb{R}^{\binom{m}{2}}$ is an ultrametric, then $-\mathbf{d} \in \text{trop}(M_{K_m})$. We now see a connection between ultrametrics and tree distances.

Lemma 4.3.9. *Every tree distance is an ultrametric plus a vector in the lineality space. Thus, the space of phylogenetic trees has the decomposition*

$$\Delta \;=\; -\,\text{trop}(M_{K_m}) + L.$$

Proof. If $-\mathbf{d} \in \text{trop}(M_{K_m})$, then $\max(d_{ij}, d_{ik}, d_{jk})$ is achieved at least twice for all i,j,k. Thus for $\lambda \gg 0$ the vector $-\mathbf{d} + \lambda\mathbf{1}$ is an ultrametric, hence a tree metric, and therefore lies in Δ. The inclusion $\supseteq$ then follows from the equality $\Delta + L = \Delta$ of (4.3.1). For the inclusion $\subseteq$, consider an arbitrary tree distance $\mathbf{d} = \mathbf{d}_\tau$. Fix a root ρ anywhere on the tree τ, and write $d_{i\rho}$ for the distance from leaf i to ρ on τ. Fix $R \gg 0$ with $R \geq d_{i\rho}$ all i. Let $\mathbf{r} = \sum_{i=1}^m (R - d_{i\rho}) \sum_{j \neq i} \mathbf{e}_{ij} \in L$. The metric $\mathbf{r} + \mathbf{d}$ is an ultrametric because every leaf has distance R from the root ρ. It thus lies in $-\,\text{trop}(M_{K_m})$. □

Toward the end of Section 2.1 we remarked that valued fields K are ultrametric spaces. We refer to [**Hol01**] for a delightful introduction to $\mathbb{Q}_p$ from this perspective. In proving Theorem 4.3.5 we may assume that the field K is algebraically closed, by Theorem 3.2.4, so its value group Γ_{val} is dense in $\mathbb{R}$. Any subset of m scalars in K defines an ultrametric on $[m]$ with coordinates in Γ_{val}. In our next proof we shall derive the converse.

First proof of $\supseteq$ in Theorem 4.3.5. Fix $\mathbf{d} \in \Delta$. Our goal is to show that $-\mathbf{d} \in \text{trop}(G^0(2,m))$. Since the Γ_{val}-valued points are dense in both polyhedral spaces, we may assume that all coordinates of $\mathbf{d}$ lie in Γ_{val}. By Lemma 4.3.9, we may assume that $\mathbf{d}$ is an ultrametric. We shall construct scalars $u_1, u_2, \ldots, u_m \in K$ such that $d_{ij} = -\text{val}(u_i - u_j)$ for all i, j. We use induction on m. The base case $m = 2$ is trivial since $d_{12} \in \Gamma_{\text{val}}$.

Let $R = \max_{ij}(d_{ij})$. There is a unique partition of $[m]$ such that $d_{ij} = R$, when i and j lie in different blocks of that partition, and $d_{ij} < R$, when i and j lie in the same block. This follows from the ultrametric property. Suppose there are r blocks. By induction on m, for each block $\sigma = \{i_1, \ldots, i_\ell\}$ there are scalars $b_{i_1}, \ldots, b_{i_\ell} \in K$, one of them zero, such that $d_{i_s i_t} = -\text{val}(b_{i_s} - b_{i_t}) < R$. We now pick arbitrary scalars $a_1, a_2, \ldots, a_r \in K$, one for each block, such that $\text{val}(a_i) = \text{val}(a_i - a_j) = -R$ for all distinct i, j. This is possible because K is algebraically closed, so the residue field $\Bbbk$ is infinite. We define $v_i = a_\sigma + b_i$ whenever $i \in \sigma$. The desired scalars $u_1, \ldots, u_m \in K$ are $u_i = v_i - v_1$. Indeed, if i, j live in the same block, then $\text{val}(u_i - u_j) = \text{val}(v_i - v_j) = \text{val}(a_\sigma + b_i - a_\sigma - b_j) = \text{val}(b_i - b_j) = -d_{ij}$. If $i \in \sigma$, $j \in \sigma'$, then $\text{val}(u_i - u_j) = \text{val}(v_i - v_j) = \text{val}((a_\sigma - a_{\sigma'}) + (b_i - b_j)) = -R$ since $\text{val}(b_i - b_j) < -R$. Now consider the plane spanned by $(1, \ldots, 1)$ and $(u_1, \ldots, u_m)$ in K^n. The corresponding point in the Grassmannian $G(2, m)$ has Plücker coordinates $u_j - u_i$. Hence $\mathbf{d} = (d_{ij})$ lies in $-\text{trop}(G^0(2, m))$. $\square$

The proof above highlights part (3) of the Fundamental Theorem 3.2.3. It explicitly constructs a $2 \times m$-matrix $U = \begin{pmatrix} 1 & 1 & \cdots & 1 \\ u_1 & u_2 & \cdots & u_m \end{pmatrix}$ over K whose 2×2-minors have the prescribed valuations. Our second proof will highlight part (2) of the Fundamental Theorem 3.2.3.

We first describe the combinatorics of the space of phylogenetic trees. Consider a tree τ on $[m]$ with e interior edges. Each edge corresponds to a split $\{I, I^c\}$ of the leaf set $[m]$ into two nonempty subsets. We write C_τ for the cone of all tree metrics on τ, where interior edges have nonnegative weights, but the weights on the m pendant edges adjacent to leaves are allowed to be negative. Then we have $C_\tau \simeq \mathbb{R}^e_{\geq 0} \times \mathbb{R}^m$. If we assign weight 1 to one edge and weight 0 to all other edges, then this determines the *split metric* $\mathbf{e}^{I,I^c} = \sum_{i \in I, j \in I^c} \mathbf{e}_{ij}$. The split metrics are linearly independent, and they generate the cone C_τ. In symbols,

$$C_\tau \;=\; \mathrm{pos}(\mathbf{e}^{I,I^c} : I \text{ is a split of } \tau) \times \mathbb{R}^m.$$

If τ is the *star tree* with exactly one interior node, then $C_{\text{star}} = L$ is the lineality space of Δ. For all other trees τ, we have $C_\tau \cap (-C_\tau) = L$.

The tree τ is trivalent if each interior node has exactly three neighbors. A trivalent tree on $[m]$ has $2m-3$ edges. Each cone $C_{\tau'}$ is contained in the $(2m-4)$-dimensional cone C_τ of some trivalent tree τ. The drop in dimension occurs because C_τ lives in $\mathbb{R}^{\binom{m}{2}}/\mathbb{R}\mathbf{1}$. By induction on m, we can see that the number of trivalent trees on $[m]$ equals

$$(2m-5)!! \;=\; (2m-5)(2m-7)\cdots(5)(3)(1).$$

Our combinatorial discussion implies the following result.

Proposition 4.3.10. *The space Δ of phylogenetic trees is the union of the $(2m-5)!!$ polyhedral cones C_τ, each of which is isomorphic to $\mathbb{R}^{m-3}_{\geq 0} \times \mathbb{R}^{m-1}$.*

For $m=5$, tree space Δ has 15 maximal cones $\mathbb{R}^2_{\geq 0} \times \mathbb{R}^4$, corresponding to the edges of the Petersen graph (Figure 4.3.2). Here is our second proof.

Second proof of $\supseteq$ in Theorem 4.3.5. Let τ be any trivalent tree on $[m]$. By Proposition 4.3.10, it suffices to show the inclusion $-C_\tau \subset \mathrm{trop}(G^0(2,m))$. Since both sets are symmetric under permutation of indices, we may assume that τ is drawn in the plane with the leaf-vertices lying on a circle and labeled $1,2,\ldots,m$ in circular order. The ideal $I_{2,m}$ is generated by the quadrics

$$\underline{p_{ik}p_{jl}} - p_{ij}p_{kl} - p_{il}p_{jk} \quad \text{for } 1 \leq i < j < k < l \leq m. \tag{4.3.5}$$

We will show that these $\binom{m}{4}$ quadrics form a Gröbner basis for $I_{2,m}$ with respect to any weight vector $-\mathbf{d}$ with $\mathbf{d}$ the relative interior of C_τ. We first show that they form a Gröbner basis for $I_{2,m}$ with the underlined monomials as initial forms. These monomials are indexed by *crossing diagonals* in the m-gon. Set the weight of p_{ij} to be the Euclidean distance between the leaf-vertices i and j of τ. By the triangle inequality, the monomials indexed by the crossing diagonals are the highest weight monomials in the quadrics (4.3.5) with respect to these weights. Let $\prec$ be any monomial term order that refines this weight order. We argue that the S-pair for any two trinomials (4.3.5) reduces to zero. If the leading monomials are relatively prime, then this is automatic by Buchberger's criterion [**CLO07**, §2.9]. Otherwise, the total number of distinct indices is at most six. Hence it suffices to check the Gröbner basis property for $m \leq 6$. This is done by a direct computation. Thus the crossing-diagonal monomials generate an initial ideal of $I_{2,m}$.

Let $\mathbf{d}$ be in the relative interior of C_τ. The initial form of (4.3.5) with respect to $-\mathbf{d}$ is the binomial $\underline{p_{ik}p_{jl}} - p_{ij}p_{kl}$, where $\{\{i,l\},\{j,k\}\}$ is the split of the subtree of τ induced on i,j,k,l. The fact that the underlined monomial is part of this binomial is a consequence of our choice of τ being drawn on the circle. The initial ideal $\mathrm{in}_\prec(\mathrm{in}_{-\mathbf{d}}(I_{2,m}))$ thus contains the crossing-diagonal monomials. It has the same Hilbert function as $I_{2,m}$ by Corollary 2.4.9, and hence these binomials form a Gröbner basis of $\mathrm{in}_{-\mathbf{d}}(I_{2,m})$. This means that, in particular, $\mathrm{in}_{-\mathbf{d}}(I_{2,m})$ is generated by the binomials $\underline{p_{ik}p_{jl}} - p_{ij}p_{kl}$. The point $(1,\ldots,1) \in (\Bbbk^*)^{\binom{m}{2}}$ lies in $V(p_{ik}p_{jl} - p_{ij}p_{kl})$ for all i,j,k,l. This means that $\mathrm{in}_{-\mathbf{d}}(I_{2,m}) \neq \langle 1 \rangle$, and we conclude $-\mathbf{d} \in \mathrm{trop}(G^0(2,m))$. □

Remark 4.3.11. This second proof also lets us compute the multiplicities on the maximal cones C_τ of $\mathrm{trop}(G^0(2,m))$. Indeed, we claim that the initial ideal $\mathrm{in}_{-\mathbf{d}}(I_{2,m})$ generated by the binomials $\underline{p_{ik}p_{jl}} - p_{ij}p_{kl}$ is prime, so all multiplicities are one. It is radical because it has a square-free initial ideal. To see that $\mathrm{in}_{-\mathbf{d}}(I_{2,m})$ is prime, we write it as the kernel of a monomial map μ_τ from $K[p_{ij} : 1 \leq i < j \leq m]$ to the auxiliary polynomial ring $K[z_e : e \text{ edge of } \tau]$. The map takes p_{ij} to the product of all variables z_e where e runs over all edges on the path from leaf i to leaf j. A monomial is *crossing-free* if it is not divisible by any underlined p-monomial. The images of the crossing-free monomials under the map μ_τ are distinct. This implies that the binomials $\underline{p_{ik}p_{jl}} - p_{ij}p_{kl}$ form a Gröbner basis for $\ker(\mu_\tau)$, and hence for $\mathrm{in}_{-\mathbf{d}}(I_{2,m})$. See [**Stu96**, Chapter 4] for details of such arguments.

We summarize our results on the Grassmannian $G(2,m)$ in a corollary:

Corollary 4.3.12. *The tropical Grassmannian* $\mathrm{trop}(G^0(2,m))$ *is the support of a pure* $(2m-4)$*-dimensional fan with* $(2m-5)!!$ *maximal cones. These cones are* $L - C_\tau$*, where* τ *runs over trivalent trees. All multiplicities are one. The* $\binom{m}{4}$ *Plücker relations* (4.3.5) *are a tropical basis of* $I_{2,m}$.

Proof. The first assertion is Theorem 4.3.5 and Proposition 4.3.10. The multiplicity of each maximal cone is one by Remark 4.3.11. The four-point condition (Lemma 4.3.6) ensures that the three-term Plücker relations (4.3.5) are a tropical basis for the Plücker ideal $I_{2,m}$. □

Remark 4.3.13. In [**BHV01**] Billera, Holmes, and Vogtmann analyze the tree space Δ from the perspective of metric geometry. They endow Δ with an intrinsic metric that differs from the extrinsic metric coming from our ambient $\mathbb{R}^{\binom{m}{2}}$. In their metric, neighboring cones C_τ and $C_{\tau'}$ always meet at right angles. This implies a strong curvature property known as CAT(0).

The tropical Grassmannian $\mathrm{trop}(G^0(r,m))$ is much more complicated for $r > 2$ than it is for $r = 2$. The following examples will give us a glimpse of this. Example 4.3.14 shows that $\mathrm{trop}(G^0(3,m))$ can depend on the residue characteristic of K, while Example 4.3.15 shows that there is no canonical simplicial fan structure on $\mathrm{trop}(G^0(3,m))$ that is as nice as Corollary 4.3.12.

Example 4.3.14. Let $r = 3$ and $m = 7$, and consider the weight vector

$$\mathbf{w} \;=\; \mathbf{e}_{124} + \mathbf{e}_{235} + \mathbf{e}_{346} + \mathbf{e}_{457} + \mathbf{e}_{156} + \mathbf{e}_{267} + \mathbf{e}_{137}.$$

This is the incidence vector for the lines in the *Fano plane*; see Example 4.2.15 and (5.3.6). In characteristic zero, the reduced Gröbner basis for the Plücker ideal $I_{3,7}$ with respect to the reverse lexicographic refinement of $\mathbf{w}$ consists of 140 quadrics, 52 cubics, and 4 quartics. One of the 52 cubics is

$$\begin{aligned} f = {} & 2p_{123}p_{467}p_{567} - p_{367}p_{567}\underline{p_{124}} - p_{167}p_{467}\underline{p_{235}} - p_{127}p_{567}\underline{p_{346}} - p_{126}p_{367}\underline{p_{457}} \\ & - p_{237}p_{467}\underline{p_{156}} + p_{134}p_{567}\underline{p_{267}} + p_{246}p_{567}\underline{p_{137}} + p_{136}\underline{p_{267}}\,\underline{p_{457}}. \end{aligned}$$

The variables with nonzero weight are underlined. The initial term of f is $\mathrm{in}_{\mathbf{w}}(f) = 2p_{123}p_{467}p_{567}$. This is a monomial, provided $2 \neq 0$, so the image of $I_{3,7}$ in $\Bbbk[p_{ijk}^{\pm 1}]$ is the unit ideal. Thus $\mathbf{w} \notin \mathrm{trop}(G^0(3,7))$ when $\mathrm{char}(\Bbbk) \neq 2$. When $\Bbbk$ has characteristic 2, the initial form $\mathrm{in}_{\mathbf{w}}(f)$ is not a monomial, and, in fact, the initial ideal $\mathrm{in}_{\mathbf{w}}(I)$ does not contain any monomial. This can be checked for $\Bbbk = \mathbb{F}_2$ using a computer algebra system such as `Macaulay2` [**M2**]; see Exercise 2.7(20). Since Gröbner algorithms never leave the field of definition, this shows it for all residue fields of characteristic two.

Now change the weight vector to $\mathbf{w}' = \mathbf{w} - \mathbf{e}_{124} = \mathbf{e}_{235} + \mathbf{e}_{346} + \mathbf{e}_{457} + \mathbf{e}_{156} + \mathbf{e}_{267} + \mathbf{e}_{137}$. Then $\mathrm{in}_{\mathbf{w}'}(f)$ is a monomial if $\mathrm{char}(K) = 2$. In characteristic zero, $\mathrm{in}_{\mathbf{w}'}(I)$ does not contain a monomial, and we have $\mathbf{w}' \in \mathrm{trop}(G^0(3,7))$ and $\mathbf{w} \notin \mathrm{trop}(G^0(3,7))$. It suffices to check this via computer over $\mathbb{Q}$. In characteristic two, we have $\mathbf{w} \in \mathrm{trop}(G^0(3,7))$ but $\mathbf{w}' \notin \mathrm{trop}(G^0(3,7))$. ◊

Example 4.3.15. The tropical Grassmannian $\mathrm{trop}(G^0(2,m))$ is the fan over a simplicial complex Σ_m that is a *flag complex*. This means that the minimal nonfaces of Σ_m have cardinality two. The vertices of Σ_m are the $2^m - m - 1$ splits, and the edges are pairs of compatible splits $\{I, I^c\}$ and $\{J, J^c\}$. Here *compatible* means $I \cap J = \emptyset$ or $I \cap J^c = \emptyset$ or $I^c \cap J = \emptyset$ or $I^c \cap J^c = \emptyset$. Facets of Σ_m are pairwise compatible collections of splits. The simplicial complex Σ_m has dimension $m - 4$. For instance, for $m = 5$ it is the Petersen graph. For $m = 6$ it has 25 vertices, 105 edges, and 105 triangles. For $m = 7$ it has 56 vertices, 490 edges, 1260 triangles, and 945 tetrahedra.

No analogous flag simplicial complex exists for $r \geq 3$. Consider the case $r = 3, m = 6$. The tropical Grassmannian $\mathrm{trop}(G^0(3,6))$ is a nine-dimensional fan in $\mathbb{R}^{\binom{6}{3}}/\mathbb{R}\mathbf{1}$. It has a unique coarsest fan structure. Modulo the lineality space, this is the fan over a three-dimensional polyhedral complex having 65 vertices, 535 edges, 1350 triangles, and 1005 facets. A classification of the facets is provided in Example 4.4.10. Of the facets,

990 are tetrahedra but 15 are bipyramids. These bipyramids show that $\text{trop}(G^0(3,6))$ does not have a canonical structures as a (flag) simplicial complex. For more detailed combinatorial information, see Figure 5.4.1 and Table 5.4.1 in Section 5.4. ◊

The Grassmannian is the first nontrivial instance of a parameter space or moduli space in algebraic geometry. Points of $G(r,m)$ are in bijection with r-dimensional subspaces of K^m, or equivalently with $(r-1)$-planes in $\mathbb{P}^{m-1}$. This bijection can be expressed as follows in terms of Plücker coordinates.

Let $\mathbf{p} \in G(r,m)$. Any index set $I = \{i_1, \ldots, i_{r+1}\}$ specifies a linear form

$$\sum_{j=1}^{r+1} (-1)^j \cdot p_{I \backslash i_j} \cdot x_{i_j}. \tag{4.3.6}$$

These linear forms were already seen in (4.1.2). We call these the *circuits* of $\mathbf{p}$. The subspace corresponding to $\mathbf{p}$ is the common zero set of all circuits. Conversely, given any r-dimensional subspace of K^m, represented as the row space of an $r \times m$-matrix B with entries in K and linearly independent rows, we can recover $\mathbf{p}$ up to scaling as the vector of maximal minors of B.

The open subset $G^0(r,m)$ parameterizes subspaces whose Plücker coordinates are all nonzero. We call such a subspace *uniform* because the corresponding rank r matroid on $[m]$ is the uniform matroid (Example 4.2.13). The following lemma underscores the importance of circuits for our study:

Lemma 4.3.16. *The circuits* (4.3.6) *of any linear subspace in* K^m *form a tropical basis for the ideal of linear forms they generate.*

In the special case when the valuation on the field K is trivial, Lemma 4.3.16 was already established in Proposition 4.1.6. The proof for arbitrary valued fields K will be given after Theorem 4.4.5. The result of Lemma 4.3.16 will be used for uniform linear spaces in the proof of Theorem 4.3.17.

The correspondence between linear subspaces and points on the Grassmannian is also true in the tropical world. If X is a uniform linear subspace of T^n, then its tropicalization $\text{trop}(X)$ is a *uniform tropicalized linear space*, or a *uniform tropicalized* $(r-1)$*-plane*, provided $\dim(X) = r-1$ as before.

Theorem 4.3.17. *The bijection between the Grassmannian* $G(r,m)$ *and the set of* r*-dimensional subspaces of* K^m *induces a bijection* $\mathbf{w} \mapsto L_{\mathbf{w}}$ *between* $\text{trop}(G^0(r,m)) \cap \Gamma_{\text{val}}^{\binom{m}{r}-1}$ *and the set of uniform tropicalized* $(r-1)$*-planes in* $\mathbb{R}^m/\mathbb{R}\mathbf{1}$.

Proof. We begin by describing the map $\mathbf{w} \mapsto L_{\mathbf{w}}$. We denote by w_J the coordinates on $\mathbb{R}^{\binom{m}{r}}/\mathbb{R}\mathbf{1}$, where J is a subset of $\{1,\ldots,m\}$ of size r. For any $I \subset \{1, \ldots, m\}$ of size $r+1$, we consider the tropical linear form

$$(4.3.7) \qquad F_I(\mathbf{u}) \;=\; \bigoplus_{i \in I} w_{I\backslash\{i\}} \odot u_i \;=\; \min_{i \in I}(w_{I\backslash\{i\}} + u_i).$$

Let $L_{\mathbf{w}}$ be the intersection of the tropical hyperplanes in $\mathbb{R}^m/\mathbb{R}\mathbf{1}$ defined by the expressions F_I as I varies over all subsets of $\{1, \ldots, m\}$ of size $r+1$.

We claim that $\mathbf{w} \mapsto L_{\mathbf{w}}$ is a bijection between $\mathrm{trop}(G^0(r,m))$ and the set of uniform tropicalized $(r-1)$-planes in $\mathbb{R}^m/\mathbb{R}\mathbf{1}$. Indeed, let $\mathbf{p} \in G^0(r,m)$ with $\mathbf{w} = \mathrm{val}(\mathbf{p})$, and let X be the linear subspace of T^m defined by $\mathbf{p}$. Since the circuits form a tropical basis (by Lemma 4.3.16), we have $L_{\mathbf{w}} = \mathrm{trop}(X)$. Hence $\mathbf{w} \mapsto L_{\mathbf{w}}$ maps onto the set of uniform tropicalized $(r-1)$-planes.

It remains to be shown that the map $\mathbf{w} \mapsto L_{\mathbf{w}}$ is injective. We do this by constructing the inverse map. Suppose we are given the tropical plane $L_{\mathbf{w}}$ as a subset of $\mathbb{R}^m/\mathbb{R}\mathbf{1}$. We need to reconstruct $\mathbf{w}$ as an element of $\mathbb{R}^{\binom{m}{r}}/\mathbb{R}\mathbf{1}$. Equivalently, for any $(r-1)$-subset J of $[m]$ and any pair $k, \ell \in [m]\backslash J$, we need to derive the real number $w_{J\cup\{\ell\}} - w_{J\cup\{k\}}$ directly from the set $L_{\mathbf{w}}$. This is equivalent because the graph that has as vertices the subsets of $[m]$ of size r and an edge connecting J' and J'' if $J' \cap J''$ has size $r-1$ is connected.

Fix a large positive number $C \gg 0$ that lies in the value group Γ_{val}. Pick $c \in K$ with $\mathrm{val}(c) = C$. Let $X \subset K^m$ be any classical linear space with $\mathrm{trop}(X) = L_{\mathbf{w}}$. Note that all Plücker coordinates $p_\bullet$ of X are nonzero.

Consider any J and k, ℓ as above. The linear space X contains a unique point $\mathbf{x} = (x_1, \ldots, x_m)$ satisfying $x_k = 1$ and $x_j = c$ for $j \in J$. From the circuit (4.3.6) for $I = J \cup \{k, \ell\}$, we obtain an identity in K of the form

$$(4.3.8) \qquad x_\ell \cdot p_{J\cup\{k\}} \;\pm\; p_{J\cup\{\ell\}} + \sum_{j \in J} \pm c \cdot p_{J\backslash\{j\}\cup\{k,\ell\}} \;=\; 0.$$

Consider the vector $\mathbf{u} = \mathrm{val}(\mathbf{x})$ in $L_{\mathbf{w}}$. It satisfies $u_k = 0$ and $u_j = C$ for $j \in J$. Since $C = \mathrm{val}(c)$ is much larger than any of the coordinates of $\mathbf{w}$, the first two terms of the identity (4.3.8) must have the same valuation:

$$w_{J\cup\{\ell\}} \;-\; w_{J\cup\{k\}} \;=\; u_\ell.$$

This shows that $\mathbf{w}$ can be recovered from the point $\mathbf{u}$, as required. □

Remark 4.3.18. A tropicalized linear space $L_{\mathbf{w}}$ is uniform if and only if its recession fan (Theorem 3.5.6) is the Bergman fan $\mathrm{trop}(U_{r,m})$ of the uniform matroid (Example 4.2.13). The $\binom{m}{r-1}$ special points $\mathbf{u}$ we constructed in the proof above correspond to the $\binom{m}{r-1}$ maximal cones of $\mathrm{trop}(U_{r,m})$.

Theorem 4.3.17 characterizes a tropical variety in $(\mathbb{R}^{\binom{m}{r}}/\mathbb{R}\mathbf{1}) \times (\mathbb{R}^m/\mathbb{R}\mathbf{1})$. Its points are the pairs $(\mathbf{w}, \mathbf{u})$ where $\mathbf{u} \in L_{\mathbf{w}}$. That tropical variety is the *universal family* over the tropical Grassmannian $\mathrm{trop}(G^0(r,m))$. A tropical basis for the universal family of r-planes is given by any tropical basis for $G^0(r,m)$ together with the $\binom{m}{r+1}$ bilinear polynomials in (4.3.6). Indeed, these circuits yield the tropical circuits F_I in (4.3.7) which cut out $L_{\mathbf{w}}$.

Example 4.3.19. Let $r=2, m=4$. The universal family over $\text{trop}(G^0(2,4))$ is a five-dimensional tropical variety with a three-dimensional lineality space that lives in the eight-dimensional ambient space $(\mathbb{R}^6/\mathbb{R}\mathbf{1}) \times (\mathbb{R}^4/\mathbb{R}\mathbf{1})$. It is cut out by the five tropical polynomials

$$\begin{array}{ccccc} w_{12} \odot w_{34} & \oplus & w_{13} \odot w_{24} & \oplus & w_{14} \odot w_{23}, \\ w_{23} \odot u_1 & \oplus & w_{13} \odot u_2 & \oplus & w_{12} \odot u_3, \\ w_{24} \odot u_1 & \oplus & w_{14} \odot u_2 & \oplus & w_{12} \odot u_4, \\ w_{34} \odot u_1 & \oplus & w_{14} \odot u_3 & \oplus & w_{13} \odot u_4, \\ w_{34} \odot u_2 & \oplus & w_{24} \odot u_3 & \oplus & w_{23} \odot u_4. \end{array} \tag{4.3.9}$$

The fibers over the three maximal cones of $\text{trop}(G^0(2,4))$ are balanced trees in $\mathbb{R}^4/\mathbb{R}\mathbf{1}$ that represent the various lines in $\mathbb{P}^3_K$. Here are the three cases:

- The tree over $\{w_{12}+w_{34} = w_{13}+w_{24} \leq w_{14}+w_{23}\}$ consists of the segment from $P_{14} = (w_{13}+w_{14}, w_{13}+w_{24}, w_{13}+w_{34}, w_{14}+w_{34})$ to $P_{23} = (w_{13}+w_{24}, w_{23}+w_{24}, w_{23}+w_{34}, w_{24}+w_{34})$ together with the four rays $P_{14}+\mathbb{R}_{\geq 0}\mathbf{e}_1$, $P_{14}+\mathbb{R}_{\geq 0}\mathbf{e}_4$, $P_{23}+\mathbb{R}_{\geq 0}\mathbf{e}_2$, $P_{23}+\mathbb{R}_{\geq 0}\mathbf{e}_3$.
- The tree over $\{w_{12}+w_{34} = w_{14}+w_{23} \leq w_{13}+w_{24}\}$ consists of the segment from $P_{13} = (w_{13}+w_{14}, w_{14}+w_{23}, w_{13}+w_{34}, w_{14}+w_{34})$ to $P_{24} = (w_{14}+w_{23}, w_{23}+w_{24}, w_{23}+w_{34}, w_{24}+w_{34})$ together with the four rays $P_{13}+\mathbb{R}_{\geq 0}\mathbf{e}_1$, $P_{13}+\mathbb{R}_{\geq 0}\mathbf{e}_3$, $P_{24}+\mathbb{R}_{\geq 0}\mathbf{e}_2$, $P_{24}+\mathbb{R}_{\geq 0}\mathbf{e}_4$.
- The tree over $\{w_{13}+w_{24} = w_{14}+w_{23} \leq w_{12}+w_{34}\}$ consists of the segment from $P_{12} = (w_{12}+w_{14}, w_{12}+w_{24}, w_{14}+w_{23}, w_{14}+w_{24})$ to $P_{34} = (w_{14}+w_{23}, w_{23}+w_{24}, w_{23}+w_{34}, w_{24}+w_{34})$ together with the four rays $P_{12}+\mathbb{R}_{\geq 0}\mathbf{e}_1$, $P_{12}+\mathbb{R}_{\geq 0}\mathbf{e}_2$, $P_{34}+\mathbb{R}_{\geq 0}\mathbf{e}_3$, $P_{34}+\mathbb{R}_{\geq 0}\mathbf{e}_4$.

Our universal family is a quotient of the six-dimensional tropical Grassmannian $\text{trop}(G^0(2,5))$, which is represented by the Petersen graph in Figure 4.3.2. Combinatorially, the map onto $\text{trop}(G^0(2,4))$ deletes the pendant edge labeled 5 in each of the 15 trivalent trees on $\{1,2,\ldots,5\}$. Algebraically, this can be seen by replacing u_i with w_{i5}. The resulting expressions in (4.3.9) are the tropicalizations of the five Plücker trinomials that generate $I_{2,5}$. ◊

4.4. Linear Spaces

In this section we finally define tropical linear spaces. To do this, we introduce a new tropical moduli space, the Dressian, which extends the tropical Grassmannian and whose points correspond to tropical linear spaces. This construction goes well beyond Theorem 4.3.17 in two different directions. First, we replace the adjective "tropicalized" with the adjective "tropical". Second, we remove the adjective "uniform" and allow arbitrary matroids M

in place of the uniform matroid. Hence the recession fan of a tropical linear space, as in Remark 4.3.18, is now allowed to be trop(M) for any matroid M. A key role will be played by matroid subdivisions of matroid polytopes.

Let $M = (E, \mathcal{B})$ be a matroid of rank r on the set $E = \{1, 2, \ldots, m\}$, as in Definition 4.2.4. For any basis $\sigma \in \mathcal{B}$ of M, we introduce a variable p_σ. The resulting Laurent polynomial ring over our field K is

$$K[\mathbf{p}_{\mathcal{B}}^{\pm 1}] \;:=\; K[\,p_\sigma^{\pm 1} : \sigma \text{ is a basis of } M].$$

We write I_M for the ideal in $K[\mathbf{p}_{\mathcal{B}}]$ which is obtained from the Plücker ideal $I_{r,m}$ by setting all variables not indexing a basis to zero. In symbols,

$$I_M \;:=\; \big(I_{r,m} + \langle p_\sigma : \sigma \text{ is not a basis of } M\,\rangle\big) \cap K[\mathbf{p}_{\mathcal{B}}^{\pm 1}].$$

The quadratic Plücker relations that generate I_M are

$$\sum_j \operatorname{sgn}(j; \sigma, \tau) \cdot p_{\sigma \cup j} \cdot p_{\tau \setminus j}, \tag{4.4.1}$$

where $\sigma, \tau \subset [m]$, $\sigma \not\subset \tau$, $|\sigma| = r - 1$, σ is independent in M, $|\tau| = r + 1$, $\operatorname{rank}(\tau) = r$ in M, and the sum is over j such that both $\sigma \cup j$ and $\tau \setminus j$ are bases of M. An easy extension of Corollary 4.3.12 shows that the quadrics (4.4.1) form a tropical basis when $r = 2$, but this fails dramatically for $r \geq 3$.

We write $T^{|\mathcal{B}|-1}$ for the torus $(K^*)^{|\mathcal{B}|}/K^*$. The variety $\mathrm{Gr}_M := V(I_M) \subset T^{|\mathcal{B}|-1}$ is the *realization space* of the matroid M. Points in Gr_M correspond to equivalence classes of $r \times m$-matrices B that realize the matroid M. Here two matrices B and B' are equivalent if $B' = g \cdot B$ for some $g \in \mathrm{GL}_r(K)$. Equivalently, Gr_M is the variety of all r-dimensional linear subspaces of K^m whose nonzero Plücker coordinates are precisely the bases of M. In particular, $\mathrm{Gr}_M = \emptyset$ if and only if the matroid M is not realizable over the field K. We note that the realization spaces Gr_M can be arbitrary varieties, due to *Mnëv's Universality Theorem*. For further reading on this see [**BS89**].

The tropicalization $\mathrm{trop}(\mathrm{Gr}_M)$ of the realization space is called the *tropical Grassmannian* of M. If $M = U_{r,m}$ is the uniform matroid, then $\mathrm{Gr}_M = G^0(r, m)$, and $\mathrm{trop}(\mathrm{Gr}_M)$ is the tropical Grassmannian we studied in Section 4.3. The ambient space for the tropical variety $\mathrm{trop}(\mathrm{Gr}_M)$ is $\mathbb{R}^{|\mathcal{B}|}/\mathbb{R}\mathbf{1}$. This is the tropicalization of the torus $T^{|\mathcal{B}|-1}$. Points in $\mathbb{R}^{|\mathcal{B}|}/\mathbb{R}\mathbf{1}$ are written as $\mathbf{w} = (w_\sigma)_{\sigma \in \mathcal{B}}$. We sometimes need to regard $\mathbf{w}$ as a vector in $\mathbb{R}^{\binom{m}{r}}$; we then set $w_\sigma = \infty$ for nonbases $\sigma \in \binom{[m]}{r} \backslash \mathcal{B}$ of M.

Fix $\sigma, \tau \subset [m]$ with $|\sigma| = r - 1$, σ is independent, $|\tau| = r + 1$, $\sigma \not\subset \tau$, and $\operatorname{rank}(\tau) = r$. The tropicalization of (4.4.1) is the tropical polynomial

$$\bigoplus_j w_{\sigma \cup j} \odot w_{\tau \setminus j}, \tag{4.4.2}$$

where j runs over indices in τ such that both $\sigma \cup j$ and $\tau \setminus j$ are bases of M.

Definition 4.4.1. The *Dressian* of the matroid M is the tropical prevariety in $\mathbb{R}^{|\mathcal{B}|}/\mathbb{R}\mathbf{1}$ obtained by intersecting the tropical hypersurfaces of the quadrics in (4.4.2). This prevariety is denoted by Dr_M. This name refers to Andreas Dress, who developed the theory of *valuated matroids* in collaboration with Walter Wenzel. The valuations on M, introduced in [**DW92**], are precisely the points $\mathbf{u}$ in Dr_M. If M is the uniform matroid $U_{r,m}$, then we write $\mathrm{Dr}(r,m) = \mathrm{Dr}_{U_{r,m}}$. While the tropical Grassmannian $\mathrm{trop}(G^0(r,m))$ depends on the residue characteristic of the field K, as seen in Example 4.2.15, the Dressian is a purely combinatorial object and is independent of K.

By definition, the tropical Grassmannian is contained in the Dressian:

$$\text{(4.4.3)} \qquad \begin{array}{ccc} & \mathrm{trop}(G^0(r,m)) \subseteq \mathrm{Dr}(r,m) & \\ \text{and} & \mathrm{trop}(\mathrm{Gr}_M) \subseteq \mathrm{Dr}_M & \text{for all matroids } M. \end{array}$$

Equality holds if and only if the quadratic Plücker relations (4.4.1) are a tropical basis. The four-point condition (Lemma 4.3.6) shows that the tropical basis property holds for $r = 2$; see Corollary 4.3.12. Since every rank-2 matroid becomes uniform after removing loops and parallel elements (circuits of size one and two), we conclude that $\mathrm{trop}(G^0(2,m)) = \mathrm{Dr}(2,m)$ and $\mathrm{trop}(\mathrm{Gr}_M) = \mathrm{Dr}_M$ for all matroids M of rank 2. We shall see in Section 5.4 that the inclusions (4.4.3) are usually strict for $r \geq 3$.

Remark 4.4.2. The various Dressians Dr_M, as M ranges over all matroids of rank r on $[m]$, fit together to form a polyhedral complex. This lives in the tropical projective space $\mathrm{trop}(\mathbb{P}^{\binom{m}{r}-1})$, which will be constructed in Chapter 6. The union of the various Grassmannians $\mathrm{trop}(\mathrm{Gr}_M)$, each sitting inside Dr_M, is the tropicalization of the classical Grassmannian in $\mathbb{P}^{\binom{m}{r}-1}$. The restriction to the torus $T^{\binom{m}{r}-1}$ corresponds to the Grassmannian $G^0(r,m)$, and similarly for $\mathrm{Dr}(r,m)$ inside $\mathrm{trop}(\mathbb{P}^{\binom{m}{r}-1})$. Extending Theorem 4.3.17, points in $\mathrm{Dr}(r,m)$ parameterize uniform tropical $(r-1)$-planes in $\mathbb{R}^m/\mathbb{R}\mathbf{1}$.

For every point $\mathbf{w}$ in Dr_M, we now construct a tropical linear space $L_{\mathbf{w}}$ as follows. Consider $\tau \subset [m]$ with $|\tau| = r+1$ and $\mathrm{rank}(\tau) = r$. Let $L_\tau(\mathbf{w})$ denote the tropical hyperplane in $\mathbb{R}^m/\mathbb{R}\mathbf{1}$ defined by the tropical polynomial

$$\text{(4.4.4)} \qquad \bigoplus_{j\in\tau} w_{\tau\backslash j} \odot u_j \;=\; \min_{j\in\tau}(w_{\tau\backslash j} + u_j).$$

In the setting of [**DW92**], these are the circuits of the valuated matroid. Our linear space is defined as the intersection of these tropical hyperplanes:

$$L_{\mathbf{w}} \;:=\; \bigcap_{\tau} L_\tau(\mathbf{w}).$$

This definition makes sense for any point $\mathbf{w} \in \mathbb{R}^{|\mathcal{B}|}/\mathbb{R}\mathbf{1}$. However, this prevariety behaves like a linear space only if $\mathbf{w}$ comes from the Dressian.

Definition 4.4.3. A *tropical linear space* in $\mathbb{R}^m/\mathbb{R}\mathbf{1}$ is a prevariety of the form $L_\mathbf{w}$, where $\mathbf{w}$ is any point in the Dressian Dr_M of a matroid M on $[m]$.

This definition is justified by the next result and the theorem thereafter.

Proposition 4.4.4. *If M is a matroid, then* $\mathrm{trop}(M)$ *is a tropical linear space. Every tropicalized linear space over K is a tropical linear space.*

Proof. Suppose that $\mathbf{w} = \mathbf{0}$ is the zero vector in $\mathbb{R}^{|\mathcal{B}|}/\mathbb{R}\mathbf{1}$. This lies in Dr_M for every matroid M. We claim that $L_\mathbf{0} = \mathrm{trop}(M)$. Indeed, the tropical linear form (4.4.4) is $\bigoplus_{j\in C} u_j$, where C is the unique circuit in τ. Since all circuits of M arise in this way, the tropical linear space $L_\mathbf{0}$ is the set described in Definition 4.2.5. The second statement follows immediately from the inclusion (4.4.3) and the characterization of the tropicalized linear space $L_\mathbf{w}$ in equation (4.3.7) from the proof of Theorem 4.3.17. □

For tropical*ized* linear spaces, the desirable properties in our next theorem can be derived from the Fundamental Theorem and the Structure Theorem. However, these hold more generally for tropical linear spaces:

Theorem 4.4.5. *Let M be a matroid of rank r on $[m]$, and let $\mathbf{w}$ be a point in its Dressian Dr_M. The tropical linear space $L_\mathbf{w}$ is a pure $(r-1)$-dimensional balanced contractible polyhedral complex in $\mathbb{R}^m/\mathbb{R}\mathbf{1}$. The recession fan of $L_\mathbf{w}$ equals* $\mathrm{trop}(M)$. *Moreover, $L_\mathbf{w}$ is a tropical cycle of degree 1, which means that, for any generic point $\mathbf{p} \in \mathbb{R}^m/\mathbb{R}\mathbf{1}$, it intersects the complementary linear space $\mathbf{p} + \mathrm{trop}(U_{m-r+1,m})$ transversally in precisely one point.*

That the circuits form a tropical basis is now a corollary to this theorem.

Proof of Lemma 4.3.16. Let X be an $(r-1)$-dimensional subspace in $\mathbb{P}^{m-1}_K$ with matroid M, and let $\mathbf{w} \in \mathrm{Gr}_M$ be the tropicalization of its Plücker coordinate vector. The tropicalized linear space $\mathrm{trop}(X \cap T^{m-1})$ is a pure $(r-1)$-dimensional tropical cycle of degree 1, by Theorem 3.3.5 and Corollary 3.6.16. Since the circuits vanish on X, we have the inclusion $\mathrm{trop}(X \cap T^{m-1}) \subseteq L_\mathbf{w}$. Both are tropical cycles of dimension $r-1$ and degree 1, and our claim states that they are equal. If not, consider any point $\mathbf{q} \in L_\mathbf{w} \backslash \mathrm{trop}(X \cap T^{m-1})$, and choose a uniform $(m-r)$-dimensional subspace P in $\mathbb{P}^{m-1}_K$ with $\mathrm{trop}(P \cap T^{m-1}) \cap L_\mathbf{w} = \{\mathbf{q}\}$. By construction, we have $\mathrm{trop}(P \cap T^{m-1}) \cap \mathrm{trop}(X \cap T^{m-1}) = \emptyset$, and this implies $P \cap X \cap T^{m-1} = \emptyset$. This is a contradiction since the $(r-1)$-dimensional subspace X and the

$(m-r)$-dimensional subspace P must meet in $\mathbb{P}_K^{m-1}$, and if P is chosen generically among uniform $(m-r)$-dimensional subspaces whose tropicalization contains $\mathbf{q}$, this intersection will be in T^{m-1}. □

The proof of Theorem 4.4.5 relies on the notion of matroid subdivisions, which we now define. A subdivision of the matroid polytope P_M is a *matroid subdivision* if its edges are translates of $\mathbf{e}_i - \mathbf{e}_j$ for some i, j. By Theorem 4.2.12, this is equivalent to saying that every cell of the subdivision is a matroid polytope. Every vector $\mathbf{w} \in \mathbb{R}^{|\mathcal{B}|}/\mathbb{R}\mathbf{1}$ induces a regular subdivision $\Delta_\mathbf{w}$ of the polytope P_M, as in Definition 2.3.8. If $\Delta_\mathbf{w}$ happens to be a matroid subdivision, then we call $\Delta_\mathbf{w}$ a *regular matroid subdivision* of P_M.

Lemma 4.4.6. *Let $M = ([m], \mathcal{B})$ be a matroid, and let $\mathbf{w} \in \mathbb{R}^{|\mathcal{B}|}/\mathbb{R}\mathbf{1}$. Then $\mathbf{w}$ lies in the Dressian Dr_M if and only if $\Delta_\mathbf{w}$ is a matroid subdivision.*

Proof. We first prove the "only-if" direction. Every edge of $\Delta_\mathbf{w}$ joins two vertices $\mathbf{e}_\sigma$ and $\mathbf{e}_{\sigma'}$ of P_M. We call $|\sigma\backslash\sigma'| = |\sigma'\backslash\sigma|$ the *length* of that edge. Our claim is that every edge of $\Delta_\mathbf{w}$ has length 1. We shall prove that $\Delta_\mathbf{w}$ has no edge of length $\ell \geq 2$. This will be done by induction on ℓ.

We start with the base case $\ell = 2$. Suppose e is an edge of length 2. Then $e = \mathrm{conv}(\mathbf{e}_{\rho ij}, \mathbf{e}_{\rho kl})$ for some $\rho \in \binom{[m]}{r-2}$ and indices i, j, k, l. The Plücker relation $p_{\rho ij}p_{\rho kl} - p_{\rho ik}p_{\rho jl} + p_{\rho il}p_{\rho jk}$ implies that the minimum of $\{w_{\rho ij} + w_{\rho kl}, w_{\rho ik} + w_{\rho jl}, w_{\rho il} + w_{\rho jk}\}$ is attained at least twice. Consider the face of the matroid polytope P_M minimizing the vector

$$-(\mathbf{e}_{\rho ij} + \mathbf{e}_{\rho ik} + \mathbf{e}_{\rho il} + \mathbf{e}_{\rho jk} + \mathbf{e}_{\rho jl} + \mathbf{e}_{\rho kl}).$$

By the basis exchange property, this is either the octahedron

$$O = \mathrm{conv}\{\mathbf{e}_{\rho ij}, \mathbf{e}_{\rho ik}, \mathbf{e}_{\rho il}, \mathbf{e}_{\rho jk}, \mathbf{e}_{\rho jl}, \mathbf{e}_{\rho kl}\},$$

or a square

$$S = \mathrm{conv}\{\mathbf{e}_{\rho ij}, \mathbf{e}_{\rho ik}, \mathbf{e}_{\rho jl}, \mathbf{e}_{\rho kl}\}.$$

The restriction of $\Delta_\mathbf{w}$ to this face is a regular subdivision of O or S. The tropical Plücker relation above implies that e is not an edge of that subdivision, and hence e is not an edge of $\Delta_\mathbf{w}$.

Next consider the case $\ell \geq 3$. Suppose that e is an edge of length ℓ in $\Delta_\mathbf{w}$. We write $e = \mathrm{conv}(\mathbf{e}_{\sigma\tau}, \mathbf{e}_{\sigma\tau'})$, where $\tau \cap \tau' = \emptyset$, $|\tau| = |\tau'| = \ell$. Let F be the face of P_M at which the linear form $\sum_{i\in\sigma} x_i - \sum_{j\notin\sigma\cup\tau\cup\tau'} x_j$ attains its maximum value $|\sigma|$. Note that $e \subset F$. We do not have $F = e$, as that would imply that e were an edge of P_M, and thus had length 1. There is thus a two-dimensional cell G of $\Delta_\mathbf{w}$ such that $G \subseteq F$ and e is an edge of

G. Let γ denote the unique path from $\mathbf{e}_{\sigma\tau}$ to $\mathbf{e}_{\sigma\tau'}$ along edges of G other than e. The linear functional that defines F ensures that any vertex $\mathbf{e}_\nu$ of F has $\sigma \subset \nu$, and $\nu \subset \sigma \cup \tau \cup \tau'$. Thus no two vertices of F are more than distance ℓ apart, so all edges of γ have length $\leq \ell$. If γ contained an edge of length ℓ, then this edge would have the form $\mathrm{conv}(\mathbf{e}_{\sigma\nu}, \mathbf{e}_{\sigma\nu'})$, where $\nu \cap \nu' = \emptyset$ and $\nu \cup \nu' = \tau \cup \tau'$. Its midpoint would coincide with the midpoint of e, contradicting the convexity of G. Hence all edges of γ have length less than ℓ. By induction, each edge of γ has length 1, so is a translate of some $\mathbf{e}_i - \mathbf{e}_j$. These edge-vectors span a two-dimensional space (the linear space parallel to G). This is only possible if the number of distinct indices i, j is at most four, and G is a triangle or a quadrilateral. In either case, e has length at most 2, and this returns us to the base case.

For the "if" direction, suppose $\mathbf{w}$ is not in the Dressian Dr_M, so it violates one of the tropical Plücker relations (4.4.2). As above, the convex hull of the $\mathbf{e}_\nu$ occurring in that tropical polynomial is a face of P_M. The regular subdivision $\Delta_{\mathbf{w}}$ restricted to that face has an edge of the form $\mathrm{conv}\{\mathbf{e}_{\sigma\cup j}, \mathbf{e}_{\tau\backslash j}\}$. Since $\sigma \not\subset \tau$ holds for every nonzero Plücker relation (4.4.2), that edge has length at least 2, and therefore $\Delta_{\mathbf{w}}$ is not a matroid subdivision. □

We now extend Corollary 4.2.11 from $\mathrm{trop}(M)$ to arbitrary tropical linear spaces $L_{\mathbf{w}}$. Fix $\mathbf{w} \in \mathrm{Dr}_M$ and $\mathbf{u} \in \mathbb{R}^m$. Consider the set of vertices $\mathbf{e}_\sigma$ of P_M for which $w_\sigma - \sum_{j\in\sigma} u_j$ is minimal. These vertices form a face of $\Delta_{\mathbf{w}}$. Since $\Delta_{\mathbf{w}}$ is a matroid subdivision, by Lemma 4.4.6, that face is the matroid polytope $P_{M_{\mathbf{u}}^{\mathbf{w}}}$ associated with some matroid $M_{\mathbf{u}}^{\mathbf{w}}$ of rank r on $[m]$.

Lemma 4.4.7. *The tropical linear space defined by a point* $\mathbf{w} \in \mathrm{Dr}_M$ *equals*

$$L_{\mathbf{w}} \;=\; \big\{\mathbf{u} \in \mathbb{R}^m/\mathbb{R}\mathbf{1} \;:\; \textit{the matroid } M_{\mathbf{u}}^{\mathbf{w}} \textit{ has no loops}\big\}.$$

Proof. The set of bases of $M_{\mathbf{u}}^{\mathbf{w}}$ is a subset of the set $\mathcal{B}$ of bases of M. Hence the circuits of $M_{\mathbf{u}}^{\mathbf{w}}$ are obtained from the circuits of M by removing elements. Each circuit of M is the support of a tropical circuit (4.4.4) indexed by a subset $\tau \subset [m]$ with $|\tau| = r+1$ and $\mathrm{rank}(\tau) = r$. It consists of those indices j for which $\tau\backslash\{j\}$ is a basis of M. The corresponding circuit of $M_{\mathbf{u}}^{\mathbf{w}}$ consists of those indices j for which the minimum in (4.4.4) is attained.

Suppose $\mathbf{u}$ is in $L_{\mathbf{w}}$. Then the minimum in (4.4.4) is attained at least twice. Hence each circuit of $M_{\mathbf{u}}^{\mathbf{w}}$ has at least two elements, so $M_{\mathbf{u}}^{\mathbf{w}}$ has no loops. Conversely, if $\mathbf{u} \notin L_{\mathbf{w}}$, then there exists $\tau \subset [m]$ with $|\tau| = r+1$ and $\mathrm{rank}(\tau) = r$ such that the minimum in (4.4.4) is attained at a unique index j. This means that j is a loop of $M_{\mathbf{u}}^{\mathbf{w}}$. □

Our description of the matroid $M^{\mathbf{w}}_{\mathbf{u}}$ by its circuits implies the identity

$$M^{\mathbf{w}}_{\mathbf{u}+\epsilon\mathbf{v}} \;=\; (M^{\mathbf{w}}_{\mathbf{u}})_{\mathbf{v}}.$$

Here $\mathbf{u},\mathbf{v}$ are any vectors in $\mathbb{R}^m$, $\epsilon > 0$ is sufficiently small, and the right-hand expression $(\cdots)_{\mathbf{v}}$ refers to the face construction of Proposition 4.2.10. This identity implies the following fact about the neighborhood of $\mathbf{u}$ in $L_{\mathbf{w}}$.

Corollary 4.4.8. *Let σ be any cell of a tropical linear space $L_{\mathbf{w}}$, and let $\mathbf{u}$ be a point in the relative interior of σ. Then*

$$\mathrm{star}_{L_{\mathbf{w}}}(\sigma) \;=\; \mathrm{trop}(M^{\mathbf{w}}_{\mathbf{u}}).$$

We are now prepared to prove our main result in this section.

Proof of Theorem 4.4.5. The recession fan of the tropical hyperplane $L_\tau(\mathbf{w})$ in (4.4.4) is the constant coefficient hyperplane $L_\tau(\mathbf{0})$ defined by the unique circuit of M that lies in τ. The recession fan of $L_{\mathbf{w}}$ is $L_{\mathbf{0}} = \mathrm{trop}(M)$. This is the intersection of the codimension-1 fans $L_\tau(0)$ for all τ.

For the first statement we use Theorem 4.2.6. For every matroid $M^{\mathbf{w}}_{\mathbf{u}}$, the tropical linear space $\mathrm{trop}(M^{\mathbf{w}}_{\mathbf{u}})$ is the support of a balanced pure simplicial fan of dimension $r-1$. These fans are all the links of $L_{\mathbf{w}}$, by Corollary 4.4.8, so $L_{\mathbf{w}}$ is balanced and pure of dimension $r-1$. To argue that $L_{\mathbf{w}}$ is contractible, we make a forward reference to the material on tropical convexity in Section 5.2. Proposition 5.2.8 tells us that $L_{\mathbf{w}}$ is tropically convex, and it is hence contractible by the last statement of Theorem 5.2.3.

It remains to be seen that $L_{\mathbf{w}}$ is a cycle of degree 1. By choosing the point $\mathbf{p}$ far away from the origin, all intersections between $L_{\mathbf{w}}$ and $\mathbf{p} + \mathrm{trop}(U_{m-r+1,m})$ will lie in unbounded cones. These cones are translates of cones in the recession fan $L_{\mathbf{0}} = \mathrm{trop}(M)$, so $L_{\mathbf{w}}$ and its recession fan have the same degree. So, it suffices to show that $\mathrm{trop}(M)$ is a cycle of degree 1.

Pick $\mathbf{p} = (p_1, p_2, \ldots, p_m)$ such that $p_1 > p_2 > \cdots > p_m$. We claim that the matroid M has a unique chain of flats $F_1 \subset F_2 \subset \cdots \subset F_{r-1}$ whose cone $\mathrm{pos}(\mathbf{e}_{F_1}, \ldots, \mathbf{e}_{F_{r-1}})$ intersects $\mathbf{p}+\mathrm{trop}(U_{m-r+1,m})$. The following construction shows that the chain exists and is unique. For $1 \le i \le r-1$, let s_i denote the smallest index such that $\{1, 2, \ldots, s_i\}$ has rank i, and let F_i be the flat spanned by $\{1, 2, \ldots, s_i\}$. Consider $t \in [m] \backslash \{s_1, s_2, \ldots, s_r\}$, and let i be the index such that $t \in F_i \backslash F_{i-1}$. (Here $F_r = [m]$.) Let $\mathbf{q}$ denote the vector obtained from $\mathbf{p}$ by adding the positive quantity $p_{s_i} - p_t$ to p_t. Then $\mathbf{q}$ is the desired intersection point. The intersection multiplicity is 1 because the coordinates of $\mathbf{q}$ are integer linear combinations of the coordinates of $\mathbf{p}$. □

Example 4.4.9. We explain the concepts in this section for the simplest nontrivial case, $r = 2$ and $m = 4$. Fix the uniform matroid $M = U_{2,4}$. Here,

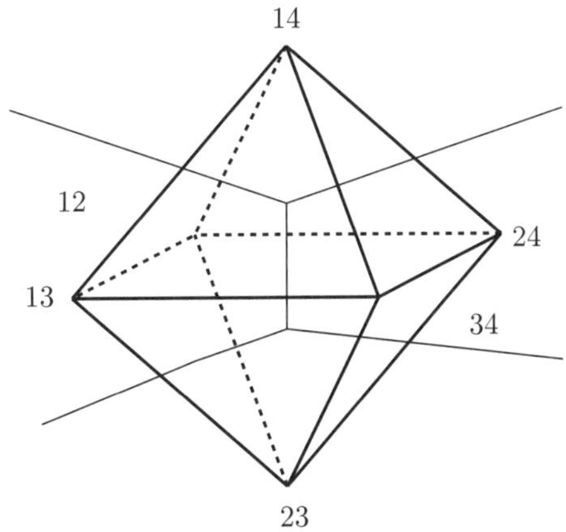

Figure 4.4.1. The hypersimplex $\Delta_{2,4}$ is a regular octahedron. Its matroid subdivisions correspond to tropical lines in 3-space.

the Dressian equals the tropical Grassmannian, by the four-point condition. Writing Δ for the space of phylogenetic trees on four taxa, we have

$$\mathrm{Dr}_M = \mathrm{Gr}_M = \mathrm{Dr}(2,4) = \mathrm{trop}(\mathrm{Gr}^0(2,4)) = -\Delta.$$

As seen in Section 4.3, this is a four-dimensional fan with three maximal cones. The cones intersect in the three-dimensional lineality space

$$L \;=\; \left\{\mathbf{w} \in \mathbb{R}^{\binom{4}{2}}/\mathbb{R}\mathbf{1} \,:\, w_{12}+w_{34} = w_{13}+w_{24} = w_{14}+w_{23}\right\}.$$

The matroid polytope P_M is the octahedron $\mathrm{conv}\{\mathbf{e}_{12}, \mathbf{e}_{13}, \mathbf{e}_{14}, \mathbf{e}_{23}, \mathbf{e}_{24}, \mathbf{e}_{34}\}$. This is shown in Figure 4.4.1.

First suppose that $\mathbf{w} \in L$. There exists $\mathbf{v} = (v_1, v_2, v_3, v_4) \in \mathbb{R}^4$ such that $w_{ij} = v_i + v_j$ for $1 \leq i < j \leq 4$. The matroid subdivision $\Delta_{\mathbf{w}}$ consists only of the octahedron and its faces, and $L_{\mathbf{w}} = \mathrm{trop}(M)$ is the star tree consisting of the four rays $-\mathbf{v} + \mathbb{R}_{\geq 0}\mathbf{e}_i$ for $i = 1,2,3,4$. The matroid $M_{\mathbf{v}}^{\mathbf{w}}$ is the original matroid $M = U_{2,4}$ which has six bases. For any $\mathbf{u}$ in the relative interior of a ray, the matroid $M_{\mathbf{u}}^{\mathbf{w}}$ has only three bases. For instance, if $\mathbf{u} \in -\mathbf{v} + \mathbb{R}_{>0}\mathbf{e}_1$, then $M_{\mathbf{u}}^{\mathbf{w}}$ has the bases $\{1,2\}$, $\{1,3\}$, and $\{1,4\}$.

Next suppose $\mathbf{w} \in \mathrm{Gr}_M \backslash L$. Up to relabeling, we can assume that $w_{12} + w_{34} = w_{13} + w_{24} < w_{14} + w_{23}$. The tropical linear space $L_{\mathbf{w}}$ is the tree in the first of the three cases of Example 4.3.19. The matroid subdivision $\Delta_{\mathbf{w}}$ cuts the octahedron into two square pyramids, namely $\mathrm{conv}\{\mathbf{e}_{12}, \mathbf{e}_{13}, \mathbf{e}_{14}, \mathbf{e}_{24}, \mathbf{e}_{34}\}$ and $\mathrm{conv}\{\mathbf{e}_{12}, \mathbf{e}_{13}, \mathbf{e}_{23}, \mathbf{e}_{24}, \mathbf{e}_{34}\}$. They correspond to the nodes P_{14} and P_{23} of the tree. The matroids $M_{P_{14}}$ and $M_{P_{23}}$ have these pyramids as matroid polytopes. They are obtained from $M = U_{2,4}$ by turning one basis (here $\{2,3\}$ or $\{1,4\}$) into a nonbasis. The bounded segment of $L_{\mathbf{w}}$ connects P_{14} and P_{23}. It is dual to the square cell $\mathrm{conv}\{\mathbf{e}_{12}, \mathbf{e}_{13}, \mathbf{e}_{24}, \mathbf{e}_{34}\}$ in $\Delta_{\mathbf{w}}$, so its matroid $M_{\mathbf{u}}^{\mathbf{w}}$ has two nonbases. The matroid on each unbounded ray has three bases and three nonbases, as before. ◊

If M is the uniform matroid $U_{r,m}$, then the matroid polytope P_M is the hypersimplex $\Delta_{r,m}$, as defined in Example 4.2.13. The corresponding Dressian $\mathrm{Dr}(r,m)$ lives in $\mathbb{R}^{\binom{m}{r}}/\mathbb{R}\mathbf{1}$. Its elements $\mathbf{w}$ define a regular matroid subdivision of $\Delta_{r,m}$. Note that a subdivision of $\Delta_{r,m}$ is a matroid subdivision if and only if every edge in the subdivision is an edge of $\Delta_{r,m}$. If $r = 2$, then all matroid subdivisions are dual to phylogenetic trees on $[m]$. These objects are familiar from Section 4.3. Here is the first really unfamiliar case.

Example 4.4.10. Let $r = 3, m = 6$, and fix the uniform matroid $M = U_{3,6}$. The five-dimensional polytope $P_M = \Delta_{3,6}$ has 20 vertices. A computation reveals that $\mathrm{Dr}(3,6) = \mathrm{Gr}(3,6)$. Modulo the lineality space, this has the structure of a four-dimensional fan with 65 rays and 1005 maximal cones. The rays come in three symmetry classes, named in [**SS04**] as follows.

Type E: 20 rays spanned by the coordinate vectors of the form $\mathbf{e}_{123}$;

Type F: 15 rays of the form $\mathbf{f}_{1234} = \mathbf{e}_{123} + \mathbf{e}_{124} + \mathbf{e}_{134} + \mathbf{e}_{234}$;

Type G: 30 rays of the form $\mathbf{g}_{123456} = \mathbf{e}_{123}+\mathbf{e}_{124}+\mathbf{e}_{345}+\mathbf{e}_{346}+\mathbf{e}_{156}+\mathbf{e}_{256}$.

We regard $\mathrm{Dr}(3,6)$ as three-dimensional polyhedral complex. Example 4.3.15 states that it has 1005 facets. The facets fall into seven symmetry classes. We label them according to which classes their vertices lie in.

Facet EEEE: 30 tetrahedra such as $\{\mathbf{e}_{123}, \mathbf{e}_{145}, \mathbf{e}_{246}, \mathbf{e}_{356}\}$;

Facet EEFF(a): 90 tetrahedra such as $\{\mathbf{e}_{123}, \mathbf{e}_{456}, \mathbf{f}_{1234}, \mathbf{f}_{3456}\}$;

Facet EEFF(b): 90 tetrahedra such as $\{\mathbf{e}_{125}, \mathbf{e}_{345}, \mathbf{f}_{1256}, \mathbf{f}_{3456}\}$;

Facet EFFG: 180 tetrahedra such as $\{\mathbf{e}_{345}, \mathbf{e}_{1256}, \mathbf{f}_{3456}, \mathbf{g}_{123456}\}$;

Facet EEEG: 180 tetrahedra such as $\{\mathbf{e}_{126}, \mathbf{e}_{134}, \mathbf{e}_{356}, \mathbf{g}_{125634}\}$;

Facet EEFG: 180 tetrahedra such as $\{\mathbf{e}_{234}, \mathbf{e}_{125}, \mathbf{f}_{1256}, \mathbf{g}_{125634}\}$;

Facet FFFGG: 15 bipyramids such as $\{\mathbf{f}_{1234}, \mathbf{f}_{1256}, \mathbf{f}_{3456}, \mathbf{g}_{123456}, \mathbf{g}_{125634}\}$.

Suppose $\mathbf{w}$ is in the relative interior of one of these seven maximal cones of $\mathrm{Dr}(3,6)$. Then $\Delta_{\mathbf{w}}$ is a finest matroid subdivision of the hypersimplex $\Delta_{3,6}$. In case EEEE that subdivision has five facets: the central facet is the matroid polytope associated with the matroid with nonbases $\{\{1,2,3\},\{1,4,5\},\{2,4,6\},\{3,5,6\}\}$. The other four matroids are $U_{3,4}$ with one of the four elements replaced by three parallel elements. The corresponding tropical plane $L_{\mathbf{w}}$ has 27 two-dimensional cells (all unbounded), 22 one-dimensional cells (four bounded), and five vertices. In the other six cases, the subdivision $\Delta_{\mathbf{w}}$ has six facets, labeled by various rank 3 matroids on $[6]$. The corresponding tropical plane $L_{\mathbf{w}}$ has 28 two-dimensional cells (one bounded), 24 one-dimensional cells (six bounded), and six vertices.

These face numbers are smaller when $\mathbf{w}$ lies on a lower-dimensional cell of $\mathrm{Dr}(3,6)$. In the most degenerate case, when $\mathbf{w}$ is in the lineality space L, the tropical plane $L_{\mathbf{w}}$ is a fan with six rays and 15 two-dimensional cones. For further information see Figure 5.4.1 and Table 5.4.1 in Section 5.4. $\diamondsuit$

The construction of matroid subdivisions to represent linear spaces is an analog to Proposition 3.1.6 for hypersurfaces. The role played by the Newton polytope of a hypersurface is now played by the matroid polytope. For hypersurfaces and linear spaces over a field K with trivial valuation, the reader should compare Proposition 3.1.10 with Proposition 4.2.10. In both cases, the tropical variety is a subfan to the normal fan of the relevant polytope. When K has a nontrivial valuation, then the tropical variety is dual to a regular subdivision of the Newton polytope or matroid polytope.

A common generalization of the Newton polytope and the matroid polytope is the *Chow polytope* which exists for an arbitrary variety $X \subseteq T^m$. The Chow polytope of X is the weight polytope of the Chow–van der Waerden form R_X associated to the projective closure $\overline{X} \subseteq \mathbb{P}^m$ of X. The tropical variety $\mathrm{trop}(X)$ is a subcomplex of the dual complex to a regular subdivision of the Chow polytope of X. We call this the *Chow complex*. In that sense, the Chow form plays a role similar to the polynomial g in (2.5.2) which was used to define the Gröbner complex of $\overline{X}$ in Section 2.5. However, R_X and g are different, and the Chow complex is different from the Gröbner complex.

Unlike the cases of hypersurfaces and linear spaces, the tropical variety $\mathrm{trop}(X)$ is in general not determined by the Chow complex. There are varieties $X, X' \subset T^m$ with the same Chow complex but $\mathrm{trop}(X) \neq \mathrm{trop}(X')$. We refer to [**KSZ92**] or [**GKZ08**, Chapter 6] for details on the Chow polytope, and to [**Fin13**] for connections to tropical geometry. Section 6 of [**Fin13**] discusses the Chow complex and contains the above example of $X, X' \subset T^m$.

Remark 4.4.11. In this section, we introduced tropical linear spaces as the balanced contractible complexes $L_{\mathbf{w}}$ associated with points $\mathbf{w}$ in a Dressian Dr_M. We showed that $L_{\mathbf{u}}$ is a cycle of degree 1 and it has recession fan $\mathrm{trop}(M)$. Either of the two latter properties actually characterizes tropical linear spaces. This was proved in [**Fin13**, Theorem 7.4].

This means that we could also have used the following as definitions:

- A *tropical linear space* is a tropical cycle of degree 1.
- A *tropical linear space* is a balanced polyhedral complex whose recession fan is (the Bergman fan of) a matroid.

While these definitions are elegant, our approach has the virtue of supplying the reader with the combinatorial tools necessary to work with linear spaces.

4.5. Surfaces

In this section we study the tropicalization of surfaces in three-dimensional space. This allows us to explore tropical shadows of classical theorems for surfaces, such as the two rulings of lines on a quadric surface and the configuration of 27 lines on a cubic surface. We shall see that these statements are not easily true in the tropical setting. A combinatorial description of surfaces of degree d that are tropically smooth appears in Theorem 4.5.2.

By a *surface in* 3*-space* we mean a variety $X = V(f)$ in the torus T^3, where $f = \sum c_{\mathbf{u}} x^{\mathbf{u}} \in K[x_1^{\pm 1}, x_2^{\pm 1}, x_3^{\pm 1}]$ is an irreducible Laurent polynomial. By Proposition 3.1.6, its tropicalization $\mathrm{trop}(X)$ is a pure two-dimensional polyhedral complex that is dual to the regular subdivision of $\mathrm{Newt}(f)$ induced by the weight vector $(\mathrm{val}(c_{\mathbf{u}}))$. In this section, a *tropical surface in* 3*-space* will be any balanced polyhedral complex of this form in $\mathbb{R}^3$. By Proposition 3.3.10 these all arise as $\mathrm{trop}(V(f))$ for some $f \in K[x_1^{\pm 1}, x_2^{\pm 1}, x_3^{\pm 1}]$.

A very first example of a tropical surface in $\mathbb{R}^3$ is the tropical plane defined by a linear polynomial $f = c_1x_1 + c_2x_2 + c_3x_3 + c_4$. If all four coefficients c_i are nonzero, then this fan is a cone over the complete graph K_4. In the notation of Example 4.2.13, the tropical plane equals $\mathrm{trop}(U_{3,4})$ but with the origin $(0, 0, 0)$ shifted to the point $(\mathrm{val}(c_4/c_1), \mathrm{val}(c_4/c_2), \mathrm{val}(c_4/c_3))$. It is instructive to verify that two tropical planes intersect in a tropical line.

We will be particularly interested in *smooth* tropical surfaces, which are those for which the regular subdivision of the Newton polytope $\mathrm{Newt}(f)$ is unimodular, so all tetrahedra have volume $1/6$. This name is justified by Proposition 4.5.1, which is true for hypersurfaces in arbitrary dimension.

A classical hypersurface $V(f) \subset T^n$ is *singular* at a point $\mathbf{y} \in V(f)$ if $(\partial f/\partial x_i)(\mathbf{y}) = 0$ for $1 \leq i \leq n$. The hypersurface $V(f)$ is *smooth* if it has no singular points. A *unimodular triangulation* of a lattice polytope in $\mathbb{R}^n$ is one for which all simplices have the same minimal volume $1/n!$.

Proposition 4.5.1. *Fix*

$$f = \sum c_{\mathbf{u}} x^{\mathbf{u}} \in K[x_1^{\pm 1}, \ldots, x_n^{\pm 1}].$$

Let $\Delta_{\mathrm{val}(c_{\mathbf{u}})}$ *be the regular subdivision of the Newton polytope* $\mathrm{Newt}(f)$ *induced by the weights* $\mathrm{val}(c_{\mathbf{u}})$. *If* $\Delta_{\mathrm{val}(c_{\mathbf{u}})}$ *is unimodular, then* $V(f) \subset T^n$ *is a smooth hypersurface.*

Proof. After multiplying by a monomial, we may assume $f \in K[x_1, \ldots, x_n]$. Let $d = \max_{c_{\mathbf{u}} \neq 0} |\mathbf{u}|$ be the maximum degree of a monomial in f. Let $g = \sum c_{\mathbf{u}} x^{\mathbf{u}} x_0^{d-|\mathbf{u}|}$ be the homogenization of f in $K[x_0, x_1, \ldots, x_n]$. A point $\mathbf{y}$ is singular on $V(f)$ if and only if $(1 : \mathbf{y})$ is singular on $V(g) \subset \mathbb{P}^n$. So, we must show that $V(g)$ has no singular point $\mathbf{z}$ in $T^n = (K^*)^{n+1}/K^*$.

Suppose that $\mathbf{z} \in V(g)$ is a singular point. Then $(\partial g/\partial x_i)(\mathbf{z}) = 0$ for $0 \leq i \leq n$, so $\sum_{i=0}^{n} a_i z_i (\partial g/\partial x_i)(\mathbf{z}) = 0$ for any $a_0, \ldots, a_n \in \mathbb{Z}$. Here multiplication of elements of K by integers a_i is the $\mathbb{Z}$-module multiplication (i.e., $3a = a + a + a$ even if $\mathrm{char}(K) = 3$). For any $\mathbf{a} \in \mathbb{Z}^{n+1}$, we set

$$\begin{aligned} W_{\mathbf{a}} &= \{\mathbf{z} \in T^n : \sum_{i=0}^{n} a_i z_i (\partial g/\partial x_i)(\mathbf{z}) = 0\} \\ &= \{\mathbf{z} \in T^n : \sum_{i=0}^{n} a_i z_i \sum_{\mathbf{u}} c_{\mathbf{u}} u_i z^{\mathbf{u}-\mathbf{e}_i} = 0\} \\ &= \{\mathbf{z} \in T^n : \sum_{\mathbf{u}} c_{\mathbf{u}} (\sum_{i=0}^{n} a_i u_i) \mathbf{z}^{\mathbf{u}} = 0\} \\ &= \{\mathbf{z} \in T^n : \sum_{\mathbf{u}} c_{\mathbf{u}} (\mathbf{a} \cdot \mathbf{u}) \mathbf{z}^{\mathbf{u}} = 0\}. \end{aligned}$$

We thus have

$$\mathbf{z} \in V(g) \cap \bigcap_{\mathbf{a} \in \mathbb{Z}^{n+1}} W_{\mathbf{a}},$$

and hence

$$\mathrm{val}(\mathbf{z}) \in \mathrm{trop}(V(g)) \cap \bigcap_{\mathbf{a} \in \mathbb{Z}^{n+1}} \mathrm{trop}(W_{\mathbf{a}}). \tag{4.5.1}$$

We will show that $V(g) \subset T^n$ is smooth by showing that this intersection of tropical hypersurfaces is empty when $\Delta_{\mathrm{val}(c_{\mathbf{u}})}$ is unimodular. Note that

$$\mathrm{trop}(W_{\mathbf{a}}) = \{\mathbf{w} : \min\big(\mathrm{val}(c_{\mathbf{u}}) + \mathrm{val}(\mathbf{a} \cdot \mathbf{u}) + \mathbf{w} \cdot \mathbf{u}\big) \text{ is achieved at least twice}\}.$$

Suppose $\mathbf{w}$ is in the right-hand side of (4.5.1). Let σ be the cell in $\Delta_{\mathrm{val}(c_{\mathbf{u}})}$ dual to the cell of $\mathrm{trop}(V(g))$ containing $\mathbf{w}$. This means that $\min(\mathrm{val}(c_{\mathbf{u}}) + \mathbf{w} \cdot \mathbf{u})$ is achieved only at lattice points $\mathbf{u} \in \sigma$. The assumption that $\Delta_{c_{\mathbf{u}}}$ is unimodular means that σ contains $\dim(\sigma) + 1$ lattice points, which are all vertices.

Choose $\mathbf{a} \in \mathbb{Z}^{n+1}$ with $\mathbf{a} \cdot \mathbf{u} = 0$ for all but one of these lattice points, and $\mathbf{a} \cdot \mathbf{u}' = 1$ for the last vector $\mathbf{u}'$. The existence of such an $\mathbf{a}$ is another consequence of σ being a unimodular simplex. Then $\mathrm{val}(\mathbf{a} \cdot \mathbf{u}') = 0$. Since $\mathbf{a} \cdot \mathbf{u} \in \mathbb{Z}$, we have $\mathrm{val}(\mathbf{a} \cdot \mathbf{u}) \geq 0$ for all $\mathbf{u}$; indeed $\mathrm{val}(\mathbf{a} \cdot \mathbf{u}) = 0$ unless $\mathrm{char}(K) = 0$ and the induced valuation on $\mathbb{Q} \subset K$ is the p-adic valuation. Thus $\mathrm{val}(c_{\mathbf{u}}(\mathbf{a} \cdot \mathbf{u})) \geq \mathrm{val}(c_{\mathbf{u}})$, so for arbitrary $\mathbf{u}$ we have

$$\begin{aligned} \mathrm{val}(c_{\mathbf{u}'}) + \mathrm{val}(\mathbf{a} \cdot \mathbf{u}') + \mathbf{w} \cdot \mathbf{u}' &= \mathrm{val}(c_{\mathbf{u}'}) + \mathbf{w} \cdot \mathbf{u} \\ &\leq \mathrm{val}(c_{\mathbf{u}}) + \mathbf{w} \cdot \mathbf{u} \\ &\leq \mathrm{val}(c_{\mathbf{u}}) + \mathrm{val}(\mathbf{a} \cdot \mathbf{u}) + \mathbf{w} \cdot \mathbf{u}. \end{aligned}$$

Since the first inequality is an equality only for $\mathbf{u} \in \sigma$ and $\mathbf{u}'$ is the only lattice point in σ for which the second point is an equality, $\min(\mathrm{val}(c_{\mathbf{u}}) + \mathrm{val}(\mathbf{a} \cdot \mathbf{u}) + \mathbf{w} \cdot \mathbf{u})$ is achieved only at $\mathbf{u}'$. Thus $\mathbf{w} \notin \mathrm{trop}(W_{\mathbf{a}})$. As the choice of $\mathbf{w}$ was arbitrary, this means that the intersection in (4.5.1) is empty, and thus $V(g)$ has no singular points. □

Smooth tropical surfaces in $\mathbb{R}^3$ are pure two-dimensional balanced polyhedral complexes, by Theorem 3.3.5. We now determine their face numbers.

Theorem 4.5.2. *Let $f \in K[x_1, x_2, x_3]$ be a polynomial of degree d whose Newton polytope is the tetrahedron $\mathrm{conv}((0,0,0),(d,0,0),(0,d,0),(0,0,d))$. If the tropical surface $S = \mathrm{trop}(V(f))$ is smooth, then it has*

d^3	*vertices,*
$2d^2(d-1)$	*edges (bounded one-dimensional cells),*
$4d^2$	*rays (unbounded one-dimensional cells),*
$d(d-1)(7d-11)/6$	*bounded two-dimensional cells, and*
$6d^2$	*unbounded two-dimensional cells.*

In particular, the Euler characteristic of a smooth tropical surface S equals

$$\chi(S) \;=\; d^3 - 2d^2(d-1) + \frac{d(d-1)(7d-11)}{6} \;=\; \frac{(d-1)(d-2)(d-3)}{6} + 1.$$

Proof. Let $\Delta_{\mathrm{val}(c_{\mathbf{u}})}$ be the unimodular triangulation of the tetrahedron $\mathrm{Newt}(f)$ induced by the coefficients $\mathrm{val}(c_{\mathbf{u}})$ of the tropical polynomial $\mathrm{trop}(f)$. Cells of dimension m in the tropical surface S are dual to simplices of $\Delta_{\mathrm{val}(c_{\mathbf{u}})}$ of dimension $3-m$. A cell is unbounded if and only if the corresponding simplex in $\Delta_{\mathrm{val}(c_{\mathbf{u}})}$ lies on the boundary of $\mathrm{Newt}(f)$. Thus to prove the proposition, we need to count the number of tetrahedra, triangles, and edges in $\Delta_{\mathrm{val}(c_{\mathbf{u}})}$, while keeping track of those on the boundary of $\mathrm{Newt}(f)$. We denote by i_m the number of m-dimensional simplices in the interior of $\Delta_{\mathrm{val}(c_{\mathbf{u}})}$, and by b_m the number of m-dimensional simplices on the boundary.

Every tetrahedron in $\Delta_{\mathrm{val}(c_{\mathbf{u}})}$ has minimal volume $1/6$. The big tetrahedron $\mathrm{Newt}(f)$ has volume $d^3/6$. Hence there are d^3 tetrahedra in the triangulation $\Delta_{\mathrm{val}(c_{\mathbf{u}})}$. Each tetrahedron has four triangular faces, which lie in two tetrahedra if they are internal and one if they are on the boundary, so $4d^3 = 2i_2 + b_2$. There are d^2 triangles on each of the four triangular faces of $\mathrm{Newt}(f)$, so $b_2 = 4d^2$, and $i_2 = 2d^3 - 2d^2$. Each boundary triangle has three edges, each of which is in two boundary triangles, so $b_1 = 3/2 b_2 = 6d^2$. There are $\binom{d+3}{3} = (d+3)(d+2)(d+1)/6$ lattice points in the tetrahedron $\mathrm{Newt}(f)$. These are the vertices in the unimodular triangulation $\Delta_{\mathrm{val}(c_{\mathbf{u}})}$. Since $\mathrm{Newt}(f)$ is homeomorphic to a ball, its Euler characteristic is one.

The Euler characteristic is the alternating sum of the face numbers, so

$$\begin{aligned} 1 &= -\,i_3 \,+\, (i_2+b_2) \,-\, (i_1+b_1) \,+\, (i_0+b_0) \\ &= -\,(d^3) \,+\, ((2d^3-2d^2)+4d^2) \,-\, (i_1+6d^2) \,+\, (d+3)(d+2)(d+1)/6 \\ &= d(d-1)(7d-11)/6+1 \,-\, i_1. \end{aligned}$$

This means that $i_1 = d(d-1)(7d-11)/6$. The face count for S now follows by dualizing. The five numbers are i_3, i_2, b_2, i_1, and b_1, in this order.

Finally, the tropical surface S is homotopic to its subcomplex of bounded faces. We can ignore the unbounded faces when computing the Euler characteristic. This gives the formula $\chi(S) = i_3 - i_2 + i_1 = \frac{(d-1)(d-2)(d-3)}{6} + 1$. □

We next examine these face numbers for tropical surfaces of low degree.

Example 4.5.3. Let S be a smooth surface of degree d as in Theorem 4.5.2.

$d=2$: Smooth tropical quadrics have eight vertices, eight edges, and one bounded 2-cell; see Figure 3.1.3. There are 16 rays, four in each coordinate direction, linked by 24 unbounded 2-cells. In each of the four planes at infinity, we see a tropical quadric as in Figure 1.3.2. An enumeration shows that these are all contractible.

$d=3$: Every smooth tropical cubic surface S is also contractible, reflecting the fact that a classical cubic in $\mathbb{P}^3$ is rational. It has 27 vertices, 36 edges, and ten bounded 2-cells. The 36 rays, nine in each coordinate direction, are linked by 54 unbounded 2-cells. This unbounded part of S represents the four elliptic curves in the planes at infinity.

$d=4$: Every smooth tropical quartic surface S is homotopic to the 2-sphere, which has $\chi(S)=2$. This sphere sits inside S. This reflects the fact that a quartic in $\mathbb{P}^3$ is a $K3$ surface. The tropical quartic surface S has 64 vertices, 96 edges, and 34 bounded 2-cells. The 64 rays, 16 in each coordinate direction, are linked by 96 unbounded 2-cells. ◊

When $d=2$, the combinatorial types of tropical surfaces of degree 2 are in bijection with the regular unimodular triangulations of the (ten lattice points in the) tetrahedron $2\Delta = \operatorname{conv}\{(0,0,0),(0,0,2),(0,2,0),(2,0,0)\}$.

Proposition 4.5.4. *There are* 192 *regular unimodular triangulations of* 2Δ, *in* 14 *symmetry classes. The unique bounded* 2*-cell of the tropical quadric is*

- *an octagon for one class with* 3 *triangulations,*
- *a heptagon for one class with* 12 *triangulations,*
- *a hexagon for three classes with* $6+12+24$ *triangulations,*
- *a pentagon for three classes with* $12+24+24$ *triangulations,*

- *a quadrilateral for five classes with* $3+12+12+12+24$ *triangulations,*
- *a triangle for one class with* 12 *triangulations.*

Proof. Each unimodular triangulation of 2Δ has exactly one interior edge e. There are three cases, namely $e = \text{conv}\{(1,0,0),(0,1,1)\}$, $e = \text{conv}\{(0,1,0),(1,0,1)\}$, or $e = \text{conv}\{(0,0,1),(1,1,0)\}$. Suppose we are in the first case. We must show that the number of these triangulations is precisely 64. This can be seen by projecting the configuration 2Δ along e. Explicitly, consider the projection $(i,j,k) \mapsto (i+j, i+k)$. The image is the planar configuration $\mathcal{P} = \{(i,j) : 0 \leq i,j \leq 2\}$. That configuration has 64 unimodular triangulations, all regular, in 14 symmetry classes. The link of the central point $(1,1)$ is an n-gon for $n \in \{3,4,5,6,7,8\}$. A triangulation of the square $\text{conv}(\mathcal{P})$ is unimodular if and only if it contains all nine points of $\mathcal{P}$ as vertices. One way to verify the enumeration of triangulations of $\mathcal{P}$ is to first construct the options for the triangles containing $(1,1)$ as a vertex, and then fill in the rest of the triangulation. The numbers for which the link is an n-gon agree with the numbers in the proposition divided by 3.

Each of these 64 planar triangulations extends uniquely to a triangulation of the tetrahedron 2Δ, which is also regular and unimodular. Indeed, each point in $\mathcal{P}$ except $(1,1)$ has a unique preimage in $2\Delta \cap \mathbb{Z}^3$. Consider the regular unimodular triangulation of $\text{conv}(\mathcal{P})$ that is induced by $\mathbf{w} \in \mathbb{R}^9$, with coordinates indexed by $\mathcal{P}$. Then we form the weight vector $\mathbf{w}' \in \mathbb{R}^{10}$ with coordinates indexed by lattice points in 2Δ by setting $w'_{(i,j,k)} = w_{(i+j,i+k)}$. The bounded two-dimensional cell of a tropical quadric is dual to the interior edge of the triangulation of 2Δ, and the number of its vertices is equal to the number of neighboring tetrahedra. This is the number of vertices of the link of $(1,1)$ in the corresponding triangulation of $\mathcal{P}$. □

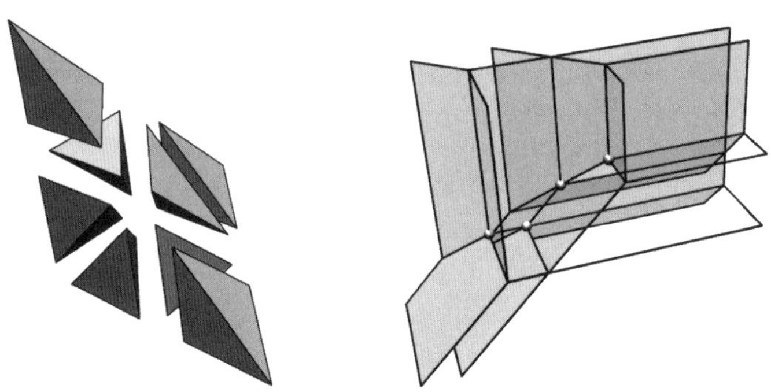

Figure 4.5.1. Triangulation of the tetrahedron 2Δ and matching tropical quadric surface. This comes from the left diagram in Figure 1.3.3.

As an example of the technique used in the proof of Proposition 4.5.4, the class of four triangulations shown on the left of Figure 1.3.3 gives 12 triangulations of 2Δ. In the resulting tropical quadrics, the bounded 2-cell is a pentagon. This triangulation and its quadric are shown in Figure 4.5.1.

In the remainder of this section we examine the tropical analogues of classical facts about lines on surfaces. We begin with surfaces of degree 2.

Quadric surfaces in $\mathbb{P}^3$ are *ruled surfaces*. Through *any* point $\mathbf{x}$ in Q there exist exactly two lines L and L' satisfying $\mathbf{x} \in L \subset Q$ and $\mathbf{x} \in L' \subset Q$. These lines come in two families, each of which covers the entire surface Q. To see this geometrically, consider the tangent plane of Q at $\mathbf{x}$. The intersection of that plane with Q is a plane conic that is singular at $\mathbf{x}$. That conic is a union $L \cup L'$ of two lines satisfying $L \cap L' = \{\mathbf{x}\}$.

We now derive a tropical version of the classical fact that every point on a quadric surface Q lies on exactly two lines of Q.

Theorem 4.5.5. *Let $\mathcal{Q}$ be a tropical quadric surface in $\mathbb{R}^3$ that is tropically smooth. Every point $\mathbf{u}$ in the relative interior of the unique bounded 2-cell of the tropical quadric $\mathcal{Q}$ lies in two tropical lines that are contained in $\mathcal{Q}$.*

Proof. The bounded 2-cell P of $\mathcal{Q}$ is dual to one of the three edges e in the proof of Proposition 4.5.4. Suppose $e = \mathrm{conv}\{(1,0,0),(0,1,1)\}$ without loss of generality, so the affine span of P is perpendicular to $(1,-1,-1)$.

Fix $\mathbf{u} \in \mathrm{relint}(P)$. Suppose L is a tropical line in $\mathcal{Q}$ containing $\mathbf{u}$. Since $\mathbf{u} \in \mathrm{relint}(P)$, the line segment or ray of L containing $\mathbf{u}$ must lie in the plane $\{v_1 - v_2 - v_3 = 0\}$. Every vertex of L is incident to a ray pointing into one of the four coordinate directions, which we take to be

$$(4.5.2) \qquad \mathbf{e}_1 = (1,0,0), \;\; \mathbf{e}_2 = (0,1,0), \;\; \mathbf{e}_3 = (0,0,1), \;\; \mathbf{e}_4 = (-1,-1,-1).$$

Since none of these directions is in the plane $\{v_1 - v_2 - v_3 = 0\}$, we conclude the following: no vertex of L lies in $\mathrm{relint}(P)$, the intersection $P \cap L$ is a line segment, and the direction of that segment is either $(1,1,0)$ or $(1,0,1)$.

We will prove that each of the two segments $P \cap (\mathbf{u} + \mathbb{R}(1,1,0))$ and $P \cap (\mathbf{u} + \mathbb{R}(1,0,1))$ extends uniquely to a tropical line L that lies on $\mathcal{Q}$. An endpoint $\mathbf{v}$ of either segment lies on an edge of P. That edge is dual to a triangle in the triangulation. The two edges of that triangle other than e connect $A_{14} = (1,0,0)$ and $A_{23} = (0,1,1)$ to a point that lies on either $B_{24} = \mathrm{conv}\{(0,2,0),(0,0,0)\}$ or $B_{13} = \mathrm{conv}\{(2,0,0),(0,0,2)\}$ or $B_{12} = \mathrm{conv}\{(2,0,0),(0,2,0)\}$ or $B_{34} = \mathrm{conv}\{(0,0,2),(0,0,0)\}$. The direction of any such edge is perpendicular to a choice of $\mathbf{e}_1, \mathbf{e}_2, \mathbf{e}_3$, or $\mathbf{e}_4$ where the ray $\mathbf{v} + \mathbb{R}_{\geq 0}\mathbf{e}_i$ lies on $\mathcal{Q}$. By attaching these four rays to $P \cap (\mathbf{u} + \mathbb{R}(1,1,0))$, we construct a tropical line that passes through $\mathbf{u}$ and lies on $\mathcal{Q}$. This works because the direction of the edge from A_{14} to B_{24} is perpendicular to $\mathbf{e}_3$, from

A_{23} to B_{24} is perpendicular to $\mathbf{e}_1$, from A_{14} to B_{13} is perpendicular to $\mathbf{e}_2$, and from A_{23} to B_{13} is perpendicular to $\mathbf{e}_4$. It also works for $P \cap (\mathbf{u} + \mathbb{R}(1,0,1))$ because the direction of the edge from A_{14} to B_{12} is perpendicular to $\mathbf{e}_3$, from A_{23} to B_{12} is perpendicular to $\mathbf{e}_4$, from A_{14} to B_{34} is perpendicular to $\mathbf{e}_2$, and from A_{23} to B_{34} is perpendicular to $\mathbf{e}_1$. □

The proof just presented gives an algorithm for constructing the two tropical lines on $\mathcal{Q}$ through a given point $\mathbf{u}$. Here is a demonstration.

Example 4.5.6. Let $\mathcal{Q}$ be the tropical quadric in $\mathbb{R}^4/\mathbb{R}\mathbf{1}$ defined by

$$2 \odot u_1^2 \oplus 0 \odot u_1u_2 \oplus 0 \odot u_1u_3 \oplus 1 \odot u_1u_4 \oplus 2 \odot u_2^2$$
$$\oplus\, 0 \odot u_2u_3 \oplus 0 \odot u_2u_4 \oplus 2 \odot u_3^2 \oplus 1 \odot u_3u_4 \oplus 3 \odot u_4^2.$$

This quadric is tropically smooth. We list the eight tetrahedra of the triangulation of 2Δ with the corresponding vertices on $\mathcal{Q} \subset \mathbb{R}^4/\mathbb{R}\mathbf{1}$:

$$\begin{array}{ll}
\{u_1u_2, u_1u_3, u_2u_3, u_2u_4\} \to (0,0,0,0), & \{u_1u_2, u_2^2, u_2u_3, u_2u_4\} \to (2,0,2,2),\\
\{u_1u_2, u_1u_3, u_1u_4, u_2u_4\} \to (0,1,1,0), & \{u_1^2, u_1u_2, u_1u_3, u_1u_4\} \to (0,2,2,1),\\
\{u_1u_3, u_1u_4, u_2u_4, u_3u_4\} \to (1,2,1,0), & \{u_1u_4, u_2u_4, u_3u_4, u_4^2\} \to (2,3,2,0),\\
\{u_1u_3, u_2u_3, u_2u_4, u_3u_4\} \to (1,1,0,0), & \{u_1u_3, u_2u_3, u_3^2, u_3u_4\} \to (2,2,0,1).
\end{array}$$

The bounded polygon P in $\mathcal{Q}$ is the convex hull of the four vertices in the left column. The remaining four bounded edges are the convex hulls of the rows of the table. This triangulation of 2Δ is obtained by cutting off the four vertices and then triangulating the octahedron that remains. It is a representative of the quadrilateral class of size three from Proposition 4.5.4. Fix the point $\mathbf{u} = (\frac{48}{109}, \frac{118}{109}, \frac{70}{109}, 0)$ in P. The two segments through $\mathbf{u}$ are

$$\operatorname{conv}\Big\{ (\frac{48}{109}, \frac{48}{109}, 0, 0), (\frac{48}{109}, \frac{157}{109}, 1, 0) \Big\}, \ \operatorname{conv}\Big\{ (1, \frac{179}{109}, \frac{70}{109}, 0), (0, \frac{70}{109}, \frac{70}{109}, 0) \Big\}.$$

The two tropical lines through $\mathbf{u}$ on $\mathcal{Q}$ are obtained by attaching the coordinate rays to the four endpoints of these two segments. ◊

Remark 4.5.7. Theorem 4.5.5 does not say that every point of $\mathcal{Q}$ lies on exactly two tropical lines in $\mathcal{Q}$. There may be infinitely many lines through a vertex of the unique bounded 2-cell of $\mathcal{Q}$. An explicit instance will be given in Example 4.5.9 after we derive the general result in Theorem 4.5.8.

A gem of nineteenth century algebraic geometry is the theorem that every smooth cubic surface $X \subset \mathbb{P}^3$ contains exactly 27 lines. This originated in a correspondence between Salmon and Cayley in the 1840s. Expositions of this result can be found, for example, in [**Rei88**, §7] or [**Har77**, §V.4]. The tropical cubic surface $\operatorname{trop}(X \cap T^3)$ contains the tropicalization of each line. These tropical lines need not be distinct, and there may exist tropical lines that do not come from classical lines. Thus, it is not clear at all whether a general tropical cubic surface should contain precisely 27 tropical lines.

As it turns out, this statement is false, in an alarming manner. Smooth tropical surfaces in $\mathbb{R}^3$ of arbitrary degree d can contain *infinitely* many lines.

Theorem 4.5.8. *For any $d \geq 2$, the scaled tetrahedron $d\Delta \subset \mathbb{R}^3$ admits regular unimodular triangulations such that all smooth tropical surfaces that are dual to these triangulations contain infinitely many tropical lines.*

Proof. In our big tetrahedron $d\Delta = \mathrm{conv}\{(0,0,0),(d,0,0),(0,d,0),(0,0,d)\}$, the small tetrahedron $\tau = \mathrm{conv}\{(0,0,0),(0,1,0),(1,d-1,0),(d-1,0,1)\}$ has unit volume. Let Σ be any regular unimodular triangulation of $d\Delta$ that contains τ as a simplex. We first prove that Σ satisfies the conclusion of the theorem, and later we show that such triangulations Σ actually exist.

Let $\mathcal{S}$ be any tropical surface of degree d dual to Σ. Then $\mathcal{S}$ is smooth since Σ is a triangulation. The tetrahedron τ corresponds to a vertex $\mathbf{t}$ on $\mathcal{S}$. We shall exhibit infinitely many tropical lines on $\mathcal{S}$ passing through $\mathbf{t}$.

The key property of the tetrahedron τ is that five of its six edges lie on the boundary of $d\Delta$. The interior edge of τ is $\mathrm{conv}\{(0,1,0),(d-1,0,1)\}$. This means that five of the six 2-cells of Σ containing $\mathbf{t}$ are unbounded. In particular, the edge $\mathrm{conv}\{(0,0,0),(0,1,0)\}$ of τ lies on an edge of $d\Delta$, and the corresponding 2-cell in $\mathcal{S}$ is the orthant $\mathbf{t} + \mathbb{R}_{\geq 0}\{\mathbf{e}_1, \mathbf{e}_3\}$. The 2-cell dual to the edge $\mathrm{conv}\{(0,0,0),(d-1,0,1)\}$ of τ has the ray $\mathbf{t} + \mathbb{R}_{\geq 0}\{\mathbf{e}_2\}$. The 2-cell dual to $\mathrm{conv}\{(1,d-1,0),(d-1,0,1)\}$ has the ray $\mathbf{t} + \mathbb{R}_{\geq 0}\{\mathbf{e}_4\}$. Here we are using the conventions from (4.5.2). Consider the tropical line whose bounded edge equals $\mathrm{conv}\{\mathbf{t}, \mathbf{t} + \lambda(\mathbf{e}_1 + \mathbf{e}_3)\}$, where $\lambda > 0$. Our construction shows that, for any $\lambda > 0$, this line lies on the tropical surface $\mathcal{S}$.

We next construct a regular unimodular triangulation Σ of $d\Delta$ containing τ. The tetrahedron τ lies in the triangular prism $d\Delta \cap \{z \leq 1\}$. It suffices to construct such a triangulation Σ' for $d\Delta \cap \{z \leq 1\}$. Indeed, any regular unimodular triangulation of $d\Delta \cap \{z \geq 1\} \simeq (d-1)\Delta$ that agrees with Σ' on the triangle $d\Delta \cap \{z = 1\}$ will give the desired Σ.

Let Σ'_0 be any regular unimodular triangulation of the big triangle $d\Delta \cap \{z = 0\}$ containing the small triangle $\mathrm{conv}\{(0,0,0),(0,1,0),(1,d-1,0)\}$. Let Σ''_0 be an arbitrary regular unimodular triangulation of $d\Delta \cap \{z = 1\}$. Fix a lifting vector for Σ'_0 with very small positive entries, and fix a lifting vector for Σ''_0 whose coordinates are very close to the values of a nonconstant linear function on $d\Delta \cap \{z = 1\}$ that attains its minimum at $(d-1,0,1)$. The concatenation of the two lifting vectors induces a regular unimodular triangulation Σ' of our prism $d\Delta \cap \{z \leq 1\}$. By construction, Σ' restricts to Σ'_0 and Σ'_1 on the two triangular faces of the prism, and τ appears in Σ'. □

Most algebraic geometers will find Theorem 4.5.8 disturbing at first glance, given that a *general* surface in $\mathbb{P}^3$ of degree $d \geq 4$ contains no lines at all. Here "general" means that the parameter space $\mathbb{P}^{\binom{d+3}{3}-1}$ of degree d surfaces, with one coordinate for each coefficient of f, has an open subset for which the corresponding surfaces contain no lines. This open set is smaller than the open set corresponding to smooth surfaces, but it is still dense in $\mathbb{P}^{\binom{d+3}{3}-1}$. This implies that "almost all" tropicalized surfaces $\mathcal{S} = \mathrm{trop}(V(f) \cap T^3)$ dual to a triangulation Σ as in Theorem 4.5.8 have infinitely many lines, but none of these lifts to a line on the classical surface.

We close this section with two explicit examples of tropical surfaces that contain infinitely many lines. More information can be found in [**Vig10**].

Example 4.5.9. Let $d = 2$, so $\mathcal{S}$ is a quadric. In the planar representation $\{(i,j) : 0 \leq i,j \leq 2\}$ used in the proof of Proposition 4.5.4, tetrahedra such as $\tau = \mathrm{conv}\{(0,0,0),(0,1,0),(1,1,0),(1,0,1)\}$ correspond to triangles such as $\mathrm{conv}\{(0,2),(1,0),(1,1)\}$. Starting from that triangle, we can create seven of the 14 combinatorial types of smooth tropical quadrics. The bounded two-dimensional cell can be a triangle, a quadrilateral, a pentagon, or a hexagon. For instance, the hexagon class of size 24 has infinitely many lines on $\mathcal{S}$ through the vertex dual to the tetrahedron τ. ◊

Example 4.5.10. Consider the case $d = 3$ when $\mathcal{S}$ is a tropical cubic. The following 27 tetrahedra form a regular unimodular triangulation Σ of 3Δ:

$$
\begin{array}{lll}
\{\mathbf{000},\mathbf{010},\mathbf{120},\mathbf{201}\}, & \{000,010,101,201\}, & \{000,001,010,101\}, \\
\{000,110,120,201\}, & \{000,100,110,201\}, & \{010,020,120,201\}, \\
\{010,020,101,201\}, & \{001,010,020,101\}, & \{020,030,120,201\}, \\
\{110,120,201,210\}, & \{020,030,101,201\}, & \{001,020,030,101\}, \\
\{030,101,111,201\}, & \{001,030,101,111\}, & \{001,011,030,111\}, \\
\{011,021,030,111\}, & \{100,110,201,210\}, & \{100,200,201,210\}, \\
\{200,201,210,300\}, & \{012,101,111,201\}, & \{012,101,102,201\}, \\
\{001,012,101,111\}, & \{001,012,101,102\}, & \{001,011,012,111\}, \\
\{011,012,021,111\}, & \{001,002,012,102\}, & \{002,003,012,102\}.
\end{array}
$$

This regular triangulation is realized by the tropical polynomial

$$
\begin{aligned}
& 14 \odot u^3 \oplus 5 \odot u^2 v \oplus \mathbf{0} \odot \mathbf{u^2 w} \oplus 8 \odot u^2 \oplus \mathbf{0} \odot \mathbf{uv^2} \\
& \quad \oplus 5 \odot uvw \oplus 1 \odot uv \oplus 22 \odot uw^2 \oplus 2 \odot uw \\
& \quad \oplus 3 \odot u \oplus 3 \odot v^3 \oplus 14 \odot v^2 w \oplus 1 \odot v^2 \\
& \quad \oplus 26 \odot vw^2 \oplus 9 \odot vw \oplus \mathbf{0} \odot \mathbf{v} \oplus 48 \odot w^3 \\
& \quad \oplus 26 \odot w^2 \oplus 5 \odot w \oplus \mathbf{0}.
\end{aligned} \tag{4.5.3}
$$

We did this computation using the technique in [**Stu96**, Proposition 8.6]. Namely, Δ is the *lexicographic triangulation* determined by the ordering

$$003 > 002 > 102 > 012 > 021 > 011 > 001 > 111 > 101 > \mathbf{201}$$
$$> 300 > 200 > 210 > 100 > 110 > 030 > 020 > \mathbf{120} > \mathbf{010} > \mathbf{000}.$$

Since τ appears (as the first) among the 27 tetrahedra of Σ, the smooth tropical cubic surface $\mathcal{S}$ has infinitely many tropical lines. $\diamondsuit$

4.6. Complete Intersections

In this section we study generic complete intersections in the torus T^n. This uses the theory of stable intersections from Section 3.6. Given n equations in n variables with generic coefficients, there are only finitely many solutions. Bernstein's Theorem expresses their number as the mixed volume of the Newton polytopes of the given equations. We will prove this in our setting. The main idea is that the tropical variety of the intersection is determined by the given tropical hypersurfaces when the coefficients are sufficiently general.

We now introduce the *mixed volume* of a collection of lattice polytopes. Recall from Section 2.3 that the *Minkowski sum* of two subsets A, B of $\mathbb{R}^n$ is $A + B = \{\mathbf{a} + \mathbf{b} : \mathbf{a} \in A, \mathbf{b} \in B\} \subseteq \mathbb{R}^n$. If A is a subset of $\mathbb{R}^n$ and $\lambda > 0$ is a real number, we can scale A to obtain $\lambda A = \{\lambda \mathbf{a} : \mathbf{a} \in A\}$. The *normalized volume* $\mathrm{vol}(P)$ of a polytope P in $\mathbb{R}^n$ is its standard Euclidean volume multiplied by $n!$. This is designed so that the smallest volume of an n-dimensional lattice polytope is 1. Given lattice polytopes $P_1, P_2, \ldots, P_r$ in $\mathbb{R}^n$ and $\lambda_1, \lambda_2, \ldots, \lambda_r \in \mathbb{R}_{\geq 0}$, the polytope λP is the Minkowski sum

$$\lambda P \;=\; \lambda_1 P_1 + \lambda_2 P_2 + \cdots + \lambda_r P_r.$$

The normalized volume of the polytope λP is a homogeneous polynomial in the parameters $\lambda_1, \ldots, \lambda_r$. This will be seen from the following construction.

Definition 4.6.1. Let $P_1, \ldots, P_r$ be lattice polytopes in $\mathbb{R}^n$, not necessarily of full dimension n. The *Cayley polytope* of the given polytopes P_i is

$$C(P_1, \ldots, P_r) \;=\; \mathrm{conv}(\mathbf{e}_1 \times P_1, \ldots, \mathbf{e}_r \times P_r) \;\subset\; \mathbb{R}^{r+n}.$$

Note that λP is affinely isomorphic to a section of the Cayley polytope:

$$(4.6.1) \qquad \lambda P \;\simeq\; C(P_1, \ldots, P_r) \cap \{x_i = \lambda_i/\ell \,:\, 1 \leq i \leq r\}.$$

Here $\ell = \sum_i \lambda_i$. We identify the two polytopes in (4.6.1). Any polyhedral subdivision of the vertex set of $C(P_1, \ldots, P_r)$ induces a subdivision of λP, by intersecting each cell with the affine subspace on the right of (4.6.1). A *mixed subdivision* of the Minkowski sum $P_1 + \cdots + P_r$ is such a subdivision for $\lambda = (1, 1, \ldots, 1)$. Given a cell Q in a subdivision of $C(P_1, \ldots, P_r)$, and $1 \leq i \leq r$, we write Q_i for the face of Q consisting of all points whose ith coordinate is 1. The cell corresponding to Q in the mixed subdivision of

λP is $Q_1 + \cdots + Q_r$. A *mixed cell* of a mixed subdivision is a cell with $\dim(Q_i) \geq 1$ for $i = 1, \ldots, r$.

Example 4.6.2. Let P_1 be the square $\text{conv}\{(0,0),(1,0),(0,1),(1,1)\}$, and let P_2 be the triangle $\text{conv}\{(0,0),(1,0),(0,1)\}$. The Minkowski sum $P_1 + P_2$ is the pentagon $\text{conv}\{(0,0),(2,0),(0,2),(2,1),(1,2)\}$. The Cayley polytope $C(P_1, P_2)$ is the following three-dimensional polytope in $\mathbb{R}^4$:

$$\text{conv}\{(1,0,0,0),(1,0,1,0),(1,0,0,1),(1,0,1,1),(0,1,0,0),(0,1,1,0),(0,1,0,1)\}.$$

One subdivision of $C(P_1, P_2)$ has the maximal cells

$$\begin{gathered}\text{conv}\{(1,0,0,0),(1,0,1,0),(1,0,0,1),(1,0,1,1),(0,1,0,0)\},\\ \text{conv}\{(1,0,1,1),(0,1,0,0),(0,1,1,0),(0,1,0,1)\},\\ \text{conv}\{(1,0,0,1),(1,0,1,1),(0,1,0,0),(0,1,0,1)\},\\ \text{conv}\{(1,0,1,0),(1,0,1,1),(0,1,0,0),(0,1,1,0)\}.\end{gathered}$$

The resulting mixed subdivision of $P_1 + P_2$ is on the left in Figure 4.6.1. Its maximal cells are $\text{conv}\{(0,0),(1,0),(0,1),(1,1)\}$, $\text{conv}\{(1,1),(2,1),(1,2)\}$, $\text{conv}\{(0,1),(1,1),(0,2),(1,2)\}$, and $\text{conv}\{(1,0),(2,0),(1,1),(2,1)\}$. The last two of these are mixed cells. In $C(P_1, P_2)$, these correspond to tetrahedra that have two vertices each on the special faces $\{x_1 = 1, x_2 = 0\}$ and $\{x_1 = 0, x_2 = 1\}$.

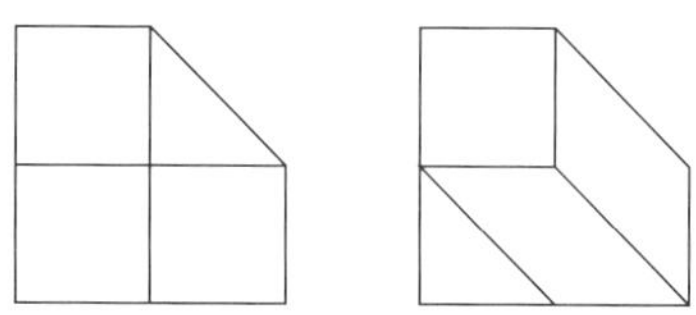

Figure 4.6.1. The two mixed subdivisions of $P_1 + P_2$ in Example 4.6.2.

Another subdivision of $C(P_1, P_2)$ has the four maximal cells

$$\begin{gathered}\text{conv}\{(1,0,0,0),(1,0,1,0),(1,0,0,1),(1,0,1,1),(0,1,0,1)\},\\ \text{conv}\{(1,0,0,0),(0,1,0,0),(0,1,1,0),(0,1,0,1)\},\\ \text{conv}\{(1,0,0,0),(1,0,1,0),(0,1,1,0),(0,1,0,1)\},\\ \text{conv}\{(1,0,1,0),(1,0,1,1),(0,1,1,0),(0,1,0,1)\}.\end{gathered}$$

This gives the mixed subdivision with cells

$$\begin{gathered}\text{conv}\{(0,1),(0,2),(1,1),(1,2)\},\\ \text{conv}\{(0,0),(1,0),(0,1)\},\\ \text{conv}\{(1,0),(0,1),(2,0),(1,1)\}, \text{ and}\\ \text{conv}\{(1,1),(2,0),(1,2),(2,1)\},\end{gathered}$$

shown on the right of Figure 4.6.1. The last two are mixed cells. ◊

Proposition 4.6.3. *Let $P_1, \ldots, P_r$ be lattice polytopes in $\mathbb{R}^n$. The normalized volume of the Minkowski sum $\lambda P = \lambda_1 P_1 + \cdots + \lambda_r P_r$ is a homogeneous polynomial in $\lambda_1, \ldots, \lambda_r$ with nonnegative integer coefficients of degree n.*

Proof. Let Σ be a triangulation of $C(P_1, \ldots, P_r)$ with all vertices at lattice points. This gives a mixed subdivision of $\lambda P \simeq C(P_1, \ldots, P_r) \cap \{x_i = \lambda_i/\ell\}$. If $\dim(C(P_1, \ldots, P_r)) = n + r - 1$, so the normalized volume is positive, each maximal simplex σ of Σ has $n + r$ vertices. Let m_i^σ denote the number of vertices of σ with ith coordinate 1. The cell τ in the mixed subdivision corresponding to σ has normalized volume $\mathrm{vol}(\sigma) \prod_{i=1}^r \big(\lambda_i^{m_i^\sigma - 1} \cdot (m_i^\sigma - 1)!\big)$. This is a monomial of degree n. Summing these monomials over all maximal simplices σ of Σ, we obtain a homogeneous polynomial of degree n with nonnegative integer coefficients. This is the normalized volume of λP. □

Definition 4.6.4. Let $P_1, \ldots, P_n$ be lattice polytopes in $\mathbb{R}^n$. Their *mixed volume* $\mathrm{MV}(P_1, \ldots, P_n)$ is the coefficient of the unique square-free monomial $\lambda_1 \lambda_2 \cdots \lambda_n$ in the polynomial $\mathrm{vol}(\lambda P)/n!$ that gives the Euclidean volume.

Example 4.6.5. Let P_1 and P_2 be the square and the triangle in Example 4.6.2. Using either mixed subdivision in Figure 4.6.1 we write the Euclidean volume of the Minkowski sum $\lambda_1 P_1 + \lambda_2 P_2$ as $\lambda_1^2 + \lambda_2^2/2 + \lambda_1\lambda_2 + \lambda_1\lambda_2$. Each of the two mixed subdivisions in Figure 4.6.1 has two mixed cells, each of volume 1. Hence the mixed volume of P_1 and P_2 equals $1 + 1 = 2$. ◊

We now derive a few general facts about the mixed volume.

Lemma 4.6.6. (1) *Given any triangulation Σ of the Cayley polytope $C(P_1, \ldots, P_n)$, the mixed volume $\mathrm{MV}(P_1, \ldots, P_n)$ is the sum of the volumes of the mixed cells of the induced subdivision of $P_1 + \cdots + P_n$.*

(2) *If Σ is any subdivision of $C(P_1, \ldots, P_n)$, then the intersection of Σ with the face $\{x_i = 1\}$ gives a subdivision Δ_i of P_i. Each cell σ of the induced mixed subdivision of $P_1 + \cdots + P_n$ has the form $\sigma = Q_1 + \cdots + Q_n$ for some cell Q_i of Δ_i, and $\mathrm{MV}(P_1, \ldots, P_n)$ is the sum of the mixed volumes $\mathrm{MV}(Q_1, \ldots, Q_n)$ over all cells σ.*

(3) *$\mathrm{MV}(P_1, \ldots, P_n)$ is positive if and only if each P_i has two vertices $\mathbf{p}_i$ and $\mathbf{q}_i$ so that the set $\{\mathbf{p}_i - \mathbf{q}_i : 1 \leq i \leq n\}$ is linearly independent in $\mathbb{R}^n$. This happens if and only if any partial Minkowski sum $P_{i_1} + \cdots + P_{i_j}$ has dimension at least j for all $1 \leq i_1 < \cdots < i_j \leq n$.*

Proof. The first part was shown in the proof of Proposition 4.6.3. For the second part, consider a refinement Σ' of Σ that is a triangulation. For each cell in Σ, this induces a refining triangulation. Every mixed cell of the subdivision of $P_1 + \cdots + P_n$ induced by Σ' is a mixed cell of the subdivision of a unique $Q_1 + \cdots + Q_n$. Hence the result follows from the first part.

We now prove the third part. If $\mathbf{p}_i, \mathbf{q}_i$ are vertices of the polytope P_i, then $\{\mathbf{p}_i - \mathbf{q}_i : 1{\leq}i{\leq}n\}$ is linearly independent if and only if the polytope $\sigma = \text{conv}\{(\mathbf{e}_i, \mathbf{p}_i), (\mathbf{e}_i, \mathbf{q}_i) : 1{\leq}i{\leq}n\}$ has dimension $2n-1$, so is a simplex. Indeed, the $2n \times 2n$-matrix with rows $(\mathbf{e}_i, \mathbf{p}_i)$ and $(\mathbf{e}_i, \mathbf{q}_i)$ is row equivalent to

$$\left(\begin{array}{c|c} I & * \\ \hline 0 & P-Q \end{array}\right),$$

where $P-Q$ has rows $\mathbf{p}_i - \mathbf{q}_i$. The polytope has dimension $2n-1$ if and only if the matrix has rank $2n$ if and only if the $\mathbf{p}_i - \mathbf{q}_i$ are linearly independent.

Given such a linearly independent collection, choose a triangulation Σ of the Cayley polytope $C(P_1, \ldots, P_n)$ that has the above simplex σ as a cell. This happens for a regular triangulation if the vertices in σ have weight 0, and all other vertices have generic positive values. Then σ contributes a term $\text{vol}(\sigma)\lambda_1 \ldots \lambda_n$ to the polynomial $\text{vol}(\lambda P)$, so the mixed volume of the P_i is positive. If no linearly independent collection exists, choose a triangulation Σ of the Cayley polytope using only the lattice points $(\mathbf{e_i}, \mathbf{u})$, for $\mathbf{u}$ a vertex of P_i, as vertices. None of the simplices of Σ have the above form with two vertices in the face $\{x_i = 1\}$ for each $1 \leq i \leq r$, so the volume polynomial has no terms of the form $\alpha\lambda_1 \cdots \lambda_n$, and the mixed volume is zero. The vectors $\{\mathbf{p}_j - \mathbf{q}_j : j \in J\}$ are parallel to the affine span of $P_J = \sum_{j \in J} P_j$. Hence if a linearly independent collection exists then $\dim(P_J) \geq |J|$ as required.

For the converse we use *Rado's Theorem on Independent Transversals.* This classical result from [**Rad42**] states the following: *if* $A_1, \ldots, A_n$ *are subsets of an* n*-dimensional vector space* V *such that* $\dim(\text{span}(\bigcup_{j \in J} A_j)) \geq |J|$ *for all subsets* $J \subseteq \{1, \ldots, n\}$, *then* V *has a basis* $\{a_1, \ldots, a_n\}$ *with* $a_i \in A_i$ *for all* i.

Suppose that our polytopes satisfy $\dim(P_J) \geq |J|$ for all J. We may assume that each P_i has $\mathbf{0}$ as a vertex, so the affine span of P_i agrees with the linear span of P_i. We apply Rado's Theorem to the nonzero vertices of $P_1, P_2, \ldots, P_n$. These span subspaces of sufficiently large dimensions, so there exist vertices $\mathbf{p}_i \in P_i$, $i = 1, \ldots, n$, that are linearly independent. If we now set $\mathbf{q}_i = \mathbf{0}$, then the n vectors $\mathbf{p}_i - \mathbf{q}_i$ are linearly independent. □

Remark 4.6.7. The mixed volume is the unique real-valued function on n-tuples of polytopes in $\mathbb{R}^n$ satisfying the following three properties:

(1) The mixed volume $\text{MV}(P, P, \ldots, P)$ is the normalized volume of P.

(2) The mixed volume is symmetric in its arguments, e.g.,

$$\text{MV}(P_1, P_2, P_3, \ldots, P_n) = \text{MV}(P_2, P_1, P_3, \ldots, P_n).$$

(3) The mixed volume is multilinear:

$$\text{MV}(aP{+}bQ, P_2, \ldots, P_n) = a\,\text{MV}(P, P_2, \ldots, P_n){+}b\,\text{MV}(Q, P_2, \ldots, P_n).$$

We refer to [**Sch93**, Chapter 5] for proofs and more details.

Our goal in this section is to connect the theory of mixed subdivisions and mixed volumes to the notion of stable intersection from Section 3.6. We begin with the remark that all regular subdivisions of the polytopes $P_1, \dots, P_r$ can be extended to mixed subdivisions of $P_1 + \cdots + P_r$. A regular subdivision of P_i is given by a weight function $\mathbf{w}_i \colon P_i \cap \mathbb{Z}^n \to \mathbb{R}$. We define a regular subdivision of the Cayley polytope $C(P_1, \dots, P_r)$ by assigning the weight $\mathbf{w}_i(\mathbf{u})$ to the vertex $(\mathbf{e}_i, \mathbf{u})$. This induces the desired mixed subdivision of $P_1 + \cdots + P_r$. Each cell in that mixed subdivision is a Minkowski sum $\sigma_1 + \cdots + \sigma_r$, where σ_i is a cell of the regular subdivision of P_i given by $\mathbf{w}_i$. Mixed subdivisions that arise in this way are *regular*.

Our first theorem concerns the stable intersection of n constant-coefficient hypersurfaces in $\mathbb{R}^n$. By Proposition 3.1.10, such a tropical hypersurface Σ_i is the $(n-1)$-skeleton of the normal fan to a lattice polytope P_i in $\mathbb{R}^n$. By Remark 3.3.11, the polytope P_i is determined by Σ_i and its multiplicities.

Theorem 4.6.8 (Tropical Bernstein). *Consider lattice polytopes $P_1, \dots, P_n$ in $\mathbb{R}^n$, with associated hypersurfaces $\Sigma_1, \dots, \Sigma_n$. Then the stable intersection*

$$\Sigma_1 \cap_{st} \Sigma_2 \cap_{st} \cdots \cap_{st} \Sigma_n \tag{4.6.2}$$

is the origin $\mathbf{0}$ with multiplicity given by the mixed volume $\mathrm{MV}(P_1, \dots, P_n)$, provided that is positive. If $\mathrm{MV}(P_1, \dots, P_n) = 0$, then (4.6.2) is the empty set.

Proof. We first assume that the stable intersection (4.6.2) is nonempty. By Theorem 3.6.10, it is a subfan of codimension n in the intersection of the Σ_i. So it is just the origin $\{\mathbf{0}\}$. We now compute the multiplicity. By Proposition 3.6.12, applied repeatedly, the stable intersection equals

$$\lim_{\epsilon \to 0} (\epsilon \mathbf{v}_1 + \Sigma_1) \cap \cdots \cap (\epsilon \mathbf{v}_n + \Sigma_n) \tag{4.6.3}$$

for sufficiently general $\mathbf{v}_1, \dots, \mathbf{v}_n \in \mathbb{R}^n$. Since (4.6.2) equals $\{\mathbf{0}\}$, the intersection in (4.6.3) is a finite collection of points. We just need to show that this number of points, counted with multiplicity, is the mixed volume.

Let $\mathbf{w} \in \bigcap(\epsilon \mathbf{v}_i + \Sigma_i)$. Since the $\mathbf{v}_i$ are generic, the point $\mathbf{w}$ lies in a maximal cone σ_i of $\epsilon \mathbf{v}_i + \Sigma_i$ for each i. This means that $Q_i = \mathrm{face}_{\mathbf{w} - \epsilon \mathbf{v}_i}(P_i)$ is an edge for each i. Write $\mathbf{q}_i \in \mathbb{Z}^n$ for a primitive vector pointing along the edge Q_i of P_i, and μ_i for the lattice length of the edge. Let $\ell_i = (\mathbf{w} - \epsilon \mathbf{v}_i) \cdot \mathbf{u}$ for $\mathbf{u} \in Q_i$. Consider the regular subdivision of the Cayley polytope $C(P_1, \dots, P_n)$ induced by the weight vector that assigns the value $\epsilon \mathbf{v}_i \cdot \mathbf{u}$ to the point $(\mathbf{e}_i, \mathbf{u})$. The cell of this subdivision that is selected by the vector $(-\ell_1, \dots, -\ell_n, \mathbf{w})$ is the convex hull of $\bigcup_{i=1}^n (\mathbf{e}_i \times Q_i)$. Hence $Q_1 + \cdots + Q_n$ is a cell of the corresponding mixed subdivision of $P_1 + \cdots + P_n$. It is a mixed cell because each Q_i is one dimensional. Since the stable intersection (4.6.3) is a finite set, the intersection of the affine spans of the σ_i is zero dimensional, so the Minkowski sum of the Q_i is n dimensional. The normalized volume of $Q_1 + \cdots + Q_n$ is the absolute value of the determinant of the matrix whose columns are the vectors $\mu_i \mathbf{q}_i$. Note from Lemma 3.4.6 that μ_i is the multiplicity of the cone of Σ_i containing $\mathbf{w} - \epsilon \mathbf{v}_i$. By Definition 3.6.11, the multiplicity of the point $\mathbf{w}$ in the stable intersection (4.6.2) equals the product of the multiplicities $\mu_1, \ldots, \mu_n$ with the lattice index

$$[N : N_{\sigma_1} + \cdots + N_{\sigma_n}] \;=\; [\mathbb{Z}^n : \mathbb{Z}\mathbf{q}_1 + \cdots + \mathbb{Z}\mathbf{q}_n] \;=\; |\det(\mathbf{q}_1, \ldots, \mathbf{q}_n)|.$$

Hence that multiplicity equals the volume of the corresponding mixed cell.

Conversely, suppose $\sum_i Q_i$ is a mixed cell of the mixed subdivision of $\sum_i P_i$ induced by assigning weight $\epsilon \mathbf{v}_i \cdot \mathbf{u}$ to the point $(\mathbf{e}_i, \mathbf{u})$ that was described above. Then $\mathrm{conv}(\mathbf{e}_i \times Q_i : 1 \le i \le n)$ is a cell of the corresponding subdivision of the Cayley polytope. This gives $Q_i = \mathrm{face}_{\mathbf{w}-\epsilon\mathbf{v}_i}(P_i)$, and hence $\mathbf{w} \in \bigcap (\epsilon \mathbf{v}_i + \Sigma_i)$. Thus the intersection points are in bijection with the mixed cells, and the sum of the multiplicities is the mixed volume.

The last paragraph also takes care of the case when the stable intersection (4.6.2) is empty. This happens if and only if $\mathrm{MV}(P_1, \ldots, P_n) = 0$, since any mixed cell gives rise to an intersection point in (4.6.3). In all cases, $\mathrm{MV}(P_1, \ldots, P_n)$ is the sum of the multiplicities of the intersection points $\mathbf{w}$ in (4.6.3), and this sum is the multiplicity of the origin $\mathbf{0}$ in (4.6.2). □

We now present a more general form of Theorem 4.6.8, where we allow tropical hypersurfaces that are not fans, and their number r is typically less than n. We fix lattice polytopes $P_1, \ldots, P_r$ in $\mathbb{R}^n$, and we write Δ_i for the regular subdivision of P_i given by the weight vector $\mathbf{w}_i$. Let Σ_i denote the tropical hypersurface in $\mathbb{R}^n$ that is dual to the subdivision Δ_i. Recall that Q_i is the cell of Δ_i selected by $\mathbf{w}$ if $\mathbf{w} \cdot \mathbf{u} + \mathbf{w}_i(\mathbf{u}) \le \mathbf{w} \cdot \mathbf{u}' + \mathbf{w}_i(\mathbf{u}')$ for all $\mathbf{u} \in Q_i$ and $\mathbf{u}' \in \Delta_i$, with a strict inequality if $\mathbf{u}' \notin Q_i$.

Theorem 4.6.9. *Let $\mathbf{w} \in \mathbb{R}^n$, and denote by Q_i the cell of Δ_i selected by $\mathbf{w}$ for $i = 1, \ldots, r$. Then $\mathbf{w}$ lies in the stable intersection $\Sigma_1 \cap_{st} \Sigma_2 \cap_{st} \cdots \cap_{st} \Sigma_r$ if and only if $\dim(\sum_{j \in J} Q_j) \ge |J|$ for all $J \subseteq \{1, \ldots, r\}$. If $\mathbf{w}$ lies in the relative interior of a maximal cell of the stable intersection, then the Q_i all lie in an r-dimensional affine subspace of $\mathbb{R}^n$ and the multiplicity of the cell containing $\mathbf{w}$ equals the r-dimensional mixed volume $\mathrm{MV}_r(Q_1, \ldots, Q_r)$.*

This can be restated informally as saying that the stable intersection is dual to the collection of mixed faces of the Minkowski sum of the P_i.

Proof. Fix $\mathbf{w} \in \mathbb{R}^n$. If σ is the smallest cell of a polyhedral complex Σ that contains $\mathbf{w}$, then we denote by $\mathrm{star}_{\mathbf{w}}(\Sigma)$ the fan $\mathrm{star}_{\Sigma}(\sigma)$. We set $\mathrm{star}_{\mathbf{w}}(\Sigma) = \emptyset$ if $\mathbf{w} \notin |\Sigma|$. By repeated application of Lemma 3.6.7, we have

$\mathrm{star}_{\mathbf{w}}(\Sigma_1 \cap_{st} \cdots \cap_{st} \Sigma_r) = \mathrm{star}_{\mathbf{w}}(\Sigma_1) \cap_{st} \cdots \cap_{st} \mathrm{star}_{\mathbf{w}}(\Sigma_2)$. Thus $\mathbf{w}$ lies in the stable intersection of the Σ_i if and only if the stable intersection of the fans $\mathrm{star}_{\mathbf{w}}(\Sigma_i)$ is nonempty.

The fan $\mathrm{star}_{\mathbf{w}}(\Sigma_i)$ consists of the codimension-1 cones of the normal fan of Q_i. We now show that the stable intersection of the fans $\mathrm{star}_{\mathbf{w}}(\Sigma_i)$ is nonempty if and only if the dimension of $Q_{i_1} + \cdots + Q_{i_j}$ is at least j for all $1 \leq i_1 < \cdots < i_j \leq r$. If this dimension is less than j for some choice, then there is an affine space L of dimension at most $j-1$ containing each of $Q_{i_1}, \ldots, Q_{i_j}$. Write $L^{\perp}$ for the orthogonal complement of the linear space parallel to L. Then $\dim(L^{\perp}) \geq n - j + 1$ and $L^{\perp}$ lies in the lineality space of $Q_{i_1}, \ldots, Q_{i_j}$. Thus by Theorem 3.6.10 the stable intersection is empty. Suppose now that the dimension of $Q_{i_1} + \cdots + Q_{i_j}$ is at least j for all $1 \leq i_1 < \cdots < i_j \leq r$. For $r+1 \leq i \leq n$ choose lattice line segments Q_i so that the property that the dimension of $Q_{i_1} + \cdots + Q_{i_j}$ is at least j holds for all $1 \leq i_1 < \cdots < i_j \leq n$. Let Δ_i be the tropical hypersurface determined by Q_i for $1 \leq i \leq n$. By Theorem 4.6.8, the stable intersection of $\Delta_1, \ldots, \Delta_n$ is the mixed volume of the Q_i, which is nonzero by part (3) of Lemma 4.6.6. Thus the stable intersection of $\Delta_1, \ldots, \Delta_r$ is also nonempty. Thus $\mathbf{w}$ lies in the stable intersection if and only if for all $1 \leq i_1 < \cdots < i_j \leq r$ the dimension of the Minkowski sum $Q_{i_1} + \cdots + Q_{i_j}$ is at least j.

When the stable intersection of the Σ_i is nonempty, it is a pure polyhedral complex of codimension r, by Theorem 3.6.10. If $\mathbf{w}$ lies in the relative interior of a maximal cell, the star of this complex at $\mathbf{w}$ is an $(n-r)$-dimensional linear space L. It lies in the lineality space of $\mathrm{star}_{\mathbf{w}}(\Sigma_i)$ for each i. Any vector in L is thus orthogonal to the affine span of the cell Q_i in the subdivision Δ_i, and so also orthogonal to the affine span of $Q_1 + \cdots + Q_r$. We may thus quotient by L to obtain r fans $\mathrm{star}_{\mathbf{w}}(\Sigma_i)/L$ in the r-dimensional vector space $\mathbb{R}^n/L$. These are given by the codimension-1 cones of the normal fans of the Q_i, now viewed as fans in $\mathbb{R}^n/L$. Theorem 4.6.8 then implies that the stable intersection is the origin $\mathbf{0}$ in $\mathbb{R}^n/L$ with multiplicity $\mathrm{MV}(Q_1, \ldots, Q_r)$. Since the mixed volume of these polytopes and the multiplicity of the stable intersection are preserved under quotienting by L, the multiplicity of the cell containing $\mathbf{w}$ equals $\mathrm{MV}(Q_1, \ldots, Q_r)$. $\square$

The previous theorem gives a purely combinatorial construction of tropical complete intersections. In what follows we shall apply this to classical polynomials $f = \sum c_{\mathbf{u}} x^{\mathbf{u}} \in K[x_1^{\pm 1}, \ldots, x_n^{\pm 1}]$ defined over a field K. The condition that f has a particular Newton polytope P is equivalent to requiring that $c_{\mathbf{u}} = 0$ for $\mathbf{u} \notin P$, and $c_{\mathbf{u}} \neq 0$ when $\mathbf{u}$ is a vertex of P. Thus the space of polynomials with Newton polytope P has the form $(K^*)^a \times \mathbb{A}^{b-a}$, where a is the number of vertices of P, and b is the number of lattice points in P.

Proposition 4.6.10. *Let $X \subset T^n_K$ be a variety, where K is a field with the trivial valuation. Fix a lattice polytope $P \subset \mathbb{R}^n$ with a vertices and b lattice points. The parameter space $(K^*)^a \times \mathbb{A}^{b-a}$ for polynomials with Newton polytope P contains a nonempty open set U such that all $f \in U$ satisfy*

$$\text{trop}(X \cap V(f)) = \text{trop}(X) \cap_{st} \text{trop}(V(f)). \tag{4.6.4}$$

Proof. We first argue that an f satisfying (4.6.4) exists. Fix any polynomial g with $P = \text{Newt}(g)$. By Theorem 3.6.1, there is a dense subset $U_g \subset T^n$ such that $\text{trop}(X \cap \mathbf{t}V(g)) = \text{trop}(X) \cap_{st} \text{trop}(V(g))$ for all $\mathbf{t} \in U_g$. Note that $\mathbf{t}V(g)$ is the variety of $f(\mathbf{x}) = g(\mathbf{t}^{-1}\mathbf{x})$. Hence $f(\mathbf{x})$ satisfies (4.6.4).

We next claim that there is an open subset U of $(K^*)^a \times \mathbb{A}^{b-a}$ such that $\text{trop}(X \cap V(f))$ is the same fan Σ for all choices of f in U. We may homogenize, so that all defining polynomials of $X \cap V(f)$ are homogeneous. When computing Gröbner bases, the Buchberger algorithm branches according to whether a leading coefficient is zero or nonzero. All such coefficients are polynomials in the coefficients $c_{\mathbf{u}}$ of $f = \sum c_{\mathbf{u}} x^{\mathbf{u}}$. Hence, requiring all leading coefficients to be nonzero defines an open subset U' of $(K^*)^a \times \mathbb{A}^{b-a}$ such that the Gröbner fan is the same for all $f \in U'$.

By Proposition 3.2.8, the tropical variety is a subfan of the Gröbner fan. For each cone σ in the Gröbner fan, let U_σ denote the set of $f \in U'$ such that σ lies in the tropical variety of $X \cap V(f)$. If U_σ is nonempty, we claim that it is open in $(K^*)^a \times \mathbb{A}^{b-a}$. To see this, we need to check whether the initial ideal $J_\sigma = \text{in}_{\mathbf{w}}(I + \langle f \rangle)$ for $\mathbf{w} \in \text{relint}(\sigma)$ contains a monomial. This happens if and only if $J_\sigma + \langle x_1 x_2 \cdots x_n z - 1 \rangle$ is the unit ideal, where z is a new variable. This is decided by running the Buchberger algorithm, and we again define U_σ by requiring that all leading coefficients occurring during that computation are nonzero. We obtain the desired Zariski open subset $U \subset (K^*)^a \times \mathbb{A}^{b-a}$ by intersecting all nonempty sets U_σ.

To complete our proof, we argue that the stable intersection $\text{trop}(X) \cap_{st} \text{trop}(V(f))$ agrees with $\text{trop}(X \cap V(f))$ for $f \in U$. This holds because the polynomial g in the first paragraph can be chosen arbitrarily in $(K^*)^a \times \mathbb{A}^{b-a}$. In particular, we can choose g in U. Then the open set of $\mathbf{t}$ for which $g(\mathbf{t}^{-1}\mathbf{x})$ lies in U is nonempty, so it intersects U_g. For $\mathbf{t}$ in this intersection and $f(\mathbf{x}) = g(\mathbf{t}^{-1}\mathbf{x})$, we thus have $\Sigma = \text{trop}(X) \cap_{st} \text{trop}(V(f))$, as desired. □

The technique in this proof relates to work of Römer and Schmitz [**RS12**] who studied the behavior of tropical varieties with respect to (classical) linear changes of coordinates. See also Exercise 6.8(11).

We now apply Proposition 4.6.10 inductively to find the tropicalization of generic complete intersections. Here we take K to be a field with the trivial valuation; the general case of valued fields will be treated later. Fix lattice polytopes $P_1, \ldots, P_r \subset \mathbb{R}^n$, where $r \leq n$, and write $\mathfrak{A}$ for the parameter space of lists $(f_1, \ldots, f_r)$ of polynomials with $\mathrm{Newt}(f_i) = P_i$. For $J \subseteq \{1, \ldots, r\}$ we write P_J for the partial Minkowski sum $\sum_{j \in J} P_j$.

Corollary 4.6.11. *Let K have the trivial valuation. There is a nonempty open set $U \subset \mathfrak{A}$ for which if $(f_1, \ldots, f_r)$ lies in U, then the tropical variety* $\mathrm{trop}(V(f_1, \ldots, f_r))$ *equals the stable intersection of the tropical hypersurfaces corresponding to $P_1, \ldots, P_r$. This is a subfan of the normal fan of $P_1 + \cdots + P_r$. It consists of $\mathbf{w} \in \mathbb{R}^n$ satisfying* $\dim(\mathrm{face}_{\mathbf{w}}(P_J)) \geq |J|$ *for all $J \subseteq \{1, \ldots, r\}$. The multiplicity of a maximal cone having $\mathbf{w}$ in its relative interior is the r-dimensional mixed volume of* $\mathrm{face}_{\mathbf{w}}(P_1), \ldots, \mathrm{face}_{\mathbf{w}}(P_r)$.

Proof. For $r = 1$, we can take $U = \mathfrak{A}$, and the result is Proposition 3.1.10. For $r \geq 2$, we are claiming that generic polynomials $f_1, \ldots, f_r$ satisfy

$$\mathrm{trop}(V(f_1, \ldots, f_r)) \;=\; \mathrm{trop}(V(f_1)) \cap_{st} \cdots \cap_{st} \mathrm{trop}(V(f_r)). \tag{4.6.5}$$

This is derived from Proposition 4.6.10 by induction on r. The description of the stable intersection on the right-hand side is then Theorem 4.6.9. □

Example 4.6.12. In this example we take $K = \mathbb{C}$ with the trivial valuation.

(1) Let $n = r = 2$, and let $P_1 = P_2 = \mathrm{conv}\{(0,0), (2,0), (0,2)\}$ so that

$$\begin{aligned} f_1 &= c_1x^2 + c_2xy + c_3y^2 + c_4x + c_5y + c_6, \\ f_2 &= d_1x^2 + d_2xy + d_3y^2 + d_4x + d_5y + d_6. \end{aligned}$$

A suitable open set U in coefficient space for Corollary 4.6.11 is

$$\det\begin{pmatrix} c_1 & c_2 & c_3 & 0 \\ 0 & c_1 & c_2 & c_3 \\ d_1 & d_2 & d_3 & 0 \\ 0 & d_1 & d_2 & d_3 \end{pmatrix} \cdot \det\begin{pmatrix} c_1 & c_4 & c_6 & 0 \\ 0 & c_1 & c_4 & c_6 \\ d_1 & d_4 & d_6 & 0 \\ 0 & d_1 & d_4 & d_6 \end{pmatrix} \cdot \det\begin{pmatrix} c_3 & c_5 & c_6 & 0 \\ 0 & c_3 & c_5 & c_6 \\ d_3 & d_5 & d_6 & 0 \\ 0 & d_3 & d_5 & d_6 \end{pmatrix} \neq 0.$$

These are Sylvester resultants (see [**CLO07**, Chapter 3]). Their nonvanishing guarantees that all intersection points of the closures of $V(f_1)$ and $V(f_2)$ in $\mathbb{P}^2$ lie in the torus. Bézout's Theorem ensures that $V(f_1, f_2) \subset T^2_{\mathbb{C}}$ consists of four points, counted with multiplicities and

$$\mathrm{trop}(V(f_1, f_2)) \;=\; \{(0,0)\} \;=\; \mathrm{trop}(V(f_1)) \cap_{st} \mathrm{trop}(V(f_2)).$$

Both of the tropical quadrics $\mathrm{trop}(V(f_i))$ have three rays, each with multiplicity 2. Their stable intersection is the origin with multiplicity 4.

(2) Let $P_1 = \cdots = P_r = \mathrm{conv}(\mathbf{0}, \mathbf{e}_1, \ldots, \mathbf{e}_n) \subseteq \mathbb{R}^n$. A polynomial f_i with Newton polytope P_i is a linear form. The coefficients of $f_1, \ldots, f_r$ form an $r \times (n+1)$-matrix. Such a matrix is in the parameter space $\mathfrak{A}$ if and only if all of its entries are nonzero.

Let U in Corollary 4.6.11 be the subset of matrices all of whose $r \times r$-minors are nonzero. In this example, the polytope analysis in Corollary 4.6.11 agrees with Example 4.2.13. The linear space $V(f_1, \dots, f_r)$ realizes the uniform matroid $U(n+1-r, n+1)$. $\diamondsuit$

Remark 4.6.13. The requirement that $f_1, \dots, f_r$ be generic is essential for Corollary 4.6.11. First, if the tropical variety is the stable intersection of r tropical hypersurfaces, it has codimension r, so Theorem 3.3.5 implies that $X = V(f_1, \dots, f_r)$ has codimension r. This means that X being a complete intersection is a necessary condition for the tropicalization of X to equal the stable intersection of the tropical hypersurfaces $\operatorname{trop}(V(f_i))$ as in (4.6.5). However, that condition is not sufficient. For instance, suppose that the coefficient matrix in part (2) of Example 4.6.12 has both zero and nonzero $r \times r$-minors. Then X is a complete intersection but (4.6.5) does not hold.

We now apply the tropical Bernstein Theorem to prove the classical Bernstein Theorem [**Ber75**], which determines the size of the variety $V(f_1, \dots, f_n)$ when $f_1, \dots, f_n$ are sufficiently generic polynomials with given Newton polytopes $P_1, \dots, P_n$. The case when $P_1 = \dots = P_n$ is due to Khovanskii and Kušnirenko [**Kuš76**]. The mixed subdivision approach of Huber and Sturmfels [**HS95**] was one of the precursors of tropical geometry.

Theorem 4.6.14 (Bernstein's Theorem). *The number of solutions in $(K^*)^n$ to a generic system of n polynomial equations $f_1 = \dots = f_n$ with given Newton polytopes $P_1, \dots, P_n$ is equal to the mixed volume* $\mathrm{MV}(P_1, \dots, P_n)$.

Proof. Let $I = \langle f_1, \dots, f_n \rangle$, and let $\Sigma_i = \operatorname{trop}(V(f_i))$ be the codimension-1-skeleton of the normal fan of P_i. By Corollary 4.6.11, $\operatorname{trop}(V(I))$ equals $\Sigma_1 \cap_{st} \cdots \cap_{st} \Sigma_n$. Also, by Theorem 4.6.8, this is the origin with multiplicity $\mathrm{MV}(P_1, \dots, P_n)$, or is empty if $\mathrm{MV}(P_1, \dots, P_n) = 0$. By definition, the multiplicity of the origin in $\operatorname{trop}(V(I))$ is the sum of the multiplicities of the minimal primes of $\operatorname{in}_{\mathbf{0}}(I)$. Since I is zero dimensional and we here take K with the trivial valuation, this is $\dim_K S/\operatorname{in}_{\mathbf{0}}(I) = \dim_K S/I$, where $S = K[x_1^{\pm 1}, \dots, x_n^{\pm 1}]$. The latter dimension is the number of solutions to $f_1 = \dots = f_n = 0$, counted with multiplicity, so the theorem follows. $\square$

Example 4.6.15. Let $f_1 = x + y + 1$ and $f_2 = 3x + 2y + 6$ in $\mathbb{Q}[x^{\pm 1}, y^{\pm 1}]$, where $\mathbb{Q}$ has the 3-adic valuation. Note that $V(f_1, f_2) = \{(-4, 3)\}$, which has valuation $(0, 1)$, so f_1 and f_2 are generic in the sense of Corollary 4.6.11.

It is instructive to revisit Lemma 4.6.6 for this example. The given Newton polytopes are $P_1 = P_2 = \operatorname{conv}\{(0,0), (1,0), (0,1)\}$. The Cayley polytope $C(P_1, P_2)$ is a triangular prism. The valuations of the coefficients of f_1 and f_2 determine the weight vectors $\mathbf{w}_1 = (0,0,0)$ and $\mathbf{w}_2 = (1,1,0)$. The regular subdivision of $C(P_1, P_2)$ given by $(\mathbf{w}_1, \mathbf{w}_2)$ has two maximal cells, one tetrahedron and one pyramid. The induced mixed subdivision of $P_1 + P_2 = \text{conv}\{(0,0),(2,0),(0,2)\}$ has two cells, one triangle, and one quadrilateral. The latter is $Q_1 + Q_2$ where $Q_1 = \text{conv}\{(0,0),(1,0),(0,1\})$ and $Q_2 = \text{conv}\{(0,0),(1,0)\}$. This cell corresponds to the stable intersection point $(0,1)$ of the two tropical lines, with multiplicity $MV(Q_1,Q_2) = 1$. ◊

Example 4.6.16. Computing the number of solutions to n generic equations in n variables arises frequently in applications. One example comes from economics, where we consider the computation of *Nash equilibria* for an n-person game where each player has two mixed strategies [**Stu02**, §6.4]. This translates mathematically into considering a system of equations $f_1 = \cdots = f_n = 0$, where f_i is a polynomial in $x_1,\ldots,x_{i-1},x_{i+1},\ldots,x_n$ that has Newton polytope P_i the ith facet of the standard n-cube, so has 2^{n-1} terms. The n-cube has normalized volume $n!$, which is also the number of permutations of an n-set. The mixed volume $MV(P_1,\ldots,P_n)$ is the number of *derangements*, which are permutations that have no fixed points. For $n = 2,3,4,5,6,\ldots$, that number equals $1,2,9,44,265,\ldots$. For instance, the mixed volume of four nonparallel facets of the four-dimensional cube is equal to 9. This is a tight upper bound for the number of isolated Nash equilibria of a four-person game where each player has two mixed strategies. See [**Stu02**, Corollary 6.9] for more information on this topic. ◊

The trivial valuation on K is needed in Corollary 4.6.11 to ensure that the tropicalization is constant on an open subset of the space of coefficients. If the valuation on K is nontrivial, then there is no unique tropicalization for the generic complete intersections with Newton polytopes $P_1,\ldots,P_r$. Even when $r = 1$, there are many different tropical hypersurfaces, arising from different regular triangulations of the same Newton polytope.

Example 4.6.17. Let $f = a+bx+cy+dxy \in K[x^{\pm 1},y^{\pm 1}]$, where $a,b,c,d \in K^*$. The combinatorial type of the tropical curve $\text{trop}(V(f))$ is determined by the sign of $\text{val}(a)-\text{val}(b)-\text{val}(c)+\text{val}(d)$. There are two types of typical behavior, arising when that quantity is either positive or negative. Indeed, the Newton polygon P of f is a square, with two regular triangulations.

A less trivial example is featured in Proposition 4.5.4: the doubled tetrahedron $P = 2\Delta$ has 192 regular triangulations, so there are 192 typical types one encounters when studying tropical quadratic surfaces in $\mathbb{R}^3$. ◊

Now let K be a field with a nontrivial valuation. While there are now many generic types of intersection, the good news is that for each of the types, the stable intersection of the tropical hypersurfaces actually coincides with the set-theoretic intersection. This is the content of the next theorem.

Theorem 4.6.18. *Let $f_1,\ldots,f_r \in K[x_1^{\pm 1},\ldots,x_n^{\pm 1}]$ be polynomials with Newton polytopes $P_1,\ldots,P_r$, and suppose the regular subdivision of the Cay-*

ley polytope $C(P_1, \ldots, P_r)$ *induced by the valuations of their coefficients is a triangulation. Then* $\{f_1, \ldots, f_r\}$ *is a tropical basis, and the tropical variety*

$$\mathrm{trop}(V(f_1, \ldots, f_r)) \;=\; \mathrm{trop}(V(f_1)) \cap \cdots \cap \mathrm{trop}(V(f_r))$$

can be computed by the combinatorial rule for the stable intersection of the tropical hypersurfaces $\mathrm{trop}(V(f_i))$ *given in Theorem* 4.6.9.

Proof. Let $\Sigma_i = \mathrm{trop}(V(f_i))$. We use the notation of Theorem 4.6.9. The subdivision Δ_i of P_i is a triangulation, by our assumption that the $\mathbf{w}_i$ given by the f_i define a regular triangulation of $C(P_1, \ldots, P_r)$. We need to prove

$$(4.6.6) \qquad \Sigma_1 \cap \Sigma_2 \cap \cdots \cap \Sigma_r \;=\; \Sigma_1 \cap_{st} \Sigma_2 \cap_{st} \cdots \cap_{st} \Sigma_r .$$

The right-hand side is always contained in the left-hand side. We must show the reverse inclusion. Let $\mathbf{w}$ be any point in the left-hand side, let σ_i be an $(n-1)$-dimensional cell of Σ_i containing $\mathbf{w}$, and let Q_i be the edge of Δ_i that is dual to σ_i. By hypothesis, $\mathrm{conv}(\mathbf{e}_i \times Q_i : i = 1, 2, \ldots, r)$ is a $(2r-1)$-simplex in the triangulation of the $(n+r-1)$-dimensional polytope $C(P_1, \ldots, P_r)$. Hence, in the corresponding mixed subdivision of $P_1 + \cdots + P_r$ in $\mathbb{R}^n$, the cell $Q_1 + \cdots + Q_r$ has codimension $n - r$ as well. This implies

$$\mathrm{codim}(\sigma_1 \cap \cdots \cap \sigma_r) \;=\; r \;=\; \sum_{i=1}^{r} \mathrm{codim}(\sigma_i).$$

We apply Theorem 3.4.12 iteratively to conclude that the relative interior of $\sigma_1 \cap \cdots \cap \sigma_r$ lies in $\Sigma_1 \cap_{st} \Sigma_2 \cap_{st} \cdots \cap_{st} \Sigma_r$. By passing to the closure, so does $\mathbf{w}$. We conclude that (4.6.6) holds, and the combinatorial rule in Theorem 4.6.9 characterizes this tropical variety along with its multiplicities. □

Remark 4.6.19. Theorem 4.6.18 should be contrasted with Corollary 4.6.11. For instance, if f and g have the same Newton polytope and their coefficients have zero valuation, then $\{f, g\}$ is almost never a tropical basis. To see this, note that $\mathrm{trop}(V(f)) = \mathrm{trop}(V(g))$, so their intersection does not even have the correct dimension unless f and g share a common factor.

A tropical complete intersection is *smooth* if the corresponding regular subdivision of the Cayley polytope $C(P_1, \ldots, P_r)$ is a unimodular triangulation. It is then given by Theorem 4.6.18. The special case of smooth surfaces ($r = 1$ and $n = 3$) was studied in Section 4.5. Generalizing Theorem 4.5.2, Steffens and Theobald [**ST10**] showed that the face numbers of a smooth tropical complete intersection are determined by the Newton polytopes. In particular, let $P_i = d_i\Delta = \mathrm{conv}(\mathbf{0}, d_i\mathbf{e}_1, \ldots, d_i\mathbf{e}_n)$. Then f_i is a dense polynomial of degree d_i, and these face numbers depend only on $d_1, \ldots, d_r$. We demonstrate this for the case of space curves ($r = 2$ and $n = 3$).

Proposition 4.6.20. *Let f and g be polynomials of degree d and e in $K[x, y, z]$ with Newton polytopes $d\Delta$ and $e\Delta$, respectively. We assume that the tropical curve $\Sigma = \mathrm{trop}(V(f, g))$ is smooth. Then Σ has*

$d^2e + de^2$	*vertices,*
$(3/2)d^2e + (3/2)de^2 - 2de$	*edges (bounded one-dimensional cells),*
$4de$	*rays (unbounded one-dimensional cells).*

The genus of the graph Σ equals $(1/2)d^2e + (1/2)de^2 - 2de + 1$.

Algebraic geometers should note that our formula for the genus agrees with that in the classical case of a complete intersection curve in $\mathbb{P}^3$. The same holds for the Euler characteristic of a surface in Theorem 4.5.2.

Proof of Proposition 4.6.20. From Sections 3.4 and 3.5, we know that Σ is a balanced connected graph. The genus of such a graph is 1 plus the number of edges minus the number of vertices, so the last sentence follows from the others.

The Cayley polytope $C(d\Delta, e\Delta)$ is four dimensional. We claim that its normalized volume equals $d^3 + d^2e + de^2 + e^3$. Indeed, this is the number of four-dimensional simplices in any unimodular triangulation of $C(d\Delta, e\Delta)$. The number of simplices that use i vertices from $d\Delta$ and $5 - i$ vertices from $e\Delta$ is $d^{i-1}e^{4-i}$. The corresponding cell in the mixed subdivision of $d\Delta + e\Delta$ is mixed if and only if $i = 2$ or $i = 3$, so the number of maximal mixed cells is $d^2e + de^2$. Applying this for the unimodular triangulation dual to Σ, we learn from the sentence after Theorem 4.6.9 that $d^2e + de^2$ is the number of vertices of Σ. The smooth tropical curve Σ has de unbounded rays pointing into each of the four coordinate directions, so the number of rays is $4de$.

To count edges, we note that Σ is a trivalent graph, so every vertex is incident to three edges or rays. Indeed, the mixed cell dual to such a vertex is a triangular prism, whose three quadrilateral faces are mixed and whose two triangle faces are not mixed. The resulting formula $3 \cdot \#\text{vertices} = 2 \cdot \#\text{edges} + \#\text{rays}$ now implies that the number of edges is as desired. □

We close this section with a brief case study of elliptic curves in 3-space that are intersections of two quadratic surfaces, and hence have degree 4.

Example 4.6.21. Let $K = \mathbb{Q}$ with the 2-adic valuation, and consider

$$\begin{aligned} f_1 &= 1024x^2 + 64xy + 8xz + 2x + 8y^2 + 2yz + y + z^2 + z + 2, \\ f_2 &= x^2 + xy + 2xz + 8x + 2y^2 + 8yz + 64y + 64z^2 + 1024z + 32768. \end{aligned}$$

Here $P_1 = P_2 = 2\Delta$ as in Proposition 4.5.4. The 2-adic valuations of the 20 coefficients of (f_1, f_2) define a regular triangulation of the four-dimensional polytope $C(P_1, P_2)$. This is a lexicographic triangulation as in Example 4.5.10. Of the 32 maximal simplices, precisely 16 give mixed cells in $C(P_1, P_2)$, and hence vertices of $X = \text{trop}(V(f_1), V(f_2)) = \text{trop}(V(f_1)) \cap \text{trop}(V(f_2))$. The tropical curve X is smooth and has genus 1. It has 16 rays and 16 bounded edges. Eight of these edges form an 8-cycle, and the other eight form four 2-chains attached to that cycle. Readers are encouraged to verify this, and to redo it for their own quadrics f_1, f_2. ◊

4.7. Exercises

(1) This exercise concerns Cramer's rule in tropical geometry.
   (a) Consider two tropical lines in the plane, given by linear polynomials $a_1 \odot x \oplus b_1 \odot y \oplus c_1$ and $a_2 \odot x \oplus b_2 \odot y \oplus c_2$. Find a formula for their intersection point in terms of $a_1, b_1, c_1, a_2, b_2, c_2$.
   (b) Consider three tropical planes in 3-space given by linear polynomials $a_i \odot x \oplus b_i \odot y \oplus c_i \odot z \oplus d_i$ for $i = 1, 2, 3$. Find a formula for their intersection point in terms of $a_1, b_1, \ldots, d_3$.
   (c) Consider two tropical planes in 3-space given by linear polynomials $a_i \odot x \oplus b_i \odot y \oplus c_i \odot z \oplus d_i$ for $i = 1, 2$. Find a formula for their intersection line in terms of $a_1, b_1, \ldots, d_2$.

(2) Let $d = 3$, let $n = 6$, and let $\mathcal{A}$ be the arrangement in $\mathbb{P}^3$ consisting of the planes spanned by the facets of a regular octahedron. Write $\mathbb{P}^3 \backslash \cup \mathcal{A}$ as a linear subvariety $V(I)$ in a torus, and determine $\text{trop}(V(I))$.

(3) In Section 4.1 we assumed that the hyperplanes in the arrangement $\mathcal{A}$ had no common intersection. Describe how to compute the tropicalization of $\mathbb{P}^d \setminus \cup \mathcal{A}$ if they do all intersect in some point $\mathbf{p} \in \mathbb{P}^d$. *Hint*: What is the image of the map from $\mathbb{P}^d \backslash \cup \mathcal{A}$ in T^n?

(4) Compute $\text{trop}(X)$ for the following varieties X defined over $\mathbb{Q}$:
   (a) $X = V(x_1 + x_2 + x_3 + x_4, x_1 + 2x_2 + 4x_3 - x_4) \subset T^4$;
   (b) $X = V(x_1 + x_2 + x_3 + x_4 + x_5, x_1 - x_2 + 3x_3 + 4x_4 + 7x_5) \subset T^5$;
   (c) $X = V(x_1 + x_2 + x_3 + x_4 + x_5, x_1 + x_2 + x_3 + 3x_4 - x_5) \subset T^5$.

   Now redo your calculation by taking $\mathbb{Q}$ not with the trivial valuation but with the 2-adic valuation. Repeat this for the 3-adic valuation.

(5) What happens in part (3) of Lemma 4.1.4 if we take an arbitrary index set L of size $d + 2$? Does this formula still give a circuit? When do different choices of L give the same linear form?

(6) Show that the two axiom systems of Definitions 4.2.2 and 4.2.3 are equivalent by constructing a rank function for every pair $(E, \mathcal{C})$ satisfying (C1) and (C2) and a set of circuits for every pair (E, ρ) satisfying (R1), (R2), and (R3).

(7) Given a classical constant-coefficient linear space X, prove that the Bergman fan of its matroid agrees with the Gröbner fan structure on $\mathrm{trop}(X)$ that is defined by the homogeneous ideal $I(X)_{\mathrm{proj}}$.

(8) Determine the graphic matroid associated with the Petersen graph. Describe the circuits, bases, rank function, and the Bergman fan.

(9) Let p_1, p_2 be points on a line L_1 in the plane $\mathbb{P}^2$. Let $\mathcal{A}$ be the arrangement in $\mathbb{P}^2$ consisting of five lines: L_1 together with two lines L_2, L_3 that intersect L_1 at p_1, and two lines L_4, L_5 that intersect L_1 in p_2, with no other triple intersections. Let $X = \mathbb{P}^2 \setminus \cup \mathcal{A}$. Compute the tropical variety $\mathrm{trop}(X) \subseteq \mathbb{R}^5/\mathbb{R}\mathbf{1}$. Show that the Bergman fan of X is not simplicial. Compare it with the fan in Theorem 4.2.6.

(10) Let M be a matroid on $[n]$. A *building set* for its lattice of flats is a set $\mathcal{G}$ of flats with the property that if F is any flat and $G_1, \ldots, G_r$ are the smallest-dimensional flats in $\mathcal{G}$ containing F, then $F = \bigcap_{i=1}^{r} G_i$. A subset σ of $\mathcal{G}$ is *nested* if any subset of σ with no pair contained one in the other has the property that the intersection of these flats does not lie in $\mathcal{G}$. The nested sets form a simplicial complex, called the *nested set complex*. This gives rise to the *nested set fan* in $\mathbb{R}^{n+1}/\mathbb{R}\mathbf{1}$ by sending a nested set σ to the cone $\mathrm{pos}(\mathbf{e}_F : F \in \sigma) + \mathbb{R}\mathbf{1} \subset \mathbb{R}^{n+1}/\mathbb{R}\mathbf{1}$. Recall that $\mathbf{e}_F = \sum_{i \in F} \mathbf{e}_i$.

(a) In Example 4.1.8 let

$$\mathcal{G} = \{\mathrm{span}(\mathbf{b}_i) : 0 \leq i \leq 4\} \cup \{\mathrm{span}(\mathbf{b}_1, \mathbf{b}_2, \mathbf{b}_3)\}.$$

Show that $\mathcal{G}$ is a building set and determine the nested set fan.

(b) Show that the support of the nested set fan equals $\mathrm{trop}(M)$ for any choice of building set $\mathcal{G}$.

(c) Show that the set of all flats of the lattice of flats is a building set. Which fan structure does this give for $\mathrm{trop}(M)$?

(d) Show that every matroid has a unique minimal building set.

(e) Find the minimal building set for the matroid of Exercise 4.7(9).

(f) What is the nested set fan for the minimal building set of the matroid of the complete graph K_4? The complete graph K_n?

Nested sets originated in the study of compactifications of hyperplane arrangements by De Concini and Procesi [**DCP95**]. See [**Fei05**] for a survey and [**FS05**] for more on tropical connections.

(11) Different graphs can have the same graphic matroid.

(a) Find two nonisomorphic graphs that have the same matroid.

(b) Despite the previous question, the tropicalization $\mathrm{trop}(M_G)$ of graphic matroid M_G still remembers some information about the graph G, such as the number of vertices and edges. Can you recover the set of circuits of a graph G from $\mathrm{trop}(M_G)$?

(12) The tropical linear space trop(M) of a matroid M is another one of the many different encodings of matroids. Verify this by describing how to recover the following information about M from trop(M):
   (a) The rank $\rho(M)$ of M;
   (b) The set of circuits of M;
   (c) The set of bases of M;
   (d) The set of independent sets of M;
   (e) Whether a subset of the ground set is a flat.

(13) Show that for any matroid M the Bergman fan on trop(M) is balanced when every maximal cone σ has weight $\text{mult}(\sigma) = 1$.

(14) The *non-Pappus matroid* is the rank 3 matroid on $\{0, 1, \ldots, 8\}$ with circuits 012, 046, 057, 136, 158, 237, 248, 345 plus every subset of size four not containing one of these eight. This matroid is not realizable over any field, as Pappus's Theorem implies that any realizable matroid with these circuits also has the circuit 678. Pappus's Theorem means that the points 6, 7, and 8 are always collinear in Figure 4.7.1. Describe the tropical linear space $\text{trop}(M) \subseteq \mathbb{R}^8$. Show directly that there is no $X \subseteq T^8$ with $\text{trop}(X) = \text{trop}(M)$.

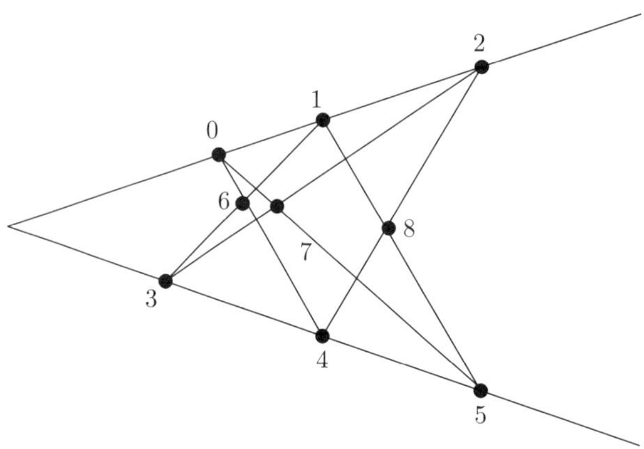

Figure 4.7.1. The non-Pappus matroid. Are points 6, 7, 8 collinear?

(15) For the tree on the right in Figure 4.3.1, express the four interior edge lengths in terms of the 21 pairwise distances d_{ij}. In other words, extend the formula for γ in (4.3.4) to a tree with seven taxa.

(16) The *f-vector* of a polyhedral complex Δ is the vector $f = (f_0, \ldots, f_d)$ where f_i is the number of cells of Δ of dimension i. Find the f-vector of the space of phylogenetic trees on m leaves for $m = 7, 8$.

(17) Compute the f-vector of the matroid polytope for the Fano matroid.

(18) Verify directly from the definition of tree metrics that the fan Δ in (4.3.1) is balanced when every maximal cone has multiplicity 1.

(19) Verify that $M_{\mathbf{w}}$ in Definition 4.2.7 is again a matroid. Determine the six initial matroids $M_{\mathbf{w}}$ of the uniform matroid $M = U_{3,6}$ given by $\mathbf{w} = (1,1,1,1,1,2)$, $\mathbf{w} = (1,1,1,1,2,2)$, $\mathbf{w} = (1,1,1,2,2,2)$, $\mathbf{w} = (1,1,2,2,2,2)$, $\mathbf{w} = (1,2,2,2,2,2)$, and $\mathbf{w} = (2,2,2,2,2,2)$.

(20) Given the arrangement $\mathcal{A}$ in Section 4.1, the vectors $\mathbf{b}_i$ are only defined up to scaling, so there are many linear spaces $X \subset T^n$ that correspond to the same hyperplane arrangement in $\mathbb{P}^d$. How is that reflected in the tropicalization of the Grassmannian $G^0(r,m)$?

(21) Verify the computation of maximal cones in $\operatorname{trop}(G^0(3,6))$ of Example 4.4.10. Pick a point $\mathbf{w}$ in the interior of your favorite cone. List all vertices, edges, and 2-cells of the tropical linear space $L_{\mathbf{w}}$.

(22) Extending Example 4.3.19, determine the universal families over the tropical Grassmannians $\operatorname{trop}(G^0(3,4))$ and $\operatorname{trop}(G^0(3,5))$.

(23) Compute the Dressian Dr_M for the non-Pappus matroid M.

(24) Determine whether the following statements are true or false.
 (a) Tropicalizing a smooth surface gives a smooth tropical surface.
 (b) A phylogenetic tree can be recovered from its tree metric.
 (c) Every tropical linear space of codimension c is the intersection of c tropical hyperplanes.
 (d) Tropical linear spaces are closed under stable intersections.
 (e) Every lattice polytope has a unimodular triangulation.

(25) Let $f, g \in \mathbb{C}[x,y,z]$ with Newton polytope $\operatorname{conv}\{(0,0,0),(2,0,0),(0,2,0),(0,0,2)\}$. Compute the tropical variety $\operatorname{trop}(V(f,g)) \subseteq \mathbb{R}^3$ when f, g are assumed to have generic coefficients. Compute explicitly the locus $U \subset \mathbb{P}^5 \times \mathbb{P}^5$ of systems (f,g) for which $\operatorname{trop}(V(f,g))$ equals this fan. What are the other possibilities for $\operatorname{trop}(V(f,g))$?

(26) What should be the definition of a smooth tropical curve in 3-space? Give an example of a curve that is smooth according to your definition but is not yet covered by Proposition 4.6.20.

(27) Let X be a hypersurface of degree d in $\mathbb{P}^4$ whose tropicalization is tropically smooth. What can you say about the number of bounded and unbounded cells in $\operatorname{trop}(X)$ of each dimension?

(28) Fix the 2-adic valuation on $\mathbb{Q}$ and the following polynomial in $\mathbb{Q}[x,y,z]$:

$$f = 16 + 2x - 2y - 31z - 16x^2 + 31xy - 2xz - 16y^2 + 2yz + 16z^2.$$

Compute the tropical surface $\mathcal{Q} = \operatorname{trop}(V(f))$, show that it is tropically smooth, and determine the two rulings of lines on $\mathcal{Q}$.

(29) Find a quadratic polynomial $f \in K[x, y, z]$ such that $\mathcal{Q} = \text{trop}(V(f))$ is in the last class in Proposition 4.5.4. Find a point $\mathbf{p}$ in the relative interior of the triangle and compute the two lines on $\mathcal{Q}$ through $\mathbf{p}$. Determine all lines on $\mathcal{Q}$ that pass through a vertex of the triangle.

(30) Find a homogeneous cubic $f \in \mathbb{Q}[x_0, x_1, x_2, x_3]$ such that $V(f)$ is smooth in $\mathbb{P}^3$ and its 27 lines are all defined over $\mathbb{Q}$. Compute and draw the p-adic tropicalizations of your 27 lines for $p = 2, 3, 5$.

(31) Consider two bivariate polynomials of the form

$$f \;=\; a_1xy + a_2x + a_3y + a_4 \;\text{ and }\; g \;=\; b_1x^3y + b_2x^3 + b_3y^3 + b_4.$$

Draw the Newton polygons of f and g and determine their mixed volume. Find precise condition on the eight coefficients under which the number of solutions in T^2 to the system of $f = g = 0$ equals the mixed volume.

(32) Three trilinear equations in three variables usually have six common solutions. Explain and prove this claim using tropical geometry. What is the (tropical) solution set for two trilinear equations?

(33) What is the maximum number of vertices of any three-dimensional polytope that is the Minkowski sum of three triangles in $\mathbb{R}^3$?

(34) In classical geometry, an irreducible quartic surface in $\mathbb{P}^3$ can have at most 16 singular points. Can this be seen in tropical geometry?

(35) Construct the triangulation promised by Theorem 4.5.8 for $d = 4$. List all 64 tetrahedra. Find an explicit realization as in (4.5.3).

(36) (a) The complex of all bounded faces in a smooth tropical cubic surface $\mathcal{S}$ in $\mathbb{R}^3$ consists of ten polygons, 36 edges and 27 vertices. Draw this complex for the specific cubic in (4.5.3).
(b) The unbounded cells of $\mathcal{S}$ form a balanced graph (at infinity) with 36 vertices and 54 edges. This graph is obtained by fusing four smooth tropical cubic curves, one for each of the four coordinate planes. Draw this graph for the cubic in (4.5.3).
(c) Connect your two pictures from parts (a) and (b). Use this to sketch a visualization of the entire cubic surface $\mathcal{S} = \text{trop}(V(f))$.

(37) Classify all possible mixed subdivisions of the pentagon $P_1 + P_2$, where P_1 and P_2 are the square and the triangle in Example 4.6.2.

(38) For the tetrahedron $P = \text{conv}\{(0,0,0), (1,0,0), (0,1,0), (0,0,1)\}$ and the triangle $Q = \text{conv}\{(2,1,0), (1,2,0), (1,1,1)\}$, determine the mixed volumes $\text{MV}(P, P, Q)$ and $\text{MV}(P, Q, Q)$.

(39) Show that if P and Q are convex lattice polygons in $\mathbb{R}^2$, then

$$\text{MV}(P, Q) \;=\; \text{vol}(P + Q) - \text{vol}(P) - \text{vol}(Q).$$

(40) Explain why the polynomials in Example 4.6.15 could not have coefficients in $\mathbb{Q}$ with the 2-adic valuation. Find two polynomials over $\overline{\mathbb{Q}}$ with a 2-adic valuation that are generic in this sense.

(41) Let f_1, f_2 be polynomials in $K[x_1^{\pm 1}, x_2^{\pm 1}]$ with Newton polygons $\mathrm{conv}((1,0),(0,1),(1,1))$ and $\mathrm{conv}((0,0),(1,0),(0,1),(1,1))$. Let $\tilde{f}_1$, $\tilde{f}_2$ be their homogenizations in $K[x_0, x_1, x_2]$. What does Bézout's Theorem say about the intersection of the curves $V(\tilde{f}_1)$ and $V(\tilde{f}_2)$? What does Bernstein's Theorem predict? Explain the difference.

(42) Consider $X = V(ax+by+cz+d, ex+fy+gz+h) \subset (\mathbb{C}^*)^3$. For what values of $a, b, \ldots, h$ is X a complete intersection? For what values is $\mathrm{trop}(X)$ equal to the stable intersection in Corollary 4.6.11?

(43) Consider three polytopes in $\mathbb{R}^3$ whose mixed volume is zero. What dimensions can these have? How about four polytopes in $\mathbb{R}^4$?

Chapter 5

Tropical Garden

After experiencing the diversity, wild beauty, and potential dangers of the tropical rain forest, we now enter the garden of tropical linear algebra.

In classical linear algebra over a field K, there are many equivalent ways to represent a d-dimensional subspace V of an n-dimensional vector space. For instance, is V the span of d linearly independent vectors, or it is the solution set of $n-d$ independent linear equations? These two notions translate to the tropical semiring, but they evolve differently. Images of tropical linear maps are *tropical polytopes*, the orchids of tropical convexity. The solution set of a finite system of tropical linear equations is a *linear prevariety*. In our garden, this remains a wallflower, in spite of its prominence in applications. We prefer to grow trees that are sturdy and balanced, so we focus on *tropicalized linear spaces* and *tropical linear spaces*. Their taxonomy is recorded in the *tropical Grassmannian* and the *Dressian*. The former arise from linear spaces over a field K with a valuation, while the latter are polyhedral complexes that share the same desirable traits. We encountered such linear spaces already in Chapter 4, and we revisit them in Section 5.4.

In Section 5.1 we investigate eigenvalues and eigenvectors in tropical linear algebra. A basic result states that every square matrix has exactly one eigenvalue. In Section 5.2 we focus on tropical convexity, which can be regarded as a shadow of classical convexity over an ordered field with a valuation. Section 5.3 explains different notions of matrix rank and how this ties in with the tropicalization of determinantal varieties. In Section 5.5 we study varieties that are parameterized by monomials in linear forms. The matroid theory of Section 4.2 makes their tropicalizations blossom.

5.1. Eigenvalues and Eigenvectors

Let A be an $n \times n$-matrix with entries in the tropical semiring $(\overline{\mathbb{R}}, \oplus, \odot)$. Here, $\overline{\mathbb{R}} = \mathbb{R} \cup \{\infty\}$. An *eigenvalue* of A is a real number λ such that

$$A \odot \mathbf{v} = \lambda \odot \mathbf{v} \tag{5.1.1}$$

for some $\mathbf{v} \in \mathbb{R}^n$. We say that $\mathbf{v}$ is an *eigenvector* of the tropical matrix A. The arithmetic operations in the equation (5.1.1) are tropical. For instance, for $n = 2$ with $A = (a_{ij})$, the left-hand side of (5.1.1) equals

$$\begin{pmatrix} a_{11} & a_{12} \\ a_{21} & a_{22} \end{pmatrix} \odot \begin{pmatrix} v_1 \\ v_2 \end{pmatrix} = \begin{pmatrix} a_{11} \odot v_1 \oplus a_{12} \odot v_2 \\ a_{21} \odot v_1 \oplus a_{22} \odot v_2 \end{pmatrix} = \begin{pmatrix} \min\{a_{11} + v_1, a_{12} + v_2\} \\ \min\{a_{21} + v_1, a_{22} + v_2\} \end{pmatrix}.$$

The right-hand side of (5.1.1) is equal to

$$\lambda \odot \begin{pmatrix} v_1 \\ v_2 \end{pmatrix} = \begin{pmatrix} \lambda \odot v_1 \\ \lambda \odot v_2 \end{pmatrix} = \begin{pmatrix} \lambda + v_1 \\ \lambda + v_2 \end{pmatrix}.$$

We represent the matrix $A = (a_{ij})$ by a weighted directed graph $G(A)$ with n nodes labeled by $[n] = \{1, 2, \ldots, n\}$. There is an edge from node i to node j if and only if $a_{ij} < \infty$, and we assign the length a_{ij} to each such edge (i, j). The *normalized length* of a directed path $i_0, i_1, \ldots, i_k$ in $G(A)$ is the sum (in classical arithmetic) of the lengths of the edges divided by the length k of the path. Thus the normalized length is $(a_{i_0 i_1} + a_{i_1 i_2} + \cdots + a_{i_{k-1} i_k})/k$. If $i_k = i_0$, then the path is a *directed cycle*, and we refer to this quantity as the normalized length of the cycle. Recall that a directed graph is *strongly connected* if there is a directed path from any vertex to any other vertex.

Theorem 5.1.1. *Let A be a tropical $n \times n$-matrix whose graph $G(A)$ is strongly connected. Then A has precisely one eigenvalue $\lambda(A)$. That eigenvalue equals the minimum normalized length of a directed cycle in $G(A)$.*

Proof. Let $\lambda = \lambda(A)$ be the minimum of the normalized lengths over all directed cycles in $G(A)$. We first prove that $\lambda(A)$ is the only possibility for an eigenvalue. Suppose that $\mathbf{z} \in \mathbb{R}^n$ is any eigenvector of A, and let γ be the corresponding eigenvalue. For any cycle $(i_1, i_2, \ldots, i_k, i_1)$ in $G(A)$ we have

$$a_{i_1 i_2} + z_{i_2} \geq \gamma + z_{i_1}, \; a_{i_2 i_3} + z_{i_3} \geq \gamma + z_{i_2},$$
$$a_{i_3 i_4} + z_{i_4} \geq \gamma + z_{i_3}, \; \ldots \;, \; a_{i_k i_1} + z_{i_1} \geq \gamma + z_{i_k}.$$

Adding the left-hand sides and the right-hand sides, we find that the normalized length of the cycle is greater than or equal to γ. In particular, we have $\lambda(A) \geq \gamma$. For the reverse inequality, start with any index i_1. Since $\mathbf{z}$ is an eigenvector with eigenvalue γ, there exists i_2 such that $a_{i_1 i_2} + z_{i_2} = \gamma + z_{i_1}$. Likewise, there exists i_3 such that $a_{i_2 i_3} + z_{i_3} = \gamma + z_{i_2}$. We continue in this manner until we reach an index i_l which was already in the sequence, say, $i_k = i_l$ for $k < l$. By adding the equations along this cycle, we find that

$$(a_{i_k i_{k+1}} + z_{i_{k+1}}) + (a_{i_{k+1} i_{k+2}} + z_{i_{k+2}}) + \cdots + (a_{i_{l-1} i_l} + z_{i_l}) \\ = (\gamma + z_{i_k}) + (\gamma + z_{i_{k+1}}) + \cdots + (\gamma + z_{i_{l-1}}).$$

We conclude that the normalized length of the cycle $(i_k, i_{k+1}, \ldots, i_l = i_k)$ in $G(A)$ is equal to γ. In particular, $\gamma \geq \lambda(A)$. This proves that $\gamma = \lambda(A)$.

It remains to prove the existence of an eigenvector. Let B be the matrix obtained from A by (classically) subtracting $\lambda(A)$ from every entry in A. All cycles in the weighted graph $G(B)$ have nonnegative length, and there exists a cycle of length zero. Using tropical matrix operations, we define

$$B^+ = B \oplus B^2 \oplus B^3 \oplus \cdots \oplus B^n.$$

This matrix is known as the *Kleene plus* of B. By Exercise 1.9(5), the entry B^+_{ij} in row i and column j of the Kleene plus B^+ is the length of a shortest path from node i to node j in the weighted directed graph $G(B)$. Since this graph is strongly connected, we have $B^+_{ij} < \infty$ for all i and j.

Fix any node j that lies on a zero-length cycle of $G(B)$. Let $\mathbf{x} = B^+_{\cdot j}$ denote the jth column vector of the matrix B^+. We have $x_j = B^+_{jj} = 0$, as there is a path from j to itself of length zero, and there are no negative weight cycles. This implies $B^+ \odot \mathbf{x} \leq B^+_{\cdot j} = \mathbf{x}$. Next note that $(B \odot \mathbf{x})_i = \min_l(B_{il} + x_l) = \min_l(B_{il} + B^+_{lj}) \geq B^+_{ij} = x_i$, since lengths of shortest paths obey the triangle inequality. In vector notation this states $B \odot \mathbf{x} \geq \mathbf{x}$. Since tropical linear maps preserve coordinatewise inequalities among vectors, we have $B^2 \odot \mathbf{x} \geq B \odot \mathbf{x}$, and $B^3 \odot \mathbf{x} \geq B^2 \odot \mathbf{x}$, etc. We conclude $B^+ \odot \mathbf{x} = B \odot \mathbf{x} \oplus B^2 \odot \mathbf{x} \oplus \cdots \oplus B^n \odot \mathbf{x} = B \odot \mathbf{x}$. This yields $\mathbf{x} \leq B \odot \mathbf{x} = B^+ \odot \mathbf{x} \leq \mathbf{x}$. This means that $B \odot \mathbf{x} = \mathbf{x}$, so $\mathbf{x}$ is an eigenvector of B with eigenvalue 0. We conclude that $\mathbf{x}$ is an eigenvector with eigenvalue λ of our matrix A:

$$A \odot \mathbf{x} = (\lambda \odot B) \odot \mathbf{x} = \lambda \odot (B \odot \mathbf{x}) = \lambda \odot \mathbf{x}.$$

This completes the proof of Theorem 5.1.1. □

It appears that the computation of the eigenvalue λ of a tropical $n \times n$-matrix requires inspecting all cycles in $G(A)$. However, this is not the case. Karp [**Kar78**] gave an efficient algorithm, based on linear programming, for computing $\lambda(A)$ from the matrix $A = (a_{ij})$. The idea is to set up the following linear program with $n+1$ decision variables $v_1, \ldots, v_n, \lambda$:

(5.1.2) maximize γ subject to $a_{ij} + v_j \geq \gamma + v_i$ for all $1 \leq i, j \leq n$.

Proposition 5.1.2 ([**Kar78**]). *The unique eigenvalue $\lambda(A)$ of the matrix $A = (a_{ij})$ coincides with the optimal value γ^* of the linear program* (5.1.2).

Proof. The dual linear program to (5.1.2) takes the form

$$\text{minimize } \sum_{i=1}^{n}\sum_{j=1}^{n} a_{ij}x_{ij} \text{ subject to } x_{ij} \geq 0 \text{ for } 1 \leq i,j \leq n,$$
$$\sum_{i,j=1}^{n} x_{ij} = 1 \text{ and } \sum_{j=1}^{n} x_{ij} = \sum_{k=1}^{n} x_{ki} \text{ for all } 1 \leq i \leq n.$$

Here the x_{ij} are the decision variables. The feasible solutions are the probability distributions (x_{ij}) on the edges of $G(A)$ that are flows in the directed graph. The vertices of the polyhedron defined by these constraints are the uniform probability distributions on the directed cycles in $G(A)$. The objective function $\sum a_{ij}x_{ij}$ of the dual linear program equals the normalized length of any such cycle. The optimal value is the minimum of these quantities over all directed cycles in $G(A)$. By strong duality, the primal linear program (5.1.2) has the same optimal value $\gamma^* = \lambda(A)$. □

We next determine the *eigenspace* of the matrix A, which is the set

$$\mathrm{Eig}(A) \quad = \quad \{\mathbf{x} \in \mathbb{R}^n : A \odot \mathbf{x} = \lambda(A) \odot \mathbf{x}\}.$$

The set $\mathrm{Eig}(A)$ is closed under tropical scalar multiplication: if $\mathbf{x} \in \mathrm{Eig}(A)$ and $c \in \mathbb{R}$, then $c \odot \mathbf{x}$ is also in $\mathrm{Eig}(A)$. We can thus identify $\mathrm{Eig}(A)$ with its image in $\mathbb{R}^n/\mathbb{R}\mathbf{1} \simeq \mathbb{R}^{n-1}$. Here $\mathbf{1} = (1, 1, \ldots, 1)$, as in Chapter 2.

Every eigenvector of the matrix A is also an eigenvector of the matrix $B = (-\lambda(A)) \odot A$ and vice versa. Hence the eigenspace $\mathrm{Eig}(A)$ is equal to

$$\mathrm{Eig}(B) \quad = \quad \{\mathbf{x} \in \mathbb{R}^n : B \odot \mathbf{x} = \mathbf{x}\}.$$

Theorem 5.1.3. *Let B_0^+ be the submatrix of B^+ given by the columns whose diagonal entry B_{jj}^+ is zero. The image of this matrix (with respect to tropical multiplication of vectors on the right) is equal to the desired eigenspace*

$$\mathrm{Eig}(A) \quad = \quad \mathrm{Eig}(B) \quad = \quad \mathrm{Image}(B_0^+).$$

Before proving Theorem 5.1.3, we present some examples of eigenspaces.

Example 5.1.4. We set $n = 4$. Each point in $\mathbb{R}^4/\mathbb{R}\mathbf{1}$ is represented by a vector in $\mathbb{R}^4$ with last coordinate zero, and we here write "Image" for the operator that computes the image in $\mathbb{R}^4/\mathbb{R}\mathbf{1}$ of a matrix with four rows.

$$\text{If } A = \begin{pmatrix} 3 & 1 & 4 & 5 \\ 5 & 2 & 4 & 2 \\ 4 & 1 & 6 & 3 \\ 2 & 6 & 3 & 6 \end{pmatrix}, \text{ then } \lambda(A) = 5/3 \text{ and } \mathrm{Eig}(A) = \mathrm{Image}\begin{pmatrix} -1/3 \\ 1/3 \\ -1/3 \\ 0 \end{pmatrix}.$$

$$\text{If } A = \begin{pmatrix} 1 & 4 & 4 & 6 \\ 1 & 1 & 1 & 2 \\ 4 & 2 & 1 & 3 \\ 6 & 3 & 6 & 4 \end{pmatrix}, \text{ then } \lambda(A) = 1 \text{ and } \mathrm{Eig}(A) = \mathrm{Image}\begin{pmatrix} -2 & 1 & 1 \\ -2 & -2 & -2 \\ -1 & -1 & -2 \\ 0 & 0 & 0 \end{pmatrix}.$$

$$\text{If } A = \begin{pmatrix} 4 & 5 & 3 & 3 \\ 3 & 5 & 4 & 6 \\ 6 & 1 & 5 & 3 \\ 5 & 5 & 2 & 5 \end{pmatrix}, \text{ then } \lambda(A) = 9/4 \text{ and } \mathrm{Eig}(A) = \mathrm{Image}\begin{pmatrix} 3/4 \\ 3/2 \\ 1/4 \\ 0 \end{pmatrix}.$$

$$\text{If } A = \begin{pmatrix} 0 & 1 & 1 & 1 \\ 1 & 0 & 1 & 1 \\ 1 & 1 & 0 & 1 \\ 1 & 1 & 1 & 0 \end{pmatrix}, \text{ then } \lambda(A) = 0 \text{ and } \mathrm{Eig}(A) = \mathrm{Image}(A).$$

For this last example, we have $A = B = B^+ = B_0^+$, and the eigenspace $\mathrm{Eig}(A)$ is a certain three-dimensional convex polytrope in $\mathbb{R}^4/\mathbb{R}\mathbf{1}$, known as the *standard polytrope*. We will discuss such polyt(r)opes in Section 5.2. ◊

Proof of Theorem 5.1.3. We saw in the proof of Theorem 5.1.1 that every column vector $\mathbf{x}$ of B_0^+ satisfies $B \odot \mathbf{x} = \mathbf{x}$. Since tropical linear combinations of eigenvectors are again eigenvectors, we have $\mathrm{Image}(B_0^+) \subseteq \mathrm{Eig}(B)$.

To prove the reverse inclusion, consider any vector $\mathbf{z} \in \mathrm{Eig}(B)$. Then $B^+ \odot \mathbf{z} = \mathbf{z}$. Let $\tilde{\mathbf{z}}$ be the vector obtained from $\mathbf{z}$ by erasing all coordinates j such that $B_{jj}^+ > 0$. We claim that $\mathbf{z} = B_0^+ \odot \tilde{\mathbf{z}}$. This will show $\mathbf{z} \in \mathrm{Image}(B_0^+)$. Recall that the directed graph $G(B)$ has minimum cycle length equal to zero. The minimum length of any path from node i to node j in $G(B)$ is the entry B_{ij}^+ of the Kleene plus B^+. That entry can be negative if $i \neq j$, but it is always nonnegative when $i = j$. Furthermore, we have $B_{ii}^+ = 0$ if and only if the node i lies on a cycle of length zero in $G(B)$.

Consider any index $i \in [n]$. We have $z_i = \min(B_{ij}^+ + z_j \, : \, j \in [n])$. If $z_i = B_{ij}^+ + z_j$ and $z_j = B_{jk}^+ + z_k$, then $B_{ij}^+ + B_{jk}^+ + z_k = z_i \leq B_{ik}^+ + z_k$. The triangle inequality $B_{ij}^+ + B_{jk}^+ \geq B_{ik}^+$ holds for the Kleene plus B^+, as B_{ij}^+ is the length of the shortest path in $G(B)$ from i to j. This implies that $z_i = B_{ik}^+ + z_k$. Continuing in this manner, we eventually revisit an index l. Writing $l = l_0, l_1, \ldots, l_s = l$ for the cycle, we have $z_l = B_{ll_1}^+ + B_{l_1 l_2}^+ + \cdots + B_{l_{s-1}l}^+ + z_l$. Thus $B_{ll}^+ = B_{ll_1}^+ + B_{l_1 l_2}^+ + \cdots + B_{l_{s-1}l}^+ = 0$, so l lies on a cycle of length zero. We then have the equality $z_i = B_{il}^+ + z_l$, which can be rewritten as $z_i = ((B_0^+) \odot \tilde{\mathbf{z}})_i$. This implies $\mathbf{z} = B_0^+ \odot \tilde{\mathbf{z}}$, and the proof is complete. □

In classical linear algebra, the eigenvalues of a square matrix are the roots of its characteristic polynomial, and we seek to extend this to tropical linear algebra. The *characteristic polynomial* of our $n \times n$-matrix A equals

$$f_A(t) \;=\; \det\big(\, A \oplus t \odot \mathrm{Id}\,\big),$$

where "det" denotes the tropical determinant. We have the following result:

Corollary 5.1.5. *The eigenvalue $\lambda(A)$ of a tropical $n \times n$-matrix A is the smallest root of its characteristic polynomial $f_A(t)$.*

Proof. Consider the expansion of the characteristic polynomial:

$$f_A(t) \;=\; t^n \,\oplus\, c_1 \odot t^{n-1} \,\oplus\, c_2 \odot t^{n-2} \,\oplus\, \cdots \,\oplus\, c_{n-1} \odot t \,\oplus\, c_n.$$

Each permutation π of $[n]$ is a disjoint union of cycles. Hence the constant term $c_n = \det(A)$ of $f_A(t)$ is the minimal length $\bigodot_{i=1}^n a_{i\pi_i}$ of any disjoint union of cycles that uses all n nodes in $G(A)$. Likewise, the coefficient c_i is the minimum over the lengths of all disjoint unions of cycles on exactly i nodes in $G(A)$. The smallest root of the tropical polynomial $f_A(t)$ equals

$$\min\{\, c_1,\, c_2/2,\, c_3/3,\, \ldots,\, c_n/n \,\}.$$

This minimum is the smallest normalized cycle length $\lambda(A)$. □

Our discussion raises the question of how the tropical eigenvalue problem is related to the classical eigenvalue problem for a matrix over a field K with a valuation. Let M be an $n \times n$-matrix with entries in K, and let $A = \mathrm{val}(M)$ be its tropicalization. If the entries in M are general enough, then the characteristic polynomial $f_A(t)$ of A coincides with the tropicalization of the classical characteristic polynomial of M. Assuming this to be the case, let us consider an arbitrary solution $(\mu, \mathbf{v})$ of the eigenvalue equation for M:

$$M \cdot \mathbf{v} \;=\; \mu \cdot \mathbf{v}.$$

This equation does not tropicalize; there will be cancellations of lowest terms in the matrix-vector product $M \cdot \mathbf{v}$, unless μ is an eigenvalue of minimal valuation $\lambda(A)$. Furthermore, the eigenvector $\mathbf{v}$ must satisfy the nontrivial combinatorial constraint imposed by Theorem 5.1.3, namely, the valuation of $\mathbf{v}$ is in the image of the matrix B_0^+. Here is an example to show this.

Example 5.1.6. Let $n = 3$, let $K = \mathbb{C}\{\{t\}\}$, and consider the matrix

$$M \;=\; \begin{pmatrix} t & 1 & t \\ 1 & t & -t^2 \\ t & t^2 & t \end{pmatrix}.$$

This matrix has three distinct eigenvalues μ in K, and we list each of them with a generator $\mathbf{v}$ for the corresponding one-dimensional eigenspace in K^3.

Eigenvalue μ	Eigenvector $\mathbf{v}$
t	$(t^2, -t, 1)^T$
$\sqrt{1+t^2-t^4}+t$	$\big(t-\sqrt{1+t^2-t^4},\, t\sqrt{1+t^2-t^4}-1,\, t(t^2-1)\big)^T$
$-\sqrt{1+t^2-t^4}+t$	$\big(t^3-\sqrt{1+t^2-t^4},\, t^4-1,\, t^2\sqrt{1+t^2-t^4}-t\big)^T$

The tropicalization of the matrix M equals $A = \left(\begin{smallmatrix} 1 & 0 & 1 \\ 0 & 1 & 2 \\ 1 & 2 & 1 \end{smallmatrix}\right)$, and the tropical characteristic polynomial $f_A(z) = z^3 \oplus 1 \odot z^2 \oplus 0 \odot z \oplus 1$ factors as

$$f_A(z) \;=\; (z \oplus 0)^2 \odot (z \oplus 1).$$

This is an identity of tropical polynomial functions. The tropical roots reflect the fact that M has two eigenvalues of valuation 0 and one eigenvalue of valuation 1. By Theorem 5.1.1, $\lambda(A) = 0$ is the only eigenvalue of the matrix A. The eigenspace $\mathrm{Eig}(A)$ is computed using Theorem 5.1.3. We have

$$B^+ \;=\; A^+ \;=\; A \oplus A^2 \oplus A^3 \;=\; \begin{pmatrix} 0 & 0 & 1 \\ 0 & 0 & 1 \\ 1 & 1 & 1 \end{pmatrix}.$$

Hence $\mathrm{Eig}(A)$ is spanned, over the tropical semiring, by the vector $(0,0,1)^T$. Equivalently, the eigenspace of A consists of the column vectors $(a, a, a+1)^T$ for all $a \in \mathbb{R}$. Each of these arises as the coordinatewise valuation of an eigenvector of the classical matrix M over the field K. For instance, the last two eigenvectors $\mathbf{v}$ listed above both have valuation $(0,0,1)^T$. ◊

In classical linear algebra, the determinant of a square matrix is the product of its eigenvalues. This is not true in tropical linear algebra, as there is only one eigenvalue. What remains true however is the geometric interpretation of the determinant as a coplanarity criterion. That result is the same both classically and tropically. We now derive the latter version.

We view the determinant of an $n \times n$-matrix as a polynomial of degree n in n^2 unknowns having $n!$ terms. The tropical hypersurface defined by that polynomial was described in Example 3.1.11; see also Remark 1.2.5. Matrices which lie on that tropical hypersurface are called *tropically singular*.

Proposition 5.1.7. *Let A be a real $n \times n$-matrix. Then A is tropically singular if and only if the rows of A lie on a tropical hyperplane in $\mathbb{R}^n/\mathbb{R}\mathbf{1}$.*

Proof. Suppose that $A = (a_{ij})$ is tropically singular. Fix an algebraically closed field K whose value group Γ_{val} contains all entries of A. By Kapranov's Theorem 3.1.3, applied to $f = \det$, there exists a singular $n \times n$-matrix U with entries in K^* for which $\mathrm{val}(U) = A$. Pick a nonzero vector in the kernel of U, and consider the classical hyperplane $H \subset K^n$ perpendicular to that vector. Then the rows of A lie in the tropical hyperplane $\mathrm{trop}(H \cap (K^*)^n)$.

For the converse, suppose that the rows $\mathbf{a}_1, \ldots, \mathbf{a}_n$ of A lie in a tropical hyperplane $\mathcal{H}$. Pick a classical vector $\mathbf{c} = (c_1, \ldots, c_n) \in (K^*)^n$ such that the linear form $\ell = c_1 x_1 + \cdots + c_n x_n$ satisfies $\mathcal{H} = \mathrm{trop}(V(\ell))$. By Kapranov's Theorem 3.1.3, now applied to ℓ, we can lift each $\mathbf{a}_i$ to a vector $\mathbf{f}_i \in (K^*)^n$ with $\ell(\mathbf{f}_i) = 0$ and $\mathrm{val}(\mathbf{f}_i) = \mathbf{a}_i$. Let F be the $n \times n$-matrix with row vectors $\mathbf{f}_1, \ldots, \mathbf{f}_n$. By construction, we have $\mathrm{val}(F) = A$ and $F \cdot \mathbf{c}^T = \mathbf{0}$. This implies that F is a classically singular, so $\mathrm{val}(F) = A$ is tropically singular. □

The spectral theory of tropical matrices is an active area of research. It has numerous applications, and offers many interesting directions for combinatorialists and geometers. We refer to the books [**BCOQ92, But10**].

5.2. Tropical Convexity

We now introduce the notions of convexity and convex polytopes in the setting of tropical geometry. The main result in this section is Theorem 5.2.19, which states that combinatorial types of tropical polytopes are in bijection with the regular subdivisions of products of two simplices.

Definition 5.2.1. A subset S of $\mathbb{R}^n$ is *tropically convex* if $\mathbf{x}, \mathbf{y} \in S$ and $a, b \in \mathbb{R}$ implies $a \odot \mathbf{x} \oplus b \odot \mathbf{y} \in S$. The *tropical convex hull* of a given subset $V \subset \mathbb{R}^n$ is the smallest tropically convex subset of $\mathbb{R}^n$ that contains V.

Any tropically convex subset S of $\mathbb{R}^n$ is closed under tropical scalar multiplication: $\mathbb{R} \odot S \subseteq S$. In other words, if $\mathbf{x} \in S$, then $\mathbf{x} + \lambda \mathbf{1} \in S$ for all $\lambda \in \mathbb{R}$. We thus identify the tropically convex set S with its image in the $(n-1)$-dimensional tropical projective torus $\mathbb{R}^n/\mathbb{R}\mathbf{1}$. A *tropical polytope* is the tropical convex hull of a finite subset V in $\mathbb{R}^n/\mathbb{R}\mathbf{1}$.

We shall see in Proposition 5.2.6 that the tropical convex hull of V coincides with the set of all tropical linear combinations

$$(5.2.1) \quad a_1 \odot \mathbf{v}_1 \oplus \cdots \oplus a_r \odot \mathbf{v}_r\,, \text{ where } \mathbf{v}_1, \ldots, \mathbf{v}_r \in V \text{ and } a_1, \ldots, a_r \in \mathbb{R}.$$

By Theorem 5.1.3, the eigenspace of a square matrix is a tropical polytope.

Remark 5.2.2. The set $\overline{\mathbb{R}}^n$ is a semimodule over the tropical semiring $(\overline{\mathbb{R}}, \oplus, \odot)$. Here $\overline{\mathbb{R}} = \mathbb{R} \cup \{\infty\}$. Tropically convex sets in $\overline{\mathbb{R}}^n/\mathbb{R}\mathbf{1}$ lift to subsemimodules, and tropical polytopes to finitely generated subsemimodules. In this section we restrict ourselves to tropical polytopes whose points have coordinates in $\mathbb{R}$. This is done to keep things simple. The theory presented here extends naturally to points with coordinates in $\overline{\mathbb{R}}$, where our ambient space $\mathbb{R}^n/\mathbb{R}\mathbf{1}$ is replaced with the tropical projective space of Chapter 6.

Tropical convexity was developed independently by several authors. Early references include [**DS04, CGQS05**]. For further reading see [**Jos**].

We shall see that every tropical polytope is a finite union of convex polytopes in the usual sense: the tropical convex hull of $V = \{\mathbf{v}_1, \ldots, \mathbf{v}_r\} \subset \mathbb{R}^n$ has a natural polyhedral cell decomposition, called the *tropical complex* generated by V. There is also a remarkable duality between tropical polytopes with r vertices in $\mathbb{R}^n/\mathbb{R}\mathbf{1}$ and tropical polytopes with n vertices in $\mathbb{R}^r/\mathbb{R}\mathbf{1}$.

We begin with pictures of tropically convex sets in the tropical plane $\mathbb{R}^3/\mathbb{R}\mathbf{1}$. A point $(x_1, x_2, x_3) \in \mathbb{R}^3/\mathbb{R}\mathbf{1}$ is represented by drawing the point

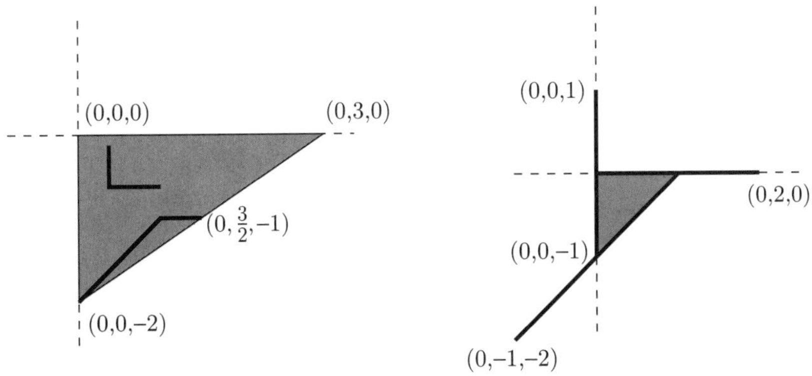

Figure 5.2.1. Tropically convex sets and tropical line segments in $\mathbb{R}^3/\mathbb{R}\mathbf{1}$.

with coordinates $(x_2 - x_1, x_3 - x_1)$ in the plane $\mathbb{R}^2$. The triangle on the left-hand side in Figure 5.2.1 is tropically convex, but it is not a tropical polytope because it is not the tropical convex hull of finitely many points. The thick edges indicate two tropical line segments. The picture on the right-hand side is a *tropical triangle*: it is the tropical convex hull of the three points $(0,0,1)$, $(0,2,0)$, and $(0,-1,-2)$ in the plane $\mathbb{R}^3/\mathbb{R}\mathbf{1}$. The thick edges represent the tropical segments connecting any two of these three points.

Tropical convex sets enjoy many of the features of ordinary convex sets:

Theorem 5.2.3. *The intersection of tropically convex sets in $\mathbb{R}^n$ is tropically convex. The projection of a tropically convex set onto a coordinate hyperplane is tropically convex. The classical hyperplane $\{ x_i - x_j = k \}$ is tropically convex. Projecting from this hyperplane to $\mathbb{R}^{n-1}$ by eliminating x_i is a tropical linear map. Tropically convex sets are contractible.*

Proof. We prove the statements in the order given. Let $\mathcal{S}$ be a collection of tropically convex subsets of $\mathbb{R}^n$ and $\mathbf{x}, \mathbf{y} \in \bigcap_{S \in \mathcal{S}} S$. Each $S \in \mathcal{S}$ contains the tropical line segment between $\mathbf{x}$ and $\mathbf{y}$, and hence so does $\bigcap_{S \in \mathcal{S}} S$.

Suppose S is a tropically convex set in $\mathbb{R}^n$. We claim that the image of S under the coordinate projection $\phi : \mathbb{R}^n \to \mathbb{R}^{n-1}$, $(x_1, x_2, \ldots, x_n) \mapsto (x_2, \ldots, x_n)$ is a tropically convex subset of $\mathbb{R}^{n-1}$. If $\mathbf{x}, \mathbf{y} \in S$, then we have

$$\phi\big(c \odot \mathbf{x} \oplus d \odot \mathbf{y}\big) \quad = \quad c \odot \phi(\mathbf{x}) \oplus d \odot \phi(\mathbf{y}).$$

This means that ϕ is a linear map in tropical arithmetic. Therefore, if S contains the tropical line segment between $\mathbf{x}$ and $\mathbf{y}$, then $\phi(S)$ contains the tropical segment between $\phi(\mathbf{x})$ and $\phi(\mathbf{y})$ and hence is tropically convex.

Most classical hyperplanes in $\mathbb{R}^n$ are not tropically convex, but we claim that classical hyperplanes of the special form $\{ x_i - x_j = k \}$ are tropically convex. If $\mathbf{x}$ and $\mathbf{y}$ lie in that hyperplane, then $x_i - y_i = x_j - y_j$, and hence $(c\odot\mathbf{x}\oplus d\odot\mathbf{y})_i - (c\odot\mathbf{x}\oplus d\odot\mathbf{y})_j = \min(x_i+c, y_i+d) - \min(x_j+c, y_j+d) = k$ for all $c, d \in \mathbb{R}$. So, the tropical segment between $\mathbf{x}$ and $\mathbf{y}$ is in $\{x_i - x_j = k\}$.

Consider the restriction of the projection map ϕ to $\{x_i - x_j = k\}$. This restriction is injective: if two points differ in the x_i coordinate, they must also differ in the x_j coordinate. It is surjective because we can recover the ith coordinate by setting $x_i = x_j + k$. Hence it is an isomorphism.

Classical convex sets are contractible because they can be retracted along line segments to any particular point. Tropically convex sets are contractible for the same reason, since tropical segments are homeomorphic to $[0,1]$. □

The relationship between classical polytopes and tropical polytopes is similar to the relationship between classical varieties and tropical varieties:

Remark 5.2.4. Let K be a real closed field with a nontrivial valuation, such as the field $K = \mathbb{R}\{\{\epsilon\}\}$ of Puiseux series with real coefficients; see [**BPR06**, Section 2.6]. Let K_+ be the subset of positive elements in K, and let $P = \operatorname{conv}(\mathbf{a}_1, \ldots, \mathbf{a}_r)$ be a classical convex polytope in $(K_+)^n$. This polytope maps to $\mathbb{R}^n/\mathbb{R}\mathbf{1}$ under the coordinatewise valuation map. The closure of the image of P under this map is the tropical convex hull of $\operatorname{val}(\mathbf{a}_1), \ldots, \operatorname{val}(\mathbf{a}_r)$. Conversely, every tropical polytope in $\mathbb{R}^n/\mathbb{R}\mathbf{1}$ can be lifted to a classical polytope P that lies in the positive orthant of K^n. We refer to [**DY07**, Section 2] for details on these constructions.

We next give a more precise description of tropical line segments.

Proposition 5.2.5. *The tropical line segment between two points* $\mathbf{x}$ *and* $\mathbf{y}$ *in* $\mathbb{R}^n/\mathbb{R}\mathbf{1}$ *is the concatenation of at most* $n-1$ *ordinary line segments. The direction of each line segment is a zero-one vector.*

Proof. After relabeling coordinates of $\mathbf{x} = (x_1, , \ldots, x_n)$ and $\mathbf{y} = (y_1, \ldots, y_n)$, and adding multiples of $\mathbf{1}$, we may assume $0 = y_1 - x_1 \leq y_2 - x_2 \leq \cdots \leq y_n - x_n$. The following points lie in the given order on the tropical segment between $\mathbf{x}$ and $\mathbf{y}$:

$$
\begin{aligned}
\mathbf{x} \;=\; (y_1 - x_1) \odot \mathbf{x} \oplus \mathbf{y} &= \big(y_1,\, y_1 - x_1 + x_2, y_1 - x_1 + x_3, \ldots, y_1 - x_1 + x_n\big) \\
(y_2 - x_2) \odot \mathbf{x} \oplus \mathbf{y} &= \big(y_1,\, y_2,\, y_2 - x_2 + x_3, \ldots, y_2 - x_2 + x_n\big) \\
(y_3 - x_3) \odot \mathbf{x} \oplus \mathbf{y} &= \big(y_1,\, y_2,\, y_3,\, \ldots, y_3 - x_3 + x_{n-1}, y_3 - x_3 + x_n\big) \\
\cdots\cdots \qquad & \qquad \cdots\cdots \qquad\qquad \cdots\cdots \\
(y_{n-1} - x_{n-1}) \odot \mathbf{x} \oplus \mathbf{y} &= \big(y_1,\, y_2,\, y_3,\, \ldots, y_{n-1}, y_{n-1} - x_{n-1} + x_n\big), \\
\mathbf{y} \;=\; (y_n - x_n) \odot \mathbf{x} \oplus \mathbf{y} &= \big(y_1,\, y_2,\, y_3,\, \ldots,\, y_{n-1},\, y_n\big).
\end{aligned}
$$

Between any two consecutive points, the tropical line segment equals an ordinary line segment of direction $(0, 0, \ldots, 0, 1, 1, \ldots, 1)$. Hence the tropical line segment between $\mathbf{x}$ and $\mathbf{y}$ is the concatenation of at most $n-1$ ordinary line segments, one for each strict inequality $y_i - x_i < y_{i+1} - x_{i+1}$. □

Proposition 5.2.5 shows an important feature of tropical convexity: segments use a limited set of directions. We next characterize convex hulls.

Proposition 5.2.6. *The smallest tropically convex subset of $\mathbb{R}^n$ containing a given set V is the set* $\mathrm{tconv}(V)$ *of all tropical linear combinations*

$$a_1 \odot \mathbf{v}_1 \oplus \cdots \oplus a_r \odot \mathbf{v}_r\,, \quad \textit{where } \mathbf{v}_1, \ldots, \mathbf{v}_r \in V \textit{ and } a_1, \ldots, a_r \in \mathbb{R}.$$

Proof. Let $\mathbf{x} = \bigoplus_{i=1}^r a_i \odot \mathbf{v}_i$ be the point in (5.2.1). If $r \leq 2$, then $\mathbf{x}$ is in the tropical convex hull of V. If $r > 2$, then we write $\mathbf{x} = a_1 \odot \mathbf{v}_1 \oplus (\bigoplus_{i=2}^r a_i \odot \mathbf{v}_i)$. The parenthesized vector lies the tropical convex hull, by induction on r, and hence so does $\mathbf{x}$. For the converse, consider any two tropical linear combinations $\mathbf{x} = \bigoplus_{i=1}^r c_i \odot \mathbf{v}_i$ and $\mathbf{y} = \bigoplus_{j=1}^r d_j \odot \mathbf{v}_j$. By the distributive law, $a \odot \mathbf{x} \oplus b \odot \mathbf{y}$ is also a tropical linear combination of $\mathbf{v}_1, \ldots, \mathbf{v}_r \in V$. Hence the set of all tropical linear combinations of V is tropically convex, so it contains the tropical convex hull of V. □

The following basic result from classical convexity holds also tropically.

Proposition 5.2.7 (Tropical Carathéodory's Theorem). *If $\mathbf{x}$ is in the tropical convex hull of a set of r points $\mathbf{v}_1, \mathbf{v}_2, \ldots, \mathbf{v}_r$ in $\mathbb{R}^n/\mathbb{R}\mathbf{1}$ with $r > n$, then $\mathbf{x}$ is in the tropical convex hull of at most n of them.*

Proof. Let $\mathbf{x} = \bigoplus_{i=1}^r a_i \odot \mathbf{v}_i$, and suppose $r > n$. For each coordinate $j \in [n]$, there exists an index $i \in \{1, \ldots, r\}$ such that $x_j = a_i + v_{ij}$. Take a subset I of $\{1, \ldots, r\}$ composed of one such i for each j. Then we also have $\mathbf{x} = \bigoplus_{i \in I} a_i \odot \mathbf{v}_i$, where I has at most n elements. □

Just as in ordinary geometry, every linear space is a convex set:

Proposition 5.2.8. *Tropical linear spaces in $\mathbb{R}^n/\mathbb{R}\mathbf{1}$ are tropically convex.*

Proof. Every tropical linear space is an intersection of tropical hyperplanes. Hence, by the first statement in Theorem 5.2.3, it suffices to show that tropical hyperplanes H are tropically convex.

Suppose that H is defined by $a_1 \odot x_1 \oplus \cdots \oplus a_n \odot x_n$, i.e., H consists of all points $\mathbf{x} = (x_1, \ldots, x_n)$ satisfying

$$(5.2.2) \quad a_i + x_i = a_j + x_j = \min\{a_k + x_k : k = 1, \ldots, n\} \text{ for some } i \neq j.$$

Let $\mathbf{x}$ and $\mathbf{y}$ be in H and consider any linear combination $\mathbf{z} = c \odot \mathbf{x} \oplus d \odot \mathbf{y}$. Let i be an index which minimizes $a_i + z_i$. We must show that this minimum is attained twice. By definition, z_i is equal to either $c + x_i$ or $d + y_i$. After permuting $\mathbf{x}$ and $\mathbf{y}$, we may assume $z_i = c + x_i \leq d + y_i$. Since, for all k, $a_i + z_i \leq a_k + z_k$ and $z_k \leq c + x_k$, it follows that $a_i + x_i \leq a_k + x_k$ for all k. Hence $a_i + x_i$ achieves the minimum of $\{a_1 + x_1, \ldots, a_n + x_n\}$. Since $\mathbf{x} \in H$, there exists an index $j \neq i$ with $a_i + x_i = a_j + x_j$. But now $a_j + z_j \leq a_j + c + x_j = c + a_i + x_i = a_i + z_i$. Since $a_i + z_i$ is the minimum of all $a_j + z_j$, the two are equal, and this minimum is obtained at least twice. □

Fix a subset $V = \{\mathbf{v}_1, \mathbf{v}_2, \ldots, \mathbf{v}_r\}$ of $\mathbb{R}^n/\mathbb{R}\mathbf{1}$, with $\mathbf{v}_i = (v_{i1}, \ldots, v_{in})$. Consider the *max-plus* tropical hyperplane with vertex at $\mathbf{v}_i$. This is the set

$$H_{\mathbf{v}_i} = \{\mathbf{x} \in \mathbb{R}^n/\mathbb{R}\mathbf{1} : \text{ the } \textit{maximum} \text{ of } x_1 - v_{i1}, \ldots, x_n - v_{in} \text{ is attained at least twice } \}.$$

The arrangement of max-plus tropical hyperplanes $H_{\mathbf{v}_1}, \ldots, H_{\mathbf{v}_r}$ divides the ambient space $\mathbb{R}^n/\mathbb{R}\mathbf{1}$ into relatively open classical convex polyhedra. We label the polyhedron containing a given point $\mathbf{x}$ in $\mathbb{R}^n/\mathbb{R}\mathbf{1}$ by a set of n subsets of $\{1, \ldots, r\}$ called its type. More formally, the *type* of $\mathbf{x}$ relative to V is the n-tuple $(S_1, \ldots, S_n)$, where

$$S_j = \{ i \in \{1, \ldots, r\} : \text{ the maximum for } H_{\mathbf{v}_i} \text{ is attained at } x_j - v_{ij} \}.$$

Thus an index i is in S_j if $v_{ij} - x_j = \min(v_{i1} - x_1, v_{i2} - x_2, \ldots, v_{in} - x_n)$. Equivalently, if we set $\lambda_i = \min\{ \lambda \in \mathbb{R} : \lambda \odot \mathbf{v}_i \oplus \mathbf{x} = \mathbf{x} \}$, then S_j is the set of all indices i such that $\lambda_i \odot \mathbf{v}_i$ and $\mathbf{x}$ have the same jth coordinate.

We say that an n-tuple of indices $S = (S_1, \ldots, S_n)$ is a *type* of V if it arises in the manner above for some $\mathbf{x}$. Note that every type satisfies $S_1 \cup S_2 \cup \cdots \cup S_n = \{1, 2, \ldots, r\}$ because the maximum for H_i is attained by some j.

Example 5.2.9. Let $r{=}n{=}3$, $\mathbf{v}_1 = (0, 0, 2)$, $\mathbf{v}_2 = (0, 2, 0)$, and $\mathbf{v}_3 = (0, 1, -2)$. The max-plus tropical lines $H_{\mathbf{v}_1}$, $H_{\mathbf{v}_2}$, and $H_{\mathbf{v}_3}$ divide the plane $\mathbb{R}^3/\mathbb{R}\mathbf{1}$ into 31 cells. There are ten two-dimensional cells (one bounded and nine unbounded), 15 edges (six bounded and nine unbounded), and six vertices. This is shown in Figure 5.2.2. The configuration $V = \{\mathbf{v}_1, \mathbf{v}_2, \mathbf{v}_3\}$ has 31 types. For instance, the point $\mathbf{x} = (0, 1, -1)$ has $\text{type}(\mathbf{x}) = (\{2\}, \{1\}, \{3\})$. Its cell is a pentagon. The point $\mathbf{x}' = (0, 0, 0)$ has $\text{type}(\mathbf{x}') = (\{1, 2\}, \{1\}, \{2, 3\})$. Its cell is a vertex. The point $\mathbf{x}'' = (0, 0, -3)$ has $\text{type}(\mathbf{x}'') = (\{1, 2, 3\}, \{1\}, \emptyset)$. Its cell is an unbounded edge. ◊

Our first application of types is the following separation theorem.

Proposition 5.2.10 (Tropical Farkas Lemma). *For all* $\mathbf{x} \in \mathbb{R}^n/\mathbb{R}\mathbf{1}$, *exactly one of the following is true:*

(i) *the point* $\mathbf{x}$ *is in the tropical polytope* $P = \text{tconv}(V)$*; or*

(ii) *there exists a tropical hyperplane that separates* $\mathbf{x}$ *from* P.

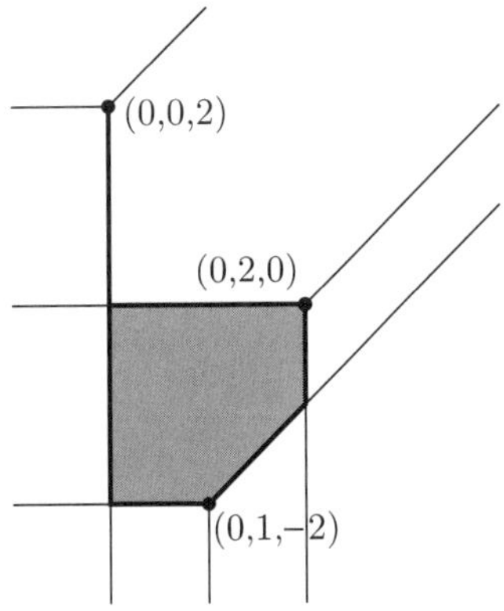

Figure 5.2.2. The polyhedral decomposition of $\mathbb{R}^3/\mathbb{R}\mathbf{1}$ from Example 5.2.9.

The phrase "separates $\mathbf{x}$ from P" is defined as follows. Suppose the hyperplane is given by the tropical linear form $a_1 \odot x_1 \oplus \cdots \oplus a_n \odot x_n$ and k is an index with $a_k + x_k = \min(a_1 + x_1, \ldots, a_n + x_n)$. Then separation means that $a_k + y_k > \min(a_1 + y_1, \ldots, a_n + y_n)$ for all $\mathbf{y} \in P$.

Proof. Let $\mathbf{x}$ be any point in $\mathbb{R}^n/\mathbb{R}\mathbf{1}$, with $\operatorname{type}(\mathbf{x}) = (S_1, \ldots, S_n)$, and let $\lambda_i = \min\{\, \lambda \in \mathbb{R} : \lambda \odot \mathbf{v}_i \oplus \mathbf{x} = \mathbf{x} \,\}$ as before. We define

$$\pi_V(x) \quad = \quad \lambda_1 \odot \mathbf{v}_1 \oplus \lambda_2 \odot \mathbf{v}_2 \oplus \cdots \oplus \lambda_r \odot \mathbf{v}_r. \tag{5.2.3}$$

Consider two cases: $\pi_V(\mathbf{x}) = \mathbf{x}$ or $\pi_V(\mathbf{x}) \neq \mathbf{x}$. The first case implies Proposition 5.2.10(i). It suffices to prove that the second case implies Proposition 5.2.10(ii). Suppose that $\pi_V(\mathbf{x}) \neq \mathbf{x}$. Since $v_{ik} + \lambda_i - x_k \geq 0$ for all k by the definition of λ_i, this means that there is $k \in \{1, \ldots, n\}$ with $v_{ik} + \lambda_i - x_k > 0$ for $i = 1, 2, \ldots, r$. Thus $S_k = \emptyset$. Fix $\varepsilon > 0$ that is less than $\min_i(v_{ik} + \lambda_i - x_k)$. We now choose our separating tropical hyperplane (5.2.2) as follows:

$$a_k \;:=\; -x_k - \varepsilon \quad \text{and} \quad a_j \;:=\; -x_j \quad \text{for } j \in \{1, \ldots, n\} \backslash \{k\}. \tag{5.2.4}$$

This satisfies $a_k + x_k = \min(a_1 + x_1, \ldots, a_n + x_n)$. Now, consider any point $\mathbf{y} = \bigoplus_{i=1}^{r} c_i \odot \mathbf{v}_i$ in $\operatorname{tconv}(V)$. Pick any index m such that $y_k = c_m + v_{mk}$. By definition of the λ_i, we have $x_l \leq \lambda_m + v_{ml}$ for all l, and there exists some j with $x_j = \lambda_m + v_{mj}$. We have $j \neq k$, as $x_k < v_{mk} + \lambda_m$. These equations and inequalities imply

$$\begin{aligned} a_k + y_k = a_k + c_m + v_{mk} &= c_m + v_{mk} - x_k - \varepsilon \;>\; c_m - \lambda_m \\ &= c_m + v_{mj} - x_j \geq y_j - x_j = a_j + y_j \geq \min(a_1 + y_1, \ldots, a_n + y_n). \end{aligned}$$

Hence, the hyperplane defined by (5.2.4) separates $\mathbf{x}$ from P as desired. $\square$

The construction in (5.2.3) defines a map $\pi_V : \mathbb{R}^n/\mathbb{R}\mathbf{1} \to P$ whose restriction to the tropical polytope $P = \operatorname{tconv}(V)$ is the identity. This map is the tropical version of the *nearest point map* onto a closed convex set.

Let $\mathcal{C}_V$ denote the polyhedral decomposition of $\mathbb{R}^n/\mathbb{R}\mathbf{1}$ defined by the various types of $\mathbf{x}$. This is the common refinement of the decompositions given by the r max-plus hyperplanes centered at the points of V. If $S = (S_1, \ldots, S_n)$ and $T = (T_1, \ldots, T_n)$ are n-tuples of subsets of $\{1, 2, \ldots, r\}$, then we write $S \subseteq T$ if $S_j \subseteq T_j$ for $j = 1, \ldots, n$. With this notation, the closed cell X_S of $\mathcal{C}_V$ indexed by the type S is

$$(5.2.5) \qquad \begin{aligned} X_S &= \{\mathbf{x} \in \mathbb{R}^n/\mathbb{R}\mathbf{1} \,:\, S \subseteq \operatorname{type}(\mathbf{x})\} \\ &= \{\mathbf{x} \,:\, x_k - x_j \leq v_{ik} - v_{ij} \ \text{ for } 1 \leq j, k \leq n \text{ and } i \in S_j\}. \end{aligned}$$

The equality of these two sets is seen by unraveling the definitions of "type". The poset structure on the cells of $\mathcal{C}_V$ is given by reverse inclusion of types. The union of the inequalities for X_S and X_T is the inequalities for $S \cup T$, so

$$X_S \cap X_T \;=\; X_{S \cup T}.$$

Proposition 5.2.11. *A cell X_S of $\mathcal{C}_V$ is bounded if and only if $S_i \neq \emptyset$ for $i = 1, \ldots, n$. The union of these cells is the tropical polytope $P = \operatorname{tconv}(V)$.*

Proof. Consider any $\mathbf{x} \in \mathbb{R}^n/\mathbb{R}\mathbf{1}$, and let $S = \operatorname{type}(\mathbf{x})$. We have seen in the proof of the Tropical Farkas Lemma (Proposition 5.2.10) that $\mathbf{x}$ lies in P if and only if no S_j is empty, so it suffices to show that this is equivalent to the polyhedron X_S being bounded. Suppose $S_j \neq \emptyset$ for all $j = 1, \ldots, n$. Then for every j and k, we can find $i \in S_j$ and $m \in S_k$, which by (5.2.5) yield the inequalities $v_{mk} - v_{mj} \leq x_k - x_j \leq v_{ik} - v_{ij}$. This implies that each $x_k - x_j$ is bounded on X_S, and hence X_S is a bounded subset of $\mathbb{R}^n/\mathbb{R}\mathbf{1}$. Conversely, suppose S_j is empty. Then the only inequalities involving x_j are of the form $x_j - x_k \leq v_j - v_k$. Consequently, if $\mathbf{x}$ is in S_j, so is $\mathbf{x} - \mu \mathbf{e}_j$ for $\mu > 0$, where $\mathbf{e}_j$ is the jth basis vector. Therefore, in this case X_S is unbounded. $\square$

The set of bounded cells X_S is called the *tropical complex* generated by V. Proposition 5.2.11 states that this is a polyhedral decomposition of the tropical polytope $P = \operatorname{tconv}(V)$. We therefore denote the tropical complex by $\mathcal{C}_P$. Equivalently, $\mathcal{C}_P$ is the subcomplex of $\mathcal{C}_V$ consisting of all bounded cells. Different sets V may have the same tropical polytope P as their convex hull, but generate different tropical complexes; the decomposition of a tropical polytope depends on the chosen V. Thus, $\mathcal{C}_P$ still depends on V.

Example 5.2.12. Let $V = \{(0,0,0), (0,1,4), (0,3,7), (0,4,5)\}$. The tropical polytope $P = \operatorname{tconv}(V)$ is shown on the left of Figure 5.2.3, together with its tropical complex. Note that the tropical convex hull does not change if we add the point $(0,3,4)$ to V, since $(0,3,4) \in P$. However the tropical complex C_P does change. This is illustrated on the right of Figure 5.2.3. $\diamondsuit$

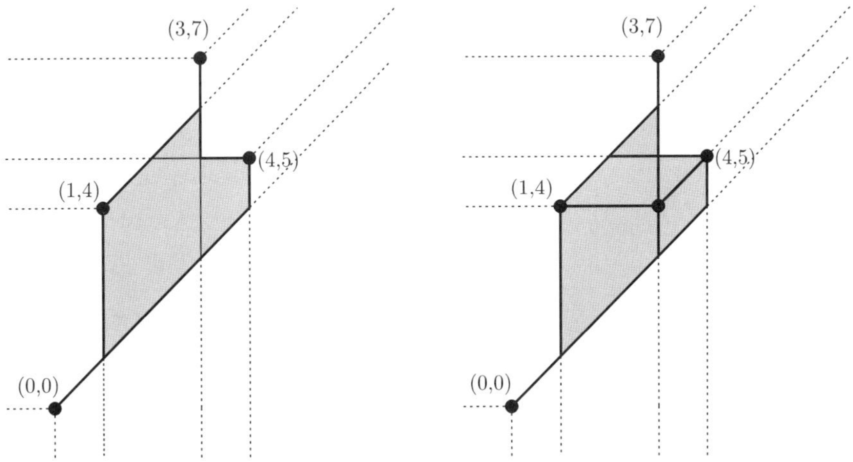

Figure 5.2.3. The two tropical complexes discussed in Example 5.2.12.

The next few results provide additional information about the classical convex polyhedron X_S in (5.2.5). Let G_S denote the undirected graph with vertices $1, \ldots, n$, where $\{j, k\}$ is an edge if and only if $S_j \cap S_k \neq \emptyset$. The polyhedron X_S in the next statement lives in the ambient space $\mathbb{R}^n/\mathbb{R}\mathbf{1}$, so its dimension d can be any integer between 0 and $n-1$.

Proposition 5.2.13. *The dimension d of the polyhedron X_S is one less than the number of connected components of G_S, and X_S is affinely and tropically isomorphic to some full-dimensional polyhedron X_T in $\mathbb{R}^{d+1}/\mathbb{R}\mathbf{1}$.*

Proof. We use induction on n. Suppose $i \in S_j \cap S_k$. Then X_S satisfies the linear equation $x_k - x_j = c$ where $c = v_{ik} - v_{ij}$. Projecting onto $\mathbb{R}^{n-1}/\mathbb{R}\mathbf{1}$ by eliminating the variable x_k, we find that X_S is isomorphic to X_T where the type T is defined by $T_\ell = S_\ell$ for $\ell \neq j$ and $T_j = S_j \cup S_k$. The cell X_T exists in the cell complex $\mathcal{C}_W$ induced by $W = \{\mathbf{w}_1, \ldots, \mathbf{w}_r\}$ with $w_{ij} = v_{ij}$ for $i \neq j$. The graph G_T is obtained from the graph G_S by contracting the edge $\{j, k\}$, and thus has the same number of connected components. Induction reduces us to the case where all of the S_j are pairwise disjoint. We must show that X_S has dimension $n-1$. Suppose not. Then X_S lies in $\mathbb{R}^n/\mathbb{R}\mathbf{1}$ but has dimension less than $n-1$. Therefore, one of the inequalities in (5.2.5) holds with equality: $x_k - x_j = v_{ik} - v_{ij}$ for all $\mathbf{x} \in X_S$. Since $x_k - x_j \leq v_{ik} - v_{ij}$ was one of the defining equalities, we have $i \in S_j$, and $x_l - x_j \leq v_{il} - v_{ij}$ for all $l \in \{1, \ldots, n\}$. Thus $x_l - x_k \leq v_{il} - v_{ik}$ for all l, and so $i \in S_k$. Hence S_j and S_k are not disjoint, which is a contradiction. $\square$

Proposition 5.2.14. *Any polytope in $\mathbb{R}^n/\mathbb{R}\mathbf{1}$ that is defined by inequalities $x_k - x_j \leq c_{jk}$ is a cell X_S in the complex $\mathcal{C}_V$ of some set $V = \{\mathbf{v}_1, \ldots, \mathbf{v}_n\}$.*

Proof. If c_{ij} does not appear in the given inequality presentation, then we get the same polytope by setting it to be a very large positive number. Define the vectors $\mathbf{v}_i$ to have coordinates $v_{ij} = c_{ij}$ for $i \neq j$, and $v_{ii} = 0$. By (5.2.5), the polytope defined by the inequalities $x_k - x_j \leq c_{jk}$ in $\mathbb{R}^n/\mathbb{R}\mathbf{1}$ is precisely the cell of type $(1, 2, \ldots, n)$ in the tropical convex hull of $\{\mathbf{v}_1, \ldots, \mathbf{v}_n\}$. □

Lemma 5.2.15. *Every bounded cell X_S in the tropical complex generated by V is itself a tropical polytope, equal to the tropical convex hull of its vertices.*

Proof. By Proposition 5.2.13, if X_S has dimension d, it is affinely and tropically isomorphic to a cell in the convex hull of a set of points in $\mathbb{R}^{d+1}/\mathbb{R}\mathbf{1}$, so it suffices to consider the full-dimensional case. The presentation (5.2.5) shows that X_S is tropically convex for all S. Therefore, it suffices to show that X_S is contained in the tropical convex hull of its vertices.

All proper faces of X_S are polytopes X_T of lower dimension, and, by induction on d, are contained in the tropical convex hull of their vertices. These vertices are among the vertices of X_S, and so each face is in the tropical convex hull. Take any point $\mathbf{x} = (x_1, \ldots, x_n)$ in the interior of X_S. We can travel in any direction from $\mathbf{x}$ while remaining in X_S. Let us travel in the $(1, 0, \ldots, 0)$ direction, until we hit the boundary, to obtain points $\mathbf{y}_1 = (x_1 + b, x_2, \ldots, x_n)$ and $\mathbf{y}_2 = (x_1 - c, x_2, \ldots, x_n)$ in the boundary of X_S. These points are in the tropical convex hull by the induction hypothesis, which means that $\mathbf{x} = \mathbf{y}_1 \oplus c \odot \mathbf{y}_2$ is also in the tropical convex hull. □

Each bounded cell X_S is a polytope both in the ordinary sense and in the tropical sense. Such objects were named *polytropes* by Joswig and Kulas [**JK10**]. By [**DS04**, Prop. 19], the number of vertices of a polytrope X_S is at most $\binom{2n-2}{n-1}$. This bound is tight for all n. For instance, the number of vertices of a three-dimensional polytrope is at most $\binom{2\cdot 4-2}{4-1} = 20$.

Proposition 5.2.16. *If P and Q are tropical polytopes in $\mathbb{R}^n/\mathbb{R}\mathbf{1}$, then $P \cap Q$ is also a tropical polytope.*

Proof. Since P and Q are tropically convex, so is $P \cap Q$. We must find a finite subset of $P \cap Q$ whose tropical convex hull is $P \cap Q$. By Proposition 5.2.11, P and Q are finite unions of bounded cells $\{X_S\}$ and $\{X_T\}$ respectively, so $P \cap Q$ is the finite union of the cells $X_S \cap X_T$. Consider any $X_S \cap X_T$. Using (5.2.5) to obtain the inequality representations of X_S and X_T, we see that this polyhedron has the form in Proposition 5.2.14. It is thus a cell X_W in some tropical complex. By Lemma 5.2.15, we can find a finite set of points whose convex hull is equal to $X_W = X_S \cap X_T$. Taking the union of these sets over all choices of S and T gives the desired finite subset of $P \cap Q$. □

Proposition 5.2.17. *Let P be a tropical polytope in $\mathbb{R}^n/\mathbb{R}\mathbf{1}$. Then there exists a unique minimal set V such that $P = \text{tconv}(V)$.*

Proof. Suppose that P has two minimal generating sets, $V = \{\mathbf{v}_1, \ldots, \mathbf{v}_m\}$ and $W = \{\mathbf{w}_1, \ldots, \mathbf{w}_r\}$. Write each element of W as $\mathbf{w}_i = \bigoplus_{j=1}^m c_{ij} \odot \mathbf{v}_j$. We claim that $V \subseteq W$. Consider $\mathbf{v}_1 \in V$ and write

$$(5.2.6)\quad \mathbf{v}_1 \;=\; \bigoplus_{i=1}^{r} d_i \odot \mathbf{w}_i \;=\; \bigoplus_{j=1}^{m} f_j \odot \mathbf{v}_j \qquad \text{where } f_j = \min_i(d_i + c_{ij}).$$

If the term $f_1 \odot \mathbf{v}_1$ does not minimize any coordinate in the right-hand side of (5.2.6), then $\mathbf{v}_1$ is a combination of $\mathbf{v}_2, \ldots, \mathbf{v}_m$, contradicting the minimality of V. However, if $f_1 \odot \mathbf{v}_1$ minimizes any coordinate in this expression, it must minimize all of them, since $(f_1 \odot \mathbf{v}_1)_j = v_{1j}$ means that $f_1 + v_{1j} = v_{1j}$, and so $f_1 = 0$. Pick any i for which $f_1 = d_i + c_{i1}$; we claim that $\mathbf{w}_i = c_{i1} \odot \mathbf{v}_1$. Indeed, if any other term in $\mathbf{w}_i = \bigoplus_{j=1}^m c_{ij} \odot \mathbf{v}_j$ contributed nontrivially to $\mathbf{w}_i$, that term would also contribute to the expression on the right-hand side of (5.2.6), which is a contradiction. Consequently, $V \subseteq W$, which means $V = W$ since both sets are minimal by hypothesis. $\square$

We now come to the connection to the product of simplices $\Delta_{n-1} \times \Delta_{r-1}$ that was promised at the beginning of this section. Every configuration $V = \{\mathbf{v}_1, \ldots, \mathbf{v}_r\}$ in $\mathbb{R}^n/\mathbb{R}\mathbf{1}$ specifies a tropical complex $\mathcal{C}_P$, which is a subdivision of the tropical polytope $P = \text{tconv}(V)$. Each cell in $\mathcal{C}_P$ is labeled by its type, which is an n-tuple of subsets of $\{1, \ldots, r\}$. Two configurations have the same *combinatorial type* if the types occurring in their tropical complexes are identical. Since X_T is a face of X_S if and only if $S \subseteq T$, this implies that the face posets of the tropical complexes are isomorphic.

Let $W = \mathbb{R}^{r+n}/\mathbb{R}(1, \ldots, 1, -1, \ldots, -1)$. The coordinates on W are denoted $(\mathbf{y}, \mathbf{z}) = (y_1, \ldots, y_r, z_1, \ldots, z_n)$. Consider the unbounded polyhedron

$$(5.2.7)\quad \mathcal{P}_V \;=\; \big\{\, (\mathbf{y}, \mathbf{z}) \in W \;:\; y_i + z_j \leq v_{ij} \;\text{ for } 1 \leq i \leq r \text{ and } 1 \leq j \leq n \,\big\}.$$

The connection to products of simplices will arise because the cone over $\Delta_{n-1} \times \Delta_{r-1}$ is dual to the recession cone of $\mathcal{P}_V$, as defined in (3.5.1). We first show that the tropical complex of V appears in the boundary of $\mathcal{P}_V$.

Lemma 5.2.18. *There is a piecewise-linear isomorphism between the tropical complex generated by V and the complex of bounded faces of the $(r+n-1)$-dimensional polyhedron $\mathcal{P}_V$. The image of a cell X_S of $\mathcal{C}_P$ under this isomorphism is the bounded face $\{y_i + z_j = v_{ij} \,:\, i \in S_j\}$ of $\mathcal{P}_V$. That bounded face maps isomorphically to X_S via projection onto the z-coordinates.*

Proof. Let F be a bounded face of $\mathcal{P}_V$, and define a subset S_j of $\{1, \ldots, r\}$ via $i \in S_j$ if $y_i + z_j = v_{ij}$ is valid on all of F. If some y_i or z_j appears in no equality, then we can subtract arbitrary positive multiples of that basis vector to obtain elements of F, contradicting the assumption that F is bounded. Therefore, each i appears in some S_j, and each S_j is nonempty.

Since every y_i appears in some equality, given a specific $\mathbf{z}$ in the projection of F onto the z-coordinates, there exists a unique $\mathbf{y}$ for which $(\mathbf{y},\mathbf{z}) \in F$, so this projection is an affine isomorphism from F to its image. We need to show that this image is equal to X_S. Let $\mathbf{z}$ be a point in the image of this projection, coming from a point $(\mathbf{y},\mathbf{z})$ in the relative interior of F. We claim that $\mathbf{z} \in X_S$. Indeed, looking at the jth coordinate of $\mathbf{z}$, we find that

$$\begin{array}{rcll} -y_i + v_{ij} & \geq & z_j & \text{for all } i, \\ -y_i + v_{ij} & = & z_j & \text{for } i \in S_j. \end{array} \tag{5.2.8}$$

The defining inequalities of X_S are $x_k - x_j \leq v_{ik} - v_{ij}$ with $i \in S_j$. Subtracting the inequality $-y_i + v_{ik} \geq z_k$ from the equality in (5.2.8) yields that this inequality is valid on $\mathbf{z}$ as well. Therefore, $\mathbf{z} \in X_S$. Similar reasoning shows that $S = \text{type}(\mathbf{z})$. We note that the relations (5.2.8) can be rewritten in terms of the tropical product of a row vector and a matrix:

$$\mathbf{z} \;=\; (-\mathbf{y}) \odot V \;=\; \bigoplus_{i=1}^{r} (-y_i) \odot \mathbf{v}_i, \tag{5.2.9}$$

where V is the $(n \times r)$-matrix with columns the $\mathbf{v}_i$. Conversely, suppose $\mathbf{z} \in X_S$. We define $\mathbf{y} = V \odot (-\mathbf{z})$. This means that

$$y_i \;=\; \min(v_{i1} - z_1, v_{i2} - z_2, \ldots, v_{in} - z_n). \tag{5.2.10}$$

We claim that $(\mathbf{y},\mathbf{z}) \in F$. Indeed, we certainly have $y_i + z_j \leq v_{ij}$ for all i and j, so $(\mathbf{y},\mathbf{z}) \in \mathcal{P}_V$. Furthermore, when $i \in S_j$, we know that $v_{ij} - z_j$ achieves the minimum in the right-hand side of (5.2.10), so that $v_{ij} - z_j = y_i$ and $y_i + z_j = v_{ij}$ is satisfied. Consequently, $(\mathbf{y},\mathbf{z}) \in F$ as desired.

It follows that the two complexes are isomorphic: if F is a face corresponding to X_S and G is a face corresponding to X_T, where S and T are both types, then X_S is a face of X_T if and only if $T \subseteq S$. By the discussion above, this is equivalent to saying that the equalities satisfied by G are a subset of the equations satisfied by F. (The former correspond to T, and the latter correspond to S.) Equivalently, F is a face of G. So X_S is a face of X_T if and only if F is a face of G, which establishes the assertion. □

The boundary complex of the polyhedron $\mathcal{P}_V$ is dual to the regular subdivision of the product of simplices $\Delta_{r-1} \times \Delta_{n-1}$ defined by the weights v_{ij}. We denote this regular polyhedral subdivision by $(\partial\mathcal{P}_V)^*$. Explicitly, a subset of vertices $(\mathbf{e}_i, \mathbf{e}_j)$ of $\Delta_{r-1} \times \Delta_{n-1}$ forms a cell of $(\partial\mathcal{P}_V)^*$ if and only if the equations $y_i + z_j = v_{ij}$ indexed by these vertices specify a face of the polyhedron $\mathcal{P}_V$. Note that $\mathcal{P}_V$ is a simple polyhedron if and only if $(\partial\mathcal{P}_V)^*$ is a triangulation of $\Delta_{r-1} \times \Delta_{n-1}$. We return to this generic situation in Theorem 5.2.22. First, however, we derive the main result of Section 5.2.

Theorem 5.2.19. *The combinatorial types of tropical complexes generated by configurations of r points in $\mathbb{R}^n/\mathbb{R}\mathbf{1}$ are in natural bijection with the regular polyhedral subdivisions of the product of two simplices $\Delta_{n-1} \times \Delta_{r-1}$.*

Proof. The poset of bounded faces of $\mathcal{P}_V$ is anti-isomorphic to the poset of interior cells of the subdivision $(\partial\mathcal{P}_V)^*$ of $\Delta_{r-1} \times \Delta_{n-1}$. Since every full-dimensional cell of $(\partial\mathcal{P}_V)^*$ is interior, the subdivision is uniquely determined by its interior cells. Hence, the combinatorial type of $\mathcal{P}_V$ is determined by the lists of facets containing each bounded face of $\mathcal{P}_V$. These lists are precisely the types of V in the tropical complex $\mathcal{C}_P$ by Lemma 5.2.18. □

Remark 5.2.20. Subdivisions of a product of simplices are related to numerous other topics in geometric combinatorics, and there are many different ways to represent such subdivisions. For further reading in this area see Section 6.2 in the textbook [**DRS10**], and the research articles [**AD09, San05**].

Theorem 5.2.19, which establishes a bijection between the tropical complexes generated by r points in $\mathbb{R}^n/\mathbb{R}\mathbf{1}$ and the regular subdivisions of a product of simplices $\Delta_{r-1} \times \Delta_{n-1}$, has some striking consequences. Notably, in the tropical world, *the row span and the column span of any matrix are equal.* This important fact is made precise in the following theorem.

Theorem 5.2.21. *Given any matrix $M \in \mathbb{R}^{r\times n}$, the tropical complex generated by its column vectors is isomorphic to the tropical complex generated by its row vectors. This isomorphism is obtained by restricting the piecewise linear maps $\mathbb{R}^n \to \mathbb{R}^r$, $\mathbf{z} \mapsto M \odot (-\mathbf{z})$ and $\mathbb{R}^r \to \mathbb{R}^n$, $\mathbf{y} \mapsto (-\mathbf{y}) \odot M$.*

Proof. By Theorem 5.2.19, the matrix M corresponds via the polyhedron $\mathcal{P}_M$ to a regular subdivision of $\Delta_{r-1} \times \Delta_{n-1}$, and the complex of interior faces of this regular subdivision is combinatorially isomorphic to both the tropical complex generated by its row vectors, which are r points in $\mathbb{R}^n/\mathbb{R}\mathbf{1}$, and the tropical complex generated by its column vectors, which are n points in $\mathbb{R}^r/\mathbb{R}\mathbf{1}$. Furthermore, Lemma 5.2.18 tells us that each bounded face of $\mathcal{P}_M$ is affinely isomorphic to its corresponding cell in both tropical complexes. Finally, in the proof of Lemma 5.2.18, we showed that any point $(\mathbf{y}, \mathbf{z})$ in a bounded face F of $\mathcal{P}_M$ satisfies $\mathbf{y} = M \odot (-\mathbf{z})$ and $\mathbf{z} = (-\mathbf{y}) \odot M$. This point projects to $\mathbf{y}$ and $\mathbf{z}$, and so the piecewise-linear isomorphism mapping these two complexes to each other is defined by the stated maps. □

Theorem 5.2.21 gives a natural bijection between the combinatorial types of tropical convex hulls of r points in $(n-1)$-space and those of tropical convex hulls of n points in $(r-1)$-space. We now discuss the generic case when the subdivision $(\partial\mathcal{P}_V)^*$ is a regular triangulation of $\Delta_{r-1} \times \Delta_{n-1}$.

Theorem 5.2.22. *For a configuration V of r points in $\mathbb{R}^n/\mathbb{R}\mathbf{1}$ with $r \geq n$ the following three conditions are equivalent.*

(1) *The regular subdivision $(\partial\mathcal{P}_V)^*$ is a triangulation of $\Delta_{r-1} \times \Delta_{n-1}$.*
(2) *No k of the points in V project into a tropical hyperplane inside a $(k-1)$-dimensional coordinate subspace, for any $2 \leq k \leq n$.*
(3) *No $k \times k$-submatrix of the $r \times n$-matrix (v_{ij}) is tropically singular, i.e., is in the tropical hypersurface of the determinant, for $2 \leq k \leq n$.*

Proof. The last equivalence follows from Proposition 5.1.7. We shall prove that (1) and (3) are equivalent. The tropical determinant of a $k \times k$-matrix M is the tropical polynomial $\bigoplus_{\sigma \in S_k} (\bigodot_{i=1}^k M_{i\sigma(i)})$. The matrix M is tropically singular if the minimum $\min_{\sigma \in S_k} (\sum_{i=1}^k M_{i\sigma(i)})$ is achieved twice.

The subdivision $(\partial\mathcal{P}_V)^*$ is a triangulation if and only if the polyhedron $\mathcal{P}_V$ is *simple*, i.e., no $r+n$ of the facets $y_i + z_j \leq v_{ij}$ meet at a single vertex. For each vertex $\mathbf{v}$, consider the bipartite graph $G_{\mathbf{v}}$ on $\{y_1, \ldots, y_n, z_1, \ldots, z_r\}$ with an edge connecting y_i and z_j if $\mathbf{v}$ lies on the corresponding facet. This graph is connected, since each y_i and z_j appears in some such inequality, and thus it will have a cycle if and only if it has at least $r+n$ edges. Consequently, $\mathcal{P}_V$ is not simple if and only if there exists some $G_{\mathbf{v}}$ with a cycle.

If there is a cycle, without loss of generality it is $y_1, z_1, y_2, z_2, \ldots, y_k, z_k$. Consider the submatrix M of (v_{ij}) given by $1 \leq i, j \leq k$. We have $y_1 + z_1 = M_{11}$, $y_2 + z_2 = M_{22}$, and so on, and also $z_1 + y_2 = M_{12}, \ldots, z_k + y_1 = M_{k1}$. Adding up these equalities yields $y_1 + \cdots + y_k + z_1 + \cdots + z_k = M_{11} + \cdots + M_{kk} = M_{12} + \cdots + M_{k1}$. Consider any element σ in the symmetric group S_k. Since $M_{i\sigma(i)} = v_{i\sigma(i)} \geq y_i + z_{\sigma(i)}$, we have $\sum M_{i\sigma(i)} \geq x_1 + \cdots + x_k + y_1 + \cdots + y_k$. Consequently, the permutations equal to the identity and to $(12 \cdots k)$ simultaneously minimize the determinant of the minor M. This logic is reversible, proving the equivalence of (1) and (3). $\square$

If the r points of V are in general position, the tropical complex they generate is a *generic tropical complex*. Such a tropical complex is dual to the cocomplex of interior faces in a regular triangulation of $\Delta_{r-1} \times \Delta_{n-1}$.

Corollary 5.2.23. *All generic tropical complexes generated by r points in $\mathbb{R}^n/\mathbb{R}\mathbf{1}$ have the same number of k-dimensional faces. That number equals*

$$\binom{r+n-k-2}{r-k-1, n-k-1, k} = \frac{(r+n-k-2)!}{(r-k-1)! \cdot (n-k-1)! \cdot k!}.$$

Proof. By Theorem 5.2.22, these objects are in bijection with regular triangulations of $P = \Delta_{r-1} \times \Delta_{n-1}$. The polytope P is unimodular, which means that all simplices formed by vertices of P are unimodular. This property implies that all triangulations of P have the same f-vector. The number of faces of dimension k of the tropical complex generated by given r points is the number of interior faces of codimension k in the corresponding triangulation. Since all triangulations of P have the same f-vector, they also have the same interior f-vector, which is obtained by (alternatingly) subtracting off the f-vectors of the induced triangulations on the proper faces of P. These proper faces are products of simplices, so all of these induced triangulations have their f-vectors independent of the original triangulation as well.

To compute this number, we consider the particular tropical complex given by $\mathbf{v}_i = (i, 2i, \ldots, ni)$ for $1 \leq i \leq r$. By Proposition 5.2.11, to count the faces of dimension k, we enumerate the types with k degrees of freedom. Consider any index i. We claim that for any $\mathbf{x}$ in $\operatorname{tconv}\{\mathbf{v}_1, \ldots, \mathbf{v}_r\}$, the set $\{j \mid i \in S_j\}$ is an interval I_i, and that if $i < m$, the intervals I_m and I_i meet in at most one point. This point is the largest element of I_m and the smallest element of I_i. Suppose $i \in S_j$ and $m \in S_l$ with $i < m$. Then $v_{ij} - x_j \leq v_{il} - x_l$ and $v_{ml} - x_l \leq v_{mj} - x_j$. Adding these inequalities yields $v_{ij} + v_{ml} \leq v_{il} + v_{mj}$, or $ij + ml \leq il + mj$. Since $i < m$, it follows that we must have $l \leq j$. Therefore, we can never have $i \in S_j$ and $m \in S_l$ with $i < m$ and $j < l$. The claim follows since the I_i cover $\{1, \ldots, n\}$.

The number of degrees of freedom of an interval set $(I_1, \ldots, I_r)$ is easily seen to be the number of indices i for which I_i and I_{i+1} are disjoint. Given this, it follows from a combinatorial counting argument that the number of interval sets with k degrees of freedom is the given multinomial coefficient. Finally, a representative for every interval set is given by $x_j = x_{j+1} - c_j$, where if S_j and S_{j+1} have an element i in common (they can have at most one), $c_j = i$, and if not, then $c_j = (\min(S_j) + \max(S_{j+1}))/2$. Therefore, each interval set is in fact a valid type, and our enumeration is complete. □

Theorem 5.2.22 implies that the number of combinatorially distinct generic tropical complexes given by r points in $\mathbb{R}^n/\mathbb{R}\mathbf{1}$ equals the number of regular triangulations of $\Delta_{r-1} \times \Delta_{n-1}$. The number of symmetry classes under the natural action of the product of symmetric groups $S_r \times S_n$ on both spaces is also the same. The symmetries in S_r correspond to permuting the points in a tropical polytope, while those in S_n correspond to permuting the coordinates. Of course, these two are dual by Theorem 5.2.21. The number of symmetry classes of regular triangulations of the polytope $\Delta_{r-1} \times \Delta_{n-1}$ is computable via Jörg Rambau's TOPCOM [**Ram02**] for small r and n:

	3	4	5	6
3	5	35	530	13621
4	35	7869		

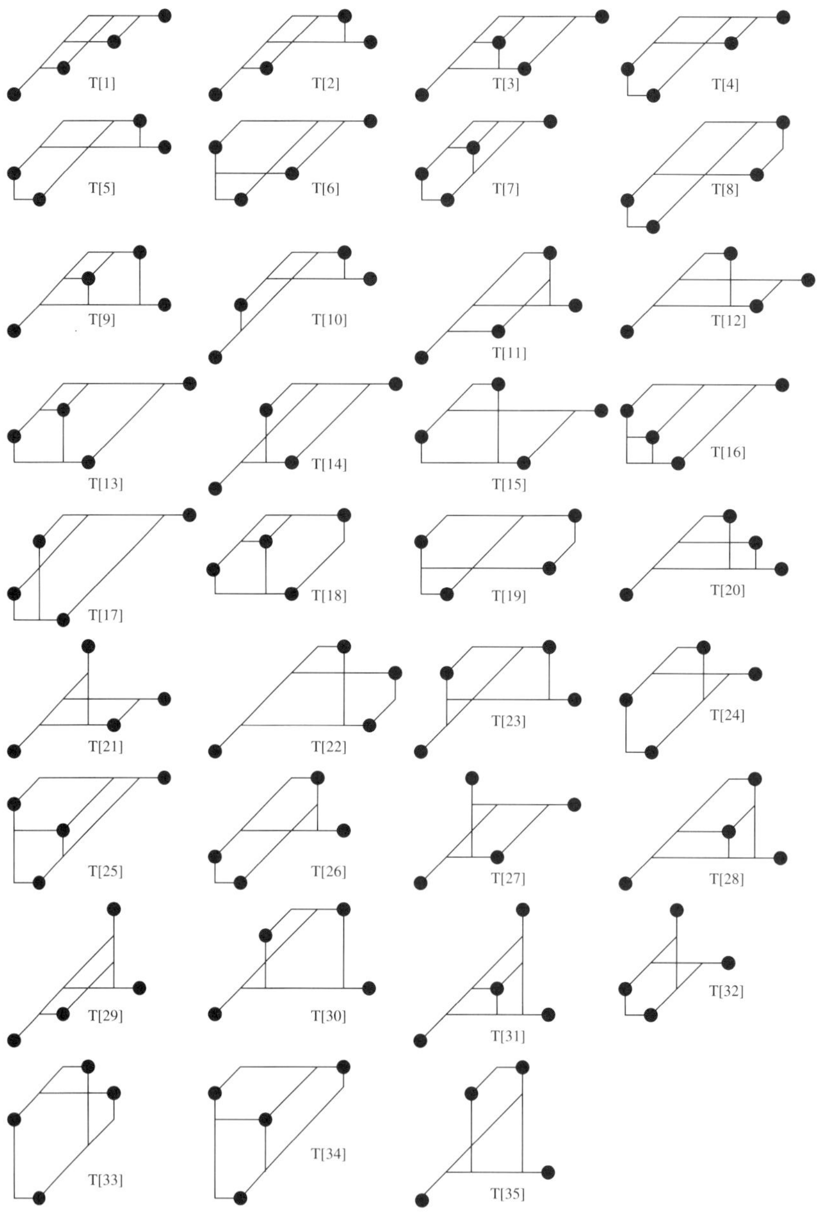

Figure 5.2.4. The 35 classes of tropical quadrilaterals in the plane.

Example 5.2.24. Let $n=3$ and $r=4$. The $(3,4)$ entry of the table above says that the five-dimensional polytope $\Delta_2\times\Delta_3$ has 35 symmetry classes of regular triangulations. These determine 35 combinatorial types of four-point configurations in $\mathbb{R}^3/\mathbb{R}\mathbf{1}$, or 35 combinatorial types of three-point configurations in $\mathbb{R}^4/\mathbb{R}\mathbf{1}$. These are shown in Figure 5.2.4 with the $\mathcal{C}_P$ they generate.

Each generic tropical complex $\mathcal{C}_P$ has ten vertices, 12 edges and three polygons. This is consistent with the formula in Corollary 5.2.23 for $k = 0, 1, 2$. ◊

We close this section with a "take-home message" from Theorem 5.2.21. In tropical geometry, the row span and the column span of any matrix can be naturally identified. That span is the tropical complex. Example 5.2.24 concerns matrices of format 3×4. The three row vectors live in $\mathbb{R}^4/\mathbb{R}\mathbf{1}$. The four column vectors live in $\mathbb{R}^3/\mathbb{R}\mathbf{1}$. They have the same tropical complex. Tropical triangles in 3-space are identified with tropical quadrilaterals in the plane. Figure 5.2.4 exhibits all the combinatorial possibilities.

5.3. The Rank of a Matrix

The rank of a matrix M is one of the most basic notions in linear algebra. It can be defined in many different ways. In particular, the following three definitions are equivalent in classical linear algebra over a field:

- The *rank* of M is the smallest positive integer r for which M can be written as the sum of r rank 1 matrices. A matrix has *rank* 1 if it is the product of a column vector and a row vector.
- The *rank* of M is the dimension of the column space. This is the smallest dimension of any linear space containing the columns of M.
- The *rank* of M is the largest positive integer r such that M has a nonsingular $r \times r$ submatrix.

In this section we examine these familiar definitions in the setting of the tropical semiring. We work in $\mathbb{R}^d$ and in the vector space $\mathbb{R}^{d\times n}$ of real $d \times n$-matrices, with the operations of addition and matrix multiplication defined tropically. All three of our definitions of matrix rank still make sense:

Definition 5.3.1. The *Barvinok rank* of a matrix $M \in \mathbb{R}^{d\times n}$ is the smallest integer r for which M can be written as the tropical sum of r rank 1 matrices. Here, we say that a $d\times n$-matrix has *rank* 1 if it is the tropical matrix product of a $d \times 1$-matrix and a $1 \times n$-matrix.

Definition 5.3.2. The *Kapranov rank* of a matrix $M \in \mathbb{R}^{d\times n}$ is the smallest integer r for which there exists a field K and a linear subspace of K^d of dimension r whose tropicalization contains the columns of M. Here K can be any field with a valuation. The attribution is a pointer to Kapranov's Theorem 3.1.3, and hence also to the Fundamental Theorem 3.2.3.

Definition 5.3.3. The *tropical rank* of a matrix $M \in \mathbb{R}^{d\times n}$ is the largest integer r such that M has a tropically nonsingular $r \times r$ submatrix. Recall from Sections 1.3 and 5.1 that a square matrix $M = (m_{ij}) \in \mathbb{R}^{r\times r}$ is *tropically singular* if the minimum in the evaluation of the tropical determinant

$$\bigoplus_{\sigma\in S_r} m_{1\sigma_1} \odot m_{2\sigma_2} \odot \cdots \odot m_{r\sigma_r} = \min(\, m_{1\sigma_1} + m_{2\sigma_2} + \cdots + m_{r\sigma_r} \,:\, \sigma \in S_r \,)$$

is attained at least twice. Here S_r denotes the symmetric group on $\{1, 2, \ldots, r\}$.

The terms *Barvinok rank*, *Kapranov rank*, and *tropical rank* were coined in [**DSS05**]. Several other notions of matrix rank are known in the literature on tropical linear algebra. In this section we limit ourselves to those in Definitions 5.3.1, 5.3.2, and 5.3.3. These are related as follows:

Theorem 5.3.4. *For every matrix M with entries in the tropical semiring,*

$$\text{(5.3.1)}\quad \textit{tropical rank}(M) \;\leq\; \textit{Kapranov rank}(M) \;\leq\; \textit{Barvinok rank}(M).$$

Both of these inequalities can be strict.

The proof of Theorem 5.3.4 consists of Propositions 5.3.15, 5.3.16, 5.3.19, and Theorem 5.3.21. As we go along, several alternative characterizations of the Barvinok, Kapranov, and tropical ranks will be offered. We shall use the fact that every $d \times n$-matrix M defines a tropically linear map $\mathbb{R}^n \to \mathbb{R}^d$. By Section 5.2, the image of M is a tropical polytope in $\mathbb{R}^d/\mathbb{R}\mathbf{1}$. We shall see that the tropical rank of M is the dimension of this tropical polytope plus one. The discrepancy among Definitions 5.3.1, 5.3.2, and 5.3.3 reflects the distinction between tropical polytopes, tropicalized linear spaces, and tropical linear spaces. A connection to Section 2.6 arises from the question whether the $(r+1) \times (r+1)$-minors of a matrix form a tropical basis. That question was answered by Shitov [**Shi13**], following earlier work of Chan, Jensen, and Rubei [**CJR11**]. This is featured in Theorem 5.3.25.

We start out by examining the Barvinok rank (Definition 5.3.1). This notion of rank arose in the context of combinatorial optimization. Barvinok, Johnson, Woeginger, and Woodroofe [**BJWW98**], building on earlier work of Barvinok, showed that for fixed r the Traveling Salesman Problem can be solved in polynomial time if the distance matrix is the tropical sum of r matrices of tropical rank 1 (with $\oplus$ as "max" instead of "min"). This motivates the definition and nomenclature of Barvinok rank as the smallest r for which $M \in \mathbb{R}^{d\times n}$ is expressible in this fashion. Since matrices of tropical rank 1 are of the form $X \odot Y^T$, for two column vectors $X \in \mathbb{R}^d$ and $Y \in \mathbb{R}^n$, this is equivalent to saying that M has a representation

$$M \quad = \quad X_1 \odot Y_1^T \;\oplus\; X_2 \odot Y_2^T \;\oplus\; \cdots \;\oplus\; X_r \odot Y_r^T. \tag{5.3.2}$$

For example, here is a 3×3-matrix that has Barvinok rank 2:

$$M \;=\; \begin{pmatrix} 0 & 4 & 2 \\ 2 & 1 & 0 \\ 2 & 4 & 3 \end{pmatrix} \;=\; \begin{pmatrix} 0 \\ 2 \\ 2 \end{pmatrix} \odot (0,4,2) \,\oplus\, \begin{pmatrix} 3 \\ 0 \\ 3 \end{pmatrix} \odot (2,1,0). \tag{5.3.3}$$

This matrix also has both tropical rank and Kapranov rank 2. The column vectors lie on the tropical line in $\mathbb{R}^3/\mathbb{R}\mathbf{1}$ defined by $2 \odot x_1 \oplus 3 \odot x_2 \oplus 0 \odot x_3$.

We next present two reformulations of Barvinok rank: in terms of tropical convex hulls, as in Section 5.2, and via tropical matrix multiplication.

Proposition 5.3.5. *For a real $d \times n$-matrix M, the following are equivalent:*

(a) *M has Barvinok rank at most r.*

(b) *The columns of M lie in the tropical convex hull of r points in $\mathbb{R}^d/\mathbb{R}\mathbf{1}$.*

(c) *There are matrices $X \in \mathbb{R}^{d\times r}$ and $Y \in \mathbb{R}^{r\times n}$ such that $M = X \odot Y$. Equivalently, M lies in the image of tropical matrix multiplication:*

$$\phi_r \;:\; \mathbb{R}^{d\times r} \times \mathbb{R}^{r\times n} \to \mathbb{R}^{d\times n} \;,\; (X,Y) \mapsto X \odot Y. \tag{5.3.4}$$

Proof. Let $M_1, \ldots, M_n \in \mathbb{R}^d$ be the column vectors of M. Let $X_1, \ldots, X_r \in \mathbb{R}^d$ be the columns of an unknown matrix $X \in \mathbb{R}^{d\times r}$ and $Y_1, \ldots, Y_r \in \mathbb{R}^n$ the rows of a matrix $Y \in \mathbb{R}^{r\times n}$. Let Y_{ij} denote the jth coordinate of Y_i. The following three algebraic identities are easily seen to be equivalent:

(a) $M \;=\; X_1 \odot Y_1 \,\oplus\, X_2 \odot Y_2 \,\oplus\, \cdots \,\oplus\, X_r \odot Y_r$;

(b) $M_j \;=\; Y_{1j} \odot X_1 \,\oplus\, Y_{2j} \odot X_2 \,\oplus\, \cdots \,\oplus\, Y_{rj} \odot X_r$ for all $j = 1, \ldots, n$;

(c) $M = X \odot Y$.

Statement (b) says that each column vector of M lies in the tropical convex hull of $X_1, \ldots, X_r$. The entries of the matrix Y are the multipliers in that tropical convex combination. This shows that the three conditions (a), (b), and (c) in the statement of the proposition are equivalent. □

We next take a closer look at the polyhedral geometry of the map ϕ_r.

Proposition 5.3.6. *The tropical matrix multiplication map ϕ_r is piecewise-linear. Its domains of linearity form a fan in $\mathbb{R}^{d\times r} \times \mathbb{R}^{r\times n}$. This fan is the common refinement of the normal fans of dn simplices of dimension $r - 1$.*

Proof. Let $U = (u_{ij})$ and $V = (v_{jk})$ be matrices of indeterminates of size $d \times r$ and $r \times n$, respectively. The entries of the classical matrix product UV are the dn quadratic polynomials $u_{i1}v_{1k} + u_{i2}v_{2k} + \cdots + u_{ir}v_{rk}$. The

Newton polytope of each quadric is an $(r-1)$-dimensional simplex P_{ik}. The (i,k)-coordinate of ϕ_r takes a pair of matrices (X,Y) to the real number $\min(x_{i1}+y_{1k},\ldots,x_{ir}+y_{rk})$. This is linear on each cone of the normal fan of P_{ik}. Hence ϕ_r is linear on each cone of the common refinement of the normal fans of the simplices P_{ik}. □

Corollary 5.3.7. *If $r=2$, then the map ϕ_2 is piecewise-linear with respect to the regions in an arrangement of dn hyperplanes in $\mathbb{R}^{d\times 2}\times\mathbb{R}^{2\times n}$.*

Proof. If $r=2$, then each P_{ik} is a line segment, so its normal fan consists of two half-spaces separated by a hyperplane. The common refinement of these dn normal fans is a hyperplane arrangement. It follows from the previous proof that ϕ_2 is piecewise linear on that hyperplane arrangement. □

Example 5.3.8. Let $d=n=3$ and $r=2$. Then ϕ_2 is linear on the regions of an arrangement of nine hyperplanes in $\mathbb{R}^{12}=\mathbb{R}^{3\times 2}\times\mathbb{R}^{2\times 3}$. A computation reveals that this arrangement has 230 maximal cones. Matrix multiplication ϕ_2 maps each of these 230 cones linearly onto an eight-dimensional cone in $\mathbb{R}^{3\times 3}$. The image of ϕ_2 is contained in the set of tropically singular 3×3-matrices, but the containment is strict. For instance, the matrix C_3 in (5.3.5) below is tropically singular but it is not in the image of ϕ_2. ◇

By Proposition 5.3.5, the set of matrices of Barvinok rank at most r is the image of the map ϕ_r. This set is the support of a polyhedral fan in $\mathbb{R}^{d\times n}$, as in Example 5.3.8. The discrepancy between Barvinok rank and Kapranov rank can be explained by the following general fact of tropical algebraic geometry. For most polynomial maps, *the image of the tropicalization is strictly contained in the tropicalization of the image*; see Remark 3.2.14.

We next demonstrate that the Barvinok rank can be much larger than the other two notions of rank. The example we consider is the $n\times n$-matrix

$$C_n \quad = \quad \begin{pmatrix} 1 & 0 & 0 & \cdots & 0\\ 0 & 1 & 0 & \cdots & 0\\ 0 & 0 & 1 & \cdots & 0\\ \vdots & \vdots & \vdots & \ddots & \vdots\\ 0 & 0 & 0 & \cdots & 1\end{pmatrix}. \tag{5.3.5}$$

This looks like the identity matrix (in classical arithmetic) but it is not the identity matrix in tropical arithmetic. That honor belongs to the $n\times n$-matrix whose diagonal entries are 0 and whose off-diagonal entries are ∞.

Proposition 5.3.9. *The Barvinok rank of the matrix C_n in (5.3.5) is the smallest positive integer r such that*

$$n \leq \binom{r}{\lfloor \frac{r}{2}\rfloor}.$$

Proof. Let r be an integer, and assume that $n \leq \binom{r}{\lfloor r/2 \rfloor}$. We first show that Barvinok rank$(C_n) \leq r$. Let $S_1, \ldots, S_n$ be distinct subsets of $\{1, \ldots, r\}$ each having cardinality $\lfloor r/2 \rfloor$. For each $k \in \{1, \ldots, r\}$, we define an $n \times n$-matrix $X_k = (x_{ij}^k)$ with entries in $\{0, 1, 2\}$ as follows:

$$x_{ij}^k = 2 \text{ if } k \in S_i \backslash S_j\,, \quad x_{ij}^k = 0 \text{ if } k \in S_j \backslash S_i, \text{ and } \quad x_{ij}^k = 1 \text{ otherwise.}$$

The matrix X_k has tropical rank 1. To see this, let $V_k \in \{0,1\}^n$ denote the row vector with ith coordinate equal to 1 if $k \in S_i$ and 0 if $k \notin S_i$. We have

$$X_k \quad = \quad V_k^T \odot (1 \odot (-V_k)).$$

To prove Barvinok rank$(C_n) \leq r$, it now suffices to establish the identity

$$C_n \quad = \quad X_1 \oplus X_2 \oplus \cdots \oplus X_r.$$

Indeed, all diagonal entries of the matrices on the right-hand side are 1, and the off-diagonal entries of the right-hand side are $\min(x_{ij}^1, x_{ij}^2, \ldots, x_{ij}^r) = 0$, because $S_j \backslash S_i$ is nonempty for $i \neq j$.

To prove the converse direction, we consider an arbitrary representation

$$C_n \quad = \quad Y_1 \oplus Y_2 \oplus \cdots \oplus Y_r,$$

where the matrices $Y_k = (y_{ij}^k)$ have tropical rank 1. For each k we set $T_k := \{(i,j) : y_{ij}^k = 0\}$. Since C_n is nonnegative, the matrices Y_k are as well. As they also have tropical rank 1, each T_k is a product $I_k \times J_k$, where I_k and J_k are subsets of $\{1, \ldots, n\}$. Moreover, we have $I_k \cap J_k = \emptyset$ because the diagonal entries of Y_k are not 0. For each $i = 1, \ldots, n$, we set

$$S_i := \{k : i \in I_k\} \quad \subseteq \quad \{1, \ldots, r\}.$$

We claim that no two of the sets $S_1, \ldots, S_n$ are contained in one another. Sperner's Theorem [**AZ04**, Ch. 23] will then imply that $n \leq \binom{r}{\lfloor r/2 \rfloor}$. To prove the claim, observe that if $S_i \subset S_j$, then y_{ij}^k cannot be 0 for any k. Indeed, if $y_{ij}^k = 0$, then $j \in J_k$, so $j \notin I_k$. But this means that $k \notin S_j$, so $k \notin S_i$, which contradicts $i \in I_k$. □

Example 5.3.10. The matrix C_6 has Barvinok rank 4. The upper bound is shown by the following decomposition into matrices of tropical rank 1:

$$\begin{pmatrix} 1&0&0&0&0&0\\ 0&1&0&0&0&0\\ 0&0&1&0&0&0\\ 0&0&0&1&0&0\\ 0&0&0&0&1&0\\ 0&0&0&0&0&1 \end{pmatrix} = \begin{pmatrix} 1&1&1&2&2&2\\ 1&1&1&2&2&2\\ 1&1&1&2&2&2\\ 0&0&0&1&1&1\\ 0&0&0&1&1&1\\ 0&0&0&1&1&1 \end{pmatrix} \oplus \begin{pmatrix} 1&1&0&0&0&1\\ 1&1&0&0&0&1\\ 2&2&1&1&1&2\\ 2&2&1&1&1&2\\ 2&2&1&1&1&2\\ 1&1&0&0&0&1 \end{pmatrix}$$
$$\oplus \begin{pmatrix} 1&0&1&0&1&0\\ 2&1&2&1&2&1\\ 1&0&1&0&1&0\\ 2&1&2&1&2&1\\ 1&0&1&0&1&0\\ 2&1&2&1&2&1 \end{pmatrix} \oplus \begin{pmatrix} 1&2&2&2&1&1\\ 0&1&1&1&0&0\\ 0&1&1&1&0&0\\ 0&1&1&1&0&0\\ 1&2&2&2&1&1\\ 1&2&2&2&1&1 \end{pmatrix}.$$

Similarly, C_{36} has Barvinok rank 8, its 35×35 submatrices have Barvinok rank 7, and its 8×8 submatrices have Barvinok rank at most 5. These claims can be verified by combinatorial computations that are based on Propositions 5.3.9 and 5.3.17. Proposition 5.3.9 also gives the asymptotics

$$\text{Barvinok rank}\,(C_n) \quad \sim \quad \log_2 n.$$

We will see in Examples 5.3.14 and 5.3.18 that the Kapranov rank and tropical rank of the matrix C_n are both 2. ◊

Recall from Definition 5.3.2 that the Kapranov rank of a matrix $M \in \mathbb{R}^{d\times n}$ is the smallest dimension of any tropicalized linear space containing the columns of M. It is not apparent in this definition that the Kapranov rank of a matrix and its transpose are the same, but this follows from our next result. Let J_r denote the ideal in $K[x_{ij}^{\pm 1}]$ that is generated by all the $(r+1)\times(r+1)$-minors of a $d \times n$-matrix of indeterminates (x_{ij}). This is the prime ideal defining the determinantal variety $V(J_r)$, which consists of all $d \times n$-matrices with entries in K^* whose (classical) rank is at most r.

Theorem 5.3.11. *For any $M = (m_{ij}) \in \mathbb{R}^{d\times n}$ the following are equivalent.*

(a) *The Kapranov rank of M is at most r.*

(b) *The matrix M lies in the tropical determinantal variety* $\mathrm{trop}(V(J_r))$.

(c) *There is a $d\times n$-matrix $F = (f_{ij})$ with nonzero entries in some field K such that the rank of F is less than or equal to r and* $\mathrm{val}(f_{ij}) = m_{ij}$ *for all i and j. We write* $\mathrm{val}(F) = M$ *and call F a* lift *of M.*

Proof. The equivalence of (b) and (c) is the Fundamental Theorem 3.2.3 applied to the ideal J_r. Indeed, a matrix over K has rank at most r if and only it it lies in the determinantal variety $V(J_r)$. To see that (c) implies (a), consider the linear subspace V of K^d spanned by the columns of F. The tropicalization of $V^0 = V \cap (K^*)^d$ is a tropicalized linear space that contains all columns of $M = \mathrm{val}(F)$. Conversely, suppose that (a) holds. Let $L = \mathrm{trop}(V^0)$ be a tropicalized linear space of dimension r containing the columns of M. By applying the Fundamental Theorem 3.2.3 to V^0, we see that each column vector of M has a preimage in $V^0 \subset (K^*)^d$ under the valuation map. Let F be the $d \times n$-matrix over K whose columns are these preimages. Then the column space of F is contained in V^0, which is r dimensional by Theorem 3.3.5, so we have $\mathrm{rank}(F) \leq r$ and $\mathrm{val}(F) = M$. □

Corollary 5.3.12. *The Kapranov rank of a matrix $M \in \mathbb{R}^{d\times n}$ is the smallest rank of any lift of M in $K^{d\times n}$.*

Example 5.3.13. The following 3×3-matrix has rank 2 over $K = \mathbb{C}\{\{t\}\}$:

$$F \quad = \quad \begin{pmatrix} 1 & t^4 & t^2 \\ t^2 & t & 1 \\ t^2+t^5 & t^4+t^6 & t^3+t^4 \end{pmatrix}.$$

We have $\mathrm{val}(F) = M$, so F is a lift of the 3×3-matrix M in (5.3.3). □

The ideal J_1 is generated by the 2×2-minors $x_{ij}x_{kl} - x_{il}x_{kj}$ of the $d \times n$-matrix (x_{ij}). Therefore, a matrix of Kapranov rank 1 must satisfy the linear equations $m_{ij} + m_{kl} = m_{il} + m_{kj}$. This happens if and only if there exist real vectors $X = (x_1, \ldots, x_d)$ and $Y = (y_1, \ldots, y_n)$ with

$$m_{ij} = x_i + y_j \text{ for all } i,j \iff m_{ij} = x_i \odot y_j \text{ for all } i,j \iff M = X^T \odot Y.$$

Conversely, if such X and Y exist, we can lift M to a matrix of rank 1 by substituting $t^{m_{ij}}$ for m_{ij}. Therefore, a matrix M has Kapranov rank 1 if and only if it has Barvinok rank 1. In general, the Kapranov rank can be much smaller than the Barvinok rank, as the following example shows.

Example 5.3.14. Let $n \geq 3$, and consider the matrix C_n in (5.3.5). Since C_n does not have Kapranov rank 1, its Kapranov rank is least 2. Fix $K = \mathbb{C}\{\{t\}\}$ and distinct complex numbers $a_3, a_4, \ldots, a_n \in \mathbb{C}$. The matrix

$$F_n \quad = \quad \begin{pmatrix} t & 1 & t+a_3 & t+a_4 & \cdots & t+a_n \\ 1 & t & 1+a_3t & 1+a_4t & \cdots & 1+a_nt \\ t-a_3 & 1 & t & t-a_3+a_4 & \cdots & t-a_3+a_n \\ t-a_4 & 1 & t-a_4+a_3 & t & \cdots & t-a_4+a_n \\ \vdots & \vdots & \vdots & \vdots & \ddots & \vdots \\ t-a_n & 1 & t-a_n+a_3 & t-a_n+a_4 & \cdots & t \end{pmatrix}$$

has rank 2 because the ith column (for $i \geq 3$) equals the first column plus a_i times the second column. Since $\mathrm{val}(F_n) = C_n$, we conclude that C_n has Kapranov rank 2. The tropicalized plane containing the columns of C_n is $\mathrm{trop}(U_{2,n})$, where $U_{2,n}$ is the uniform matroid as in Example 4.2.13. ◊

We remark that Corollary 5.3.12 is usually not an effective method for computing the Kapranov rank of a matrix. Instead, it is preferable to use Theorem 5.3.11(b) and test membership in $\text{trop}(V(J_r))$ using Gröbner bases.

The following proposition establishes half of Theorem 5.3.4.

Proposition 5.3.15. *The Kapranov rank of any matrix $M \in \mathbb{R}^{d \times n}$ is less than or equal to the Barvinok rank of M, and this inequality can be strict.*

Proof. Fix a valued field K with an infinite residue field. Suppose that M has Barvinok rank r. Write $M = M_1 \oplus \cdots \oplus M_r$, where each M_i has Barvinok rank 1. Then M_i has Kapranov rank 1, so there exists a rank 1 matrix F_i over K with $\text{val}(F_i) = M_i$. By multiplying the matrices F_i by suitable scalars with valuation 0, we can choose F_i such that the sum $F = F_1 + \cdots + F_r$ satisfies $\text{val}(F) = M$. This uses the hypothesis that the residue field of K is infinite. The matrix F has rank at most r. Theorem 5.3.11 implies that M has Kapranov rank at most r. Example 5.3.14 shows that the inequality can be strict. □

Our next step is to prove the first inequality in Theorem 5.3.4.

Proposition 5.3.16. *The tropical rank of any matrix $M \in \mathbb{R}^{d \times n}$ is less than or equal to the Kapranov rank of M.*

Proof. If the matrix M has a tropically nonsingular $r \times r$-submatrix, then any lift of M must have the corresponding $r \times r$-submatrix nonsingular over the field K. Consequently, no lift of M to the field K can have rank less than r. By Theorem 5.3.11, the Kapranov rank of M must be at least r. □

We now present a combinatorial formula for the tropical rank of a zero-one matrix, or any matrix which has only two distinct entries. We define the *support* of a vector in tropical space $\mathbb{R}^d$ as the set of its 0 coordinates. We define the *support poset* of a matrix M to be the set of all unions of supports of column vectors of M. This set is partially ordered by inclusion.

Proposition 5.3.17. *The tropical rank of a zero-one matrix M with no column of all ones equals the maximum length of a chain in its support poset.*

Proof. We can assume that every union of supports of columns of M is the support of a column. Indeed, the tropical sum of a set of columns is a vector whose support is the union of supports, and appending that vector as a new column to M does not change the tropical rank. This follows from the fact that if a square matrix has a column that is a tropical sum of other columns, then that matrix is tropically singular. Therefore, if there is a chain of length r in the support poset, we may assume that there is a set of r columns with supports properly contained in one another. Since there is no column of ones, we can extract an $r \times r$-submatrix with zeros on and below the diagonal and ones above the diagonal. This is tropically nonsingular.

Conversely, suppose there is a tropically nonsingular $r \times r$-submatrix N. We claim that the support poset of N has a chain of length r, from which it follows that the support poset of M also has a chain of length r. Assume without loss of generality that the unique minimum permutation sum is obtained in the diagonal. This minimum sum cannot be more than one, because if n_{ii} and n_{jj} are both one, then changing them for n_{ij} and n_{ji} does not increase the sum. If the minimum is zero, orienting an edge from i to j if entry ij of N is zero yields an acyclic digraph, which admits an ordering. Rearranging the rows and columns according to this ordering yields a matrix with ones above the diagonal and zeros on and below the diagonal. The tropical sum of the last i columns then produces a vector with zeros exactly in the last i positions. Hence, there is a proper chain of supports of length r.

If the minimum permutation sum in N is one, then let n_{ii} be the unique diagonal entry equal to one. The ith row in N must consist of all ones: if n_{ij} is zero, then changing n_{ij} and n_{ji} for n_{ii} and n_{jj} does not increase the sum. Changing this row of ones to a row of zeros does not affect the support poset of N, and it yields a nonsingular zero-one matrix with minimum sum zero to which we can apply the argument in the previous paragraph. □

Example 5.3.18. The tropical rank of the matrix C_n in Proposition 5.3.9 equals 2, since all its 3×3-submatrix are tropically singular, while the principal 2×2-submatrices are not. The supports of its columns are all the sets of cardinality $n - 1$. The support poset consists of these and the whole set $\{1, \ldots, n\}$. The maximal chains in the poset have length 2. □

Matroid theory allows us to construct matrices whose tropical and Kapranov ranks disagree. To explain this approach, we need a definition. The *cocircuit matrix* of a matroid M, denoted $\mathbf{C}(M)$, has rows indexed by the elements of the ground set of M and columns indexed by the cocircuits of M, that is, the circuits of the *dual matroid* M^*. By definition, M^* is the matroid whose bases are the set complements of the bases of M. The matrix $\mathbf{C}(M)$ has a zero in entry (i, j) if the ith element is in the jth cocircuit, and a one otherwise. In other words, $\mathbf{C}(M)$ is the zero-one matrix whose columns have the cocircuits of M as supports. Here, as before, the support of a column is its set of zeros. As an example, the matrix C_n in Proposition 5.3.9 is the cocircuit matrix of the uniform matroid of rank 2 with n elements. Similarly, the cocircuit matrix of the uniform matroid $U_{n,r}$ has size $n \times \binom{n}{r-1}$ and its columns are all the zero-one vectors with exactly $r - 1$ ones. The following result shows that its tropical and Kapranov ranks equal r.

Proposition 5.3.19. *The tropical rank of the cocircuit matrix* $\mathbf{C}(M)$ *is the rank of the matroid* M.

Proof. The complements of the cocircuits are the *hyperplanes* of the matroid M. The hyperplanes are the maximal proper flats. The flats of M are the intersections of hyperplanes, so the complements of flats are unions of cocircuits. This implies that the support poset of $\mathbf{C}(M)$ is anti-isomorphic to the lattice of flats of M. The rank of M is the length of any maximal chain of flats. Equivalently, the rank of M is the maximum length of a chain in the support poset of $\mathbf{C}(M)$. Proposition 5.3.17 now implies the claim. □

Example 5.3.20. Let M be the Fano matroid in Figure 4.2.1. Its cocircuit matrix is the following 7×7-matrix, which has tropical rank 3:

$$\mathbf{C}(M) = \begin{pmatrix} 1 & 1 & 1 & 0 & 0 & 0 & 0 \\ 1 & 0 & 0 & 1 & 1 & 0 & 0 \\ 1 & 0 & 0 & 0 & 0 & 1 & 1 \\ 0 & 1 & 0 & 1 & 0 & 1 & 0 \\ 0 & 0 & 1 & 0 & 1 & 1 & 0 \\ 0 & 0 & 1 & 1 & 0 & 0 & 1 \\ 0 & 1 & 0 & 0 & 1 & 0 & 1 \end{pmatrix}. \tag{5.3.6}$$

Theorem 5.3.21. *The Kapranov rank of* $\mathbf{C}(M)$ *is equal to the rank of* M *if and only if the matroid* M *is realizable over some field.*

Proof. Let M be a matroid of rank r on $\{1,\ldots,d\}$ which has n cocircuits. We first prove the "only-if" direction. Suppose that $F \in K^{d\times n}$ is a rank-r lift of the cocircuit matrix $\mathbf{C}(M)$. For each row $\mathbf{f}_i$ of F, let $\mathbf{v}_i \in \Bbbk^d$ be its image in the residue field $\Bbbk$. We claim that $V = \{\mathbf{v}_1,\ldots,\mathbf{v}_d\}$ realizes M. First note that V has rank at most r since every K-linear relation among the vectors f_i translates into a $\Bbbk$-linear relation among the $\mathbf{v}_i$. Our claim says that $\{i_1,\ldots,i_r\}$ is a basis of M if and only if $\{\mathbf{v}_{i_1},\ldots,\mathbf{v}_{i_r}\}$ is a basis of $\Bbbk^d$. Suppose that $I = \{i_1,\ldots,i_r\}$ is a basis of M. Consider the $r\times r$-submatrix of $\mathbf{C}(M)$ whose rows are indexed by I and whose columns are indexed by the *basic cocircuits* for the basis I, i.e., the cocircuits of M that are disjoint from $I\backslash\{i_j\}$ for $j = 1,2,\ldots,r$. We can order the rows and columns of that $r\times r$-submatrix so that all entries one are strictly above the diagonal. The lifted submatrix over $\Bbbk$ is lower-triangular with nonzero entries along the diagonal. Hence $\mathbf{v}_{i_1},\ldots,\mathbf{v}_{i_r}$ are linearly independent, and, since $\operatorname{rank}(V) \le r$, they must be a basis. We also conclude that $\operatorname{rank}(V) = r$. If $I = \{i_1,\ldots,i_r\}$ is not a basis in M, then there exists a cocircuit disjoint from I; this means that some column of $\mathbf{C}(M)$ has all ones in rows $i_1,\ldots,i_r$. Therefore, $\mathbf{f}_{i_1},\ldots,\mathbf{f}_{i_r}$ all have positive valuation in that coordinate, which means that $\mathbf{v}_{i_1},\ldots,\mathbf{v}_{i_r}$ are all zero in that coordinate. Since the cocircuit is not empty, not all vectors $\mathbf{v}_j$ have an entry of zero in that coordinate, and so $\{\mathbf{v}_1, \ldots, \mathbf{v}_r\}$ cannot be a basis. This shows that V realizes M over $\Bbbk$, which proves the "only-if" direction.

For the "if" direction, we assume that M is realizable over some field $\Bbbk$. We may assume that $\Bbbk$ is infinite and we fix $K = \Bbbk(t)$ with its usual valuation. Fix a matrix $A \in \Bbbk^{d \times n}$ such that the rows of A realize M and the sets of nonzero coordinates along the columns of A are the cocircuits of M. We may assume that no row of A is zero. Suppose $\{1, \ldots, r\}$ is a basis of M, and let A' be the submatrix of A consisting of the first r rows. Write

$$A \;=\; \begin{pmatrix} \mathbf{I}_r \\ C \end{pmatrix} \cdot A',$$

where $\mathbf{I}_r$ is the identity matrix and $C \in \Bbbk^{(d-r)\times r}$. Since $\Bbbk$ is infinite, there exists an $r \times n$-matrix A'' with entries in $\Bbbk$ such that all entries of the product $\binom{\mathbf{I}_r}{C} \cdot A''$ are nonzero scalars in $\Bbbk$. We now define

$$F \;=\; \begin{pmatrix} \mathbf{I}_r \\ C \end{pmatrix} \cdot (A' + tA'') \quad \in K^{d \times n}.$$

This matrix has rank r and $\mathrm{val}(F) = \mathbf{C}(M)$. This completes the proof. □

Corollary 5.3.22. *Let M be a matroid that is not realizable over any field. Then the Kapranov rank of the matrix $\mathbf{C}(M)$ exceeds its tropical rank.*

This furnishes many examples of matrices whose Kapranov rank exceeds their tropical rank. For instance, if M is the non-Pappus matroid in Figure 4.7.1 then $\mathbf{C}(M)$ has tropical rank 3 and Kapranov rank 4. In Example 5.3.20, the Kapranov rank of $\mathbf{C}(M)$ is 3 because the Fano matroid M is realizable over the two-element field. A variant of the Kapranov rank is to fix the characteristic of the residue field. For example, the Kapranov rank of $\mathbf{C}(M)$ would be 4 if one restricts to only fields of characteristic 0.

One can also get examples where the difference between the two ranks is arbitrarily large. Indeed, given matrices A and B, we can construct the matrix

$$M := \begin{pmatrix} A & \infty \\ \infty' & B \end{pmatrix},$$

where ∞ and ∞' denote matrices of the appropriate dimensions and whose entries are sufficiently large. Appropriate choices of these large values will ensure that the tropical and Kapranov ranks of M are the sums of those of A and of B. For other constructions see [**Shi13**], and Exercise 5.6(10).

Recall from Definition 5.3.3 that the tropical rank of a matrix is the size of the largest nonsingular square submatrix. Another characterization is:

Theorem 5.3.23. *The tropical rank of a matrix $M \subset \mathbb{R}^{d \times n}$ equals 1 plus the dimension of the tropical convex hull in $\mathbb{R}^d/\mathbb{R}\mathbf{1}$ of the n columns of M.*

Proof. Let $V = \{\mathbf{v}_1, \ldots, \mathbf{v}_n\}$ be the set of columns of M and $P = \text{tconv}(V)$. Let r be the tropical rank of M: there exists a tropically nonsingular $r \times r$-submatrix M' of M, but all larger square submatrices are tropically singular. We first show that $\dim(P) \geq r - 1$. Deleting the rows outside M' means projecting P into $\mathbb{R}^r/\mathbb{R}\mathbf{1}$, and deleting the columns outside M' means passing to a tropical subpolytope P' of the image. Both operations can only decrease the dimension, so it suffices to show $\dim(P') \geq r - 1$. Hence, we can assume that M is itself a tropically nonsingular $r \times r$-matrix. Also, without loss of generality, we can assume that the minimum over $\sigma \in S_r$ of

$$f(\sigma) \quad = \quad \sum_{i=1}^{r} v_{\sigma(i),i} \tag{5.3.7}$$

is uniquely achieved when σ is the identity id. We now claim that the cell $X_{(\{1\},\ldots,\{r\})}$ in P' has dimension $r-1$. The inequalities defining this cell are $x_k - x_j \leq v_{jk} - v_{jj}$ for $j \neq k$. Suppose that this cell is not full dimensional. By Farkas's Lemma, some nonnegative linear combination of the inequalities $x_k - x_j \leq v_{jk} - v_{jj}$ has the form $0 \leq c$ for some nonpositive real c. This implies that some other $\sigma \in S_r$ has $f(\sigma) \leq f(\text{id})$, a contradiction.

For the converse, suppose $\dim(P) \geq r$. Pick a region X_S of dimension r. By Proposition 5.2.13, the graph G_S has $r+1$ connected components, so we can pick $r+1$ elements of $\{1, \ldots, n\}$ of which no two appear in a common S_j. Assume without loss of generality that this set is $\{1, \ldots, r+1\}$, so that $i \in S_j$ if and only if $i = j$, for $1 \leq i, j \leq r+1$. We claim that the square submatrix consisting of the first $r+1$ rows and columns of M is tropically nonsingular. Note that

$$f(\sigma) - f(\text{id}) \quad = \quad \sum_{i=1}^{r+1} v_{\sigma(i),i} - \sum_{i=1}^{r+1} v_{ii} \quad = \quad \sum_{i=1}^{r+1} \big(v_{\sigma(i),i} - v_{ii}\big).$$

Whenever $\sigma(i) \neq i$, we have $v_{\sigma(i),i} - v_{ii} > 0$ since $i \in S_i$ and $i \notin S_{\sigma(i)}$. Therefore, if σ is not the identity, we have $f(\sigma) - f(\text{id}) > 0$, and id is the unique permutation in S_{r+1} minimizing the expression (5.3.7). So M has tropical rank at least $r+1$. This is a contradiction. We conclude $\dim(P) = r - 1$. □

We close this section by discussing the connection with tropical bases.

Corollary 5.3.24. *Fix positive integers d, n, r. The $(r+1) \times (r+1)$-minors of a $d \times n$-matrix of variables (x_{ij}) are a tropical basis if and only if every $d \times n$-matrix M of tropical rank at most r has Kapranov rank at most r.*

Proof. By Definition 5.3.3, the set of matrices of tropical rank at most r is the intersection of the tropical hypersurfaces given by the $(r+1) \times (r+1)$-minors. By Theorem 5.3.11, the set of matrices of Kapranov rank at most r

is the tropical variety $\mathrm{trop}(V(J_r))$. The former set contains the latter, and equality holds if and only if the minors form a tropical basis for J_r. □

The following theorem gives a complete characterization of all triples (d, n, r) for which the condition in Corollary 5.3.24 is satisfied.

Theorem 5.3.25 (Shitov's Theorem). *The set of $(r+1) \times (r+1)$-minors of a $d \times n$-matrix of indeterminates (x_{ij}) is a tropical basis if and only if*

$$r \leq 2 \quad \textit{or} \quad r+1 = \min\{d,n\} \quad \textit{or} \quad (\, r=3 \textit{ and } \min\{d,n\} \leq 6\,). \tag{5.3.8}$$

For the first two cases in (5.3.8), the tropical basis property was proved in [**DSS05**]. That article also showed it fails when $4 \leq r \leq \min\{d,n\} - 3$. Chan, Jensen, and Rubei [**CJR11**] established the tropical basis property for $r = 3$ and $\min\{d,n\} = 5$. This left only the case $\min\{d,n\} = 6$, which was settled in [**Shi13**]. In that article, Shitov established the tropical basis property for $r = 3$, and he gave a counterexample for $r = 4$. We named the theorem after Shitov because [**Shi13**] concluded this topic. In Exercise 5.6(10) our readers are invited to verify his $r = 4$ example.

5.4. Arrangements of Trees

What follows is a continuation of Section 4.4. Section 5.4 offers an in-depth study of two-dimensional tropical linear spaces and their parameter spaces. These tropical planes will be represented by arrangements of trees, just like classical planes were represented (in Section 4.1) by arrangements of lines. This makes sense in the tropical garden. After all, our trees represent lines.

Recall that the *tropical Grassmannian* $\mathrm{trop}(\mathrm{Gr}_M)$ parameterizes tropicalized linear spaces supported on a given matroid M, and the *Dressian* Dr_M parameterizes tropical linear spaces supported on M. The former is a tropical variety, but the latter is only a prevariety: it even fails to be pure dimensional (by Theorem 5.4.1). We have the inclusion $\mathrm{trop}(\mathrm{Gr}_M) \subseteq \mathrm{Dr}_M$.

The tropical Grassmannian $\mathrm{trop}(\mathrm{Gr}_M)$ comes with a *Gröbner fan structure*, as it is the tropical variety defined by the homogeneous ideal I_M in Section 4.4. The Dressian Dr_M has two natural fan structures. The *secondary fan structure* arises from Lemma 4.4.6, with $\mathbf{w} \sim \mathbf{w}'$ if and only if their matroid subdivisions agree: $\Delta_{\mathbf{w}} = \Delta_{\mathbf{w}'}$. In the *Plücker fan structure*, $\mathbf{w} \sim \mathbf{w}'$ whenever they attain the same minima in the quadrics (4.4.2).

In this section, which is based on [**HJJS09**], we focus on the case when M is a uniform matroid of rank 3. We wish to understand the inclusion

$$\mathrm{trop}(G^0(3,n)) \;\subseteq\; \mathrm{Dr}(3,n). \tag{5.4.1}$$

This inclusion is an equality when $n \leq 6$, but, as we shall see, it is strict when $n \geq 7$. The case $n = 6$ was covered in Example 4.4.10. Points in

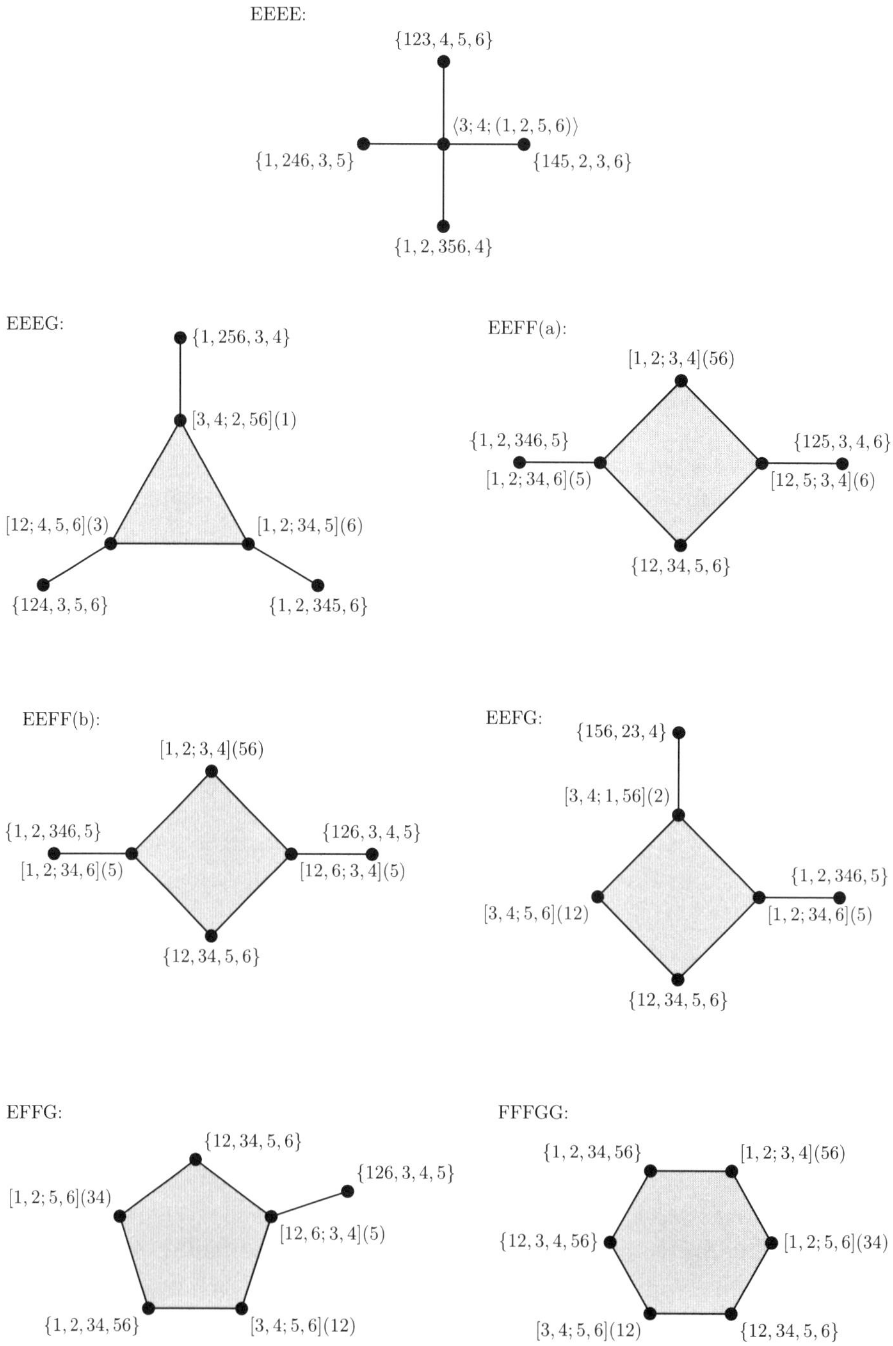

Figure 5.4.1. Seven types of generic tropical planes given by $\mathrm{trop}(\mathrm{G}^0(3,6))$.

$\mathrm{Dr}(3,6) = \mathrm{trop}(G^0(3,6))$ correspond to tropical planes in $\mathbb{R}^6/\mathbb{R}\mathbf{1}$. There are seven generic combinatorial types, labeled EEEE, EEFF(a), ..., FFFGG. Figure 5.4.1 visualizes these seven types. Each picture shows the polyhedral complex of bounded cells in that tropical plane. So, the picture is dual to a matroid subdivision $\Delta_{\mathbf{w}}$ of the hypersimplex $\Delta_{3,6}$. The nodes represent the maximal cells of $\Delta_{\mathbf{w}}$, which are matroid polytopes of graphic matroids (of rank 3 on six elements). The edges indicate their adjacency relations in $\Delta_{\mathbf{w}}$. Our notation for the node labels will be explained later in Figure 5.4.7.

Let us pause to remind the reader that the hypersimplex $\Delta_{3,n}$ is an $(n-1)$-dimensional polytope with $\binom{n}{3}$ vertices. We saw in Example 4.2.13 that $\Delta_{3,n}$ is the matroid polytope of the uniform matroid $U_{3,n}$. Points in the Dressian $\mathrm{Dr}(3,n)$ determine matroid subdivisions of $\Delta_{3,n}$ by Lemma 4.4.6.

We now consider the smallest case when the inclusion (4.4.3) is strict, namely $n = 7$. The following theorem was found by an explicit computation, first reported in [**HJJS09**]. Subsequently, in [**HJS14**], also the Dressian $\mathrm{Dr}(3,8)$ was computed. These computations are hard. Our aim here is not to explain how this was done, but what the output means for the theory.

Theorem 5.4.1. *The tropical Grassmannian* $\mathrm{trop}(G^0(3,7))$ *with Gröbner fan structure is the fan over a five-dimensional simplicial complex with f-vector*

$$(721, 16800, 124180, 386155, 522585, 252000).$$

The Dressian $\mathrm{Dr}(3,7)$ *with Plücker fan structure is a nonsimplicial fan. The underlying polyhedral complex is six dimensional and has the f-vector*

$$(616, 13860, 101185, 315070, 431025, 211365, 30).$$

In both cases, the reduced homology is free abelian and concentrated in dimension 5*:*

$$\begin{aligned} H_*(\mathrm{trop}(G^0(3,7));\mathbb{Z}) &= H_5(\mathrm{trop}(G^0(3,7));\mathbb{Z}) = \mathbb{Z}^{7470}, \\ H_*(\mathrm{Dr}(3,7);\mathbb{Z}) &= H_5(\mathrm{Dr}(3,7);\mathbb{Z}) = \mathbb{Z}^{7440}. \end{aligned}$$

The statement about $\mathrm{trop}(G^0(3,7))$ requires the underlying field to have characteristic different from 2. Indeed, the $30 = 7470 - 7440$ extra homology cycles correspond to the 30 relabelings of the Fano matroid (4.2.1). The statement about the homology refers to the underlying simplicial complex (respectively, the polyhedral complex), which is the link of the fan at the origin. The algebro-geometric meaning of these homology groups will be discussed at the end of Section 6.7. In that context, the above Betti number 7470 may serve as a nontrivial example to illustrate Hacking's result in [**Hac08**].

The symmetric group S_7 acts naturally on $\mathrm{trop}(\mathrm{G}^0(3,7))$ and on $\mathrm{Dr}(3,7)$, and it makes sense to count their cells up to this symmetry. The face numbers of the underlying polyhedral complexes modulo symmetry are

$$\begin{aligned} f(\mathrm{trop}(\mathrm{G}^0(3,7)) \bmod S_7) &= (6,37,140,296,300,125) \quad \text{and} \\ f(\mathrm{Dr}(3,7) \bmod S_7) &= (5,30,107,217,218,94,1)\,. \end{aligned}$$

The tropical Grassmannian $\mathrm{trop}(\mathrm{G}^0(3,7))$ has 125 orbits of five-dimensional simplices. These are merged to 94 orbits of five-dimensional polytopes in the Dressian $\mathrm{Dr}(3,7)$. One of these cells is not a facet because it lies in the unique cell of dimension 6 (corresponding to the Fano plane). This means that $\mathrm{Dr}(3,7)$ has $93+1=94$ maximal cells up to the S_7-symmetry.

Each point $\mathbf{w}$ in $\mathrm{Dr}(3,n)$ determines a tropical plane $L_{\mathbf{w}}$ in $\mathbb{R}^n/\mathbb{R}\mathbf{1}$ and conversely, by Theorem 4.4.5. The cells of $\mathrm{Dr}(3,n)$ modulo S_n correspond to combinatorial types of tropical planes in $\mathbb{R}^n/\mathbb{R}\mathbf{1}$. Facets of $\mathrm{Dr}(3,n)$ correspond to *generic planes*, and there is a census of these for small n:

Corollary 5.4.2. *The number of combinatorial types of generic planes in* $\mathbb{R}^n/\mathbb{R}\mathbf{1}$ *is equal to* $1,1,7,94$ *for* $n=4,5,6,7$, *respectively.*

Proof. The generic plane in 3-space $\mathbb{R}^4/\mathbb{R}\mathbf{1}$ is the cone over the complete graph K_4. Planes in 4-space are parameterized by the Petersen graph $\mathrm{Dr}(3,5) = \mathrm{trop}(G^0(3,5))$. The unique generic type is dual to the trivalent tree with five leaves. The seven types of generic planes in 5-space were listed in Example 4.4.10. Drawings of their bounded parts are given in Figure 5.4.1. Their unbounded cells are represented by the tree arrangements in Table 5.4.1. The number 94 for $n=7$ is derived from Theorem 5.4.1. □

As the number n grows, the Dressian $\mathrm{Dr}(3,n)$ becomes much larger than $\mathrm{trop}(G^0(3,n))$. The dimension of that tropical Grassmannian is $3n-9$, so it grows linearly in n. On the other hand, for the Dressian we have

Theorem 5.4.3. *The dimension of the Dressian* $\mathrm{Dr}(3,n)$ *is of order* $\Theta(n^2)$.

Proof. For the proof of the upper bound we refer to [**HJJS09**, Theorem 3.6]. In what follows we prove a lower bound that is quadratic in n. This will be done by identifying cells of sufficiently large dimension in $\mathrm{Dr}(3,n)$.

The *generalized Fano matroid* F_r is a matroid of rank 3 on 2^r-1 elements. That matroid is defined as follows. Its bases are the noncollinear triples of points in the $(r-1)$-dimensional projective space over the field with two elements, denoted $\mathbb{P}^{r-1}_{\mathbb{F}_2}$. Thus the number of bases of F_r equals

$$\beta_r \;:=\; \frac{1}{6}(2^r-1)(2^r-2)(2^r-4)\,.$$

Equivalently, the number of nonbases of F_r of size 3 is

$$\nu_r \;:=\; \binom{2^r-1}{3} - \beta_r \;=\; \frac{1}{6}(2^r-1)(2^r-2)\,.$$

In this manner, the vertices of the hypersimplex $\Delta_{3,2^r-1}$ are partitioned into bases and nonbases of the matroid F_r. We claim that the nonbases form a stable set (i.e., no two adjacent) in the edge graph of $\Delta_{3,2^r-1}$. Indeed, the nonbases are precisely the collinear triples in $\mathbb{P}^{r-1}_{\mathbb{F}_2}$. Two distinct lines in $\mathbb{P}^{r-1}_{\mathbb{F}_2}$ share at most one point, and hence the two corresponding vertices of $\Delta_{3,2^r-1}$ differ in more than two coordinates, which means that they are not connected by an edge of $\Delta_{3,2^r-1}$.

The quadratic lower bound for $\dim(\mathrm{Dr}(3,n))$ is now derived as follows. For any given n, let r be the unique natural number satisfying $2^r - 1 \leq n < 2^{r+1}$. Then the generalized Fano matroid F_r yields a stable set of size $\nu_r = 1/6(2^r-1)(2^r-2) \geq n^2/24 - n/12$ in the edge graph of the hypersimplex $\Delta_{3,n}$. The latter inequality follows from $2^r - 1 \geq n/2$.

Let σ denote the cone of nonnegative vectors in $\mathbb{R}^{\binom{n}{3}}$ whose support is contained in the above stable set. For $\mathbf{w}$ in relint(σ), the subdivision $\Delta_{\mathbf{w}}$ cuts off certain vertices from $\Delta_{3,n}$, namely those indexed by the support of $\mathbf{w}$. Here a vertex being "cut off" means that $\Delta_{3,n}$ is divided by the hyperplane passing through its neighboring vertices. No two such hyperplanes intersect in the interior of $\Delta_{3,n}$. No new edges get created, so the subdivision $\Delta_{\mathbf{w}}$ is a matroid subdivision. (This was first shown in [**HJ08**, Lemma 7.4]). This means that the cone σ is contained in the Dressian $\mathrm{Dr}(3,n)$. Since σ has dimension $n^2/24 - n/12$, this number is a lower bound for $\dim(\mathrm{Dr}(3,n))$. □

It is instructive to review the above argument for $r = 3$ and $n = 2^r - 1 = 7$. The matroid F_3 is the Fano plane, shown in Figure 4.2.1. The seven bases of F_3 define a six-dimensional simplex in the Dressian $\mathrm{Dr}(3,7)$ when regarded as a polyhedral complex as in Theorem 5.4.1. For dimension reasons, that simplex cannot be in $\mathrm{trop}(G^0(3,7))$, so the inclusion (5.4.1) is strict for $n = 7$. Note that all 30 six-dimensional cells of $\mathrm{Dr}(3,7)$ come from the Fano matroid F_3 by relabeling. They form a single orbit under the S_7 action, since the automorphism group $\mathrm{GL}_3(\mathbb{F}_2)$ of F_3 has order $168 = 5040/30$. Each of the 30 Fano simplices of $\mathrm{Dr}(3,7)$ intersects $\mathrm{trop}(G^0(3,7))$ its boundary. Hence each Fano simplex cancels precisely one homology cycle of $\mathrm{trop}(G^0(3,7))$. This explains the difference 30 in the homology ranks in Theorem 5.4.1.

We now come to the title theme of this section, namely arrangements of trees. In Section 4.1 we identified d-dimensional linear varieties in a torus with complements of hyperplane arrangement in $\mathbb{P}^d$. If $d = 1$, then the arrangement consists of points in $\mathbb{P}^1$, and these correspond to the leaves in the trees of Section 4.3. We now consider the case of planes ($d = 2$), where the arrangement consists of lines in $\mathbb{P}^2$. Each line tropicalizes to a tree. Thus the study of tropical planes leads us naturally to tree arrangements.

Let $n \geq 4$, and consider an n-tuple of metric trees $T = (T_1, T_2, \ldots, T_n)$, where T_i has the set of leaves $[n]\backslash\{i\}$. Here we are using the term *metric tree* to refer to a phylogenetic tree (a tree with n labeled leaves and no vertices of degree 2) together with a fixed tree metric on that tree. Thus, each edge in the metric tree T_i has a fixed nonnegative length. We write $\delta_i : ([n]\backslash\{i\}) \times ([n]\backslash\{i\}) \to \mathbb{R}_{\geq 0}$ for the tree metric defined by T_i. We call the n-tuple T of metric trees a *metric tree arrangement* if

$$\delta_i(j,k) \;=\; \delta_j(k,i) \;=\; \delta_k(i,j) \tag{5.4.2}$$

for all $i, j, k \in [n]$ pairwise distinct. The following definition is more subtle.

Definition 5.4.4. Consider $T = (T_1, T_2, \ldots, T_n)$, where T_i is a phylogenetic tree whose set of leaves is $[n]\backslash\{i\}$. Thus each T_i is just a combinatorial tree, without any metric. We say that T is an *abstract tree arrangement* if either

- $n = 4$; or
- $n = 5$, and T is the set of quartets of a tree with five leaves; or
- $n \geq 6$, and $(T_1\backslash i, \ldots, T_{i-1}\backslash i, T_{i+1}\backslash i, \ldots, T_n\backslash i)$ is an abstract tree arrangement on $n-1$ elements for each $i \in [n]$.

Here $T_j\backslash i$ denotes the tree on $[n]\backslash\{i,j\}$ obtained from T_j by deleting leaf i.

We wish to emphasize once more that this definition is very subtle. It will take the reader some time to unravel and digest this recursive definition.

Example 5.4.5. We apply Definition 5.4.4 to construct all arrangements of trivalent trees for $n \leq 6$. There are three trivalent trees on $[4]$, so there are $3^5 = 243$ 5-tuples $(T_1, T_2, T_3, T_4, T_5)$ where T_i is a trivalent tree on $[5]\backslash\{i\}$. Of these 243, precisely 15 are abstract tree arrangements. Likewise, there are $15^6 = 11390625$ tuples $(T_1, T_2, \ldots, T_6)$ where T_i is a trivalent tree on $[6]\backslash\{i\}$. Precisely 1005 of these pass the test in the recursive step. (Needless to say, these results are the output of a computer program.) For all other 6-tuples, at least one of the six deletions is not one of the 15 abstract tree arrangements for $n = 5$, but is instead one of the 228 other 5-tuples. The 1005 abstract tree arrangements we found for $n = 6$ are listed in Table 5.4.1. $\diamondsuit$

Proposition 5.4.6. *Each metric tree arrangement gives rise to an abstract tree arrangement if we simply disregard the edge lengths.*

Proof. The statement is trivial for $n = 4$. For $n = 5$, the proof goes as follows. Let $T = (T_1, T_2, T_3, T_4, T_5)$ be a metric tree arrangement. We define a metric $\mathbf{d}$ on $[5] = \{1,2,3,4,5\}$ by setting $d_{lm} := \delta_i(j,k) = \delta_j(k,i) = \delta_k(i,j)$ whenever $\{i,j,k,l,m\} = [5]$. Both the triangle inequality and the four point condition are satisfied for $\mathbf{d}$ because $\delta_i, \delta_j, \delta_k$ are tree metrics. Hence,

Table 5.4.1. The 1005 abstract trivalent tree arrangements for $n = 6$ fall into seven symmetry classes. These correspond to the seven types of tropical planes in 5-space. Each tree is denoted $ab\,c\,de$ as in Figure 5.4.2, and the names for the types match Example 4.4.10 and Figure 5.4.1.

Type	Tree 1	Tree 2	Tree 3	Tree 4	Tree 5	Tree 6	Orbit
EEEE	23 6 45	13 5 46	12 4 56	15 3 26	14 2 36	24 1 35	30
EEEG	26 5 34	16 5 34	14 2 56	13 2 56	12 3 46	12 3 45	240
EEFF(a)	25 6 34	15 6 34	12 5 46	12 5 36	12 6 34	12 5 34	90
EEFF(b)	25 6 34	15 6 34	12 6 45	12 6 35	12 6 34	12 5 34	90
EEFG	25 6 34	15 6 34	24 1 56	23 1 56	12 6 34	12 5 34	360
EFFG	34 2 56	34 1 56	12 6 45	12 6 35	12 6 34	12 5 34	180
FFFGG	34 2 56	34 1 56	12 4 56	12 3 56	12 6 34	12 5 34	15

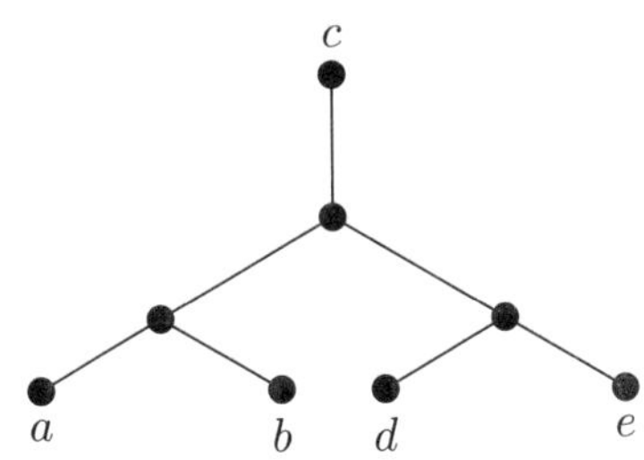

Figure 5.4.2. We use the notation $ab\,c\,de$ for this tree on five labeled leaves.

by Lemma 4.3.6, $\mathbf{d}$ is a tree metric. This means that there is a metric tree with five leaves with metric $\mathbf{d}$; this is the tree required for the $n = 5$ case in Definition 5.4.4. For $n \geq 6$, the statement holds for the following reason: whenever T is a metric tree arrangement, then so is $(T_1\backslash i, \ldots, T_{i-1}\backslash i, T_{i+1}\backslash i, \ldots, T_n\backslash i)$. □

The converse to Proposition 5.4.6 is not true: for $n \geq 9$, there exist abstract arrangements of n trees that do not come from any metric tree arrangement. We will get to this in Example 5.4.12 and Figure 5.4.6.

The hypersimplex $\Delta_{d,n}$ is the intersection of the cube $[0,1]^n$ with the hyperplane $\sum x_i = d$. Its facets correspond to facets of $[0,1]^n$. We call the facet $\{x_i = 0\} \simeq \Delta_{d,n-1}$ the *ith deletion facet* of $\Delta_{d,n}$, and the facet $\{x_i = 1\} \simeq \Delta_{d-1,n-1}$ the *ith contraction facet*. These names make sense: if M is any matroid on $[n]$ of rank d, then the intersection of P_M with the ith deletion (contraction) facet is the matroid polytope of $M\backslash i$ (respectively, M/i).

Lemma 5.4.7. *Each matroid subdivision Σ of the hypersimplex $\Delta_{3,n}$ defines an abstract arrangement $T(\Sigma)$ of n trees. Moreover, if the matroid subdivision Σ is regular, then $T(\Sigma)$ supports a metric tree arrangement.*

Proof. Each of the n contraction facets of $\Delta_{3,n}$ is isomorphic to $\Delta_{2,n-1}$. Hence Σ induces matroid subdivisions on n copies of $\Delta_{2,n-1}$. By [**Kap93**, Theorem 1.3.6] each matroid subdivision is dual to a phylogenetic tree. Write T_i for the phylogenetic tree induced by the ith contraction facet. We claim that $T = (T_1, \dots, T_n)$ is an abstract tree arrangement. This is trivial for $n = 4$ and formal for $n \geq 6$. For $n = 5$, the subdivision is induced by some vector $\mathbf{w} \in \text{trop}(G^0(3,5)) = \text{Dr}(3,5)$, so T is a metric tree arrangement, and we are in the situation discussed next.

Let Σ be regular with weights $\mathbf{w} \in \text{Dr}(3,n)$. The contractions of $\mathbf{w}$ give weights on each contraction facet $\Delta_{2,n-1}$. These are tree distances, where (5.4.2) is now the coordinate w_{ijk} of $\mathbf{w}$. After adding a multiple of $\mathbf{1}$ to $\mathbf{w}$, all n tree distances T_i are tree metrics, so T is a metric tree arrangement. □

Theorem 5.4.8. *Let Σ and $\bar{\Sigma}$ be matroid subdivisions of $\Delta_{3,n}$ such that Σ refines $\bar{\Sigma}$ and they agree on the boundary of $\Delta_{3,n}$. Then Σ and $\bar{\Sigma}$ are equal.*

Proof. Suppose that Σ strictly refines $\bar{\Sigma}$. Then there is a codimension 1 cell F of Σ which is not a cell in $\bar{\Sigma}$. Let $\bar{F}$ be the unique full-dimensional cell of $\bar{\Sigma}$ that contains F. In particular, F is not contained in the boundary of $\Delta_{3,n}$. Then F is a rank 3 matroid polytope $F = P_M$ of codimension 1. By [**FS05**, Proposition 2.4], the matroid M is a disjoint union $M = M_1 \cup M_2$, where M_i is a matroid of rank i. Geometrically, we have $F \cong P_{M_1} \times P_{M_2}$. The affine span H of F is defined by the equation $\sum_{i \in I} x_i = 1$, where I is the set of elements of M_1. These are all parallel in M because $\text{rank}(M_1) = 1$.

Since $\bar{F}$ is divided by H, there exist vertices $\mathbf{v}, \mathbf{w}$ of $\bar{F}$ on either side of H. Now $\bar{F}$ is also a matroid polytope of some matroid $\bar{M}$ containing M as a submatroid. Up to relabeling, we can assume $\mathbf{v} = \mathbf{e}_{12i}$ and $\mathbf{w} = \mathbf{e}_{345}$. Then $\{1,2,i\}$ and $\{3,4,5\}$ are bases of $\bar{M}$ which are not bases of M, where $1, 2 \in I$ and $i, 3, 4, 5 \notin I$. If $i \notin \{3,4,5\}$, then we can replace i in the basis $\{1,2,i\}$ by some $j \in \{3,4,5\}$ to obtain a new basis of $\bar{M}$. We can assume that $i = 5$ or $j = 5$. Hence $\{1,2,5\}$ and $\{3,4,5\}$ are bases of $\bar{M}$ that are not bases of M. Notice that $\mathbf{e}_{125}$ and $\mathbf{e}_{345}$ are still on different sides of H as $\mathbf{e}_{12i}$ and $\mathbf{e}_{125}$ are connected by an edge and $\{1,2,5\}$ is not a basis of M.

As $\text{rank}(M_i) \leq 2$, both M_1 and M_2 are realizable as affine point configurations over $\mathbb{R}$. Hence we can draw M as a point configuration (with multiple points) in the affine plane. By the description given above, the first five points look like one of the two configurations shown in Figure 5.4.3.

Consider the intersection of $\Delta_{3,n}$ with the affine space defined by $x_5 = 1$ and $x_6 = x_7 = \cdots = x_n = 0$. This gives us an octahedron $C \cong \Delta_{2,4}$ in the boundary of $\Delta_{3,n}$. The intersection $S = F \cap C$ is a square; in Figure 5.4.3 it is the convex hull of the four points $\mathbf{e}_{135}$, $\mathbf{e}_{145}$, $\mathbf{e}_{235}$, and $\mathbf{e}_{245}$. In particular, the square S is a cell of Σ. However, since $\mathbf{e}_{125}$ and $\mathbf{e}_{345}$ are vertices of

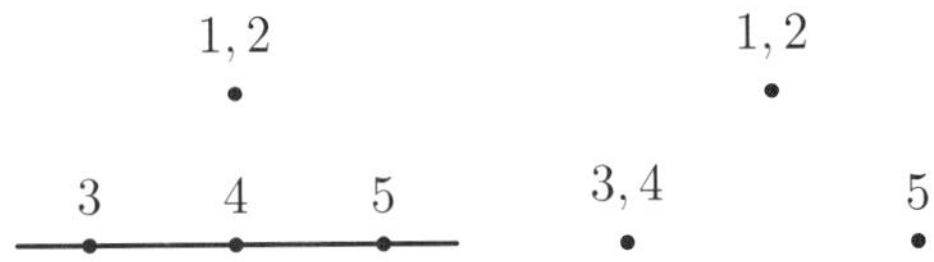

Figure 5.4.3. Two point configurations in the Euclidean plane.

$\bar{F} = P_{\bar{M}}$ as discussed above, C is a cell of $\bar{\Sigma}$. We conclude that the square S is a cell of Σ but not a cell of $\bar{\Sigma}$. By construction $S \subset C$ is contained in the boundary of $\Delta_{3,n}$. This yields the desired contradiction, as Σ and $\bar{\Sigma}$ induce the same subdivision on the boundary. □

Two metric tree arrangements are *equivalent* if they induce the same abstract tree arrangement.

Theorem 5.4.9. *Equivalence classes of arrangements of* n *metric trees are in bijection with regular matroid subdivisions of the hypersimplex* $\Delta_{3,n}$*. The secondary fan structure on* $\mathrm{Dr}(3, n)$ *equals the Plücker fan structure.*

Proof. Each regular matroid subdivision gives rise to a metric tree arrangement by Lemma 5.4.7. The harder direction is to construct regular matroid subdivisions from metric tree arrangements. We shall do this by induction on n. The hypersimplex $\Delta_{3,4}$ is a 3-simplex, and $\mathrm{Dr}(3, 4)$ is a single point modulo the lineality space (in both fan structures). The hypersimplex $\Delta_{3,5}$ is isomorphic to $\Delta_{2,5}$, and $\mathrm{Dr}(3, 5) = \mathrm{trop}(G^0(3, 5)) \cong \mathrm{trop}(G^0(2, 5))$ is the Petersen graph (Figure 4.3.2). In this case, the result can be verified directly. This establishes the basis of our induction. We now assume $n \geq 6$.

Let T be an arrangement of n tree metrics $\delta_1, \delta_2, \ldots, \delta_n$. In view of the axiom (5.4.2), the following map $\pi : [n]^3 \to \mathbb{R} \cup \{\infty\}$ is well defined:

$$-\pi_{ijk} \;=\; \begin{cases} \delta_i(j, k) = \delta_j(k, i) = \delta_k(i, j) & \text{if } i, j, k \text{ are pairwise distinct,} \\ \infty & \text{otherwise.} \end{cases}$$

We must show that the minimum of $\{\pi_{hij} + \pi_{hkl}, \pi_{hik} + \pi_{hjl}, \pi_{hil} + \pi_{hjk}\}$ is attained at least twice, for any pairwise distinct $h, i, j, k, l \in [n]$. Now, since $n \geq 6$, each 5-tuple in $[n]$ is already contained in some deletion, and hence the desired property is satisfied by induction. We conclude that the restriction of the map π to increasing triples $i < j < k$ is a finite tropical Plücker vector, that is, it is an element of $\mathrm{Dr}(3, n)$. By Lemma 4.4.6, the map π defines a matroid subdivision $\Sigma(T)$ of the hypersimplex $\Delta_{3,n}$.

Let T' be an arrangement that is equivalent to T. The maps π and π' associated with T and T' lie in the same cone of the Plücker fan structure on $\mathrm{Dr}(3, n)$. What we must prove is that they are also in the same cone of the secondary fan structure. Equivalently, we must show $\Sigma(T') = \Sigma(T)$.

Suppose the secondary fan structure on $\mathrm{Dr}(3,n)$ is strictly finer than the Plücker fan structure. Pick a regular matroid subdivision Σ of $\Delta_{3,n}$ whose secondary cone $S(\Sigma)$ lies strictly in the corresponding cone $P(\Sigma)$ of tropical Plücker vectors. Pick a point $\mathbf{w}$ in the boundary of $S(\Sigma)$ which is in the interior of $P(\Sigma)$. Then Σ strictly refines $\bar{\Sigma} = \Delta_{\mathbf{w}}$. By induction we can assume that Σ and $\bar{\Sigma}$ induce the same subdivision on the boundary of $\Delta_{3,n}$. Theorem 5.4.8 now implies $\Sigma = \bar{\Sigma}$, a contradiction. □

We next introduce some special tree arrangements which can be represented by pictures as in Figures 5.4.4 or 5.4.6. The *vertex figure* of the $(n-1)$-dimensional polytope $\Delta_{3,n}$ at any vertex is the $(n-2)$-dimensional polytope $\Delta_2 \times \Delta_{n-4}$. By Theorem 5.2.19, regular subdivisions of $\Delta_2 \times \Delta_{n-4}$ correspond to tropical complexes generated by $n-3$ points in the plane. We claim that each regular subdivision of $\Delta_2 \times \Delta_{n-4}$ extends to a unique regular matroid subdivision of $\Delta_{3,n}$. This extension can be constructed as follows using tree arrangements. Let $L_1, L_2, \ldots, L_{n-3}$ be the lines dual to the given points, and let L_x, L_y, L_z be three boundary lines at infinity in the x-, y-, and z-directions of $\mathbb{R}^3/\mathbb{R}\mathbf{1}$. Each of these n tropical lines translates into a tree. The tree for L_x is obtained by branching off the leaves $\{1, 2, \ldots, n-3\}$ on the path between leaves y and z, in the order in which the L_j intersect L_x. The trees for L_y and L_z are analogous. The tree for L_i consists of the three rays, marked by leaves x, y, and z at infinity. Along each ray, we branch off additional leaves j for each line L_j that intersects the line L_i in that ray. This branching takes place in the order in which the lines L_j intersect L_i. In this manner, every arrangement of $n-3$ tropical lines in $\mathbb{R}^3/\mathbb{R}\mathbf{1}$ determines n trees, each with $n-1$ labeled leaves.

Consider $n-3$ copies of the standard triangle Δ_2. Their Minkowski sum is the scaled triangle $(n-3)\Delta_2$, and their Cayley polytope is $C(\Delta_2, \ldots, \Delta_2) = \Delta_2 \times \Delta_{n-4}$; see Definition 4.6.1. Arbitrary subdivisions of $\Delta_2 \times \Delta_{n-4}$ correspond to mixed subdivisions of $(n-3)\Delta_2$. These need not be regular. If the mixed subdivision comes from a triangulation of $\Delta_2 \times \Delta_{n-4}$, then its pieces are lozenges and unit upward triangles. A *lozenge* is a parallelogram that is the union of one upward triangle and one downward triangle. We call a mixed cell *even* if it can be tiled by lozenges only. Those that need an upward triangle in any tiling are *odd*. A counting argument now reveals that each mixed subdivision of $(n-3)\Delta_2$ contains up to $n-3$ odd polygonal cells.

Proposition 5.4.10. *Each subdivision of $\Delta_2 \times \Delta_{n-4}$, or mixed subdivision of the triangle $(n-3)\Delta_2$, determines an abstract arrangement of n trees.*

Proof. Let Σ be a triangulation of $\Delta_2 \times \Delta_{n-4}$. The corresponding mixed subdivision $M(\Sigma)$ of $(n-3)\Delta_2$ has exactly $n-3$ odd cells, all upward triangles, and the even cells are lozenges. Placing a labeled node into each upward triangle defines a tree in the graph dual to $M(\Sigma)$. Each of its three branches uses the edges in $M(\Sigma)$ which are in the parallelism class as one fixed edge of that upward triangle. Two opposite edges in a lozenge are parallel, and the *parallelism* we refer to is the transitive closure of this relation. Each parallelism class extends to the boundary of $(n-3)\Delta_2$. Doing so for all the upward triangles, the dual graph of $M(\Sigma)$ decomposes into n trees $T_1, \ldots, T_n$, where T_i has leaves $[n]\backslash\{i\}$. We claim that this is an abstract tree arrangement. This is trivial for $n = 4$, it holds for $n = 5$ because all triangulations of $\Delta_2 \times \Delta_1$ are regular, and it is formal for $n \geq 6$.

Next suppose that Σ is not a triangulation, so $M(\Sigma)$ is a coarser mixed subdivision of $(n-3)\Delta_2$. We shall associate a tree arrangement with $M(\Sigma)$. Pick any triangulation Σ' which refines Σ. The above procedure maps Σ' to a tree arrangement $T(\Sigma')$. Then, as Σ' refines Σ, one can contract edges in the trees of $T(\Sigma')$. In this way one also arrives at another arrangement of n trees. Three of them come from the boundary of $(n-3)\Delta_2$. The $n-3$ nonboundary trees are assigned surjectively to the $\leq n-3$ odd cells. The resulting $T(\Sigma)$ might depend on the choice of the triangulation Σ'. □

Example 5.4.11. Let $n = 6$ and consider the two mixed subdivisions of $3\Delta_2$ shown in Figure 5.4.4. The left-hand one is a lozenge tiling which encodes a regular triangulation of $\Delta_2 \times \Delta_2$. It corresponds to the abstract tree arrangement FFFGG in Table 5.4.1. The mixed subdivision of $3\Delta_2$ on the right in Figure 5.4.4 coarsens the lozenge tiling on the left. It corresponds to

$$34\,2\,56\,,\ 34\,1\,56\,,\ 12\,(456)\,,\ 12\,(356)\,,\ 12\,6\,34\,,\ 12\,5\,34\,.$$

The tree $ab\,(cde)$ is obtained from $ab\,c\,de$ by contracting the interior edge between c and de. Odd cells (shaded in Figure 5.4.4) correspond to trees. ◊

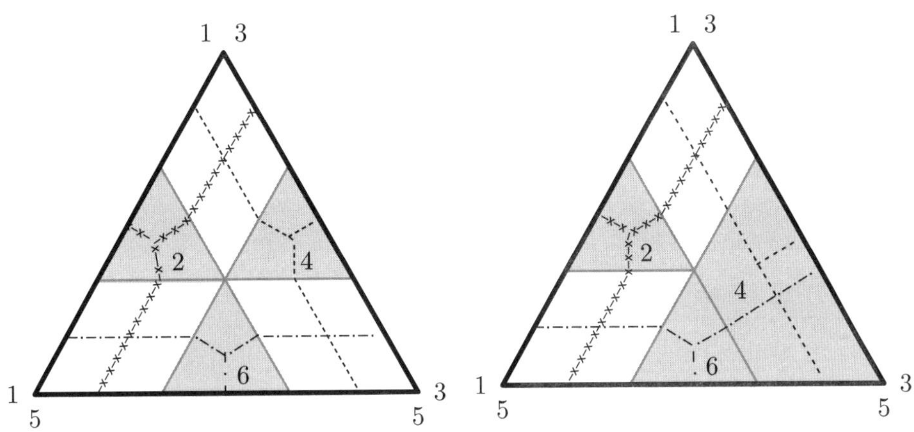

Figure 5.4.4. Mixed subdivisions of $3\Delta_2$ and arrangements of six trees.

Example 5.4.12. The lozenge tiling of $6\Delta_2$ in Figure 5.4.6 encodes a nonregular matroid subdivision. This picture translates into the abstract arrangement of nine trees in Table 5.4.2. The corresponding matroid subdivision of $\Delta_{3,9}$ is not regular, because the underlying triangulation of $\Delta_2 \times \Delta_5$ is not regular. We conclude that the Dressian $\mathrm{Dr}(3,9)$ has no cell for the abstract tree arrangement in Figure 5.4.6. This example is due to Santos; he proved its nonregularity in [**San05**, Proposition 4.1]. See also [**AD09**] where this serves as an example of a *tropical oriented matroid* that is not realizable. ◊

Table 5.4.2. Abstract arrangement of nine caterpillar trees on eight leaves encoding a nonregular matroid subdivision of $\Delta_{3,9}$; see Figure 5.4.6. The notation for caterpillar trees is explained in Figure 5.4.5.

Tree 1: C(24, 6598, 37)	Tree 2: C(14, 5768, 39)	Tree 3: C(17, 5846, 29)
Tree 4: C(12, 6579, 38)	Tree 5: C(26, 4198, 37)	Tree 6: C(14, 5729, 38)
Tree 7: C(13, 5894, 26)	Tree 8: C(15, 7346, 29)	Tree 9: C(15, 7468, 23)

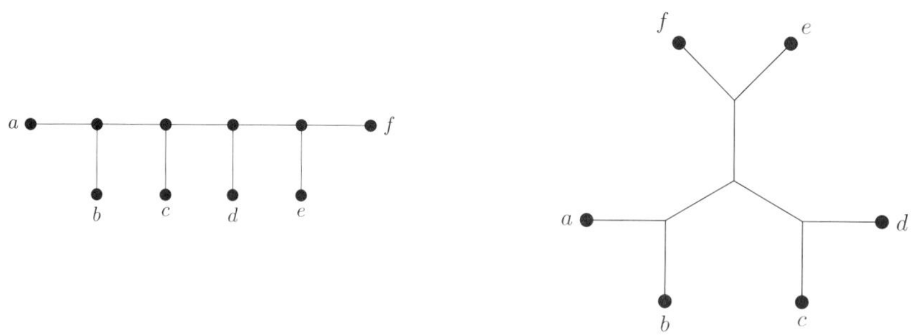

Figure 5.4.5. Caterpillar tree $\mathrm{C}(ab, cd, ef)$ and snowflake tree $\mathrm{S}(ab, cd, ef)$.

We next answer the question of [**HJJS09**]: *How do we draw a tropical plane*? Tropical planes are contractible polyhedral surfaces dual to regular matroid subdivisions of $\Delta_{3,n}$. Consider any point $\mathbf{w}$ in the Dressian $\mathrm{Dr}(3,n)$. The associated tropical plane $L_\mathbf{w}$ in $\mathbb{R}^n/\mathbb{R}\mathbf{1}$ is defined by the tropical circuits

$$w_{ijk} \odot x_l \ \oplus \ w_{ijl} \odot x_k \ \oplus \ w_{ikl} \odot x_j \ \oplus \ w_{jkl} \odot x_i.$$

The first answer to our question is *draw the metric tree arrangement*. The correspondence is given by Theorems 4.4.5 and 5.4.9. We may also disregard the metrics and just draw the abstract tree arrangements. For instance, consider Table 5.4.1 for $n = 6$. One translates the seven rows into seven pictures of tree arrangements. For example, the arrangement of type FFFGG in Table 5.4.1 is the picture on the left-hand side in Figure 5.4.4.

The second answer to our question is *draw and label the bounded cells*, as in Figure 5.4.1. Each vertex of a tropical plane $L_\mathbf{w}$ is labeled by a connected matroid of rank 3. Its matroid polytope is a maximal cell in the

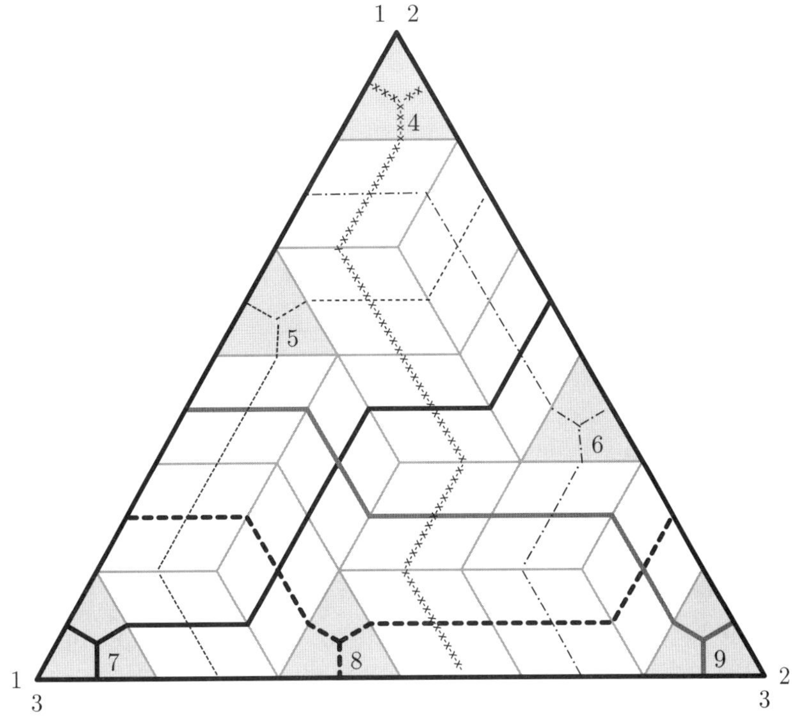

Figure 5.4.6. Abstract arrangement of nine caterpillar trees on eight leaves encoding a matroid subdivision of $\Delta_{3,9}$ which is not regular (Table 5.4.2). Trees 1, 2, and 3 are the edges of the triangle. The leaves of each tree are attached at the intersection points with the other eight trees.

matroid subdivision of $\Delta_{3,n}$ given by $L_\mathbf{w}$. For $n = 6$ only three classes of matroids occur as node labels of generic planes. These matroids are denoted $\{A, B, C, D\}$, $[A, B, C, D](E)$, or $\langle A; a; (b, c, d, e)\rangle$. Here capital letters are nonempty subsets of the set $\{1, 2, 3, 4, 5, 6\}$, and lower-case letters are elements thereof. All three matroids are graphical. The corresponding graphs are shown in Figure 5.4.7. An edge labeled with a set of l points should be considered as l parallel edges each labeled with one element of the set.

The graph for the matroid $\langle A; b; (c, d, e, f)\rangle$ is the complete graph K_4. The set A is a singleton. This matroid occurs in the unique orbit (of type EEEE) with no bounded 2-cell. The two-dimensional pictures in Figure 5.4.1 use only the matroids $\{A, B, C, D\}$ and $[A, B; C, D](E)$ for their labels.

The third answer to our question is the synthesis of the previous two: *draw both* the bounded complex and the tree arrangement. The two pictures can be connected, by linking each node of $L_\mathbf{w}$ to the adjacent unbounded rays and 2-cells. This leads to an accurate diagram of the tropical planes $L_\mathbf{w}$.

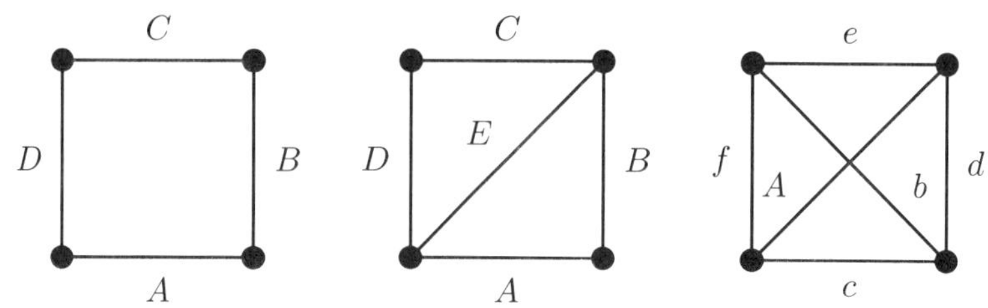

Figure 5.4.7. The graphic matroids corresponding to the labels $\{A, B, C, D\}$, $[A, B; C, D](E)$, and $\langle A; b; (c, d, e, f)\rangle$ used in Figure 5.4.1.

The analogous complete description for $n = 7$ was given in [**HJJS09**]. The $211365 + 30$ maximal cells of the Dressian $\mathrm{Dr}(3, 7)$ correspond to arrangements of seven trivalent trees. It was found that for $n = 7$, there is no difference between abstract tree arrangements and metric tree arrangements: nothing like Example 5.4.12 exists in this case. To draw the arrangements, one uses the *caterpillar* and the *snowflake* trees, as in Figure 5.4.5. Caterpillars exist for all $n \geq 5$, and are encoded using notation as in Figure 5.4.2.

This concludes our combinatorial study of the structure of tropical linear spaces. The next section will be concerned with tropical varieties that are obtained by applying a classical linear map to a tropical linear space.

5.5. Monomials in Linear Forms

In this section we present an application of tropical linear spaces to a special class of algebraic varieties, namely, those that admit a parameterization by products of linear forms. Varieties in this class include discriminants, resultants, and some classical moduli spaces. In this section we use the A-discriminants of [**GKZ08**] as our primary example. Readers interested in the connection to moduli spaces can find additional information in [**RSS14**].

To keep things simple, we now assume that K is a field with the trivial valuation. In particular, all tropical linear spaces to be encountered here have the form $\mathrm{trop}(M)$ where M is realizable matroid, as in Section 4.1.

Suppose we are given two matrices. The first is an $n \times d$-matrix $B = (b_{ij})$ with entries in K. The second is an $m \times n$-matrix $C = (c_{ij})$ with integer entries. The first matrix B specifies n linear forms in $K[x_1, \ldots, x_d]$:

$$\ell_i(x) = b_{i1}x_1 + b_{i2}x_2 + \cdots + b_{id}x_d \qquad \text{for } i = 1, 2, \ldots, n. \tag{5.5.1}$$

The second matrix C specifies m Laurent monomials in $K[y_1^{\pm 1}, \ldots, y_n^{\pm 1}]$:

$$y_1^{c_{j1}} y_2^{c_{j2}} \cdots y_n^{c_{jn}} \qquad \text{for } j = 1, 2, \ldots, m. \tag{5.5.2}$$

We now substitute $y_i = \ell_i(x)$ into (5.5.2). Our data specify a rational map

$$\psi : \mathbb{A}^d \dashrightarrow \mathbb{A}^m \quad \text{with coordinates} \quad \psi_j(x) = \ell_1(x)^{c_{j1}} \ell_2(x)^{c_{j2}} \cdots \ell_n(x)^{c_{jn}}.$$

Let Y denote the closure of the image of ψ. This is an affine variety in $\mathbb{A}^m$. We wish to compute the tropical variety $\operatorname{trop}(Y \cap (K^*)^m)$ in $\mathbb{R}^m$. To this end, we consider the matroid M on the ground set $[n] = \{1, 2, \ldots, n\}$ defined by the rows of the matrix B. This matroid has rank at most d, and the rank is exactly d if the columns of B are linearly independent.

Let $\operatorname{trop}(M)$ be the tropical linear space for this matroid, with a fixed polyhedral fan structure as in Theorem 4.2.6. By Exercise 4.7(13), we have $\operatorname{mult}(\sigma) = 1$ for all maximal cones σ of $\operatorname{trop}(M)$. The matrix C defines a (classical) linear map $\mathbb{R}^n \to \mathbb{R}^m$. We apply this map to $\operatorname{trop}(M)$.

Theorem 5.5.1. *The tropicalization of the variety Y is the balanced fan*

$$\operatorname{trop}(Y) \quad = \quad C \cdot \operatorname{trop}(M). \tag{5.5.3}$$

Proof. The image of the linear map given by the matrix B is a linear subspace X of K^n. As in Section 4.1, we regard X as a hyperplane arrangement complement, embedded in the torus T^n. By Theorem 4.1.11 and Definition 4.2.5, we have $\operatorname{trop}(X) = \operatorname{trop}(M)$. We now apply Corollary 3.2.13 where ϕ is the monomial map defined by the matrix C. Then $\operatorname{trop}(\phi)$ is the linear map defined by C, and the desired identity follows directly from (3.2.2). □

Remark 5.5.2. We know from Lemma 3.6.3 that $C \cdot \operatorname{trop}(M)$ has the structure of a balanced fan with the multiplicities given by (3.6.2). These multiplicities are, in fact, the multiplicities on $\operatorname{trop}(Y)$, up to a factor recording the degree of the map ψ. This is proved in [**ST08**]. Their proof uses the method of geometric tropicalization, which will appear in Section 6.5.

In some applications of Theorem 5.5.1, the rows of the matrix C have all the same sum. In that case, the m monomials in (5.5.2) have the same degree and we can regard ψ as a rational map between projective spaces:

$$\psi : \mathbb{P}^{d-1} \dashrightarrow \mathbb{P}^{m-1}.$$

Theorem 5.5.1 remains valid when the valuation on the field K is nontrivial. In that case, the linear space X is defined over that K and its tropicalization $\operatorname{trop}(X)$ is a tropical linear space as in Sections 4.4 and 5.4. To keep things simple, however, we have here restricted to the constant-coefficient case.

Example 5.5.3. Let $d = m = 3$, let $n = 5$, and take the same matrix twice:

$$B^T = C = \begin{pmatrix} 1 & -2 & 1 & 0 & 0 \\ 0 & 1 & -2 & 1 & 0 \\ 0 & 0 & 1 & -2 & 1 \end{pmatrix}.$$

Writing (u, v, w) for the coordinates on the image, the map $\psi : \mathbb{A}^3 \dashrightarrow \mathbb{A}^3$ is given by the following three Laurent monomials in five linear forms:

$$u = \frac{x_1(x_1 - 2x_2 + x_3)}{(x_2 - 2x_1)^2}, \quad v = \frac{(x_2 - 2x_1)(x_2 - 2x_3)}{(x_1 - 2x_2 + x_3)^2}, \quad w = \frac{(x_1 - 2x_2 + x_3)x_3}{(x_2 - 2x_3)^2}.$$

The image of ψ is the surface Y in $\mathbb{A}^3$ that is defined by the polynomial

$$\begin{aligned}\underline{256u^3v^4w^3} &- 192u^2v^3w^2 - 128u^2v^2w^2 + 144u^2v^2w \\ &+ 144uv^2w^2 - \underline{27u^2v^2} - 6uv^2w - \underline{27v^2w^2} - 80uvw \\ &+ 18uv + \underline{16uw} + 18vw - \underline{4u} - \underline{4v} - \underline{4w} + \underline{1}.\end{aligned}$$

The Newton polytope of this polynomial is combinatorially isomorphic to the three-dimensional cube, with vertices corresponding to the underlined terms. We now derive the normal fan of this polytope from Theorem 5.5.1.

The linear space $X = \text{image}(B)$ is three dimensional in $\mathbb{A}^5$. Since all ten 3×3-minors of B are nonzero, the matroid M is the uniform matroid $U_{3,5}$. Hence $\text{trop}(X) = \text{trop}(U_{3,5})$ is the cone over the complete graph K_5, by Example 4.2.13. Its image under the linear map $C : \mathbb{R}^5 \to \mathbb{R}^3$ is the two-dimensional fan formed by the ten cones spanned by any two columns of C. Five of the rays of $\text{trop}(Y)$ are spanned by the columns of C, but there are more rays since the graph K_5 is not planar: any drawing of K_5 on the 2-sphere must have crossing edges. In our situation, exactly one pair of edges crosses; namely, the cones spanned by the first two and the last two columns of C intersect in the ray $\mathbb{R}_{\geq 0}(0,1,0)^T$. This is the sixth ray of $\text{trop}(Y)$. The resulting graph on the 2-sphere is the edge graph of an octahedron. The corresponding fan in $\mathbb{R}^3$ is the normal fan of a 3-cube.

The surface Y is a dehomogenized version of the discriminant in Example 3.3.3. Indeed, consider the monomial substitution that is obtained from C by labeling the rows and columns by u, v, w and a, b, c, d, e, respectively:

$$u = \frac{ac}{b^2}, \quad v = \frac{bd}{c^2}, \quad w = \frac{ce}{d^2}. \tag{5.5.4}$$

Making this substitution in the equation of Y and clearing denominators yields the discriminant of a binary quartic, displayed in Example 3.3.3. $\Diamond$

Example 5.5.3 is an instance of the construction of *tropical discriminants* by Dickenstein et al. [**DFS07**]. We now explain this result in general. Let $A = (\mathbf{a}_1, \mathbf{a}_2, \ldots, \mathbf{a}_n)$ be an $r \times n$-matrix with nonnegative integer entries such that $\text{rank}(A) = r$ and all column sums of A are equal. For any vector of coefficients $\mathbf{c} = (c_1, c_2, \ldots, c_n) \in K^n$, we have a homogeneous polynomial

$$f_{\mathbf{c}} \;=\; c_1 x^{\mathbf{a}_1} + c_2 x^{\mathbf{a}_2} + \cdots + c_n x^{\mathbf{a}_n}.$$

Here $x^{\mathbf{a}_j} = x_1^{a_{1j}} x_2^{a_{2j}} \cdots x_r^{a_{rj}}$. Consider the hypersurface $V(f_{\mathbf{c}})$ defined by this polynomial in the torus $T^{r-1} \subset \mathbb{P}^{r-1}$. This hypersurface is smooth for general $\mathbf{c}$. We are interested in those special $\mathbf{c}$ for which $V(f_{\mathbf{c}})$ has a singular point in T^{r-1}. The closure of the set of such $\mathbf{c}$ is a proper irreducible subvariety in K^n. This variety is called the *A-discriminant* and is denoted Δ_A.

Example 5.5.4. Let $r = 2$, let $n = 3$, and fix the matrix $A = \left(\begin{smallmatrix} 2 & 1 & 0 \\ 0 & 1 & 2 \end{smallmatrix}\right)$. The polynomial $f_{\mathbf{c}}$ is the binary quadric $c_1x_1^2 + c_2x_1x_2 + c_3x_2^2$. Its variety $V(f_{\mathbf{c}})$ is a hypersurface in $\mathbb{P}^1$. For general $\mathbf{c} = (c_1, c_2, c_3)$, this hypersurface is a set of two distinct points in T^1 given by the quadratic formula we all memorized in high school. Here, $V(f_{\mathbf{c}})$ having a singular point means that two roots of $f_{\mathbf{c}}$ coincide, so the discriminant of the binary form is zero. We conclude that the A-discriminant for the given 2×3-matrix is $\Delta_A = V(4c_1c_3 - c_2^2)$. ◊

The tropical discriminant $\text{trop}(\Delta_A)$ lives in $\mathbb{R}^n/\mathbb{R}\mathbf{1}$. Our goal is to compute this tropical variety using Theorem 5.5.1. To this end, we present a parameterization of Δ_A via products of linear forms. That parameterization is known as the *Horn uniformization*. We learned it from [**GKZ08**].

Set $d = n - r$, and let B be an $n \times d$-matrix whose columns span the kernel of A as a $\mathbb{Z}$-module. The rows of B form a *Gale transform* of the columns of A. We set $C = B^T$ as in Example 5.5.3. Also, let $\gamma : K^n \dashrightarrow K^d$ denote the monomial map given by the rows of C, as in (5.5.4).

Proposition 5.5.5. *With $C = B^T$ as above, the A-discriminant Δ_A is the inverse image of the variety Y in Theorem* 5.5.1 *under the monomial map γ.*

Proof. Consider the d-dimensional subspace $X = \text{image}(B) = \text{kernel}(A)$ of K^n. A vector $\mathbf{c}$ lies in X precisely when the hypersurface $V(f_{\mathbf{c}})$ is singular at the point $(1 : 1 : \cdots : 1)$ in $\mathbb{P}^{d-1}$. This implies that $V(f_{\mathbf{c}})$ is singular at $(x_1^{-1} : x_2^{-1} : \cdots : x_r^{-1})$ if and only if $(c_1x^{\mathbf{a}_1}, c_2x^{\mathbf{a}_2}, \ldots, c_nx^{\mathbf{a}_n})$ lies in X. The vectors of this form in $(K^*)^n$ are precisely those that are mapped into Y under the monomial map $(K^*)^n \to (K^*)^d$ given by C. Hence, the A-discriminant Δ_A is the closure of the set of all vectors $\mathbf{c} \in (K^*)^n$ that are mapped into Y under the map γ. This is precisely what was claimed. □

Example 5.5.6. We continue with Example 5.5.4. The kernel of A is the row span of $C = B^T = \begin{pmatrix} 1 & -2 & 1 \end{pmatrix}$. The map $\psi : K^1 \dashrightarrow K^1$ is the constant function that takes t to $(t)^1 \cdot (-2t)^{-2} \cdot (t)^1 = 1/4$. The discriminant Δ_A is the inverse image of the point $Y = \{1/4\}$ under $\gamma : K^3 \dashrightarrow K^1, \mathbf{c} \mapsto c_1c_3/c_2^2$. ◊

Corollary 5.5.7. *Modulo its lineality space, the tropical A-discriminant is*

$$\text{trop}(\Delta_A) \;=\; \text{trop}(\text{kernel}(A)) \;+\; \text{row-space}(A). \tag{5.5.5}$$

Proof. By Theorem 5.5.1, we have $\text{trop}(Y) = C \cdot \text{trop}(\text{row-space}(B^T)) = B^T \cdot \text{trop}(\text{kernel}(A))$. If follows from Proposition 5.5.5 that the tropical A-discriminant $\text{trop}(\Delta_A)$ is the inverse image of $\text{trop}(Y)$ under the linear map B^T. The kernel of that map is the row space of A, so (5.5.5) follows. □

Example 5.5.8. We continue with Example 5.5.4. The tropicalization of the kernel of A is the classical line spanned by $\mathbf{1} = (1, 1, 1)$ in $\mathbb{R}^3$. This is contained in the row space of A. Hence the tropical A-discriminant $\mathrm{trop}(\Delta_A) = \mathrm{trop}(V(4c_1c_3 - c_2^2)) \subset \mathbb{R}^3$ coincides with the row-space of A. ◊

The formulas (5.5.3) and (5.5.5) allow us to compute the tropicalizations of some interesting nonlinear varieties Y using matroid theory, namely $\mathrm{trop}(Y)$ is the image of $\mathrm{trop}(M)$ under the linear map C, where M is the matroid of B. This was stated with multiplicities in Theorem 5.5.1, and that works for Corollary 5.5.7 as well. Here is an example to check this.

Example 5.5.9. Example 3.3.3 concerns the A-discriminant for

$$A = \begin{pmatrix} 0 & 1 & 2 & 3 & 4 \\ 4 & 3 & 2 & 1 & 0 \end{pmatrix}.$$

We saw that 11 of the 12 maximal cones σ in $\mathrm{trop}(\Delta_A)$ have multiplicity 1, while one has multiplicity 2. We also see this in Example 5.5.3 via (3.6.2) and (5.5.5). After refining $\Sigma = \mathrm{trop}(U_{3,5})$, as required for (3.6.2), each of the 12 maximal cones σ' in $\Sigma' = \mathrm{trop}(\Delta_A)$ is the image of a unique cone σ in Σ, identified by a 3×2-submatrix of $B^T = C$. The lattice index $[N' : \phi(N_\sigma)]$ in (3.6.2) is the gcd of the three 2×2-minors of that submatrix. For nine of the ten pairs of columns of B^T, that gcd is 1. The exception is the pair $\{2, 4\}$, for which the gcd is 2. The resulting image is the unique multiplicity two cone in Σ'. ◊

For most matrices A, the A-discriminant Δ_A is a hypersurface. It is a delicate problem to characterize those special cases where this fails. We can approach this question tropically. The vector $\mathbf{1} = (1, 1, \ldots, 1)$ is in the kernel of the matrix B^T, hence B^T defines a classically linear map $\mathbb{R}^n/\mathbb{R}\mathbf{1} \to \mathbb{R}^r$. The tropical linear space $\mathrm{trop}(\mathrm{kernel}(A))$ lives in $\mathbb{R}^n/\mathbb{R}\mathbf{1}$ where it has dimension $r - 1 = n - d - 1$. Hence the tropical discriminant (5.5.5) is a balanced fan of dimension at most $r-1$ in $\mathbb{R}^r$. It has dimension exactly $r-1$ when B^T is injective on at least one of the maximal cones of $\mathrm{trop}(\mathrm{kernel}(A))$. Here, for maximal cones, we can use either those in the order complex of Definition 4.1.9 or those in the Bergman fan of Corollary 4.2.11.

Suppose now that Δ_A is a hypersurface. We identify this hypersurface with its unique (up to sign) irreducible defining polynomial $\Delta_A \in \mathbb{Z}[c_1, c_2, \ldots, c_n]$. The balanced weighted fan (5.5.5) consists of the codimension-1 cones in the normal fan of the Newton polytope $P = \mathrm{Newt}(\Delta_A)$. According to Proposition 3.3.10 and Remark 3.3.11, this Newton polytope can be uniquely recovered from the tropical hypersurface $\mathrm{trop}(\Delta_A)$:

Corollary 5.5.10. *Using the formula* (5.5.5), *we can test whether* Δ_A *is a hypersurface, and, if yes, we can derive its Newton polytope from the tropical hypersurface* $\mathrm{trop}(\Delta_A)$ *using the algorithm in the proof of Proposition* 3.3.10.

See [**DFS07**, Theorem 1.2] for an explicit combinatorial formula, derived from this approach, for all extreme monomials of the A-discriminant Δ_A.

We conclude our discussion of A-discriminants by emphasizing the importance of matroid theory. The material in Sections 4.1 and 4.2 enables us to compute the right-hand side in the equation (5.5.5) of Corollary 5.5.7.

We wish to also point out a connection to Proposition 4.5.1. Consider a tropical hypersurface that is tropically smooth. The coefficient vector $\mathbf{c}$ of such a hypersurface is a point in $\mathbb{R}^n \backslash \text{trop}(\Delta_A)$, so, by the Fundamental Theorem 3.2, all of its classical lifts $V(f_{\mathbf{c}})$ are smooth hypersurface over K.

One way to interpret the Horn uniformization (Proposition 5.5.5) is that Δ_A is the Hadamard product of the kernel of A with the toric variety defined by A. It therefore makes sense to briefly talk about Hadamard products under tropicalization. Let X and Y be subvarieties of the torus T^n over the valued field K. Their *Hadamard product* $X \star Y$ is the set of all vectors $(x_1y_1, x_2y_2, \ldots, x_ny_n)$ where $(x_1, x_2, \ldots, x_n) \in X$ and $(y_1, y_2, \ldots, y_n) \in Y$. Equivalently, $X \star Y$ is the image of $X \times Y$ under the monomial map from $T^{2n} = T^n \times T^n$ to T^n given by multiplying corresponding coordinates. Its Zariski closure $\overline{X \star Y}$ is a closed subvariety of the torus T^n. The tropicalization of this subvariety can be computed combinatorially as follows.

Proposition 5.5.11. *The tropicalization of the Hadamard product of two varieties in T^n is the Minkowski sum of their tropicalizations. In symbols,*

$$\text{trop}(\overline{X \star Y}) \quad = \quad \text{trop}(X) + \text{trop}(Y). \tag{5.5.6}$$

Proof. It is easy to see, using either of the characterizations in Theorem 3.2.3, that tropicalization commutes with direct products of varieties:

$$\text{trop}(X \times Y) \quad = \quad \text{trop}(X) \times \text{trop}(Y). \tag{5.5.7}$$

We now apply Corollary 3.2.13 where ϕ is the monomial map $T^{2n} \to T^n$ given by multiplying corresponding coordinates. Its tropicalization $\text{trop}(\phi)$ is the linear map that adds two vectors. This gives the desired conclusion:

$$\begin{aligned} \text{trop}(\overline{\phi(X \times Y)}) &= \text{trop}(\phi)\big(\text{trop}(X \times Y)\big) \\ &= \text{trop}(\phi)\big(\text{trop}(X) \times \text{trop}(Y)\big) \;=\; \text{trop}(X) + \text{trop}(Y). \end{aligned}$$

The equation (5.5.6) holds not just set-theoretically, but also as weighted balanced fans. To get weights on the Minkowski sum, we use (3.6.2) to push forward the product weights on (5.5.7) to (5.5.6) under the map $\text{trop}(\phi)$. □

5.6. Exercises

(1) The eigenspaces of the second and the fourth matrix in Example 5.1.4 are tropical polytopes of dimension 2 and 3, respectively. Draw these objects and determine their tropical complexes.

(2) A *polytrope* is a subset of $\mathbb{R}^n/\mathbb{R}\mathbf{1}$ that is both tropically convex and classically convex. Show that every polytrope arises as the eigenspace of a tropical $n \times n$-matrix.

(3) Determine the eigenvalue and all eigenvectors of the tropical matrix

$$A = \begin{pmatrix} 4 & 4 & 5 \\ 1 & 3 & 2 \\ 1 & 3 & 4 \end{pmatrix}.$$

What is the determinant of this matrix? Compute the image of the tropical linear map $\mathbb{R}^3/\mathbb{R}\mathbf{1} \to \mathbb{R}^3/\mathbb{R}\mathbf{1}$ that is defined by A.

(4) Find the image, eigenvalue, and eigenspace of the $n\times n$-matrix whose diagonal entries are 1 and whose off-diagonal entries are 0.

(5) Prove that the three different definitions of type given in Section 5.2 are equivalent.

(6) Consider the nearest point map $\mathbb{R}^3/\mathbb{R}\mathbf{1} \to P$, defined in (5.2.3), for tropical quadrilateral P in Figure 5.2.3. Give an explicit piecewise linear formula for this map, and show what it does in the picture.

(7) Consider the two point configurations given by a matrix, namely, by the rows and by the columns. How are their types related?

(8) Verify the duality of tropical complexes in Theorem 5.2.21 for the configuration in Example 5.2.9. Draw both tropical triangles and describe the corresponding triangulation of $\Delta_2 \times \Delta_2$.

(9) Pick three random 3×4-matrices with real entries. For each of your matrices, locate the combinatorial type of its tropical column span among the 35 pictures in Figure 5.2.4.

(10) The following matrix is due to Shitov [**Shi13**, Example 1.9]:

$$\begin{pmatrix} 0 & 0 & 4 & 4 & 4 & 4 \\ 0 & 0 & 2 & 4 & 1 & 4 \\ 4 & 4 & 0 & 0 & 4 & 4 \\ 2 & 4 & 0 & 0 & 2 & 4 \\ 4 & 4 & 4 & 4 & 0 & 0 \\ 2 & 4 & 1 & 4 & 0 & 0 \end{pmatrix}.$$

Show that this 6×6-matrix has tropical rank 4 but Kapranov rank 5.

(11) What is the maximal number of vertices of a four-dimensional polytrope? Answer this question for both classical and tropical vertices.

(12) Find the maximal Barvinok rank of any 5×5-matrix whose entries are 0 or 1. Do you have a conjecture for $n \times n$-matrices over $\{0, 1\}$?

(13) Consider a general arrangement of k tropical hyperplanes in $\mathbb{R}^n$. How many connected components does the complement of such an arrangement have?

(14) Compute the tropicalization of the row-space over $\mathbb{C}\{\{t\}\}$ of the matrix F_n in Example 5.3.14. Draw a picture of this tropical linear space. How does your answer depend on the choice of $a_3, a_4, \ldots, a_n$?

(15) In classical linear algebra, the rank of a matrix can drop by at most 1 when deleting one row of the matrix. Is the same true for Barvinok rank? Is it true for tropical rank? For Kapranov rank?

(16) Describe the hyperplane arrangement referred to in Corollary 5.3.7. Find a formula (in terms of d and n) for its number of regions.

(17) The tropical determinantal variety $\operatorname{trop}(V(J_r))$ is defined via a field. Does this polyhedral complex depend on the characteristic? What does this mean for the three characterizations in Theorem 5.3.11?

(18) Classically, a convex polyhedron in standard form is given as the set of nonnegative points in an affine-linear subspace L of $\mathbb{R}^n$. Tropicalize this definition. What is the set of "positive points" in $\operatorname{trop}(L)$? (*Hint*: [**AKW06**]). Show that your set is tropically convex.

(19) Write out explicitly the cocircuit matrix $\mathbf{C}(M)$ when M is the non-Pappus matroid in Figure 4.7.1. Determine the tropical rank of M.

(20) Can tropical polytopes be represented as intersections of tropical half-spaces? How would you define facets of a tropical polytope?

(21) Draw the seven generic tropical planes in 5-spaces by augmenting the seven pictures in Figure 5.4.1 with the seven "trees at infinity" given by the seven rows of Table 5.4.1.

(22) Formulate and prove Theorem 5.5.1 for the more general case when the matrix B has entries in a field with nontrivial valuation.

(23) According to Theorem 5.4.1, the Dressian $\operatorname{Dr}(3,7)$ has cells that are not simplices. Identify such a cell and explain how it gets subdivided in $\operatorname{trop}(G^0(3,7))$. Draw the corresponding tree arrangements. Draw the bounded cells (as in Figure 5.4.1) of your tropical planes. (*Hint*: First better understand the cell FFFGG in Example 4.4.10.)

(24) Given an example of two closed subvarieties X and Y in a torus T^n such that the Hadamard product $X \star Y$ is not closed.

(25) Tropicalize the variety X of pairs of intersecting lines in 3-space. This is a subvariety of $\mathbb{P}^5 \times \mathbb{P}^5$, as the two lines can be given by their Plücker vectors $\mathbf{p} = (p_{12}, p_{13}, p_{14}, p_{23}, p_{24}, p_{34})$ and $\mathbf{q} = (q_{12}, q_{13}, q_{14}, q_{23}, q_{24}, q_{34})$.
   (a) Show that X is a complete intersection of codimension 3.

(b) Compute trop(X) from the prime ideal I_X, e.g., using `Gfan`.
(c) Show that X can be defined by the parametric representation

$$p_{ij} = x_i x_j (y_i - y_j) \text{ and } q_{ij} = x_i x_j (z_i - z_j) \quad \text{for } 1 \leq i < j \leq 4.$$

(d) Compute trop(X) using Theorem 5.5.1.
(e) Prove: The three generators of I_X are a tropical basis.

(26) Prove Theorem 1.5.2.

(27) Let $f_1, f_2, f_3, g_1, g_2, g_3$ be polynomials in one variable, and let $X \subset T^3$ be the surface given by the parameterization

$$x = f_1(s)g_1(t), \quad y = f_2(s)g_2(t), \quad z = f_3(s)g_3(t).$$

Explain how to find trop(X) and how to find the Newton polytope of X. (*Hint*: Hadamard product of two curves as in Theorem 1.5.2.) Carry out your algorithm for one example.

(28) Prove that $\text{trop}(G^0(n-2, n)) = \text{Dr}(n-2, n)$ for all $n \geq 3$.

(29) Show that $\text{Dr}(4, 6)$ has 105 maximal cells, in two symmetry classes. Describe the two corresponding tropical 3-plane $L_{\mathbf{w}}$ in $\mathbb{R}^6/\mathbb{R}\mathbf{1}$. In each case, draw the complex of bounded faces of $L_{\mathbf{w}}$.

(30) Let X be the variety in matrix space $(K^*)^{m \times n}$ defined by the parameterization $x_{ij} = a_i b_j (c_i + d_j)$ via monomials in linear forms. Do you know this variety? How would you compute trop(X)?

(31) Find a 3×6-matrix with entries in the Puiseux series field $\mathbb{C}\{\{t\}\}$ whose row space has its tropicalization of type FFFGG. This refers to the notation used in Example 4.4.10, Figure 5.4.1, and Table 5.4.1.

Chapter 6

Toric Connections

The theory of toric varieties is one of the main interfaces between combinatorics and algebraic geometry. In this chapter we will see how the tropical connection between these fields is intimately connected with the toric one.

A toric variety is a variety with an action of the algebraic torus T^n on it that contains a dense copy of T^n. It decomposes into a union of T^n-orbits. We shall tropicalize this notion to obtain a tropical toric variety. The tropicalization of a projective toric variety is homeomorphic to the associated lattice polytope. If we tropicalize subvarieties of a projective toric variety, then we obtain compact objects which live inside that polytope.

Tropical geometry answers some (a priori nontropical) toric questions. Given a subvariety Z of a toric variety, we see in Section 6.3 how its tropicalization $\mathrm{trop}(Z)$ records the torus orbits of the toric variety that intersect Z.

A normal toric variety is determined by the combinatorial data of a rational polyhedral fan. For $Y \subset T^n$, a choice of fan structure on $\mathrm{trop}(Y)$ then determines a toric variety with torus T^n. The closure of Y in this toric variety is a compactification of Y. This extends the story begun in Section 1.8. Conversely, a good choice of compactification of $Y \subset T^n$ leads to a computation of $\mathrm{trop}(Y)$. Degenerations of Y are also controlled by the tropical variety $\mathrm{trop}(Y)$. We study these in Section 6.6 before turning to the tropical and toric approaches to intersection theory in the last section.

In this chapter we assume familiarity with modern algebraic geometry. In particular, we assume the basics of toric geometry, as in the books [**CLS11**], [**Ful93**], or [**Oda88**], and just briefly review notation in Section 6.1.

6.1. Toric Background

A normal toric variety is defined by a rational fan Σ in $N_{\mathbb{R}} = N \otimes \mathbb{R} \cong \mathbb{R}^n$ for a lattice $N \cong \mathbb{Z}^n$. Since Σ is rational, each ray has a primitive generator in N. The lattice dual to N is $M = \mathrm{Hom}(N, \mathbb{Z})$. The real vector space dual to $N_{\mathbb{R}}$ is $M_{\mathbb{R}} = M \otimes \mathbb{R} \cong \mathbb{R}^n$. We will work with toric varieties X_Σ over a field K with a valuation. The torus T^n of X_Σ is $N \otimes K^* \cong \mathrm{Hom}(M, K^*) \cong (K^*)^n$. We denote by $\Sigma(d)$ the set of d-dimensional cones of the fan Σ.

Each cone $\sigma \in \Sigma$ determines a local chart $U_\sigma = \mathrm{Spec}(K[\sigma^\vee \cap M])$, where $\sigma^\vee = \{\mathbf{u} \in M : \mathbf{u} \cdot \mathbf{v} \geq 0 \text{ for all } \mathbf{v} \in \sigma\}$ is the dual cone. For the cone $\sigma = \{\mathbf{0}\}$, we have $\sigma^\vee = M_{\mathbb{R}}$, so $K[\sigma^\vee \cap M] = K[M]$. This is the Laurent polynomial ring, and $U_\sigma \cong T^n$. Every affine normal toric variety has the form U_σ for some cone $\sigma \subset N_{\mathbb{R}}$. The cone σ also determines a T^n-orbit $\mathcal{O}_\sigma \cong (K^*)^{n-\dim(\sigma)}$. The closure in X_Σ of the orbit $\mathcal{O}_\sigma$ is denoted by $V(\sigma)$.

Example 6.1.1. (1) Let Σ be the fan in $\mathbb{R}^n$ with $n+1$ rays spanned by $\mathbf{e}_1, \ldots, \mathbf{e}_n$ and $\mathbf{e}_0 = -\sum_{i=1}^n \mathbf{e}_i$, and cones spanned by $\{\mathbf{e}_i : i \in \sigma\}$ for any proper subset $\sigma \subset \{0, \ldots, n\}$. The case $n = 2$ is shown in Figure 6.1.1. Then $X_\Sigma \cong \mathbb{P}^n$. The orbit indexed by σ consists of the points $(x_0 : \cdots : x_n) \in \mathbb{P}^n$ with $x_i = 0$ for $i \in \sigma$ and $x_i \neq 0$ for $i \notin \sigma$.

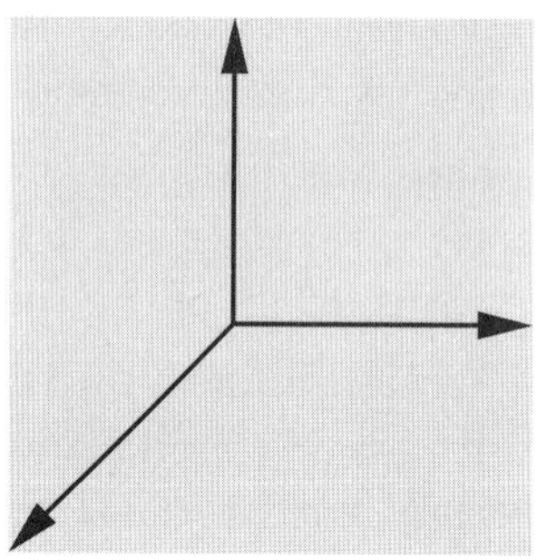

Figure 6.1.1. The fan of $\mathbb{P}^2$.

(2) Let Σ be the fan in $\mathbb{R}^2$ with rays $(1,0), (1,1), (0,1)$ and maximal cones $\mathrm{pos}\{(1,0),(1,1)\}$ and $\mathrm{pos}\{(0,1),(1,1)\}$. The toric surface X_Σ is the blow-up of $\mathbb{A}^2$ at the origin.

The only smooth affine toric varieties are the products $\mathbb{A}^d \times T^{n-d}$. These correspond to d-dimensional cones $\sigma \subset N_{\mathbb{R}}$ generated by part of a basis for the lattice N. In general, a toric variety X_Σ is smooth if and only if every cone $\sigma \in \Sigma$ is generated by part of a basis for N. We call such Σ a *smooth fan*. Resolution of singularities for toric varieties is a combinatorial operation, and works in arbitrary characteristic. Specifically, given any fan

Σ, there is a smooth fan $\widetilde{\Sigma}$ that refines Σ, and the refinement of fans induces a proper birational map $\pi\colon X_{\widetilde{\Sigma}} \to X_\Sigma$. See [**Ful93**, Section 2.2] for details.

Toric varieties also have a global quotient description that generalizes the construction of projective space $\mathbb{P}^n$ as the quotient of $\mathbb{A}^{n+1} \setminus \{0\}$ by K^*. We recall this here in the case that the fan Σ is smooth. Number the rays of Σ from 1 to s, and let $S = K[x_1, \ldots, x_s]$ be the polynomial ring with one generator for each ray. The ring S is graded by the class group of X_Σ. This is the group $A_{n-1}(X_\Sigma)$, which can be defined as the cokernel of the following map. Identify $\mathbb{Z}^s$ with the group of torus-invariant Weil divisors. This has basis $D_1, \ldots, D_s$, where D_i is the divisor corresponding to the ith ray of Σ. We map M to $\mathbb{Z}^s$ given by taking $\mathbf{u} \in M$ to $\sum_{i=1}^{s} (\mathbf{u} \cdot \mathbf{v}_i) D_i$, where $\mathbf{v}_i$ is the first lattice point on the ith ray of Σ, and D_i is the torus-invariant divisor corresponding to that ray. Thus $A_{n-1}(X_\Sigma)$ is given by the exact sequence

$$0 \to M \cong \mathbb{Z}^n \xrightarrow{V} \mathbb{Z}^s \xrightarrow{\deg} A_{n-1}(X_\Sigma) \to 0, \tag{6.1.1}$$

where V is the $s \times n$-matrix whose ith row is $\mathbf{v}_i$. The hypothesis that Σ is smooth ensures that $A_{n-1}(X_\Sigma)$ is torsion-free. We grade S by setting $\deg(x_i) = [D_i] = \deg(\mathbf{e}_i) \in A_{n-1}(X_\Sigma)$. The graded ring S is the *Cox homogeneous coordinate ring* $\mathrm{Cox}(X_\Sigma)$ of the toric variety X_Σ.

Applying the functor $\mathrm{Hom}(-, K^*)$ to the exact sequence (6.1.1) gives

$$\mathrm{Hom}(M, K^*) = T^n \xleftarrow{V^T} \mathrm{Hom}(\mathbb{Z}^s, K^*) \cong (K^*)^s \longleftarrow H \leftarrow 0. \tag{6.1.2}$$

The group $H = \mathrm{Hom}(A_{n-1}(X_\Sigma), K^*)$ is also a torus, since $A_{n-1}(X_\Sigma)$ is torsion-free when Σ is smooth. The sequence (6.1.2) is also exact. Here V^T denotes the map $(K^*)^s \to T$ that takes $t = (t_1, \ldots, t_s)$ to $(t^{(V^T)_1}, \ldots, t^{(V^T)_n})$, where $t^{(V^T)_i} = \prod_{j=1}^{s} t_j^{V_{ji}}$. The inclusion of H into $(K^*)^s$ gives an action of H on $\mathbb{A}^s$. The *irrelevant ideal* in $\mathrm{Cox}(X_\Sigma)$ is

$$B = \big\langle \prod_{\mathbf{v}_i \notin \sigma} x_i : \sigma \in \Sigma \big\rangle. \tag{6.1.3}$$

Those familiar with combinatorial commutative algebra [**MS05**, Chapter 6] will note that this is the Alexander dual of the Stanley–Reisner ideal of the simplicial complex corresponding to Σ. The torus H acts on $\mathbb{A}^s \setminus V(B)$, and

$$X_\Sigma = (\mathbb{A}^s \setminus V(B))/H. \tag{6.1.4}$$

For any cone $\sigma \in \Sigma$, consider the restriction of the action by H to the coordinate subspace $\{\mathbf{x} \in \mathbb{A}^s : x_i = 0 \text{ for } \mathbf{v}_i \in \sigma\}$ with $V(B)$ removed. The quotient of that action is the closed orbit $V(\sigma)$ corresponding to σ.

The torus orbit $\mathcal{O}_\sigma$ is the quotient modulo H of the set

$$\{\mathbf{x} \in \mathbb{A}^s : x_i = 0 \text{ for } \mathbf{v}_i \in \sigma \text{ and } x_i \neq 0 \text{ for } \mathbf{v}_i \notin \sigma\} \setminus V(B).$$

When X_Σ is (quasi)projective, the quotient is a geometric invariant theory (GIT) construction of X_Σ as a quotient $\mathbb{A}^s /\!\!/_\alpha H$. Here α is a character of the torus H that is very ample when regarded as an element of $A_{n-1}(X_\Sigma)$. See [**Dol03**, Chapter 12] or [**CLS11**, Chapter 5] for more information about GIT and toric varieties.

Example 6.1.2. (1) Let $X_\Sigma = \mathbb{P}^n$. The Cox ring is the usual homogeneous coordinate ring $S = K[x_0, x_1, \ldots, x_n]$. The grading is also the standard grading $\deg(x_i) = 1 \in \mathbb{Z} \cong A_{n-1}(\mathbb{P}^n)$. The irrelevant ideal B is the usual irrelevant ideal $\mathfrak{m} = \langle x_0, \ldots, x_n\rangle$, and $V(B)$ is the origin in $\mathbb{A}^{n+1}$. This means that the quotient description is the familiar construction of $\mathbb{P}^n$ as $(\mathbb{A}^{n+1}\backslash\{0\})/K^*$. In this sense the Cox construction of toric varieties generalizes that of $\mathbb{P}^n$.

(2) Let $X_\Sigma = (\mathbb{P}^1)^3$. This has fan $\Sigma \subset \mathbb{R}^3$ with rays through the vectors $\pm\mathbf{e}_1, \pm\mathbf{e}_2, \pm\mathbf{e}_3$, and eight maximal cones $\mathrm{pos}(\epsilon_1\mathbf{e}_1, \epsilon_2\mathbf{e}_2, \epsilon_3\mathbf{e}_3)$, where $(\epsilon_1, \epsilon_2, \epsilon_3) \in \{-1, 1\}^3$. The Cox ring is $S = K[x_1, y_1, x_2, y_2, x_3, y_3]$, where x_i corresponds to the ray through $\mathbf{e}_i$, and y_i corresponds to the ray through $-\mathbf{e}_i$. The class group of $(\mathbb{P}^1)^3$ is isomorphic to $\mathbb{Z}^3$, and the grading is given by $\deg(x_i) = \deg(y_i) = \mathbf{e}_i \in \mathbb{Z}^3$. This gives an action of $H \cong (K^*)^3$ on $\mathbb{A}^6$ via $(t_1, t_2, t_3)\cdot(x_1, y_1, x_2, y_2, x_3, y_3) = (t_1x_1, t_1y_1, t_2x_2, t_2y_2, t_3x_3, t_3y_3)$. The irrelevant ideal is

$$\begin{aligned} B &= \langle x_1x_2x_3, x_1x_2y_3, x_1y_2x_3, x_1y_2y_3, y_1x_2x_3, y_1x_2y_3, y_1y_2x_3, y_1y_2y_3\rangle \\ &= \langle x_1, y_1\rangle \cap \langle x_2, y_2\rangle \cap \langle x_3, y_3\rangle. \end{aligned}$$

We have $\mathbb{A}^6 \setminus V(B) \cong (\mathbb{A}^2 \setminus (0,0))^3$, and so $(\mathbb{A}^6 \setminus V(B))/H \cong (\mathbb{P}^1)^3$.

Every subvariety Y of a smooth toric variety X_Σ arises, under the quotient construction (6.1.4), from an H-invariant subvariety of $\mathbb{A}^s$ that is not contained in $V(B)$. The ideal I_Y of such a subvariety lives in $\mathrm{Cox}(X_\Sigma) = K[x_1, \ldots, x_s]$ and is homogeneous with respect to the H-grading. It can be assumed to be B-saturated: $(I_Y : B^\infty) = I_Y$. Conversely, every radical ideal in $K[x_1, \ldots, x_s]$ that has these two properties specifies a subvariety of X_Σ. In the next section, this description will be used to compute the tropicalization of a subvariety Y in a toric variety X_Σ. To do this, we compute the tropical variety in $\mathbb{R}^s$ from the ideal as in Theorem 6.2.15(2), and then we take the quotient modulo the additive version (6.2.1) of H.

Example 6.1.3. Consider the plane in T^3 defined by the equation $x+y+z = 1$, and let $\overline{Y}$ denote its closure in $X_\Sigma = (\mathbb{P}^1)^3$. The corresponding ideal in the Cox ring $S = K[x_1, x_2, y_1, y_2, z_1, z_2]$ is the principal prime ideal

$$I_{\overline{Y}} = \langle x_1y_2z_2 + x_2y_1z_2 + x_2y_2z_1 - x_2y_2z_2\rangle.$$

The open surface $\overline{Y} \cap T^3$ is the complement of four lines in a plane $\mathbb{P}^2$, as in Proposition 4.1.1. Its boundary $\overline{Y}\backslash T^3$ consists of six irreducible curves in

$(\mathbb{P}^1)^3$. These curves are defined by the six minimal primes of the ideal

$$(6.1.5) \qquad \big(I_Y + \langle x_1x_2y_1y_2z_1z_2\rangle : B^\infty\big). \qquad \diamondsuit$$

6.2. Tropicalizing Toric Varieties

We saw in Chapter 3 that the tropicalization of a subvariety of a torus T^n is a polyhedral complex that lives in $\mathbb{R}^n = \operatorname{trop}(T^n)$. We now extend the notion of tropicalization from T^n to an arbitrary toric variety X_Σ. This will allow us to tropicalize subvarieties of X_Σ. We use coordinate-free language for the theory but translate this into coordinates to calculate examples. The tropicalization map $T^n \to N_\mathbb{R}$ sends $\mathbf{v} \otimes a \in N \otimes K^*$ to $\mathbf{v} \otimes \operatorname{val}(a)$. Equivalently, writing $T^n = \operatorname{Hom}(M, K^*)$, this map sends $\phi\colon M \to K^*$ to $\operatorname{val} \circ \phi\colon M \to \mathbb{R}$ in $N_\mathbb{R}$. We begin with a few simple motivating examples.

One of the simplest toric varieties is the affine line $\mathbb{A}^1$. We consider $\mathbb{A}^1_K$ where K is an algebraically closed field with a nontrivial valuation $\operatorname{val}\colon K \to \mathbb{R} \cup \{\infty\}$. Generalizing the characterization in part (3) of Theorem 3.2.3, we consider $\overline{\mathbb{R}} = \mathbb{R} \cup \{\infty\} = \{\operatorname{val}(x) : x \in \mathbb{A}^1\}$ to be the tropicalization of $\mathbb{A}^1$. This is the union of $\mathbb{R} = \operatorname{trop}(T^1)$, and a point $\{\infty\}$, which we regard as the tropicalization of the origin. The toric variety $\mathbb{A}^1$ has two torus orbits: T^1 and the origin, and its tropicalization is the union of the two tropicalizations. More generally, the tropicalization of affine n-space $\mathbb{A}^n$ is $(\overline{\mathbb{R}})^n$. Again, this is the union of the tropicalizations of the 2^n torus orbits on $\mathbb{A}^n$.

The tropical affine line $\overline{\mathbb{R}} = \mathbb{R} \cup \{\infty\}$ is a semigroup under the usual addition (with $a + \infty = \infty$ for all $a \in \overline{\mathbb{R}}$). This is the *multiplicative* semigroup structure on $\overline{\mathbb{R}}$, here regarded as the tropical semiring with operations minimum and addition. This is sometimes denoted by $\mathbb{T}$ in the literature (often by authors who use the maximum convention). We place a topology on $\overline{\mathbb{R}}$ that extends the usual topology on $\mathbb{R}$, by taking intervals of the form (a, b) for $a, b \in \mathbb{R}$ and $(a, \infty]$ for $a \in \mathbb{R}$ to be a basis for the topology.

We next tropicalize the projective line. There are two natural ways to do this. The first is to follow the construction of $\mathbb{P}^1$ as the union of two copies of $\mathbb{A}^1$. This suggests gluing together two copies of $\overline{\mathbb{R}}$. In the classical construction the two copies of $\mathbb{A}^1$ are glued by identifying x and x^{-1}. Tropically, we identify two copies of $\overline{\mathbb{R}}$ by identifying x and $-x$ for $x \in \mathbb{R}$. Classically, $\mathbb{P}^1$ is the union of three torus orbits: the torus T^1 and two torus-fixed points. The tropical projective line also has this property; it is the union of the tropical torus $\mathbb{R} = \operatorname{trop}(T^1)$ with two copies of ∞. This is illustrated in Figure 6.2.1. We give $\operatorname{trop}(\mathbb{P}^1)$ the quotient topology coming from identifying the two copies of $\overline{\mathbb{R}}$ over the common open set $\mathbb{R}$. Note that this is homeomorphic to the interval $[0, 1]$ in the standard topology on $\mathbb{R}$.

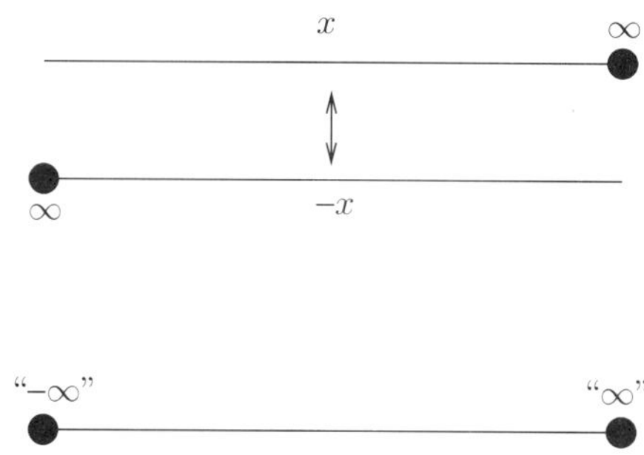

Figure 6.2.1. The tropical projective line.

The other approach is to follow the classical construction of $\mathbb{P}^1$ as the quotient of $\mathbb{A}^2$, with the origin removed, by K^*. We interpret 0 here as the additive identity, so tropically $\mathbb{A}^2\backslash\{(0,0)\}$ becomes $\overline{\mathbb{R}}^2\backslash\{(\infty,\infty)\}$. The diagonal multiplicative action of K^* on $\mathbb{A}^2$ becomes the action of $\mathbb{R}$ on $\overline{\mathbb{R}}^2$ given by translation: $a\cdot(x,y) = (a\odot x, a\odot y) = (a+x, a+y)$. We can thus define the tropical projective line to be $(\overline{\mathbb{R}}^2\backslash\{(\infty,\infty)\})/\mathbb{R}$. This is the union of the quotient vector space $\mathbb{R}^2/\mathbb{R}(1,1)$ and the two points $(\{\infty\}\times\mathbb{R})/\mathbb{R}$ and $(\mathbb{R}\times\{\infty\})/\mathbb{R}$. This coincides with the first description of tropical $\mathbb{P}^1$.

The general definition of a tropical toric variety can be also be viewed in these two ways. We first give the construction of a tropical toric variety from its fan and then show the equivalence with the quotient construction. For a cone τ of a fan we write $\operatorname{span}(\tau)$ for the subspace of $N_{\mathbb{R}}$ spanned by τ.

Definition 6.2.1. Let Σ be a rational polyhedral fan in $N_{\mathbb{R}}$. For each cone $\sigma\in\Sigma$, we consider the $(n-\dim(\sigma))$-dimensional vector space $N(\sigma) = N_{\mathbb{R}}/\operatorname{span}(\sigma)$. As a set, the tropical toric variety X_Σ^{trop} is the disjoint union

$$X_\Sigma^{\mathrm{trop}} = \coprod_{\sigma\in\Sigma} N(\sigma).$$

To place a topology on X_Σ^{trop}, we associate to each cone $\sigma\in\Sigma$ the space

$$U_\sigma^{\mathrm{trop}} = \operatorname{Hom}(\sigma^\vee\cap M, \overline{\mathbb{R}})$$

of semigroup homomorphisms from $(\sigma^\vee\cap M, +)$ to $(\overline{\mathbb{R}},\odot)$. Note that if $\phi(\mathbf{u})=\infty$ for some $\phi\in U_\sigma^{\mathrm{trop}}$, then $\phi(\mathbf{u}+\mathbf{v})=\infty$ for all $\mathbf{v}\in\sigma^\vee\cap M$. Also, if $\phi(\mathbf{u}+\mathbf{v})=\infty$, then at least one of $\phi(\mathbf{u})$ and $\phi(\mathbf{v})$ equals ∞. Thus the set $\{\mathbf{u} : \phi(\mathbf{u})\neq\infty\}$ equals $\sigma^\vee\cap\tau^\perp\cap M$ for some face τ of σ. Here $\tau^\perp = \{\mathbf{u}\in M_{\mathbb{R}} : \mathbf{u}\cdot\mathbf{v} = 0 \text{ for all } \mathbf{v}\in\tau\}$. The map ϕ induces a group homomorphism $\phi\colon M\cap\tau^\perp\to\mathbb{R}$, which in turn induces a group homomorphism $\tilde{\phi}\colon M\to\mathbb{R}$ with $\tilde{\phi}(\mathbf{v})=0$ for $\mathbf{v}\notin\tau^\perp$. Hence ϕ induces an element of $N(\tau)$, and all elements of $N(\tau)$ arise in this way. Thus $U_\sigma^{\mathrm{trop}} = \coprod_{\tau\preceq\sigma} N(\tau)$.

We place the pointwise-convergence topology on the set U_σ^{trop}. This is the topology induced from the product topology on products of $\overline{\mathbb{R}}$, where we identify U_σ^{trop} as a subset of the product space $(\overline{\mathbb{R}})^{\sigma^\vee \cap M}$. Explicitly, U_σ^{trop} is the subset of those maps from the infinite set $\sigma^\vee \cap M$ to $\overline{\mathbb{R}}$ that are homomorphisms of semigroups. If τ is a face of a cone $\sigma \in \Sigma$, then $\sigma^\vee \cap M$ is a subsemigroup of $\tau^\vee \cap M$, and the map $U_\tau^{\text{trop}} \to U_\sigma^{\text{trop}}$ given by $p \mapsto p|_{\sigma^\vee \cap M}$ is injective. This follows from the fact that $\tau^\vee \cap M = \sigma^\vee \cap M + (\tau^\perp \cap M)$, and for all $\mathbf{u} \in \tau^\perp \cap M$ we can write $\mathbf{u} = \mathbf{u}_1 - \mathbf{u}_2$ with $\mathbf{u}_1, \mathbf{u}_2 \in \sigma^\vee \cap \tau^\perp$. If $p \colon \tau^\vee \cap M \to \overline{\mathbb{R}}$ is a homomorphism, then $p(\mathbf{u}) + p(\mathbf{u}_2) = p(\mathbf{u}_1)$. So, if $p(\mathbf{u}_2) \neq \infty$, we have $p(\mathbf{u}) = p(\mathbf{u}_1) - p(\mathbf{u}_2)$. If $p(\mathbf{u}_2) = \infty$, then since $-\mathbf{u}_2 \in \tau^\perp$ we have $p(\mathbf{0}) = p(\mathbf{u}_2) + p(-\mathbf{u}_2)$, so $p(\mathbf{0}) = \infty$. We conclude that $p(\mathbf{u}') = \infty$ for all $\mathbf{u}' \in \tau^\vee \cap M$, so p is the constant ∞ function. Thus the map p is determined by $p|_{\sigma^\vee \cap M}$. This means that the restriction is injective.

The image of U_τ^{trop} in U_σ^{trop} consists of all maps p with $p(\mathbf{u}) \neq \infty$ for all $\mathbf{u} \in \tau^\perp$, plus the constant ∞ map. This means that the inclusion $\tau \to \sigma$ identifies U_τ^{trop} with an open subset of U_σ^{trop}. The tropical toric variety X_Σ is then obtained by gluing the spaces U_σ^{trop} for $\sigma \in \Sigma$ along common faces.

For more on this definition, see [**Kaj08**] or [**Rab12**].

Example 6.2.2. (1) Let $\sigma = \text{pos}(\mathbf{e}_1, \ldots, \mathbf{e}_n) \subset N_\mathbb{R} \cong \mathbb{R}^n$. Then $\sigma^\vee = \text{pos}(\mathbf{e}_1, \ldots, \mathbf{e}_n)$ in $M_\mathbb{R}$, and so $U_\sigma^{\text{trop}} = \text{Hom}(\mathbb{N}^n, \overline{\mathbb{R}}) = \overline{\mathbb{R}}^n$. Since $U_\sigma \cong \mathbb{A}^n$, this agrees with our calculation above.

(2) Let $\sigma = \text{pos}((0,1),(2,-1)) \subset N_\mathbb{R} \cong \mathbb{R}^2$. Then $\sigma^\vee = \text{pos}\{(1,0),(1,2)\}$ in $M_\mathbb{R}$, so $\sigma^\vee \cap M$ is the semigroup generated by $(1,0)$, $(1,1)$ and $(1,2)$. Note that any semigroup homomorphism $\phi : \sigma^\vee \cap M \to \overline{\mathbb{R}}$ must have $\phi((1,0)) + \phi((1,2)) = 2\phi((1,1))$. We can thus identify $U_\sigma^{\text{trop}} = \text{Hom}(\sigma^\vee \cap M, \overline{\mathbb{R}})$ with $\{(a,b,c) \in \overline{\mathbb{R}}^3 : a + c = 2b\}$. ◊

Remark 6.2.3. The semigroup $\sigma^\vee \cap M$ is finitely generated; see, for example, [**Ful93**, p. 12] or [**CLS11**, Proposition 1.2.17]. Choose m generators, and consider the linear relations among them. These relations realize $U_\sigma^{\text{trop}} = \text{Hom}(\sigma^\vee \cap M, \overline{\mathbb{R}})$ as a subset of $\overline{\mathbb{R}}^m$. The induced topology on U_σ^{trop} coming from that of $\overline{\mathbb{R}}^m$ equals the more intrinsic one described above.

Example 6.2.4. Let Σ be the fan in $N_\mathbb{R} \cong \mathbb{R}^n$ defining projective space $\mathbb{P}^n$ that was described in part (1) of Example 6.1.1. The tropical toric variety $\text{trop}(\mathbb{P}^n)$ is the union of $\binom{n+1}{k}$ copies of $\mathbb{R}^{n-k}$ for $0 \leq k \leq n$; one copy of $\mathbb{R}^{n-k}$ for each k-dimensional cone of Σ. It is obtained by gluing together $n+1$ copies of $\overline{\mathbb{R}}^n$. The case $n = 2$ is shown in Figure 6.2.2. ◊

Remark 6.2.5. The concept of *tropical convexity*, introduced in Section 5.2, extends naturally from $\mathbb{R}^n/\mathbb{R}\mathbf{1}$ to $\text{trop}(\mathbb{P}^{n-1})$. If S is any tropically

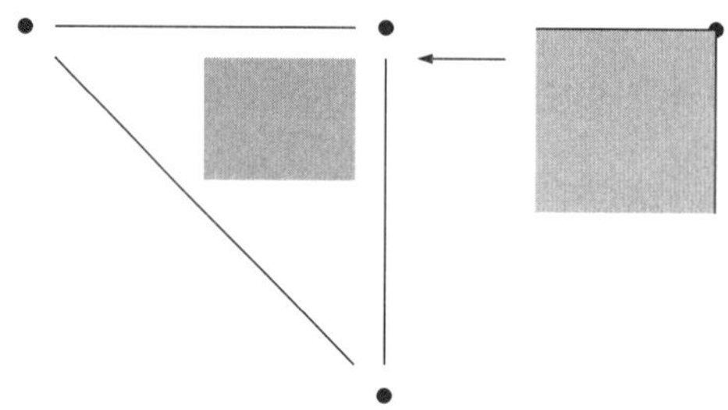

Figure 6.2.2. The tropical projective plane.

convex subset of $\mathbb{R}^n/\mathbb{R}\mathbf{1}$, then its compactification $\overline{S}$ is convex in trop$(\mathbb{P}^{n-1})$. In particular, by Proposition 5.2.8, every tropical linear space trop$(\overline{Y})$ is tropically convex. Moreover, trop$(\overline{Y})$ has the structure of a tropical polytope in trop$(\mathbb{P}^{n-1})$: it is the convex hull of its *cocircuit vectors*. This is related to Proposition 5.3.19 and is proved in [**JSY07**, Theorem 14]. The cocircuit vectors arise from orbits $\mathcal{O}_\sigma$ that intersect $\overline{Y}$ in a single point. See [**JSY07**] for more on tropical convexity and its connection to *affine buildings*.

The quotient description of a toric variety also tropicalizes naturally. Recall from Section 6.1 that if X_Σ is a simplicial toric variety with s rays, then $X_\Sigma = (\mathbb{A}^s \backslash V(B))/H$, where B is the irrelevant ideal of (6.1.3), and $H = \mathrm{Hom}(A_{n-1}(X_\Sigma), K^*)$. The exact sequence (6.1.2) gives an embedding of H into the torus T^s of $\mathbb{A}^s$. This tropicalizes to

$$\text{(6.2.1)} \qquad \mathrm{trop}(H) \;=\; \mathrm{Hom}(A_{n-1}(X_\Sigma), \mathbb{R}) \;=\; \ker(V^T),$$

where V is the matrix in (6.1.1), and V^T is regarded as a map from $\mathbb{R}^s$ to $\mathbb{R}^n$. The H-action on $(K^*)^s$ by multiplication tropicalizes to an additive action of $\ker(V^T)$ on $\mathbb{R}^s$. This action extends to $\overline{\mathbb{R}}^s$ by setting $a + \infty = \infty$ for $a \in \mathbb{R}$. The quotient description of toric varieties tropicalizes as follows.

Proposition 6.2.6. *Let X_Σ be a simplicial toric variety with s rays and Cox irrelevant ideal $B \subset K[x_1, \ldots, x_s]$. The tropical toric variety equals*

$$\mathrm{trop}(X_\Sigma) \;=\; \big(\mathrm{trop}(\mathbb{A}^s) \backslash \mathrm{trop}(V(B))\big)/\,\mathrm{trop}(H).$$

Proof. Fix $\sigma \in \Sigma$, and let $V_\sigma = \{\mathbf{x} \in \overline{\mathbb{R}}^s : x_i \neq \infty \text{ for } i \notin \sigma\}$. We first claim that $V_\sigma/\,\mathrm{trop}(H) = U_\sigma^{\mathrm{trop}}$. By [**Ful93**, p. 53] and [**Cox95**, Section 2],

$$U_\sigma \;=\; \mathrm{Spec}(K[\sigma^\vee \cap M]) \;=\; \mathrm{Hom}(\sigma^\vee \cap M, K) \;=\; \Big(\mathbb{A}^s \setminus V\Big(\prod_{i \notin \sigma} x_i\Big)\Big)/H.$$

Hence $\mathrm{trop}(U_\sigma) = \mathrm{Hom}(\sigma^\vee \cap M, \overline{\mathbb{R}}) = U_\sigma^{\mathrm{trop}}$, and the claim follows.

Since $V_\sigma = \mathrm{trop}(\mathbb{A}^s) \backslash \mathrm{trop}(V(\prod_{i \notin \sigma} x_i))$, we have $V_\sigma \cap \mathrm{trop}(V(B)) = \emptyset$, and $\mathrm{trop}(\mathbb{A}^s) \backslash \mathrm{trop}(V(B)) = \bigcup_{\sigma \in \Sigma} V_\sigma$. As the overlaps induce the appropriate gluing, this proves the desired identity. □

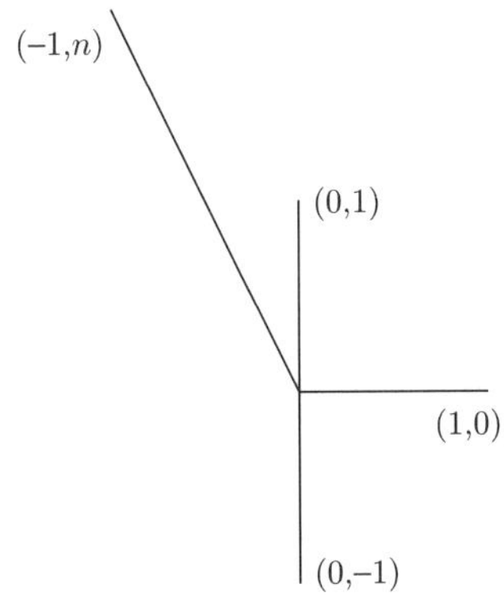

Figure 6.2.3. The fan for the Hirzebruch surface $\mathbb{F}_n$.

Example 6.2.7. (1) Let X_Σ be the Hirzebruch surface $\mathbb{F}_n$. The fan Σ has four rays and four two-dimensional cones, as in Figure 6.2.3.

The Cox ring of $\mathbb{F}_n$ is $K[x_1,x_2,x_3,x_4]$, with irrelevant ideal $B = \langle x_1,x_3\rangle \cap \langle x_2,x_4\rangle$. This implies $\operatorname{trop}(\mathbb{A}^4)\setminus\operatorname{trop}(V(B)) = \{\mathbf{x}\in\overline{\mathbb{R}}^4 : (x_1 \text{ or } x_3\neq\infty) \text{ and } (x_2 \text{ or } x_4\neq\infty)\}$. The torus $H\cong(K^*)^2$ acts on $\mathbb{A}^4$ by $(t_1,t_2)\cdot(x_1,x_2,x_3,x_4) = (t_1x_1, t_1^{-n}t_2x_2, t_1x_3, t_2x_4)$, so $\operatorname{trop}(H) = \operatorname{span}\{(1,-n,1,0),(0,1,0,1)\} \subseteq \mathbb{R}^4$. The tropical Hirzebruch surface $\operatorname{trop}(\mathbb{F}_n)$ is then the union of nine orbits:

(a) $\mathbb{R}^2 \cong \mathbb{R}^4/\operatorname{trop}(H)$;
(b) $\mathbb{R} \cong \{(\infty,x_2,x_3,x_4) : x_2,x_3,x_4 \in \mathbb{R}\}/\operatorname{trop}(H) \cong \{(\infty,0,0, x_4-x_2-nx_3)\}$, and the three other analogous orbits; and
(c) four points $\{(\infty,\infty,x_3,x_4) : x_3,x_4\in\mathbb{R}\}/H = \{(\infty,\infty,0,0)\}$, $\{(\infty,0,0,\infty)\}$, $\{(0,\infty,\infty,0)\}$, and $\{(0,0,\infty,\infty)\}$.

(2) Let $X_\Sigma = (\mathbb{P}^1)^3$. In part (2) of Example 6.1.2, we saw that this toric threefold is the quotient of $\mathbb{A}^6\setminus V(B)$ by a three-dimensional torus H. The tropicalization of H is $\operatorname{span}\{(1,1,0,0,0,0),(0,0,1,1,0,0),(0,0,0,0,1,1)\}$. We obtain the tropicalization of $(\mathbb{P}^1)^3$ as the quotient of $\{\mathbf{x}\in\overline{\mathbb{R}}^6 : (x_1\neq\infty \text{ or } x_2\neq\infty) \text{ and } (x_3\neq\infty \text{ or } x_4\neq\infty) \text{ and } (x_5\neq\infty \text{ or } x_6\neq\infty)\}$ by $\operatorname{trop}(H)$. This has one three-dimensional orbit $\mathbb{R}^3$, and six two-dimensional orbits with representatives of the form $(\infty,0,x_3,0,x_5,0)$. There are twelve one-dimensional orbits with representatives $(\infty,0,\infty,0,x_5,0)$, and eight zero-dimensional orbits of the form $(\infty,0,\infty,0,\infty,0)$. ◊

If $\overline{Y}$ is a subvariety of a toric variety X_Σ, then its tropicalization $\operatorname{trop}(\overline{Y})$ lives in $\operatorname{trop}(X_\Sigma)$. For each torus orbit $\mathcal{O}_\sigma$ of X_σ, we have $\operatorname{trop}(\mathcal{O}_\sigma) = N(\sigma)$. Set $Y_\sigma = \overline{Y}\cap\mathcal{O}_\sigma$. Then $\operatorname{trop}(Y_\sigma)$ is a balanced complex in $N(\sigma)$. When $\sigma = \{0\}$, we have $Y_\sigma = \overline{Y}\cap T$, and $\operatorname{trop}(Y_\sigma)\subseteq N_\mathbb{R}\cong\mathbb{R}^n$ as before. Identifying $\coprod_{\sigma\in\Sigma} N(\sigma)$ with $\operatorname{trop}(X_\Sigma)$ gives a construction of $\operatorname{trop}(\overline{Y})$ inside $\operatorname{trop}(X_\Sigma)$.

This can be carried out in coordinates, starting with $X_\Sigma = \mathbb{A}^s$, where $\operatorname{trop}(Y)$ is the closure of $\{(\operatorname{val}(x_1), \ldots, \operatorname{val}(x_s)) : (x_1, \ldots, x_s) \in Y\}$. Zero coordinates go to $\infty \in \overline{\mathbb{R}}$. We can pass to quotients via Proposition 6.2.6.

Example 6.2.8. (1) Let $\overline{Y} = V(x+y+z) \subseteq \mathbb{P}^2$. The tropical variety $\operatorname{trop}(\overline{Y})$ is the union of the standard tropical line and three extra points, one in each of the three copies of $\mathbb{R}$ in the boundary of $\operatorname{trop}(\mathbb{P}^2)$. We draw $\operatorname{trop}(\mathbb{P}^2)$ as a closed triangle. See Figure 6.2.4.

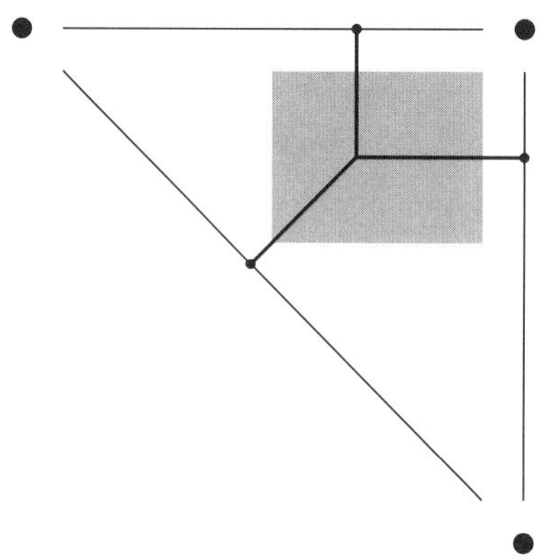

Figure 6.2.4. A compactified tropical line.

(2) The curves Y in Example 3.1.8 live in the two-dimensional torus over $\mathbb{C}\{\{t\}\}$. We consider their closures $\overline{Y}$ in the affine plane $\mathbb{A}^2_{\mathbb{C}\{\{t\}\}}$. Each $\operatorname{trop}(\overline{Y})$ is obtained from the picture in Figure 3.1.2 by adding a point at the end of each ray in northern or eastern direction.

(3) Let $B = \langle x_1, y_1\rangle \cap \langle x_2, y_2\rangle \cap \langle x_3, y_3\rangle \subset \mathbb{C}[x_1, y_1, x_2, y_2, x_3, y_3]$. The variety $V(B) \subseteq \mathbb{A}^6$ is a union of three four-dimensional linear spaces. The tropical variety $\operatorname{trop}(B) \subset \operatorname{trop}(\mathbb{A}^6) = \overline{\mathbb{R}}^6$ is the union of their tropicalizations: $\{(\infty, \infty, a, b, c, d) : a, b, c, d \in \overline{\mathbb{R}}\}$, $\{(a, b, \infty, \infty, c, d) : a, b, c, d \in \overline{\mathbb{R}}\}$, and $\{(a, b, c, d, \infty, \infty) : a, b, c, d \in \overline{\mathbb{R}}\}$. ◊

Example 6.2.9. Let $Y \subset T^3$ be defined by the equation $x + y + z = 1$, with closure $\overline{Y}$ in $(\mathbb{P}^1)^3$ as in Example 6.1.3. The tropical toric variety $\operatorname{trop}\big((\mathbb{P}^1)^3\big) = (\operatorname{trop}(\mathbb{P}^1))^3$ is a three-dimensional cube. It contains $\operatorname{trop}(\overline{Y})$ as a compact balanced surface. The subset $\operatorname{trop}(Y)$ is a tropical linear space as in Section 4.2: it is a two-dimensional fan with four rays and six maximal cones. The boundary $\operatorname{trop}(\overline{Y})\backslash \operatorname{trop}(Y)$ consists of nine edges and seven vertices. It is a subdivision of the graph K_4. In total, $\operatorname{trop}(\overline{Y})$ has eight vertices, thirteen edges, and six two-cells. ◊

The Fundamental Theorem 3.2.3 generalizes easily to this setting. We first extend the concept of tropical hypersurfaces from $\mathbb{R}^n$ to $\overline{\mathbb{R}}^n$.

Definition 6.2.10. Let $f = \sum c_{\mathbf{u}} x^{\mathbf{u}} \in K[x_1, \ldots, x_n]$. Here $\mathbf{u}$ runs over $\mathbb{N}^n$ but only finitely many $c_{\mathbf{u}}$ are nonzero. The tropical polynomial $\operatorname{trop}(f)$ is given by $\operatorname{trop}(f)(\mathbf{w}) = \min\{\operatorname{val}(c_{\mathbf{u}}) + \mathbf{w} \cdot \mathbf{u} : \mathbf{u} \in \mathbb{N}^n\}$. We follow the convention here that if $w_i = \infty$, then the term $w_i u_i$ equals ∞ if $u_i \neq 0$, and equals 0 if $u_i = 0$. The tropical polynomial $\operatorname{trop}(f)$ can be viewed as a function from $\overline{\mathbb{R}}^n$ to $\overline{\mathbb{R}}$. The *extended tropical hypersurface* of f is then

$$\operatorname{trop}(V(f)) = \{\mathbf{w} \in \overline{\mathbb{R}}^n : \text{the minimum in } \operatorname{trop}(f) \text{ is achieved at least twice}\}.$$

We use the same notation for both the tropical hypersurface in $\mathbb{R}^n$ and the extended tropical hypersurface in $\overline{\mathbb{R}}^n$ to avoid excessive ornamentation. Note that the intersection of the latter hypersurface with $\mathbb{R}^n$ equals the former.

Example 6.2.11. If $f = x + y + 1 \in \mathbb{C}[x, y]$, then $\operatorname{trop}(V(f))$ in $\overline{\mathbb{R}}^2$ is the standard tropical line with the points $(\infty, 0)$ and $(0, \infty)$ added.

If $f = xy$, then $\operatorname{trop}(V(f)) = \{\infty\} \times \overline{\mathbb{R}} \cup \overline{\mathbb{R}} \times \{\infty\}$ consists of the two boundary lines. Indeed, if f is any monomial, then the minimum in $\operatorname{trop}(f)(\mathbf{w})$ is only achieved twice when $\mathbf{w} \in \overline{\mathbb{R}}^n$ has at least one infinite coordinate. ◊

Lemma 6.2.12. *The extended tropical hypersurface* $\operatorname{trop}(V(f)) \subset \overline{\mathbb{R}}^n$ *of a polynomial* $f \in K[x_1, \ldots, x_n]$ *is a closed subset of* $\overline{\mathbb{R}}^n = \operatorname{trop}(\mathbb{A}^n)$.

Proof. Let $\mathbf{w}$ be an arbitrary point in $\overline{\mathbb{R}}^n \setminus \operatorname{trop}(V(f))$. There exists a term $c_{\mathbf{u}} x^{\mathbf{u}}$ in f with $\operatorname{trop}(f)(\mathbf{w}) = \operatorname{val}(c_{\mathbf{u}}) + \mathbf{w} \cdot \mathbf{u} < \infty$. By comparing that term with all the other terms of f, we can find $\epsilon > 0$ small enough and $L \gg 0$ large enough such that the following set U is disjoint from $\operatorname{trop}(V(f))$:

$$U = \{\mathbf{v} \in \overline{\mathbb{R}}^n : |v_i - w_i| < \epsilon \text{ if } w_i < \infty \text{ and } v_i > L \text{ if } w_i = \infty\}.$$

Indeed, for $\mathbf{w}' \in U$, we have $\operatorname{trop}(f)(\mathbf{w}') = \operatorname{val}(c_{\mathbf{u}}) + \mathbf{w}' \cdot \mathbf{u}$, and this is the only term in $\operatorname{trop}(f)$ achieving the minimum in $\operatorname{trop}(f)(\mathbf{w}')$. The set U is open. This proves that $\operatorname{trop}(V(f))$ is a closed subset of $\overline{\mathbb{R}}^n$. □

We next extend the definition of initial forms and initial ideals in Section 2.4 so as to allow ∞ as a coordinate in $\mathbf{w}$. Write $\overline{\Gamma}_{\text{val}} = \Gamma_{\text{val}} \cup \{\infty\}$ for the image of $\operatorname{val}\colon K \to \mathbb{R} \cup \{\infty\}$. We assume that there is a splitting $\Gamma_{\text{val}} \to K^*$ of the valuation map given by $w \mapsto t^w$. Note that this splitting extends to a semigroup homomorphism from $\overline{\Gamma}_{\text{val}}$ to K by sending ∞ to 0.

Definition 6.2.13. Let $\mathbf{w} \in \overline{\mathbb{R}}^n$ and $f = \sum c_{\mathbf{u}} x^{\mathbf{u}} \in K[x_1, \ldots, x_n]$. If $\operatorname{trop}(f)(\mathbf{w}) < \infty$, then the *initial form* of f is the polynomial

$$\operatorname{in}_{\mathbf{w}}(f) = \sum_{\mathbf{u}:\operatorname{val}(c_{\mathbf{u}}) + \mathbf{w} \cdot \mathbf{u} = \operatorname{trop}(f)(\mathbf{w})} \overline{t^{-\operatorname{val}(c_{\mathbf{u}})} c_{\mathbf{u}}} x^{\mathbf{u}} \in \Bbbk[x_1, \ldots, x_n].$$

As before, $\Bbbk$ is the residue field of K, and the sum is over $\mathbf{u} \in \mathbb{N}^n$ with $c_{\mathbf{u}} \neq 0$. If $\operatorname{trop}(f)(\mathbf{w}) = \infty$, then $\operatorname{in}_{\mathbf{w}}(f) = 0$. The *initial ideal* of an ideal $I \subseteq K[x_1, \ldots, x_n]$ is the ideal $\operatorname{in}_{\mathbf{w}}(I) = \langle \operatorname{in}_{\mathbf{w}}(f) : f \in I \rangle$ in $\Bbbk[x_1, \ldots, x_n]$.

Example 6.2.14. Let $K = \mathbb{C}$ with the trivial valuation, and $f = xy + 3x + 4y - 2 \in \mathbb{K}[x, y]$. For $\mathbf{w} = (\infty, 3)$ we have $\operatorname{in}_{\mathbf{w}}(f) = -2$, for $\mathbf{w} = (\infty, 0)$ we have $\operatorname{in}_{\mathbf{w}}(f) = 4y - 2$, and for $\mathbf{w} = (\infty, \infty)$ we have $\operatorname{in}_{\mathbf{w}}(f) = -2$. ◊

Fix $\sigma \subset \{1, \ldots, n\}$. For a vector $\mathbf{w}$ in $\mathbb{R}^{n-|\sigma|}$ with coordinates indexed by $\{i : i \notin \sigma\}$, we write $\mathbf{w} \times \infty^\sigma$ for the vector in $\mathbb{R}^n$ with ith coordinate w_i if $i \notin \sigma$, and ∞ otherwise. For a subset Σ of $\mathbb{R}^{n-|\sigma|}$ we write $\Sigma \times \infty^\sigma$ for the set $\{\mathbf{w} \times \infty^\sigma : \mathbf{w} \in \Sigma\}$. Our next result is the extension of the Fundamental Theorem 3.2.3 from the torus T^n to the affine space $\mathbb{A}^n$:

Theorem 6.2.15 (Extended Fundamental Theorem). *Let Y be a subvariety of $\mathbb{A}^n$, and let I be its ideal in $S = K[x_1, \ldots, x_n]$. Then the following subsets of $\overline{\mathbb{R}}^n = \operatorname{trop}(\mathbb{A}^n)$ coincide:*

(1) $\bigcap_{f \in I} \operatorname{trop}(V(f))$;

(2) *the set of all vectors $\mathbf{w} \in \overline{\mathbb{R}}^n$ for which $\operatorname{in}_{\mathbf{w}}(I) \subseteq \Bbbk[x_1, \ldots, x_n]$ does not contain a monomial; and*

(3) *the set*

$$\bigcup_{\sigma \subseteq \{1, \ldots, n\}} \operatorname{trop}(Y \cap \mathcal{O}_\sigma) \times \infty^\sigma,$$

where $\mathcal{O}_\sigma = \{\mathbf{x} \in \mathbb{A}^n : x_i = 0 \text{ for } i \in \sigma, \text{ and } x_j \neq 0 \text{ for } j \notin \sigma\}$.

Proof. For $\sigma \subset \{1, \ldots, n\}$, we regard $Y_\sigma = Y \cap \mathcal{O}_\sigma$ as a subvariety of $\mathcal{O}_\sigma \cong (K^*)^{n-|\sigma|}$. Set $I_\sigma = (I + \langle x_j : j \in \sigma \rangle) \cap K[x_i : i \notin \sigma]$. For $f = \sum c_{\mathbf{u}} x^{\mathbf{u}} \in I$, we consider the subsum $f_\sigma = \sum_{\operatorname{supp}(\mathbf{u}) \cap \sigma = \emptyset} c_{\mathbf{u}} x^{\mathbf{u}}$. Note that $I_\sigma = \langle f_\sigma : f \in I \rangle$.

For any polynomial $f \in S$, the tropical hypersurface $\operatorname{trop}(V(f))$ equals $\bigcup_{\sigma \subseteq \{1, \ldots, n\}} \operatorname{trop}(V(f_\sigma)) \times \infty^\sigma$. Indeed, if $f = \sum c_{\mathbf{u}} x^{\mathbf{u}}$ and $\mathbf{w} \in \operatorname{trop}(V(f))$ with $\sigma = \{i : w_i = \infty\}$, then $\operatorname{trop}(f)(\mathbf{w}) = \min(\operatorname{val}(c_{\mathbf{u}}) + \mathbf{w} \cdot \mathbf{u})$ is achieved at a term in f_σ if $f_\sigma \neq 0$, in which case $\operatorname{trop}(f)(\mathbf{w}) = \operatorname{trop}(f_\sigma)(\mathbf{w})$. Conversely, for any $\mathbf{w} \in \operatorname{trop}(V(f_\sigma)) \subseteq \mathbb{R}^{n-|\sigma|}$, we have $\mathbf{w} \times \infty^\sigma \in \operatorname{trop}(V(f))$, as the minimum in $\operatorname{trop}(f)(\mathbf{w} \times \infty^\sigma) = \operatorname{trop}(f_\sigma)(\mathbf{w})$ is achieved at least twice for coordinates not in σ. This shows that for $\mathbf{w} \in \mathbb{R}^{n-|\sigma|}$, we have $\operatorname{in}_{\mathbf{w} \times \infty^\sigma}(f) = \operatorname{in}_{\mathbf{w}}(f_\sigma)$. This is a monomial if and only if the minimum in $\operatorname{trop}(f)(\mathbf{w})$ is achieved once, so $\mathbf{w} \times \infty^\sigma \notin \operatorname{trop}(V(f))$. Since $\operatorname{in}_{\mathbf{w}}(I)$ is generated by $\operatorname{in}_{\mathbf{w}}(f)$

for $f \in I$, this shows the equivalence of (1) and (2). We also have

$$\begin{aligned}\bigcap_{f\in I}\operatorname{trop}(V(f)) &= \bigcap_{f\in I}\bigcup_{\sigma\subseteq\{1,\dots,n\}}\operatorname{trop}(V(f_\sigma))\times\infty^\sigma\\ &= \bigcup_{\sigma\subseteq\{1,\dots,n\}}\bigcap_{f\in I}\operatorname{trop}(V(f_\sigma))\times\infty^\sigma\\ &= \bigcup_{\sigma\subseteq\{1,\dots,n\}}\Big(\bigcap_{g\in I_\sigma}\operatorname{trop}(V(g))\Big)\times\infty^\sigma\\ &= \bigcup_{\sigma\subseteq\{1,\dots,n\}}\operatorname{trop}(Y_\sigma)\times\infty^\sigma.\end{aligned}$$

This shows the equivalence of sets (1) and (3), so completes the proof. □

Theorem 6.2.15 generalizes to an arbitrary toric variety X_Σ. Recall that the torus $H = \operatorname{Hom}(A_{n-1}(X_\Sigma), K^*)$ acts naturally on the Cox ring $K[x_1, \dots, x_s]$, and that $\mathbf{v}_i$ is the first lattice point on the ith ray of Σ.

Corollary 6.2.16. *Let Y be a subvariety of a smooth toric variety X_Σ, and let I be its homogeneous B-saturated ideal in the Cox ring $K[x_1, \dots, x_s]$ of X_Σ. Then the following subsets of $\overline{\mathbb{R}}^s \setminus \operatorname{trop}(V(B))$ coincide:*

(1) $\bigcap_{f\in I}\operatorname{trop}(V(f)) \setminus \operatorname{trop}(V(B))$;

(2) *the set of all $\mathbf{w} \in \overline{\mathbb{R}}^s \setminus \operatorname{trop}(V(B))$ such that $\operatorname{in}_{\mathbf{w}}(I)$ does not contain a monomial.*

The quotient of this set by $\operatorname{trop}(H)$ equals

$$\operatorname{trop}(Y) = \bigcup_{\sigma\in\Sigma}\operatorname{trop}(Y\cap\mathcal{O}_\sigma).$$

Proof. The equivalence of (1) and (2) follows immediately from the corresponding equivalence in Theorem 6.2.15. The second claim is a consequence of the Cox construction of X_Σ and part (3) of Theorem 6.2.15, as we now explain. By the Cox construction, we have $X_\Sigma = (\mathbb{A}^s\backslash V(B))/H$, and $\mathcal{O}_\sigma = \{\mathbf{x} \in \mathbb{A}^s\backslash V(B) : x_i = 0 \text{ when } \mathbf{v}_i \in \sigma \text{ and } x_i \neq 0 \text{ when } \mathbf{v}_i \notin \sigma\}/H$.

The set (1) = (2) consists of all points $\mathbf{w}$ in $\operatorname{trop}(V(I)) \subset \overline{\mathbb{R}}^s$ that do not lie in $\operatorname{trop}(V(B))$. This is the set of $\mathbf{w} \in \operatorname{trop}(V(I))$ for which there exists $\sigma \in \Sigma$ with $w_i < \infty$ whenever $\mathbf{v}_i \notin \sigma$. For such a $\mathbf{w}$, set $\tau = \{i : w_i = \infty\}$. The assumption $\mathbf{w} \notin \operatorname{trop}(V(B))$ means that $\tau^\Sigma = \operatorname{pos}(\mathbf{v}_i : i \in \tau)$ is a face of σ, and thus a cone of Σ. Let $\mathcal{O}^\tau = \{\mathbf{x} \in \mathbb{A}^s : x_i = 0 \text{ for } i \in \tau \text{ and } x_i \neq 0 \text{ for } i \notin \tau\}$. Thus $\mathbf{w} \in \operatorname{trop}(V(I) \cap \mathcal{O}^\tau) \times \infty^\tau$. By part (3) of Theorem 6.2.15 we have $\mathbf{w} + \operatorname{trop}(H) \in (\operatorname{trop}(Y \cap \mathcal{O}_{\tau^\Sigma}) \subseteq (\overline{\mathbb{R}}^s \setminus \operatorname{trop}(V(B)))/\operatorname{trop}(H)$. Conversely, given a point $\mathbf{y} \in Y \cap \mathcal{O}_\sigma$ for a cone $\sigma \in \Sigma$, we can choose a lift $\mathbf{y}' \in \mathbb{A}^s$ with $y_i' = 0$ when $\mathbf{v}_i \in \sigma$ and $y_i' \neq 0$ when

$\mathbf{v}_i \notin \sigma$. Then $\text{val}(\mathbf{y}')_i = \infty$ when $\mathbf{v}_i \in \sigma$, and $\text{val}(\mathbf{y}')_i < \infty$ when $\mathbf{v}_i \notin \sigma$. Thus $\mathbf{w} = \text{val}(\mathbf{y}') \in (\text{trop}(\mathcal{O}_\sigma) \times \infty^\sigma) \setminus \text{trop}(V(B))$. Hence, the quotient by $\text{trop}(H)$ of the subset of (1) consisting of those $\mathbf{w}$ with $w_i = \infty$ if and only if $\mathbf{v}_i \in \sigma$ equals $\text{trop}(Y \cap \mathcal{O}_\sigma)$. The result now follows. □

Tropicalization commutes with toric morphisms, in the following sense:

Corollary 6.2.17. *Let $\pi \colon X_\Sigma \to X_\Delta$ be a morphism of toric varieties, given by a map of fans $\pi \colon \Sigma \to \Delta$, and let $\text{trop}(\pi) : \text{trop}(X_\Sigma) \to \text{trop}(X_\Delta)$ be the induced map. If Y is a subvariety of X_Σ, then $\text{trop}(\pi(Y)) = \text{trop}(\pi)(\text{trop}(Y))$.*

Proof. This follows from the Extended Fundamental Theorem 6.2.15, along with Corollary 6.2.16 and Corollary 3.2.13. □

Our final result in this section says that tropicalization commutes with taking closures in toric varieties.

Theorem 6.2.18. *Let $Y \subseteq T$, and let $\overline{Y}$ be the closure of Y in a toric variety X_Σ. Then $\text{trop}(\overline{Y})$ is the closure of $\text{trop}(Y) \subset \mathbb{R}^n$ in $\text{trop}(X_\Sigma)$.*

Proof. Since $Y \subseteq \overline{Y}$, we have $\text{trop}(Y) \subseteq \text{trop}(\overline{Y})$. We denote by s the number of rays of Σ. Let I be the ideal of $\overline{Y}$ in $\text{Cox}(X_\Sigma) = K[x_1, \ldots, x_s]$. For $f \in I$, the extended tropical hypersurface $\text{trop}(V(f))$ is a closed subset of $\overline{\mathbb{R}}^s$ by Lemma 6.2.12. This means $\bigcap_{f \in I} \text{trop}(V(f))$ is a closed subset of $\overline{\mathbb{R}}^s$. This makes $(\bigcap_{f \in I} \text{trop}(V(f)) \setminus \text{trop}(V(B)))/H$ a closed subset of $\text{trop}(X_\Sigma)$, as Proposition 6.2.6 shows that the topology on $\text{trop}(X_\Sigma)$ is the quotient topology from the quotient construction. By Corollary 6.2.16 this equals $\text{trop}(\overline{Y})$, so $\text{trop}(\overline{Y})$ is a closed subset of $\text{trop}(X_\Sigma)$ containing $\text{trop}(Y)$.

To show that $\text{trop}(\overline{Y})$ is the closure of $\text{trop}(Y)$, we again make use of the Cox construction. Note that the ideal $I' = IK[x_1^{\pm 1}, \ldots, x_s^{\pm 1}]$ of $V(I) \cap (K^*)^s$ satisfies $I = I' \cap K[x_1, \ldots, x_s]$. The proof is by induction on s. When $s = 1$, we either have $\text{trop}(V(I)) = \overline{\mathbb{R}}$ or $\text{trop}(V(I))$ is a finite set of points. In the first case the claim is true, while in the second $\text{trop}(V(I')) \subset \mathbb{R}$ is a finite set of points, so $V(I')$ is a finite set of points by Lemma 3.3.9. This means that the closure $V(I) \subset \mathbb{A}^1$ does not add any points, so in particular $0 \notin V(I)$, and thus $\text{trop}(V(I)) = \text{trop}(V(I'))$ is the closure as required.

Suppose the claim is true for $s - 1$. Given $\mathbf{w} \in \text{trop}(V(I)) \setminus \text{trop}(V(I'))$, we must show that every open set in $\overline{\mathbb{R}}^s$ containing $\mathbf{w}$ intersects $\text{trop}(V(I'))$. We may assume that $\mathbf{w} \in \overline{\Gamma}_{\text{val}}^s$. By Theorem 6.2.15, we have $\mathbf{w} = \text{val}(\mathbf{y})$ for some $\mathbf{y} \in V(I)$. Let $\sigma = \{i : w_i = \infty\} = \{i : y_i = 0\}$. It suffices to show that for all $m \gg 0$ there is $\mathbf{w}' \in \text{trop}(V(I')) \subset \mathbb{R}^s$ with $w_i' = w_i$ for $i \notin \sigma$, and $w_i' > m$ for $i \in \sigma$. Without loss of generality we may assume that $s \in \sigma$. We first note that for all $m > 0$ there is $\mathbf{w}_m \in \text{trop}(V(I'))$ with $(\mathbf{w}_m)_s > m$.

Indeed, if not, by Corollary 6.2.17 the projection of $\operatorname{trop}(V(I))$ to the last coordinate is a finite set, so the projection of $\operatorname{trop}(V(I'))$ is also finite, and hence, by Lemma 3.3.9, the projection of $V(I')$ to the last coordinate is finite. Since the projection of $V(I)$ is the closure of the projection of $V(I')$, there would be no point in $V(I)$ with last coordinate zero, and thus, by the Extended Fundamental Theorem 6.2.15, no $\mathbf{w} \in \operatorname{trop}(V(I))$ with $w_s = \infty$.

Choose $\mathbf{y} \in V(I) \subset (K^*)^s$ with $\operatorname{val}(\mathbf{y}) = \mathbf{w}_m$. Let $I_m = I|_{x_s=y_s} \subset K[x_1, \ldots, x_{s-1}]$, and write $\pi_s : \overline{\mathbb{R}}^s \to \overline{\mathbb{R}}^{s-1}$ for the projection onto the first $s-1$ coordinates. By Theorems 3.4.12 and 6.2.15, we know that $\operatorname{trop}(V(I_m)) \subseteq \overline{\mathbb{R}}^{s-1}$ equals $\pi_s(\operatorname{trop}(V(I)) \cap \{\mathbf{w}' : w'_s = (\mathbf{w}_m)_s\})$. Thus, in particular, $\pi_s(\mathbf{w}) \in \operatorname{trop}(V(I_m))$. By induction this means that there is $\mathbf{w}' \in \operatorname{trop}(V(I_m)) \cap \mathbb{R}^{s-1}$ with $w'_i > m$ for $i \in \sigma \backslash \{s\}$ and $w'_i = w_i$ for $i \notin \sigma$, and so there is also $\tilde{\mathbf{w}} \in \operatorname{trop}(V(I)) \subset \overline{\mathbb{R}}^s$ with $\pi_s(\tilde{\mathbf{w}}) = \mathbf{w}'$ and $\tilde{w}_s = (\mathbf{w}_m)_s$. By construction, we have $\tilde{w}_i = w_i$ for $i \notin \sigma$, and $\tilde{w}_i > m$ for $i \in \sigma$, so the claim follows. □

6.3. Orbits

Let $T^n = (K^*)^n$, and let Y be a subvariety of T^n. Fix a toric variety X_Σ, and let $\overline{Y}$ be the closure of Y in X_Σ. We emphasize that we do not assume that X_Σ is a complete toric variety, so the support $|\Sigma|$ of Σ need not be all of $\mathbb{R}^n$. The following is a natural question in the context of toric geometry:

Question 6.3.1. Which T^n-orbits of X_Σ does $\overline{Y}$ intersect?

We illustrate this for a line in the plane, and for a plane in 3-space.

Example 6.3.2. Let $Y = V(x + y + 1) \subset (K^*)^2$.

(1) Let $X_\Sigma = \mathbb{P}^2$, with torus $T^2 = \{(x : y : 1) : x, y \in K^*\}$ and homogeneous coordinates $(x : y : z)$. Then $\overline{Y} = V(x + y + z) = Y \cup \{(1 : -1 : 0), (1 : 0 : -1), (0 : 1 : -1)\}$. The closure $\overline{Y}$ thus intersects all T-orbits of $\mathbb{P}^2$ except for the three T-fixed points.

(2) Let $X_\Sigma = (\mathbb{P}^1)^2$, with torus $T^2 = \{((x{:}1), (y{:}1)) : x, y \in K^*\}$ and homogeneous coordinates $((x_1{:}x_2), (y_1{:}y_2))$. Then $\overline{Y}$ is the subvariety of $(\mathbb{P}^1)^2$ defined by the equation $x_1y_2 + x_2y_1 + x_2y_2 = 0$. Thus $\overline{Y} = Y \cup \{((-1{:}1), (0{:}1)), ((0{:}1), (-1{:}1)), ((1{:}0), (1{:}0))\}$. The closure $\overline{Y}$ intersects four of the nine torus orbits of $(\mathbb{P}^1)^2$, namely T^2, $\{((a{:}1), (1{:}0)) : a \in K^*\}$, $\{((1{:}0), (a{:}1)) : a \in K^*\}$, and $((1{:}0), (1{:}0))$. The corresponding cones are shown in Figure 6.3.1.

Example 6.3.3. Let $Y = V(x + y + z + 1) \subset (K^*)^3$ and $X_\Sigma = (\mathbb{P}^1)^3$, as in Examples 6.1.3 and 6.2.9. The compact surface $\overline{Y}$ intersects three of the six two-dimensional orbits, six of the twelve one-dimensional orbits, and

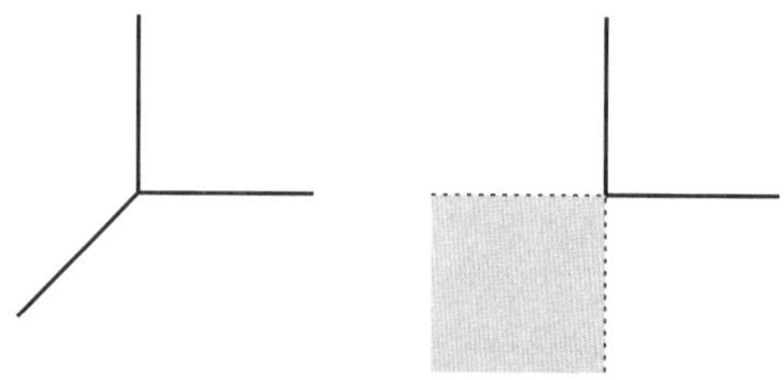

Figure 6.3.1. Torus orbits intersecting the curve $\overline{Y}$ in Example 6.3.2.

four of the eight zero-dimensional orbits on X_Σ. This can be verified by computations in $S = \mathrm{Cox}(X_\Sigma)$ using the description of Example 6.1.3. ◊

Let $\mathrm{trop}(Y_{\mathrm{triv}})$ denote the tropicalization of Y with respect to the trivial valuation on K. Tropical geometry answers Question 6.3.1 as follows:

Theorem 6.3.4. *Fix a toric variety X_Σ with torus T^n. Let Y be a subvariety of T^n, and let $\overline{Y}$ be its closure in X_Σ. For any $\sigma \in \Sigma$, we have $\overline{Y} \cap \mathcal{O}_\sigma \neq \emptyset$ if and only if $\mathrm{trop}(Y_{\mathrm{triv}})$ intersects the relative interior of the cone σ.*

We first consider the case where the toric variety X_Σ is affine space $\mathbb{A}^n$. Here, we give two proofs of the result: one using the tropical toric varieties of the previous section and also a direct one based on commutative algebra.

Proposition 6.3.5. *Let $Y \subset T^n$ be a subvariety, and let $\overline{Y}$ be the closure of Y in $\mathbb{A}^n$. Then $\mathbf{0} \in \overline{Y}$ if and only if $\mathrm{trop}(Y_{\mathrm{triv}}) \cap \mathbb{R}^n_{>0} \neq \emptyset$.*

First proof. By Theorems 3.4.12 and 6.2.18, the tropical variety $\mathrm{trop}(\overline{Y}_{\mathrm{triv}})$ is the closure of $\mathrm{trop}(Y_{\mathrm{triv}}) \subset \mathbb{R}^n$ in $\mathrm{trop}(\mathbb{A}^n) = \overline{\mathbb{R}}^n$. Thus $(\infty, \ldots, \infty) \in \mathrm{trop}(\overline{Y}_{\mathrm{triv}})$ if and only if for all $m > 0$ there is $\mathbf{w} \in \mathrm{trop}(Y_{\mathrm{triv}})$ with $w_i > m$ for all i. Since the tropicalization is with respect to the trivial valuation, $\mathrm{trop}(Y_{\mathrm{triv}})$ is a fan, so this occurs if and only if $\mathrm{trop}(Y_{\mathrm{triv}}) \cap \mathbb{R}^n_{>0} \neq \emptyset$. □

Second proof. Let $I = I_Y \subseteq K[x_1^{\pm 1}, \ldots, x_n^{\pm 1}]$. The ideal of $\overline{Y}$ in $S = K[x_1, \ldots, x_n]$ is $I_{\mathrm{aff}} = I \cap S$. Suppose first that $\mathbf{0} \notin \overline{Y}$. Then there is $f \in I_{\mathrm{aff}}$ of the form $f = 1 + g$, with $g \in \langle x_1, \ldots, x_n \rangle$. But then for $\mathbf{w} \in \mathbb{R}^n$ with $w_i > 0$ for $1 \leq i \leq n$, we have $\mathrm{in}_{\mathbf{w}}(f) = 1$. Since $f \in I$ when viewed as a Laurent polynomial, this means $\mathrm{in}_{\mathbf{w}}(I) = \langle 1 \rangle$, so $\mathbf{w} \notin \mathrm{trop}(Y)$.

Conversely, suppose that $\mathbf{0} \in \overline{Y}$. This implies $\dim(Y) > 0$ as if $\dim(Y) = 0$, then $\overline{Y} = Y \subset (K^*)^n$. We now proceed by induction on $\dim(Y)$. If $Y = \bigcup_i V_i$ is an irreducible decomposition of Y, then $\overline{Y} = \bigcup_i \overline{V_i}$, and thus $\mathbf{0}$ lies in the closure of one of the irreducible components V_i of Y. Since we also have $\mathrm{trop}(Y) = \bigcup_i \mathrm{trop}(V_i)$, we may assume that Y is irreducible. If $\dim(Y) > 1$, we choose a polynomial $h \in S$ with $h \notin \sqrt{I_{\mathrm{aff}} + \langle x_i \rangle}$ for any

i. This means that $\overline{Y} \cap V(h)$ does not contain the intersection of $\overline{Y}$ with any coordinate hyperplane, and every irreducible component of $\overline{Y} \cap V(h)$ intersects T^n. We again restrict to an irreducible component containing $\mathbf{0}$. Let I' be its ideal, and let $Y' = V(I') \subseteq T^n$, where we view I' as an ideal in $K[x_1^{\pm 1}, \dots, x_n^{\pm 1}]$. Then $\mathbf{0} \in \overline{Y'}$, and $\dim(Y') < \dim(Y)$, so by induction $\text{trop}(Y') \cap \mathbb{R}^n_{>0} \neq \emptyset$. Since $\text{trop}(Y') \subset \text{trop}(Y)$ the result follows.

This reduces the proof to the base case of the induction: $\dim(Y) = \dim(I) = 1$. We again assume that Y, and thus $\overline{Y}$, is irreducible. Let J be the integral closure of I_{aff}, and consider the ideal $J_{\mathfrak{m}} \subset S_{\mathfrak{m}}$, where $\mathfrak{m} = \langle x_1, \dots, x_n \rangle$. Note that J does not contain any variable x_i; otherwise x_i would satisfy a monic polynomial for which all but the leading coefficient lie in I, which would imply $x_i \in I_{\text{aff}}$. Since J is a prime of dimension 1, $\mathfrak{m}$ is a codimension-1 ideal in $S_{\mathfrak{m}}/J_{\mathfrak{m}}$. Thus by Serre's condition R1, the ring $R = S_{\mathfrak{m}}/J_{\mathfrak{m}}$ is a discrete valuation ring (see [**Eis95**, Theorem 11.5]) with maximal ideal $\mathfrak{m}$. The completion $\hat{R}$ of R at $\mathfrak{m}$ is then a complete regular local ring. There is thus an isomorphism $\pi \colon \hat{R} \to K[[t]]$ for some parameter t (see [**Eis95**, Proposition 10.16]). Let $p_j = \pi(x_j) \in K[[t]]$ for $1 \leq j \leq n$.

The map $S \to S/J$ induces a map $K[[x_1, \dots, x_n]] \to \hat{R}$, which does not send any variable x_i to 0 and contains in its kernel all $f \in I_{\text{aff}}$ when the polynomial f is viewed as a power series. This means that $f(p_1, \dots, p_n) = 0$ for $f \in I_{\text{aff}}$, and $p_i \neq 0$ for all i. Let L be the field of generalized power series with coefficients in L (see Example 2.1.7). We thus have $(p_1, \dots, p_n) \in V(I_{\text{aff}}) \subseteq T^n_L$, and therefore $(\text{val}(p_1), \dots, \text{val}(p_n)) \in \text{trop}(V(I))$. Since the isomorphism $\pi \colon \hat{R} \to K[[t]]$ must take the maximal ideal to the maximal ideal, each p_j lies in $\langle t \rangle$, so $\text{val}(p_j) > 0$, and thus $\text{trop}(V(I)) \cap \mathbb{R}^n_{>0} \neq \emptyset$. □

Remark 6.3.6. The closure of Y in $\mathbb{A}^n$ depends on how T^n is embedded into $\mathbb{A}^n$. For example, consider $Y = V(t_1 + t_2 + 1) \subset T^2$. For the standard embedding $i : T^2 \to \mathbb{A}^2$ given by $i(t_1, t_2) = (x, y)$, we have $\overline{Y} = V(x+y+1)$. But, if $i : T^2 \to \mathbb{A}^2$ is given by $i(t_1, t_2) = (t_2/t_1, t_1)$, then $\overline{Y} = V(y+xy+1)$.

Proof of Theorem 6.3.4. The special case $X_\Sigma = \mathbb{A}^n$ is Proposition 6.3.5. Next suppose that Σ is a cone σ generated by d elements in a basis for $N \simeq \mathbb{Z}^n$, so

$$X_\Sigma = U_\sigma \cong \mathbb{A}^d \times T^{n-d}.$$

Let $I \subset K[x_1^{\pm 1}, \dots, x_n^{\pm 1}]$ be the ideal of $Y \subset T^n$. We identify $T^{n-d} = \{\mathbf{t} \in T^n : t_1 = \cdots = t_d = 1\}$. Let

$$\tilde{Y} = \{\mathbf{t} \cdot \mathbf{y} : \mathbf{t} \in T^{n-d} \text{ and } \mathbf{y} \in Y\},$$

and $\tilde{I} = I \cap \Bbbk[x_1^{\pm 1}, \dots, x_d^{\pm 1}]$. Then $\tilde{Y}/T^{n-d} = V(\tilde{I}) \subset T^d$. Similarly,

$$\overline{Y} = V(I \cap \Bbbk[x_1, \dots, x_d, x_{d+1}^{\pm 1}, \dots, x_n^{\pm 1}]).$$

Let $Y' = \{\mathbf{t}\cdot\mathbf{y} : \mathbf{t} \in T^{n-d} \text{ and } \mathbf{y} \in \overline{Y}\}$. Then $Y'/T^{n-d} = V(I \cap K[x_1, \ldots, x_d])$. Then $\overline{Y} \cap \mathcal{O}_\sigma \neq \emptyset$ if and only if $\mathbf{0} \in Y'/T^{n-d}$, and $\operatorname{trop}(Y_{\mathrm{triv}})$ intersects the relative interior of σ if and only if $\operatorname{trop}((\tilde{Y}/T^{n-d})_{\mathrm{triv}})$ intersects the interior of the positive orthant in $\mathbb{R}^d$. The theorem in this case thus follows from Proposition 6.3.5.

We now consider the general case of an arbitrary toric variety X_Σ. Choose a toric resolution of singularities $\pi : X_{\tilde{\Sigma}} \to X_\Sigma$, where $\tilde{\Sigma}$ is a smooth fan that refines Σ. We also denote by π the map of fans $\pi\colon \tilde{\Sigma} \to \Sigma$.

Let $\tilde{Y}$ be the closure of Y in $X_{\tilde{\Sigma}}$. This is the strict transform of $\overline{Y}$. Suppose first that $\overline{Y}$ intersects an orbit $\mathcal{O}_\sigma$ of X_σ. Then there is some $\sigma' \in \tilde{\Sigma}$ with $\pi(\operatorname{relint}(\sigma')) \subseteq \operatorname{relint}(\sigma)$, and $\tilde{Y} \cap \mathcal{O}_{\sigma'} \neq \emptyset$. This means that the closure $Y_{\sigma'}$ of Y in $U_{\sigma'}$ intersects $\mathcal{O}_{\sigma'}$. Since $U_{\sigma'} \cong \mathbb{A}^d \times T^{n-d}$, the previous paragraph implies that $\operatorname{trop}(Y_{\mathrm{triv}}) \cap \operatorname{relint}(\sigma') \neq \emptyset$, and thus $\operatorname{trop}(Y_{\mathrm{triv}}) \cap \operatorname{relint}(\sigma) \neq \emptyset$. Conversely, suppose that $\operatorname{trop}(Y_{\mathrm{triv}}) \cap \operatorname{relint}(\sigma) \neq \emptyset$ for some cone $\sigma \in \Sigma$. Then there is $\sigma' \in \tilde{\Sigma}$ with $\operatorname{trop}(Y_{\mathrm{triv}}) \cap \operatorname{relint}(\sigma') \neq \emptyset$, so the closure $Y_{\sigma'}$ of Y in $U_{\sigma'}$ intersects $\mathcal{O}_{\sigma'}$ by the argument at the start of the proof, and thus $\tilde{Y}$ intersects $\mathcal{O}_{\sigma'}$. This implies that $\overline{Y}$ intersects $\mathcal{O}_\sigma$, as required. □

Theorem 6.3.4 has the following interpretation: Given a subvariety Y of a torus $T^n = (K^*)^n$, its tropicalization $\operatorname{trop}(Y)$ gives us information about the closure of Y in *any* toric compactification of T^n. In particular, it suggests that we use a fan structure on $\operatorname{trop}(Y)$ itself as an economical way of compactifying Y. We saw a first glimpse of this in Section 1.8, and we shall develop such compactifications systematically in the next section.

In the remainder of this section we present a detailed example that illustrates various concepts introduced so far in Chapter 6. We shall highlight relations to some of the constructions seen earlier in this book.

Example 6.3.7. We fix the four-dimensional projective space $X_\Sigma = \mathbb{P}^4$ over the field of rational numbers $K = \mathbb{Q}$ with the 2-adic valuation. The fan Σ consists of $31 = 1 + 5 + 10 + 10 + 5$ cones in $\mathbb{R}^5/\mathbb{R}\mathbf{1} \simeq \mathbb{R}^4$. Let $\overline{Y} \simeq \mathbb{P}^2$ be the projective plane inside $\mathbb{P}^4$ that consists of all vectors in the kernel of

$$\begin{bmatrix} 1 & 1 & 2 & 4 & 8 \\ 8 & 4 & 2 & 1 & 1 \end{bmatrix}. \tag{6.3.1}$$

For a cone σ in Σ, we have $\overline{Y} \cap \mathcal{O}_\sigma \neq \emptyset$ if and only if $\dim(\sigma) \leq 2$. For each two-dimensional cone σ, the intersection consists of a unique point. Namely, identifying $\sigma = \text{pos}\{\mathbf{e}_i, \mathbf{e}_j\}$ with $\{i, j\}$, these ten special points on $\overline{Y}$ are

σ	$\overline{Y} \cap \mathcal{O}_\sigma$	$\text{trop}(\overline{Y} \cap \mathcal{O}_\sigma)$
$\{0,1\}$	$(0:0:2:-7:3)$	$(\infty:\infty:1:0:0)$
$\{0,2\}$	$(0:4:0:-31:15)$	$(\infty:2:\infty:0:0)$
$\{0,3\}$	$(0:14:-31:0:6)$	$(\infty:1:0:\infty:1)$
$\{0,4\}$	$(0:2:-5:2:0)$	$(\infty:1:0:1:\infty)$
$\{1,2\}$	$(4:0:0:-63:31)$	$(2:\infty:\infty:0:0)$
$\{1,3\}$	$(2:0:-9:0:2)$	$(1:\infty:0:\infty:1)$
$\{1,4\}$	$(6:0:-31:14:0)$	$(1:\infty:0:1:\infty)$
$\{2,3\}$	$(31:-63:0:0:4)$	$(0:0:\infty:\infty:2)$
$\{2,4\}$	$(15:-31:0:4:0)$	$(0:0:\infty:2:\infty)$
$\{3,4\}$	$(3:-7:2:0:0)$	$(0:0:1:\infty:\infty)$

The projective space $\text{trop}(\mathbb{P}^4)$ can be thought of as a four-dimensional simplex. The tropical plane $\text{trop}(\overline{Y})$ is a balanced polyhedral complex that intersects the faces of dimension at least 2 of that simplex. The ten two-dimensional faces intersect $\text{trop}(\overline{Y})$ in the ten points (given by representatives in $\overline{\mathbb{R}}^5$) that are listed in the third column above.

Consider next the five facets of the simplex $\text{trop}(\mathbb{P}^4)$. These correspond to codimension-1 orbits $\mathcal{O}_{\{\mathbf{e}_i\}}$ on $\mathbb{P}^4$. The intersection $\overline{Y} \cap \mathcal{O}_{\{\mathbf{e}_i\}}$ is a line in $\mathbb{P}^3$ with four points removed. Its tropicalization is a trivalent tree with four leaves. We can record such a tree by giving the partition $\{\{i, j\}, \{k, l\}\}$, also called a *split*, induced on the leaves of the tree by removing the unique internal edge. The five trees in the boundary of $\text{trop}(\mathbb{P}^4)$ are found to be:

- in the facet dual to $\sigma = \{\mathbf{e}_0\}$, the tree has the split $[\{1,2\},\{3,4\}]$;
- in the facet dual to $\sigma = \{\mathbf{e}_1\}$, the tree has the split $[\{0,2\},\{3,4\}]$;
- in the facet dual to $\sigma = \{\mathbf{e}_2\}$, the tree has the split $[\{0,1\},\{3,4\}]$;
- in the facet dual to $\sigma = \{\mathbf{e}_3\}$, the tree has the split $[\{0,1\},\{2,4\}]$;
- in the facet dual to $\sigma = \{\mathbf{e}_4\}$, the tree has the split $[\{0,1\},\{2,3\}]$.

This is an *abstract tree arrangement* with $n = 5$ as in Definition 5.4.4. By Theorem 5.4.9, this data determines a coarsest matroid subdivision of the four-dimensional hypersimplex $\Delta_{3,5}$. It has three maximal cells, which are all matroid polytopes. In the notation of Figure 5.4.7, these are the matroids

$$\{\{0,1\},2,3,4\} \quad \text{and} \quad [0,1;3,4](2) \quad \text{and} \quad \{0,1,2,\{3,4\}\}. \tag{6.3.2}$$

The very affine surface $Y = \overline{Y} \cap T^4$ is $\mathbb{P}^2$ minus five lines. Its tropicalization $\text{trop}(Y)$ is a uniform tropicalized two-dimensional plane $L_\mathbf{w} \subset \mathbb{R}^5/\mathbb{R}\mathbf{1}$ as in Theorem 4.3.17. Its tropical Plücker vector $\mathbf{w} \in \text{trop}(G^0(3,5))$ is read off from the ten $2{\times}2$-minors of (6.3.1) by dualizing and taking the 2-adic valuation:

$$\begin{aligned} &w_{012} = 2, \quad w_{013} = 1, \quad w_{014} = 1, \quad w_{023} = 0, \quad w_{024} = 0, \\ &w_{034} = 1, \quad w_{123} = 0, \quad w_{124} = 0, \quad w_{134} = 1, \quad w_{234} = 2. \end{aligned}$$

The plane $L_{\mathbf{w}}$ consists of 15 unbounded polygons, ten unbounded edges, and two bounded edges. It has three vertices, labeled by the matroids in (6.3.2).

We can also construct $L_{\mathbf{w}} = \text{trop}(Y)$ as a complete intersection of two tropical hyperplanes in $\mathbb{R}^5/\mathbb{R}\mathbf{1}$. The rows (6.3.1) satisfy the hypotheses of Theorem 4.6.18, with $r = 2, n = 4$, and $P_1 = P_2$ the standard four-dimensional simplex. Hence $\text{trop}(Y)$ is the fan derived in Theorem 4.6.9. We use the notation $[Q_1, Q_2]$ to denote simplices in the triangulation of the five-dimensional Cayley polytope $C(P_1, P_2)$. Here Q_i is face of P_i. The triangulation has five five-dimensional simplices. Three of these are mixed:

$$[\{0,1\},\{1,2,3,4\}] \quad \text{and} \quad [\{0,1,2\},\{2,3,4\}] \quad \text{and} \quad [\{0,1,2,3\},\{3,4\}].$$

These correspond to the vertices of $\text{trop}(Y)$, in the order given in (6.3.2):

$$(0:0:1:2:2) \quad \text{and} \quad (1:1:0:1:1) \quad \text{and} \quad (2:2:1:0:0).$$

The triangulation of $C(P_1, P_2)$ has twelve mixed four-dimensional simplices. Two are dual to the bounded edges of $\text{trop}(Y)$: they are $[\{0,1\},\{2,3,4\}]$ and $[\{0,1,2\},\{3,4\}]$. The other ten mixed four-dimensional simplices correspond to the nodes in the five trees: $[\{1,2,3\},\{3,4\}], \ldots, [\{0,1\},\{1,2,3\}]$. Further, the triangulation has ten mixed tetrahedra corresponding to the two-dimensional cells of $\text{trop}(Y)$. They are labeled $[\{i,j\},\{k,l\}]$. The list is

σ	mixed cell
{0,1}	[{2,3},{3,4}]
{0,3}	[{1,2},{2,4}]
{1,2}	[{0,3},{3,4}]
{1,4}	[{0,2},{2,3}]
{2,4}	[{0,1},{1,3}]

σ	mixed cell
{0,2}	[{1,3},{3,4}]
{0,4}	[{1,2},{2,3}]
{1,3}	[{0,2},{2,4}]
{2,3}	[{0,1},{1,4}]
{3,4}	[{0,1},{1,2}]

In Remark 6.2.5 we noted that a tropical linear space is the tropical convex hull of its cocircuit vectors. Hence our tropical plane in $\text{trop}(\mathbb{P}^4)$ is a tropical polygon with ten vertices, labeled by $\sigma = \{0,1\}, \{0,2\}, \ldots, \{3,4\}$:

$$L_{\mathbf{w}} \;=\; \text{trop}(\overline{Y}) \;=\; \text{tconv}\big\{(\infty{:}\infty{:}1{:}0{:}0),\, (\infty{:}2{:}\infty{:}0{:}0),\, \ldots,\, (0{:}0{:}1{:}\infty{:}\infty)\big\}.$$

In summary, $\text{trop}(\overline{Y})$ is a compact contractible balanced polyhedral complex in $\text{trop}(\mathbb{P}^4)$. It is the union of the pieces $\text{trop}(\overline{Y} \cap \mathcal{O}_\sigma)$. Here σ runs over the $16 = 1+5+10$ cones of dimension 0, 1, and 2 in Σ. We obtain a polyhedral complex structure on $\text{trop}(\overline{Y})$ by taking the union over these various orbits. That complex has $23 = 3 + 5\cdot 2 + 10 \cdot 1$ vertices, $47 = 12 + 5 \cdot (5+2)$ edges, and 25 polygons (20 triangles, four quadrilaterals, and one pentagon). ◇

6.4. Tropical Compactifications

Throughout this book we studied varieties Y embedded in a torus T^n. Section 6.3 was about the closure $\overline{Y}$ inside a toric variety X_Σ with torus T^n. In this section we focus on special choices of Σ and the resulting properties of $\overline{Y}$. We begin by explaining how to speak of the tropicalization of Y without reference to the embedding. This relies on the existence of an *intrinsic torus* into which Y embeds. We shall use the following result of Samuel [**Sam66**].

Lemma 6.4.1. *Let R be a finitely generated K-algebra that is an integral domain, and let R^* be the multiplicative group of units of R. Then the quotient group R^*/K^* is free abelian and finitely generated.*

Proof. Fix an embedding of $Y = \mathrm{Spec}(R)$ into some affine space $\mathbb{A}^m$, and let $\overline{Y}$ be the closure of Y in $\mathbb{P}^m$. We first consider the case that $\overline{Y}$ is normal. Consider the group homomorphism from R^* to the group $\mathrm{Div}\,\overline{Y}$ of Weil divisors on $\overline{Y}$ given by sending $f \in R^*$ to the divisor $\mathrm{div}(f)$ it determines. Its kernel is K^*, as any other unit defines a nontrivial divisor on $\overline{Y}$. Since $f \in R^*$, the divisor $\mathrm{div}(f)$ is supported on the boundary $\overline{Y}\backslash Y$. This means that the image of this group homomorphism is contained in the free abelian group generated by the finitely many divisorial components of $\overline{Y}\backslash Y$. Since every subgroup of a finitely generated free abelian group is finitely generated and free abelian, the group R^*/K^* is finitely generated and free abelian.

If $\overline{Y}$ is not normal, we consider the normalization map $\phi\colon \widetilde{Y} \to \overline{Y}$. We have $\phi^{-1}(Y) = \mathrm{Spec}(\widetilde{R})$, where $\widetilde{R}$ is the integral closure of R. This is an extension of R, so R^*/K^* is a subgroup of $\widetilde{R}^*/K^*$. We know that $\widetilde{R}^*/K^*$ is finitely generated and free abelian. Hence so is its subgroup R^*/K^*. □

Definition 6.4.2. Let Y be a subvariety of a torus T^n. We call Y a *very affine* variety. By Lemma 6.4.1, the group $K[Y]^*/K^*$ is isomorphic to $\mathbb{Z}^m$ for some m. The *intrinsic torus* of Y is the torus $T_{\mathrm{in}} := \mathrm{Hom}(K[Y]^*/K^*, K^*)$.

Every very affine variety Y embeds into its intrinsic torus T_{in}. This embedding is not canonical, but depends on a small choice. We now explain this. Let $f_1, \ldots, f_m$ be Laurent polynomials in n variables whose images generate the group $K[Y]^*/K^* \simeq \mathbb{Z}^m$, so $T_{\mathrm{in}} \simeq (K^*)^m$. An embedding $Y \hookrightarrow T_{\mathrm{in}}$ is given by $\mathbf{y} \mapsto (f_1(\mathbf{y}), \ldots, f_m(\mathbf{y}))$. This embedding is not unique because we can replace $f_1, \ldots, f_m$ by another sequence that generates $K[Y]^*/K^*$. For example, we can multiply each f_i by a different scalar from K^*.

Example 6.4.3. (1) Let $Y = V(x + y + 1) \subset (K^*)^2$. The units of $K[Y] \cong K[x^{\pm 1}, y^{\pm 1}]/\langle x + y + 1\rangle$ have the form $ax^u y^v$ for $a \in K^*$ and $u, v \in \mathbb{Z}$. Hence $K[Y]^*/K^* \cong \mathbb{Z}^2$ under the isomorphism that takes $ax^u y^v$ to (u, v). The intrinsic torus T_{in} equals $\mathrm{Hom}(\mathbb{Z}^2, K^*) \cong (K^*)^2$, and the embedding of Y into T_{in} is the original embedding.

(2) Fix the surface $Y = V(x_1x_3 - x_2^2, x_2x_4 - x_3^2, x_1x_4 - x_2x_3) \subset (K^*)^4$. We have $K[Y] \cong K[y_1^{\pm 1}, y_2^{\pm 1}]$ under the map

$$x_1 \mapsto y_1,\ x_2 \mapsto y_1y_2,\ x_3 \mapsto y_1y_2^2,\ x_4 \mapsto y_1y_2^3.$$

Thus $Y \cong (K^*)^2$, so Y is its own intrinsic torus. The tropicalization is $\operatorname{trop}(Y) = \{\mathbf{w} \in \mathbb{R}^4 : w_1 + w_3 = 2w_2 \text{ and } w_2 + w_4 = 2w_3\}$. Note that this is isomorphic to $\mathbb{R}^2 = \operatorname{trop}((K^*)^2)$.

(3) Let $H_0, \ldots, H_n$ be hyperplanes in $\mathbb{P}^d$, with linear forms $\ell_0, \ldots, \ell_n \in K[x_0, \ldots, x_d]$. Let Y be the arrangement complement as in Section 4.1. The map (4.1.1) that takes Y into $(K^*)^{n+1}/K^* \cong (K^*)^n$ is the embedding of Y into its intrinsic torus T_{in}. Indeed, setting $\ell_0 = x_0$, so that $Y \subset \mathbb{A}^n$, we have $K[Y] = K[x_1, \ldots, x_d][\ell_1'^{-1}, \ldots, \ell_n'^{-1}]$, where $\ell_i' = \ell_i|_{x_0=1}$. The group $T_{\text{in}} = K[Y]^*/K^*$ is generated by $\ell_1', \ldots, \ell_n'$, and the embedding $Y \hookrightarrow T_{\text{in}}$ coincides with (4.1.1).

(4) Let $Y = V(x^3 + y^3 - 2x^2y - 2x + 1) \subseteq (\mathbb{C}^*)^2$. The tropicalization of Y is the standard tropical line, with multiplicities changed from 1 to 3. This is not the embedding of Y into T_{in}. The units of $\mathbb{C}[Y] = \mathbb{C}[x^{\pm 1}, y^{\pm 1}]/\langle x^3 + y^3 - 2x^2y - 2x + 1\rangle$ include $1 - x + y$ in addition to x and y: $(1-x+y)(x^{-1}y^{-1}(1-x-y-x^2+xy+y^2)) = 1$ in $\mathbb{C}[Y]$. We use this to re-embed Y into a larger torus, by setting $z = 1 - x + y$. This gives $Y = V(x^3 + y^3 - 2x^2y - 2x + 1, z + x - y - 1) \subset (\mathbb{C}^*)^3$.

The next proposition shows that the scenarios in Example 6.4.3 are representative of the general case. We say that $\phi : T^m \to T^n$ is a *monomial map* if $\phi(t_1, \ldots, t_m) = (c_1\mathbf{t}^{\mathbf{a}_1}, \ldots, c_n\mathbf{t}^{\mathbf{a}_n})$ for some $\mathbf{a}_1, \ldots, \mathbf{a}_n \in \mathbb{Z}^m$ and $\mathbf{c} = (c_1, \ldots, c_n) \in (K^*)^n$. The map ϕ is a morphism of affine varieties. If $c_1 = \cdots = c_n = 1$, then ϕ is a morphism of tori, and $\phi = \phi_A$ where A is the $m \times n$ matrix with ith column $\mathbf{a}_i$ as in Corollary 3.2.13.

Proposition 6.4.4. *Let $j : Y \to T^n$ be a closed embedding, and let $i : Y \to T^m$ be the embedding of Y into its intrinsic torus. Then there is a monomial map $\phi \colon T^m \to T^n$ as above for which the following diagram commutes.*

$$\begin{array}{ccc} Y & \xrightarrow{\ i\ } & T^m \\ & \searrow^{j} & \downarrow{\scriptstyle \phi} \\ & & T^n \end{array}$$

Write $\operatorname{trop}(\phi) \colon \mathbb{R}^n \to \mathbb{R}^m$ *for the affine map given by* $\mathbf{x} \mapsto A^T\mathbf{x} + \operatorname{val}(\mathbf{c})$. *The tropicalization of $Y \subset T^n$ is the image under the affine map* $\operatorname{trop}(\phi)$ *of the tropicalization of the embedding of Y into its intrinsic torus.*

Proof. The embedding j expresses the coordinate ring of Y as $K[Y] \cong K[x_1^{\pm 1}, \ldots, x_n^{\pm 1}]/I$ for some ideal I. Choose Laurent polynomials $f_1, \ldots, f_m$

in $K[x_1^{\pm 1}, \ldots, x_n^{\pm 1}]$ whose images in $K[Y]$ freely generate the group $K[Y]^*/K^*$. Since the x_i are units themselves, we can find an integer matrix $A = (a_{ij})$ and scalars $c_i \in K^*$ such that $x_i \equiv c_i f_1^{a_{i1}} \cdots f_m^{a_{im}}$ modulo I for $i = 1, \ldots, n$. The corresponding monomial map ϕ satisfies $j = \phi \circ i$. The last sentence follows from Corollary 3.2.13 and the discussion at the end of Section 2.6. □

Remark 6.4.5. Given any very affine variety Y, we may now speak of *the* tropicalization of Y when referring to $\mathrm{trop}(Y \hookrightarrow T_{\mathrm{in}})$. Proposition 6.4.4 tells us that any other embedding of Y into a torus can be recovered from the intrinsic one, and so by Corollary 3.2.13 the tropicalization of any embedding $Y \subset T^n$ can be recovered from the tropicalization of $Y \subset T_{\mathrm{in}}$. Also note that if a group G acts on Y, then the action extends to $K[Y]^*$, and so to the intrinsic torus. This means that a shadow of G (which may be trivial) acts on the tropicalization of Y. See Exercise 6.8(12).

We now discuss how tropical geometry can be used to compactify subvarieties of tori. This will complete the journey we began in Section 1.8. For the rest of the section we assume that the valuation on K is trivial.

Definition 6.4.6. A variety $\overline{Y}$ defined over K is *complete* if it is universally closed, so the projection map $p\colon \overline{Y} \times Z \to Z$ is closed for every variety Z. This notion plays the role for algebraic geometry of compactness in topology.

A variety being complete is a synonym for it being "proper over Spec(K)", or simply "proper" if the context is clear. The definition of a proper morphism includes the criteria that it be separated and of finite type; these are automatic for varieties over K. Recall from [**CLS11**, Theorem 3.4.6] that a toric variety X_Σ is complete if and only if the fan Σ is complete, so $|\Sigma| = \mathbb{R}^n$. We shall use the following consequences of the definition of completeness in the situation when Y is a subvariety of T^n:

(a) If X_Σ is a complete toric variety with torus T^n, then the closure $\overline{Y}$ of Y in X_Σ is complete.

(b) In this case, the intersection of $\overline{Y}$ with any torus orbit closure $V(\sigma)$ is again complete.

(c) Let X_Σ be a toric variety with torus T^n, and assume that the closure $\overline{Y}$ of Y in X_Σ is complete. Let $\Sigma' \subseteq \mathbb{R}^n$ be a fan that contains Σ as a subfan. Then the closure of Y in $X_{\Sigma'}$ equals $\overline{Y}$.

Proposition 6.4.7. *Let Y be a d-dimensional irreducible subvariety of T^n, and let $\overline{Y}$ be its closure in a toric variety X_Σ. Fix the trivial valuation on K.*

(1) *The variety $\overline{Y}$ is complete if and only if* $\mathrm{trop}(Y) \subseteq |\Sigma|$.

(2) *Suppose the equivalent conditions in* (1) *hold. Then* $\mathrm{trop}(Y) = |\Sigma|$ *if and only if $\overline{Y} \cap \mathcal{O}_\sigma$ is nonempty and pure of dimension $d - \dim(\sigma)$ for all $\sigma \in \Sigma$.*

Proof. For the "only-if" direction in (1), suppose $\overline{Y}$ is complete, but $\operatorname{trop}(Y) \not\subseteq |\Sigma|$. Choose a complete fan Σ' that contains Σ as a subfan. This exists by [**Ewa96**, Theorem III.2.8]. Fix a cone σ of $\Sigma' \backslash \Sigma$ that has a point of $\operatorname{trop}(Y)$ in its relative interior. Since $\overline{Y}$ is complete, by fact (c) above, the closure $\overline{Y}'$ of Y in $X_{\Sigma'}$ equals $\overline{Y}$. By Theorem 6.3.4 we know that $\overline{Y}'$ intersects the torus orbit $\mathcal{O}_\sigma$ of $X_{\Sigma'}$. However, this contradicts $\overline{Y} \subseteq X_\Sigma$, since $\mathcal{O}_\sigma \cap X_\Sigma = \emptyset$.

For the "if" direction in (1), suppose $\operatorname{trop}(Y) \subseteq |\Sigma|$, and fix a complete fan Σ' as above. By Theorem 6.3.4, the closure $\overline{Y}'$ of Y in $X_{\Sigma'}$ does not intersect any orbit $\mathcal{O}_\sigma$ with $\sigma \in \Sigma' \backslash \Sigma$. Hence $\overline{Y}'$ is contained in X_Σ, and thus equals the closure $\overline{Y}$ of Y in X_Σ. Since $X_{\Sigma'}$ is complete, so is $\overline{Y}' = \overline{Y}$.

For the "only if" direction in (2), suppose $\operatorname{trop}(Y) = |\Sigma|$, and consider $\sigma \in \Sigma$. Let $\overline{Y}$ be the closure of Y in X_Σ, let Z be an irreducible component of $\overline{Y} \cap \mathcal{O}_\sigma$ with the reduced scheme structure, and let $\overline{Z}$ be the closure of Z in $V(\sigma)$. By part (1), $\overline{Y}$ is complete, and hence $\overline{Y} \cap V(\sigma)$ is complete, by fact (b). As $\overline{Z}$ is the closure of Z in $\overline{Y} \cap V(\sigma)$, it is also complete. The tropical variety of $Z \subseteq \mathcal{O}_\sigma$ is contained in the fan of $V(\sigma)$ in the quotient space $N(\sigma) = N_\mathbb{R}/\operatorname{span}(\sigma)$. This fan is pure of dimension $d - \dim(\sigma)$, since $\operatorname{trop}(Y) = |\Sigma|$, so $\dim(Z) \leq d - \dim(\sigma)$. Since toric varieties are Cohen–Macaulay (see [**CLS11**, 9.2.9]), $\mathcal{O}_\sigma$ is locally set-theoretically cut out by $\dim(\sigma)$ equations. This means that $\overline{Y} \cap \mathcal{O}_\sigma$, and thus Z, has codimension at most $\dim(\sigma)$. We conclude that $\dim(Z) = d - \dim(\sigma)$, as required.

For the "if" direction, suppose that $\overline{Y} \cap \mathcal{O}_\sigma$ is nonempty and pure of dimension $d - \dim(\sigma)$ for all $\sigma \in \Sigma$. We have $\operatorname{trop}(Y) \subseteq |\Sigma|$, by assumption (1). Since $\overline{Y}$ has dimension d, we have $\dim(\sigma) \leq d$ for all $\sigma \in \Sigma$. Theorem 6.3.4 implies that $\operatorname{trop}(Y)$ intersects the relative interior of every $\sigma \in \Sigma$. By the Structure Theorem 3.3.5, $\operatorname{trop}(Y)$ is the support of a pure d-dimensional fan. Hence the fan Σ is also pure of dimension d. Suppose there is $\sigma \in \Sigma$ with $\dim(\sigma) = d$ and $\sigma \not\subseteq \operatorname{trop}(Y)$. Then $\operatorname{trop}(Y) \cap \sigma$ is properly contained in σ, and is the support of a pure d-dimensional fan Σ_σ. Thus there must be a $(d-1)$-dimensional cone τ of Σ_σ that lives in only one d-dimensional cone of Σ_σ. This contradicts the balancing condition, so we conclude $|\Sigma| = \operatorname{trop}(Y)$. □

Remark 6.4.8. We do not require X_Σ to be complete here; it is possible for the closure $\overline{Y}$ of Y in X_Σ to be proper even though X_Σ is not. A simple example is given by considering the noncomplete toric variety $\mathbb{P}^2 \backslash \{(1{:}0{:}0), (0{:}1{:}0), (0{:}0{:}1)\}$, and $Y = V(x+y+1) \subset T^2$. Then Y is isomorphic to $\mathbb{P}^1$ with three points removed, and $\overline{Y} \cong \mathbb{P}^1$, which is complete.

Remark 6.4.9. One consequence of Proposition 6.4.7 is that if Y is a subvariety of T^n and Σ is a fan with $\operatorname{trop}(Y) = |\Sigma|$, then the boundary $\overline{Y} \backslash Y$

added in the compactification $\overline{Y}$ of Y in the toric variety X_Σ is divisorial. This means that every irreducible component has codimension 1 in $\overline{Y}$. In addition, these boundary components have *combinatorial normal crossings*: any nonempty intersection of l components has codimension l in $\overline{Y}$. These facts will be used to compute the tropical variety in Section 6.5.

Let $Y \subset T^n$ be a subvariety. A *tropical compactification* of Y is its closure $\overline{Y}$ in a toric variety X_Σ with $|\Sigma| = \text{trop}(Y)$, as in Proposition 6.4.7. Returning to Example 1.8.1, we now show how this works for plane curves.

Example 6.4.10. Let $Y = V(1+2x-3y+5xy) \subset T^2$. The tropical curve $\text{trop}(Y)$ consists of the four coordinate rays, each with multiplicity 1.

Consider first the closure $\overline{Y}_1 = V(z^2+2xz-3yz+5xy)$ of Y in $\mathbb{P}^2$. This projective curve intersects the line $x=0$ in two points: $(0:1:0)$ and $(0:1:3)$; it intersects the line $y=0$ in two points: $(1:0:0)$ and $(1:0:2)$; and intersects the line $z=0$ in $(1:0:0)$ and $(0:1:0)$. Now let X_Σ be $\mathbb{P}^2$ with the three torus-invariant points removed. The closure $\overline{Y}_2$ of Y in X_Σ is thus $\overline{Y}_1$ with two points removed, which is not complete. This is as expected, as $\text{trop}(Y)$ is not contained in $|\Sigma| = \text{pos}\{(1,0)\} \cup \text{pos}\{(0,1)\} \cup \text{pos}\{(-1,-1)\}$.

Consider next the closure $\overline{Y}_3$ of Y in $\mathbb{P}^1 \times \mathbb{P}^1$. This is the curve defined by the bihomogeneous polynomial $x_0y_0+2x_1y_0-3x_0y_1+5x_1y_1$. The intersection of $\overline{Y}_3$ with the torus-invariant divisor $\{x_0=0\}$ is the point $(0:1)\times(5:-2)$; the intersection with $\{x_1=0\}$ is the point $(1:0)\times(3:1)$; the intersection with $\{y_0=0\}$ is the point $(5:3)\times(0:1)$; and the intersection with $\{y_1=0\}$ is the point $(2:-1)\times(1:0)$. Let X_Σ be $\mathbb{P}^1\times\mathbb{P}^1$ with the four torus-fixed points removed. Since $\overline{Y}_3$ does not contain any of the torus-fixed points of $\mathbb{P}^1\times\mathbb{P}^1$, the closure $\overline{Y}_4$ of Y in X_Σ equals $\overline{Y}_3$, which is complete. This is again as expected, as $\text{trop}(Y)$ equals $|\Sigma|$, the union of the coordinate rays.

The toric surface X_Σ is the union of five torus orbits; the dense orbit, and four one-dimensional orbits. The intersection of $\overline{Y}_4$ with the dense orbit is Y, which is codimension zero in $\overline{Y}_4$. The intersection of $\overline{Y}_4$ with each one-dimensional orbit is a point, which is codimension 1 in $\overline{Y}_4$. ◊

Example 6.4.11. Let $f = 3x_1x_3+5x_2x_3-x_1+2x_2-x_3+7 \in \mathbb{C}[x_1^{\pm 1}, x_2^{\pm 1}, x_3^{\pm 1}]$, and $Y = V(f) \subset T^3$. The Newton polytope of f is a triangular prism, so $\text{trop}(Y)$ is the fan over the edge graph of a bipyramid. The closure $\overline{Y}_1$ of Y in $\mathbb{P}^3$ is the quadric $V(3x_1x_3+5x_2x_3-x_0x_1+2x_0x_2-x_0x_3+7x_0^2)$. The intersection of $\overline{Y}_1$ with the orbit closure $\{x_0=x_3=0\}$ in $\mathbb{P}^3$ is that entire orbit closure, so has codimension 1 in $\overline{Y}_1$, rather than the expected codimension of 2. Indeed, the smallest toric subvariety of $\mathbb{P}^3$ containing $\overline{Y}_1$ is $\mathbb{P}^3 \backslash \{(1{:}0{:}0{:}0)\}$, and $\text{trop}(Y)$ does not equal the support of the fan of this toric variety. On the other hand, let X_Σ be the toric variety obtained by removing the torus-fixed points from $\mathbb{P}^2\times\mathbb{P}^1$. That toric threefold has Cox ring $K[x_0, x_1, x_2, y_0, y_1]$, where $y_1 = x_3$. The closure $\overline{Y}_2$ of Y in X_Σ is defined by the homogeneous ideal $\langle 3x_1y_1 + 5x_2y_1 - x_1y_0 + 2x_2y_0 - y_1x_0 + 7x_0y_0 \rangle$. We see that $\overline{Y}_2$ does not contain any of the torus-fixed points of $\mathbb{P}^2 \times \mathbb{P}^1$, but it does intersect every other torus orbit. This is consistent with the fact that trop(Y) is precisely the union of all two-dimensional cones in Σ. $\Diamond$

An interesting example of a tropical compactification is the moduli space $\overline{M}_{0,n}$ of stable genus zero curves with n marked points. This is the Deligne–Mumford compactification of the moduli space $M_{0,n}$, which parameterizes ways to arrange n distinct labeled points on $\mathbb{P}^1$. We first recall these spaces.

By elementary projective geometry, any three distinct points in $\mathbb{P}^1$ can be mapped to any other three distinct points via a unique automorphism of $\mathbb{P}^1$. Given a collection of n distinct labeled points, we may thus assume that points $1, 2$, and 3 are $\infty, 0$, and 1. More formally, they are $(1:0), (0:1)$, and $(1:1)$. This means that $M_{0,3}$ is a point and $M_{0,4}$ is $\mathbb{P}^1 \backslash \{0, 1, \infty\}$. In general, $M_{0,n}$ is $(\mathbb{P}^1 \backslash \{0, 1, \infty\})^{n-3}$ with the diagonals $\{x_i = x_j\}$ removed:

$$\begin{aligned} M_{0,n} &= (\mathbb{P}^1 \backslash \{0,1,\infty\})^{n-3} \setminus \text{diagonals} \\ &= (\mathbb{C}^* \backslash \{1\})^{n-3} \setminus \text{diagonals} \\ &= \mathbb{P}^{n-3} \setminus \big\{ x_i = 0, x_i = x_j : 0 \leq i, j \leq n-3 \big\}. \end{aligned}$$

We have realized $M_{0,n}$ as the complement of $\binom{n-1}{2} = n - 2 + \binom{n-2}{2}$ hyperplanes in $\mathbb{P}^{n-3}$. Following Section 4.1, this defines a closed embedding of $M_{0,n}$ into $T^{\binom{n-1}{2}-1}$ where the defining equations are linear. Explicitly, the morphism $M_{0,n} \to T^{\binom{n-1}{2}-1}$ from (4.1.1) is given by $\mathbf{x} \mapsto B^T\mathbf{z}$, where B is the $(n-2) \times \binom{n-1}{2}$-matrix with first $n-2$ columns an identity matrix and the remaining $\binom{n-2}{2}$-columns of the form $\mathbf{e}_i - \mathbf{e}_j$ with $0 \leq i < j \leq n-3$. For example, for $n = 5$, we have the 3×6-matrix B of Example 4.1.2. The kernel of B does not change if we add a last row with first $n-2$ entries -1 and all other entries 0. The columns of this new matrix B' are then precisely the simple roots $\{\mathbf{e}_i - \mathbf{e}_j : 0 \leq i < j \leq n-2\}$ of the root system A_{n-2}. By Example 4.2.14, the associated matroid is the matroid of the complete graph K_{n-1}. By Section 4.3, the corresponding tropical variety is the space Δ of phylogenetic trees. The toric variety X_Δ has dimension $\binom{n-1}{2} - 1$.

The closure of $M_{0,n} \subset T^{\binom{n-1}{2}-1}$ in X_Δ is the Deligne–Mumford moduli space $\overline{M}_{0,n}$. See [**Tev07**, Theorem 5.5] or [**GM10**, Theorem 5.7]. This is the moduli space of *stable* genus zero curves with n distinct marked points. A stable genus zero curve is a tree of $\mathbb{P}^1$s intersecting in nodes for which every copy of $\mathbb{P}^1$ contains at least three nodes or marked points. See Figure 6.4.1 for an example. Such diagrams are dual to pictures of trees as in Figure 4.3.1.

We summarize this example in the following theorem.

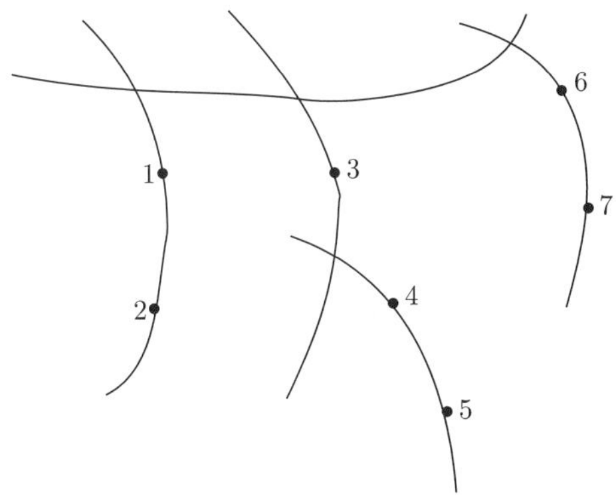

Figure 6.4.1. A stable genus 0 curve with seven marked points.

Theorem 6.4.12. *The moduli space* $M_{0,n}$ *is the variety in the torus* $T^{\binom{n-1}{2}-1} \subset \mathbb{P}^{\binom{n-1}{2}-1}$ *defined by the homogeneous linear ideal*

$$I_{0,n} = \langle z_{ij} - z_{1j} + z_{1i} : 2 \leq i < j \leq n-1 \rangle \subset K[z_{ij} : 1 \leq i < j \leq n-1]. \tag{6.4.1}$$

Its tropicalization $\mathrm{trop}(M_{0,n}) \subset \mathbb{R}^{\binom{n-1}{2}-1}$ *is the space* Δ *of phylogenetic trees on* n *leaves from Section* 4.3. *The closure of* $M_{0,n}$ *in the corresponding toric variety* X_Δ *equals the Deligne–Mumford compactification* $\overline{M}_{0,n}$.

This tropical compactification is a special case of the setup in Section 4.1. If $\mathcal{A}$ is any arrangement of $n+1$ hyperplanes in $\mathbb{P}^d$, then the complement $Y = \mathbb{P}^d \backslash \mathcal{A}$ is a very affine variety. By part (3) of Example 6.4.3, the embedding of Y into the torus T^n is the embedding into the intrinsic torus. As discussed in Chapter 4, there are several different fan structures on $\mathrm{trop}(Y) \subset \mathbb{R}^n$. A choice of *building set* $\mathcal{G}$ (see Exercise 4.7(10) of Chapter 4) for the lattice of flats of $\mathcal{A}$ determines a fan structure Σ with associated simplicial complex the *nested set complex*. The tropical compactification of Y using this fan structure is the wonderful compactification of Y due to De Concini and Procesi. For a proof and more details we refer to [**Tev07**, §4] and [**FS05**].

It is sometimes useful to refine a given fan structure on a tropical variety. Recall that a morphism $\psi : X \to Y$ is *flat* if for every point $\mathfrak{p} \in X$ the local ring $\mathcal{O}_{X,\mathfrak{p}}$ is a flat $\mathcal{O}_{Y,\psi(\mathfrak{p})}$-module. If X and Y are affine with $X = \mathrm{Spec}(A)$ and $Y = \mathrm{Spec}(B)$, then ψ is flat if and only if the map $\psi^*: B \to A$ makes A into a flat B-module. This means that the right exact functor $- \otimes_B A$ is exact.

Flatness is a niceness property that guarantees that the fibers of ψ share many numerical invariants. See [**Vak13**, Chapter 24] for a summary of such properties. A morphism ψ is *faithfully flat* if ψ is flat and surjective.

Definition 6.4.13. Fix a subvariety $Y \subset T^n$ and a fan Σ with $|\Sigma| = \operatorname{trop}(Y)$ in $\mathbb{R}^n$. The closure $\overline{Y}$ of Y in X_Σ is *flat tropical* if $\overline{Y}$ is complete and the multiplication map $\psi : T \times \overline{Y} \to X_\Sigma$ given by $(\mathbf{t}, \mathbf{x}) \mapsto \mathbf{tx}$ is faithfully flat.

The notion of a compactification being flat tropical is due to Tevelev. His original paper [**Tev07**] does not use the prefix "flat"; we add it here to distinguish from tropical compactifications for which $|\Sigma| = \operatorname{trop}(Y)$ is the only condition. The requirement in Definition 6.4.13 that $\overline{Y}$ is complete implies $\operatorname{trop}(Y) \subseteq |\Sigma|$, by Proposition 6.4.7. The condition that ψ is surjective is equivalent to requiring that $\overline{Y}$ intersects every torus orbit of X_Σ and so, by Theorem 6.3.4, to requiring that $\operatorname{trop}(Y)$ intersects the relative interior of every cone of Σ. We now show that if this holds, then $|\Sigma| = \operatorname{trop}(Y)$ and that any refinement of Σ also induces a flat tropical compactification:

Proposition 6.4.14. *Let $Y \subset T^n$ be a subvariety, and let $\Sigma \subset \mathbb{R}^n$ be a fan for which the closure $\overline{Y}$ in X_Σ is a flat tropical compactification. Any refinement Σ' of Σ also has this property. In addition, the support $|\Sigma|$ equals* $\operatorname{trop}(Y)$ *when K is given the trivial valuation.*

Proof. Let $\pi \colon X_{\Sigma'} \to X_\Sigma$ be the toric morphism induced by the refinement Σ' of Σ. Since the support $|\Sigma'|$ equals $|\Sigma|$, and the latter contains $\operatorname{trop}(Y)$, the closure $\overline{Y}'$ is complete by part (1) of Proposition 6.4.7. To show that $\overline{Y}'$ is a flat tropical compactification, we thus only need to show that the multiplication morphism $\psi' \colon \overline{Y}' \times T \to X_{\Sigma'}$ is faithfully flat. The pullback of a faithfully flat morphism is faithfully flat (see, for example [**Vak13**, §24.5.1]). Hence it suffices to show that ψ' is the pullback $\pi^*(\psi)$ of the multiplication map on X_Σ, so $\overline{Y}' \times T = (\overline{Y} \times T) \times_{X_\Sigma} X_{\Sigma'}$, as in the following diagram.

$$\begin{array}{ccc} \overline{Y}' \times T & \xrightarrow{\psi'=\pi^*(\psi)} & X_{\Sigma'} \\ \big\downarrow & & \big\downarrow \pi \\ \overline{Y} \times T & \xrightarrow{\quad\psi\quad} & X_\Sigma \end{array}$$

Since π is the identity on T, the restriction of $\pi^*(\psi)$ to $Y \times T$ equals ψ'. It thus suffices to show that $Z := (\overline{Y} \times T) \times_{X_\Sigma} X_{\Sigma'}$ is reduced and irreducible. Consider the map $\pi^*(\psi) \colon Z \to X_{\Sigma'}$, which is flat as noted above. The preimage of $T^n \subset X_{\Sigma'}$ is $Y \times T^n$, so is in particular reduced and irreducible. Restricting to an affine open set, we need to show the following: if $\phi : \operatorname{Spec}(A) \to \operatorname{Spec}(B)$ is flat and surjective, $\operatorname{Spec}(B)$ is reduced and irreducible, and the preimage in $\operatorname{Spec}(A)$ of some open set $U \subset \operatorname{Spec}(B)$ is reduced and irreducible, then $\operatorname{Spec}(A)$ is reduced and irreducible. Algebraically, this means showing that if $\phi^* \colon B \to A$ is a injection that makes A into a flat B-module, B is a domain, and there is $f \in B$ for which A_f is a domain, then A is a domain. This follows from applying the exact functor $- \otimes_B A$ to the sequence $0 \to B \to B_f$. It shows that A includes into the domain A_f, so is itself a domain.

We now show that the support $|\Sigma|$ of Σ equals $\mathrm{trop}(Y)$. Since $\overline{Y}$ is complete, we know that $\mathrm{trop}(Y) \subseteq |\Sigma|$. Suppose there exists a vector $\mathbf{v} \in (|\Sigma| \cap \mathbb{Q}^n) \setminus \mathrm{trop}(Y)$. By assumption, $\mathrm{trop}(Y)$ intersects the relative interior of every cone of Σ, so $\mathbf{v}$ does not lie on a ray of Σ. Form the *stellar subdivision* Σ' of Σ using the ray $\mathbf{v}$. See, for example, [**CLS11**, §1.1], where this is called the star subdivision of Σ at $\mathbf{v}$. The fan Σ' refines Σ, so by above the closure $\overline{Y}'$ of Y in $X_{\Sigma'}$ is a flat tropical compactification. But this means that $\mathrm{trop}(Y)$ intersects the relative interior of every cone of Σ', so contains the ray through $\mathbf{v}$, which is a contradiction. Thus $\mathrm{trop}(Y) = |\Sigma|$. □

We next discuss some consequences of a compactification being flat tropical that will be useful in Section 6.7. Recall that a local ring $(R, \mathfrak{m})$ of Krull dimension d is *Cohen–Macaulay* if there is a regular sequence $r_1, \ldots, r_d$ in $\mathfrak{m}$. This means that $\langle r_1, \ldots, r_d \rangle \neq R$, the element r_1 is a nonzerodivisor on R, and r_i is a nonzerodivisor on $R/\langle r_1, \ldots, r_{i-1} \rangle$ for all $i > 1$. A variety X is Cohen–Macaulay at a point $\mathbf{p} \in X$ if the local ring $\mathcal{O}_{X,\mathbf{p}}$ of X at $\mathbf{p}$ is Cohen–Macaulay. If $X = V(I)$ is affine, where $I \subseteq S := K[x_1, \ldots, x_n]$, then the local ring $\mathcal{O}_{X,\mathbf{p}}$ is the localization $(S/I)_{\mathfrak{p}}$, where $\mathfrak{p} = I(\mathbf{p}) \subseteq S$. The condition that X is Cohen–Macaulay at a point $\mathbf{p}$ is weaker than the requirement that X be smooth at $\mathbf{p}$, or locally a complete intersection, but still places some strong conditions on X. In particular, the aspect that we will use in Section 6.7 is that it simplifies the intersection theory of X.

Proposition 6.4.15. *Let $\overline{Y} \subset X_\Sigma$ be a flat tropical compactification of a d-dimensional variety $Y \subset T^n$, with X_Σ smooth. Let $\sigma \in \Sigma$ with $\dim(\sigma) = d$, and fix a point $\mathbf{p} \in \overline{Y} \cap \mathcal{O}_\sigma$. Then $\overline{Y}$ is Cohen–Macaulay at $\mathbf{p}$.*

Proof. Consider the restriction $(\overline{Y} \cap U_\sigma) \times T^n \to U_\sigma$ of the multiplication map ψ to the affine chart U_σ. Write $\overline{Y} \cap U_\sigma = \mathrm{Spec}(R)$, so $R = K[\sigma^\vee \cap M]/I$ for some ideal I. Since X_Σ is smooth, the orbit closure $V(\sigma) \subset U_\sigma$ (which equals $\mathcal{O}_\sigma$) is defined by a regular sequence $f_1, \ldots, f_d \in K[\sigma^\vee \cap M]$. Since the multiplication map ψ is flat, $R \otimes K[M]$ is a flat $K[\sigma^\vee \cap M]$-module. This implies that the images $g_i = \psi^*(f_i)$ form a regular sequence in $R \otimes K[M]$. Indeed, the fact that f_{i+1} is a nonzerodivisor on $A = K[\sigma^\vee \cap M]/\langle f_1, \ldots, f_i \rangle$ means that $0 \to A \to A_{f_{i+1}}$ is exact. Since tensoring with $R \otimes K[M]$ is an exact functor, the natural map from $(R \otimes K[M])/\langle g_1, \ldots, g_i \rangle$ to $((R \otimes K[M])/\langle g_1, \ldots, g_i \rangle)_{g_{i+1}}$ is an injection, and so g_{i+1} is a nonzerodivisor on $(R \otimes K[M])/\langle g_1, \ldots, g_i \rangle$. Thus the g_i form a regular sequence.

The support of the subscheme defined by $\langle g_1, \ldots, g_d \rangle$ is a union of sets of the form $\{\mathbf{q}\} \times T^n$ for finitely many points $\mathbf{q}$, one of which equals $\mathbf{p}$. This follows from part (2) of Proposition 6.4.7. Choose $\mathbf{t} = (t_1, \ldots, t_n) \in T^n$ for which $(\mathbf{p}, \mathbf{t})$ lies outside any embedded component of the scheme defined by $\langle g_1, \ldots, g_d \rangle \subseteq R \otimes K[M]$, and let $x_1, \ldots, x_n$ be the coordinates on $K[M]$. Then $g_1, \ldots, g_d, x_1 - t_1, \ldots, x_n - t_n$ is a regular sequence on $R \otimes K[M]$. After localizing at the ideal $\mathfrak{p}'$ in $R \otimes K[M]$ of the point $(\mathbf{p}, \mathbf{t})$, we may permute the order of this regular sequence to obtain that $g_1, \ldots, g_d$ is a regular sequence on $(R \otimes K[M])_{\mathfrak{p}'} / \langle x_1 - t_1, \ldots, x_n - t_n \rangle \cong R_{\mathfrak{p}}$, where $\mathfrak{p}$ is the ideal of the point $\mathbf{p}$. Since $d = \dim(\sigma) = \dim(Y) = \dim(R)$, we conclude that $R_{\mathfrak{p}}$ is Cohen–Macaulay, and thus that $\overline{Y}$ is Cohen–Macaulay at $\mathbf{p}$. □

If the given variety Y admits a flat tropical compactification, then we can find one with the toric variety X_Σ smooth. This is done by toric resolution of singularities. A consequence of Proposition 6.4.15 is that not every fan Σ with support $|\Sigma| = \mathrm{trop}(Y)$ is flat tropical, as the following example shows.

Example 6.4.16. Let $\overline{Y} \subset \mathbb{P}^n$ be an irreducible d-dimensional projective variety that is not Cohen–Macaulay at a point $\mathbf{p} = (p_0 : \cdots : p_n) \in \overline{Y}$. Choose coordinates on $\mathbb{P}^n$ so that $p_0 = \cdots = p_{d-1} = 0$, $p_i \neq 0$ for $i \geq d$, and no point of $\overline{Y}$ has more than d coordinates equal to zero. This can be achieved by choosing d general hyperplanes passing through $\mathbf{p}$, and $n+1-d$ general hyperplanes that do not pass through $\mathbf{p}$, and changing coordinates so that these are the coordinate hyperplanes. See Exercise 6.8(11). Let $Y = \overline{Y} \cap T^n$. We claim that $\mathrm{trop}(Y)$ has the same support as the *d-skeleton* of the fan of $\mathbb{P}^n$, i.e., the fan whose maximal cones are spanned by d-tuples $\{\mathbf{e}_{i_1}, \ldots, \mathbf{e}_{i_d}\}$ of coordinate rays. Indeed, by Theorem 6.3.4, since $\overline{Y}$ does not intersect torus orbits of $\mathbb{P}^n$ with more than d zero coordinates, $\mathrm{trop}(Y)$ is contained in the d-skeleton. However, the d-dimensional variety $\overline{Y}$ intersects all torus orbits with fewer than d zero coordinates. Hence, again by Theorem 6.3.4, $\mathrm{trop}(Y)$ intersects the relative interior of every d-dimensional cone in the fan of $\mathbb{P}^n$. The toric variety defined by the d-skeleton is $\mathbb{P}^n$ with every torus orbit of codimension larger than d removed, so it is smooth. By construction, the point $\mathbf{p}$ lies in $\overline{Y} \cap \mathcal{O}_\sigma$, where σ is a d-dimensional cone. If $\overline{Y}$ were a flat tropical compactification, Proposition 6.4.15 would imply that $\overline{Y}$ were Cohen–Macaulay at $\mathbf{p}$. Hence $\overline{Y}$ is not a flat tropical compactification. ◇

Whether a compactification is flat tropical depends on the variety itself and not just on the tropical variety as a set. Indeed, any generic complete intersection of $n - d$-hypersurfaces in $\mathbb{P}^n$ has the same tropicalization as $\mathrm{trop}(Y)$ in Example 6.4.16. By choosing all but one of the hypersurfaces to be hyperplanes, we may ensure that the degree of the complete intersection equals the degree of $\overline{Y}$, and thus that the multiplicities also coincide.

We next show that every subvariety $Y \subset T^n$ has a flat tropical compactification $\overline{Y}$. We give the field K the trivial valuation, so trop(Y) can be given the structure of a polyhedral fan. The key idea is to choose the fan Σ' on trop(Y) to come from the Gröbner fan of the homogenization of the ideal of Y; this is a fan by Corollary 2.5.12.

Proposition 6.4.17. *Let Y be a subvariety of T^n with ideal $I = I(Y) \subseteq K[x_1^{\pm 1}, \ldots, x_n^{\pm 1}]$. We assume that Y is not fixed by any subtorus of T^n. Let I_{proj} be the ideal of the closure of Y in $\mathbb{P}^n$, and let Σ be the Gröbner fan of I_{proj}, regarded as a fan in $\mathbb{R}^n/\mathbb{R}\mathbf{1}$. Let $\overline{Y}$ be the closure of Y in X_Σ. Then $\overline{Y}$ is a flat tropical compactification of Y.*

Proof. Let P be the Hilbert polynomial of I_{proj}. The *Hilbert scheme* $\text{Hilb}_P(\mathbb{P}^n)$ parameterizes subschemes of $\mathbb{P}^n$ with Hilbert polynomial P or, equivalently, homogeneous ideals in $K[x_0, \ldots, x_n]$ with Hilbert polynomial P that are saturated with respect to the irrelevant ideal $\mathfrak{m} = \langle x_0, \ldots, x_n \rangle$. The torus $T^n \cong (K^*)^{n+1}/K^*$ of $\mathbb{P}^n$ acts on $\text{Hilb}_P(\mathbb{P}^n)$ by setting $\lambda \cdot I = \langle f(\lambda_0 x_0, \ldots, \lambda_n x_n) : f \in I \rangle$. The Hilbert scheme has a universal family $\mathcal{U} \subset \text{Hilb}_P(\mathbb{P}^n) \times \mathbb{P}^n$. The projection $\pi \colon \mathcal{U} \to \text{Hilb}_P(\mathbb{P}^n)$ is flat, and the fiber over the point corresponding to a subscheme of $\mathbb{P}^n$ is that subscheme.

Let Z be the closure of the T^n-orbit of the point $\tilde{Y}$ in $\text{Hilb}_P(\mathbb{P}^n)$ corresponding to I_{proj}. The assumption that Y is not fixed by any subtorus of T^n means that the orbit of I_{proj} is isomorphic to T^n. The normalization $\tilde{Z}$ of Z is thus a normal toric variety with torus T^n. The limit of a one-parameter subtorus of T^n labeled by $\mathbf{w} \in N \cong \mathbb{Z}^n$ is the point of $\text{Hilb}_P(\mathbb{P}^n)$ corresponding to the subscheme of $\mathbb{P}^n$ defined by the initial ideal $\text{in}_{\mathbf{w}}(I_{\text{proj}})$. This means that the Gröbner fan Σ refines the fan of $\tilde{Z}$. Let Σ' be the subfan of the fan of $\tilde{Z}$ consisting of all cones such that the initial ideal of I_{proj} contains no monomial. By Proposition 6.4.14 it thus suffices to show that the closure $\overline{Y}'$ of Y in $\tilde{Z} = X_{\Sigma'}$ is a flat tropical compactification.

By Corollaries 2.5.12 and 3.5.5, the support $|\Sigma'|$ equals trop(Y), so we need only prove that the multiplication map is flat. The composition of the normalization map and the embedding gives a morphism $X_{\Sigma'} \to \text{Hilb}_P(\mathbb{P}^n)$. Let $U' \subset X_{\Sigma'} \times \mathbb{P}^n$ be the pullback of the universal family $\mathcal{U} \subset \text{Hilb}_P(\mathbb{P}^n) \times \mathbb{P}^n$ over $\text{Hilb}_P(\mathbb{P}^n)$ via this morphism. Let U be the intersection of U' with $X_{\Sigma'} \times T^n$. This gives a flat morphism $U \to X_{\Sigma'}$. To finish the proof, we show that $U \cong \overline{Y}' \times T^n$, and the map $U \to X_{\Sigma'}$ is the multiplication map.

We denote by $[W]$ the point in $\text{Hilb}_P(\mathbb{P}^n)$ corresponding to a subscheme $W \subset \mathbb{P}^n$. By construction $U = \{([W], \mathbf{w}) \in X_{\Sigma'} \times T^n : [W] \in X_{\Sigma'}, \mathbf{w} \in W \cap T^n\}$. This is isomorphic to $\{(\mathbf{w}^{-1}[W], \mathbf{w}) : [W] \in X_{\Sigma'}, \mathbf{w} \in W \cap T^n\}$. Note that $\mathbf{w}^{-1}[W] = [\mathbf{w}^{-1}W]$, and $(1 : \cdots : 1) \in \mathbf{w}^{-1}W$. So it suffices to show that $\overline{Y} = \{[W] \in X_\Sigma : (1 : \cdots : 1) \in W\}$. The containment $\subseteq$

follows from the observation that $(1 : \cdots : 1) \in y^{-1}\tilde{Y}$ for any $y \in Y$, since the right-hand set is closed. This means that $\overline{Y}$ is an irreducible component of this set. It remains to see that this set is irreducible, so equals $\overline{Y}$. In Proposition 6.4.14 we showed that if the preimage of an open set under a flat surjective morphism is irreducible, then the source of the morphism is irreducible. Applying this to the torus T^n the claim follows. □

Remark 6.4.18. If Y is fixed by a subtorus T' of T^n, then the proof of Proposition 6.4.17 actually constructs a flat tropical compactification of the quotient Y/T'. To get a flat tropical compactification of Y, we combine this with a compactification of the subtorus factor.

The Gröbner fan may be finer than the fan of $\tilde{Z}$ constructed in the proof of Proposition 6.4.17. This is because two distinct initial ideals may have the same saturation with respect to the irrelevant ideal $\langle x_0, \ldots, x_n \rangle$, and so describe the same subscheme of $\mathbb{P}^n$. One can also replace the Hilbert scheme $\mathrm{Hilb}_P(\mathbb{P}^n)$ by the multigraded Hilbert scheme [**HS04**] that parameterizes ideals with a fixed Hilbert function. In that case the fan constructed would be precisely the Gröbner fan.

In this setting, we can use initial ideals to identify the piece $\overline{Y} \cap \mathcal{O}_\sigma$ added in the compactification for any $\sigma \in \Sigma$. Let $T_\sigma = N_\sigma \otimes K^* \subset T^n$. Note that $\mathcal{O}_\sigma \cong T^n/T_\sigma$. Fix $\mathbf{w} \in \mathrm{relint}(\sigma)$. The subscheme Y^σ of T^n defined by $\mathrm{in}_{\mathbf{w}}(I_Y)$ has an action of T_σ, as the initial ideal is homogeneous with respect to the grading induced by N_σ. We use here that the valuation on K is trivial, so $K = \Bbbk$. The subscheme $\overline{Y} \cap \mathcal{O}_\sigma$ is then Y^σ/T_σ.

The compactification $\overline{Y}$ used in Proposition 6.4.17 depends on the choice of embedding of T^n into $\mathbb{P}^n$, which is induced from the choice of coordinates on T^n. This is shown in Example 3.2.9. In that example, the fan Σ that works for Proposition 6.4.17 is the coarsest fan structure on the set $\mathrm{trop}(Y)$. However no such coarsest fan may exist, as seen in Example 3.5.4. To compute the fan Σ for any given Y, one can use the software `Gfan` [**Jen**].

We finish this section with one further niceness condition.

Definition 6.4.19. Let $Y \subset T^n$ be a subvariety, and let $\overline{Y}$ be a tropical compactification obtained by taking the closure of Y in a toric variety X_Σ. The compactification $\overline{Y}$ is *schön* if $\overline{Y} \cap \mathcal{O}_\sigma$ is smooth for every torus orbit $\mathcal{O}_\sigma$ in X_Σ.

The case when $Y = V(f)$ is a hypersurface in T^n has been well studied for decades. Here Σ is the normal fan of the Newton polytope of the Laurent polynomial f. The compactification of Y is *schön* precisely when Y is *nondegenerate with respect to its Newton boundary*. Much of the geometry of such hypersurfaces is determined by the geometry of the toric variety X_Σ. Examples include the relationship between the Milnor number of a hypersurface singularity and its Newton polytope given by Kushnirenko [**Kou76**] and the computation of the Hodge numbers by Danilov and Khovanskii [**DK86**].

The following theorem, whose proof we omit, summarizes further properties that come from requiring a tropical compactification $\overline{Y}$ to be *schön*.

Theorem 6.4.20. (1) *If $Y \subset T^n$ has a* schön *compactification, then any tropical compactification of Y is* schön.

(2) *A* schön *compactification of Y is regularly embedded, normal, and has toroidal singularities.*

(3) *If the field K has characteristic zero, then any projective variety Z contains a Zariski open subset Y with a* schön *compactification $\overline{Y}$.*

The construction of Y and $\overline{Y}$ in part (3) requires that a resolution of singularities exists for Z. This is the only reason for the characteristic zero requirement. For proofs of these results see [**Tev07**], [**Tev14**], and [**LQ11**].

6.5. Geometric Tropicalization

We saw in Section 6.4 how, given a subvariety $Y \subset T^n$, the tropical variety determines a good choice of compactification of Y. We now explore the converse, and see how a nice compactification of Y determines $\mathrm{trop}(Y)$. Throughout this section we assume that the field K has the trivial valuation, and all varieties are taken to be irreducible. A key idea is the characterization of $\mathrm{trop}(Y)$ in terms of *divisorial valuations*, given in Proposition 6.5.4. This can be thought of as a fourth part of the Fundamental Theorem 3.2.3.

Fix a subvariety $Y \subset T^n$. Its function field $K(Y)$ is the field of fractions of the coordinate ring $K[Y]$. Let Y' be a variety birational to Y. This means that Y and Y' contain open sets on which they are isomorphic. Examples of such Y' are compactifications of Y. In what follows we assume that Y' is normal and $\mathbb{Q}$-factorial. The second of these conditions means that a multiple of every Weil divisor is a Cartier divisor, so every codimension-1 subvariety of Y' is locally defined by a single equation. The assumption that Y' is birational to Y means that $K(Y')$ is isomorphic to $K(Y)$. This is also the field of fractions of the coordinate ring $K[Z]$ of any affine chart Z of Y'.

Every irreducible divisor D on Y' determines a valuation on $K(Y)$, as we now recall. Since Y' is normal, the coordinate ring $K[Z]$ of any affine chart Z is normal. Choose a chart Z that intersects D, and let $P \subset K[Z]$ be the prime ideal defining $D \cap Z$. By Serre's condition R1 (see [**Eis95**, Theorem 11.5]), since D has codimension 1 and $K[Z]$ is normal, the localization $K[Z]_P$ is a DVR. Write $\mathrm{val}_D : K(Y) \to \mathbb{Z}$ for the associated discrete valuation on the quotient field $K(Y)$ of $K[Z]_P$. We call such valuations *divisorial valuations* on $K(Y)$. Note that the valuation val_D is trivial on the field K.

Example 6.5.1. Let $\overline{Y} = V(x_0+x_1+x_2+x_3) \subset \mathbb{P}^3$, and $Y = \overline{Y} \cap (K^*)^3$. Then $K(\overline{Y}) = K(Y)$ is the field of rational functions in two variables. The difference $\overline{Y} \backslash Y$ consists of four lines, L_0, L_1, L_2, L_3, where $L_i = \overline{Y} \cap \{x_i = 0\}$. To compute the divisorial valuation on $\overline{Y}$ corresponding to L_1, we consider the affine chart $Z = \{x_0 \neq 0\}$ with $K[Z] = K[y_1, y_2, y_3]/\langle 1+y_1+y_2+y_3 \rangle$ given by $y_i = x_i/x_0$ for $i = 1, 2, 3$. On the chart Z, the divisor L_1 has the equation $y_1 = 0$. The valuation val_{L_1} is computed on a rational function f in the quotient field of $K[Z]$ by writing $f = y_1^m f'$ where $m \in \mathbb{Z}$ and f' has neither numerator nor denominator divisible by y_1. With this, $\mathrm{val}_{L_1}(f) = m$. To see this explicitly, note that the localization $K[Z]_{\langle y_1 \rangle}$ is a regular local ring of dimension 1, so it is a discrete valuation ring. The maximal ideal is generated by y_1, so the valuation has the desired property. For the line L_0 we need to choose a different affine chart to compute the divisorial valuation. For example, we may take $Z' = \{x_1 \neq 0\}$, which has coordinate ring $K[y_0', y_2', y_3']/(y_0' + 1 + y_2' + y_3')$, where $y_i' = x_i/x_1$. The valuation val_{L_0} is given by the exponent of the largest power of y_0' dividing the function. ◊

Definition 6.5.2. Let $\overline{Y}$ be a normal $\mathbb{Q}$-factorial compactification of $Y \subset T^n$. Recall that an element $m \in M$ induces an element of $\mathrm{Hom}(T^n, K^*)$, and thus an element of $K(Y)$. A divisorial valuation val_D on $K(Y)$ determines an element $[\mathrm{val}_D] \in N \otimes \mathbb{R} \cong \mathrm{Hom}(M, \mathbb{R})$ by setting $[\mathrm{val}_D](m) = \mathrm{val}_D(m)$ for $m \in M$. When we choose coordinates for T^n, so $K[T^n] \cong K[x_1^{\pm 1}, \ldots, x_n^{\pm 1}]$, we get an integer vector $[\mathrm{val}_D] = (\mathrm{val}_D(x_1), \ldots, \mathrm{val}_D(x_n))$ in $\mathbb{R}^n$.

Example 6.5.3. We continue Example 6.5.1. We take y_1, y_2, y_3 as generators for the group of units $K[Y]^*/K^* \cong \mathbb{Z}^3$. Then $[\mathrm{val}_{L_1}] = (1, 0, 0) \in \mathbb{R}^3$. By the same argument, $[\mathrm{val}_{L_2}] = (0, 1, 0)$ and $[\mathrm{val}_{L_3}] = (0, 0, 1)$. Note that $y_1 = 1/y_0'$, $y_2 = y_2'/y_0'$, and $y_3 = y_3'/y_0'$. Hence $[\mathrm{val}_{L_0}] = (-1, -1, -1)$. ◊

Proposition 6.5.4. *Let Y be a subvariety of T^n. The tropical variety* $\mathrm{trop}(Y)$ *is the closure of the following subset in* $N_{\mathbb{R}} \cong \mathbb{R}^n$*:*

$$\big\{[c\,\mathrm{val}_D] : c \in \mathbb{Q}_{\geq 0} \text{ and } \mathrm{val}_D \text{ is a divisorial valuation on } K(Y)\big\}. \tag{6.5.1}$$

Proof. Let $I \subseteq K[x_1^{\pm 1}, \ldots, x_n^{\pm 1}]$ be the ideal of Y, so $K(Y)$ is the quotient field of $K[x_1^{\pm 1}, \ldots, x_n^{\pm 1}]/I$. Let val_D be a divisorial valuation on $K(Y)$, and write $[\mathrm{val}_D] = \mathbf{w} = (w_1, \ldots, w_n) \in \mathbb{R}^n$ where $w_i = \mathrm{val}_D(x_i)$. We first show that $\mathbf{w} \in \mathrm{trop}(Y)$. For $f = \sum_{\mathbf{u}} c_{\mathbf{u}} x^{\mathbf{u}} \in I$, the image of f in $K(Y)$ is zero. Hence $\mathrm{val}_D(\sum c_{\mathbf{u}} x^{\mathbf{u}}) > \min_{\mathbf{u}}(\mathrm{val}_D(c_{\mathbf{u}} x^{\mathbf{u}})) = \min_{\mathbf{u}}(\sum_i u_i \,\mathrm{val}_D(x_i)) = \min_{\mathbf{u}}(\mathbf{w} \cdot \mathbf{u}) = \mathrm{trop}(f)(\mathbf{w})$, since K has the trivial valuation. Since $\mathrm{val}(a + b) = \min(\mathrm{val}(a), \mathrm{val}(b))$ if $\mathrm{val}(a) \neq \mathrm{val}(b)$, this means that the minimum in $\mathrm{trop}(f)(\mathbf{w})$ must be achieved at least twice. As f is arbitrary, this implies $\mathbf{w} = [\mathrm{val}_D] \in \mathrm{trop}(Y)$. The equality $[c\,\mathrm{val}_D] = c[\mathrm{val}_D]$ then implies that $\mathrm{trop}(Y)$ contains the closure of (6.5.1).

We now show the opposite inclusion. Since $\operatorname{trop}(Y)$ is the support of a rational polyhedral fan, $\operatorname{trop}(Y) \cap \mathbb{Q}^n$ is dense in $\operatorname{trop}(Y)$. It thus suffices to show that every $\mathbf{w} \in \operatorname{trop}(Y) \cap \mathbb{Q}^n$ has the form $[c \operatorname{val}_D]$ where $\operatorname{val}_D : K(Y) \to \mathbb{R}$ is the divisorial valuation corresponding to a Cartier divisor D on some normal variety birational to Y. After scaling, we may assume that $\mathbf{w} \in \mathbb{Z}^n$ with $\gcd(w_i) = 1$. After a change of coordinates on T^n we may then assume $\mathbf{w} = \mathbf{e}_1$. Let Σ be the fan in $N_{\mathbb{R}}$ consisting of just the ray spanned by $\mathbf{e}_1$, so $X_\Sigma = \mathbb{A}^1 \times T^{n-1}$. Let $\overline{Y}$ be the closure of Y in the toric variety X_Σ. We denote by D_1 the torus-invariant divisor defined by $x_1 = 0$ on X_Σ. By Theorem 6.3.4, since $\mathbf{e}_1 \in \operatorname{trop}(Y)$, the variety $\overline{Y}$ intersects D_1. As $\overline{Y}$ is irreducible, the intersection $\overline{Y} \cap D_1 = \overline{Y} \cap \{x_1 = 0\}$ is a hypersurface in $\overline{Y}$.

Let $\nu\colon Y' \to \overline{Y}$ be the normalization of $\overline{Y}$. This is a birational morphism, so a Cartier divisor on Y' gives rise to a divisorial valuation on $K(Y)$. Let D' be an irreducible component of $\nu^{-1}(\overline{Y} \cap D_1)$. We claim that $\mathbf{e}_1 = [c \operatorname{val}_{D'}]$ for some $c > 0$. Indeed, for $2 \leq i \leq n$ the coordinate function x_i is invertible on $\mathbb{A}^1 \times T^{n-1}$, so its restriction to $\overline{Y}$ is also invertible. The same holds for the pullback $\nu^*(x_i)$ on Y'. This means that $\operatorname{val}_{D'}(\nu^*(x_i)) = 0$. The coordinate x_1 vanishes on D_1, so the pullback $\nu^*(x_1)$ vanishes on D', and thus $\lambda := \operatorname{val}_{D'}(\nu^*(x_1)) > 0$. Using the coordinates $x_1, \ldots, x_n$ for T^n, we get $[\operatorname{val}_{D'}] = \lambda \mathbf{e}_1$. Setting $c = 1/\lambda$, we thus have $\mathbf{w} = \mathbf{e}_1 = [c \operatorname{val}_{D'}]$. □

Remark 6.5.5. The result of Proposition 6.5.4 that every $\mathbf{w} \in \operatorname{trop}(Y) \cap \mathbb{Q}^n$ has the form $[c \operatorname{val}_D]$ for a divisorial valuation on $K(Y)$ is also a consequence of the fact that the tropicalization map $\operatorname{trop}\colon Y \to N_{\mathbb{R}}$ extends to a map from the *Berkovich analytification* Y^{an} to $N_{\mathbb{R}}$. Berkovich spaces are beyond the scope of this book, but have many important connections to tropical geometry. Baker's article [**Bak08a**] is an excellent first introduction. For the extension of tropicalization to analytic spaces, see [**Pay09a**].

We will use the characterization of $\operatorname{trop}(Y)$ of Proposition 6.5.4 to show how a sufficiently nice compactification of Y determines $\operatorname{trop}(Y)$. The connection is via a simplicial complex that comes from the compactification.

Definition 6.5.6. Let $Y \subset T^n$ be a variety, and let $\overline{Y}$ be a compactification of Y, so $\overline{Y}$ is a complete variety containing Y. The *boundary* of $\overline{Y}$ is the set $\partial \overline{Y} = \overline{Y} \backslash Y$. Throughout this section we shall assume that the boundary $\partial \overline{Y}$ is *divisorial*, meaning that it is a union of codimension-1 subvarieties of $\overline{Y}$. Let $D_1, \ldots, D_l$ be the irreducible components of $\partial \overline{Y}$.

The boundary $\partial \overline{Y}$ is a *combinatorial normal crossings divisor* if, for any subset $\sigma \subseteq \{1, \ldots, l\}$, the intersection $\bigcap_{i \in \sigma} D_i$ has codimension $|\sigma|$ in $\overline{Y}$. The pair $(\overline{Y}, \partial \overline{Y})$ is then called a *combinatorial normal crossings (cnc) pair.* If, in addition, this intersection is transverse, the boundary is *simple normal crossings*, and the pair is a *simple normal crossings (snc) pair.*

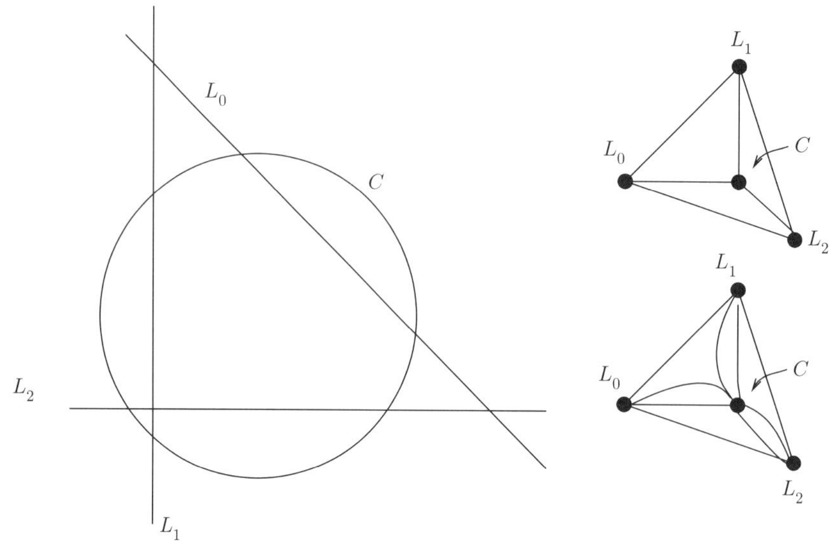

Figure 6.5.1. A compactification and its boundary complex.

The *boundary complex* $\Delta(\partial\overline{Y})$ of the pair $(\overline{Y}, \partial\overline{Y})$ is a simplicial complex with one vertex v_i for each divisor D_i. A subset $\sigma = \{v_{i_1}, \dots, v_{i_j}\}$ is a simplex in $\Delta(\partial\overline{Y})$ whenever the intersection $D_{i_1} \cap \dots \cap D_{i_j}$ is nonempty.

Example 6.5.7. (1) Let $\overline{Y} = V(x_0+x_1+x_2+x_3) \subset \mathbb{P}^3$ be a copy of $\mathbb{P}^2$ inside $\mathbb{P}^3$ as in Examples 6.5.1 and 6.5.3, and let $Y = \overline{Y} \cap T^n$. The boundary $\partial\overline{Y}$ consists of the four lines $L_i = V(x_0+x_1+x_2+x_3, x_i)$. Any two of these lines intersect in one point, but the intersection of any three is empty. The boundary complex Δ has four vertices, and one edge for any pair of vertices, so is the complete graph K_4.

(2) For $i = 0, 1, 2$, let L_i be the coordinate line $\{\mathbf{x} \in \mathbb{P}^2 : x_i = 0\}$, and let C be a general conic in $\mathbb{P}^2$ with equation $f \in K[x_0, x_1, x_2]$ of degree 2. Here "general" means that C does not contain any of the three torus fixed points of $\mathbb{P}^2$. Let $Y = \mathbb{P}^2 \backslash (L_0 \cup L_1 \cup L_2 \cup C) = (K^*)^2 \setminus C$. Then Y can be embedded into T^3 via the map $(x_0 : x_1 : x_2) \mapsto (x_1/x_0, x_2/x_0, f(x_0, x_1, x_2)/x_0^2)$. For the compactification $\overline{Y} = \mathbb{P}^2$ of Y, the boundary complex has four vertices v_0, v_1, v_2, v_C. There is an edge between each pair of these vertices. The boundary complex $\Delta(\partial\overline{Y})$ is again K_4, as shown on the top right of Figure 6.5.1. ◊

Remark 6.5.8. The notion of snc pairs is ubiquitous in the algebraic geometry literature, while the notion of combinatorial normal crossings was developed by Tevelev in [**Tev07**]. The fact that cnc suffices for many "niceness" properties is an important feature of the toric-tropical interplay.

Many authors prefer a more refined version of the boundary complex, where the simplicial complex is replaced by a *Delta-complex*. Following [**Hat02**, Chapter 2], in a Delta-complex different faces of a simplex are allowed to coincide, and there can be multiple simplices with the same set of vertices. The Delta-complex associated to the pair $(\overline{Y}, \partial\overline{Y})$ has one vertex for each component D_i in the boundary, and a simplex $\sigma = \{i_1, \ldots, i_j\}$ for each irreducible component of $D_{i_1} \cap \cdots \cap D_{i_j}$ whenever this is nonempty. This is shown in the lower-right diagram of Figure 6.5.1 for part (2) of Example 6.5.7. We place the additional genericity assumption here that the conic C is not tangent to any of the coordinate lines. The Delta-complex is a graph with four vertices and nine edges, while our $\Delta(\partial\overline{Y})$ is a graph with four vertices and six edges. The edges representing the intersections of the lines with the conic have been split into two edges each, corresponding to the two intersection points of C with each line.

The Delta-complex remembers more information about the compactification than the simplicial complex. See [**Hac08**] or [**Pay13**] for some examples. We here use the simpler version as that suffices for Theorem 6.5.11.

Our next goal is to explain how the boundary complex of a nice compactification $\overline{Y}$ of Y determines $\operatorname{trop}(Y)$. We first explain how to construct a fan containing $\operatorname{trop}(Y)$ from *any* pair $(\overline{Y}, \partial\overline{Y})$ with divisorial boundary. Recall that for a set $B \subset \mathbb{R}^n$ the cone over B is the set $\{\lambda \mathbf{b} \colon \mathbf{b} \in B, \lambda \geq 0\}$.

Proposition 6.5.9. *Let $\overline{Y}$ be a complete variety containing a very affine subvariety Y with divisorial boundary $\partial\overline{Y}$. Let $\pi : \Delta(\partial\overline{Y}) \to N_{\mathbb{R}}$ be the map defined by sending v_i to $[\operatorname{val}_{D_i}]$ and extending linearly on every simplex. Then the cone over the image of π contains the tropical variety* $\operatorname{trop}(Y)$.

Proof. By Proposition 6.5.4, we must show that if val_D is a divisorial valuation on $L = K(Y)$, then $[\operatorname{val}_D]$ lies in the cone over the image of π. Let R_L be the valuation ring of L with respect to val_D. Since the valuation on K is trivial, we have $K \subset R_L$, so there is a morphism $\operatorname{Spec}(R_L) \to \operatorname{Spec}(K)$. By the valuative criterion for properness, there is a unique morphism $\phi\colon \operatorname{Spec}(R_L) \to \overline{Y}$ that makes the following diagram commute.

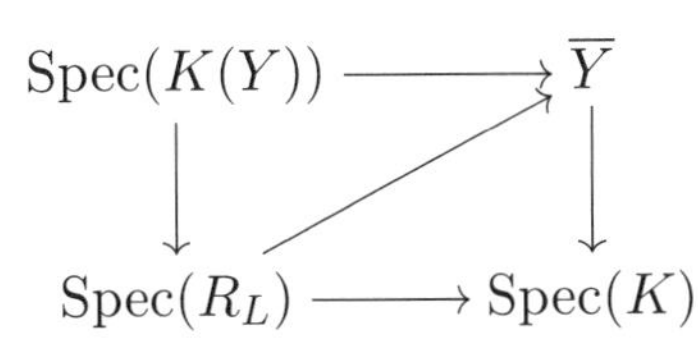

The valuation ring R_L has exactly two prime ideals, namely 0 and $\mathfrak{m}_L = \{a \in R_L : \operatorname{val}_D(a) > 0\}$. Let s be the closed point of $\operatorname{Spec}(R_L)$, corresponding to $\mathfrak{m}_L$. Let $D_1, \ldots, D_l$ be the components of $\partial\overline{Y}$, and let $\sigma = \{i : \phi(s) \in D_i\}$.

We now show that $[\mathrm{val}_D]$ lies in the cone spanned by the $[\mathrm{val}_{D_j}]$ with $j \in \sigma$. If not, since $\mathrm{pos}([\mathrm{val}_{D_j}] : j \in \sigma)$ is a rational polyhedral cone, there would be $\mathbf{m} = (m_1, \ldots, m_n) \in \mathbb{Z}^n$ with $\sum_{i=1}^n m_i \,\mathrm{val}_D(x_i) < 0$ and $\sum_{i=1}^n m_i \,\mathrm{val}_{D_j}(x_i) > 0$ for all $j \in \sigma$. Let $u = \prod_{i=1}^n x_i^{m_i} \in K(Y)$. Since the x_i are units in $K[Y]$, we have $u \in K[Y]^*$. The choice of $\mathbf{m}$ means that $\mathrm{val}_D(u) < 0$ and $\mathrm{val}_{D_j}(u) > 0$ for $j \in \sigma$. Let $B \subset \overline{Y}$ be the locus where the rational function u has poles, so u is regular on $\overline{Y}\backslash B$. Since u is a unit on Y, we have $B \subseteq \partial\overline{Y}$. Since $\mathrm{val}_{D_j}(u) > 0$ for $j \in \sigma$, we have $B \cap D_j = \emptyset$ for $j \in \sigma$, so the image of the function ϕ: $\mathrm{Spec}(R_L) \to \overline{Y}$ lies in $\overline{Y}\backslash B$. Thus the pullback $\phi^*(u)$ of the regular function u on $\overline{Y}\backslash B$ is regular on $\mathrm{Spec}(R_L)$. This means that $u \in R_L$, so $\mathrm{val}_D(u) \geq 0$, contradicting our assumption. We conclude that $[\mathrm{val}_D]$ lies in the cone spanned by the $[\mathrm{val}_{D_i}]$. □

Example 6.5.10. Let Y be the complement in $\mathbb{P}^2$ of the three coordinate lines and an irreducible conic C with equation $f \in K[x_0, x_1, x_2]$. Then Y is the subvariety of $(K^*)^3$ defined by the equation $y - f/x_0^2 \in K[x_1/x_0, x_2/x_0, y]$, so $\mathrm{trop}(Y)$ is a two-dimensional polyhedral fan in $\mathbb{R}^3$.

Consider $\overline{Y} = \mathbb{P}^2$ as a compactification for Y. The boundary $\partial\overline{Y}$ has four components: the three coordinate lines D_0, D_1, D_2, and the conic D_C. The divisorial valuation vectors are formed by applying the four divisorial valuations to the coordinates $x_1/x_0, x_2/x_0, y$ for $(K^*)^3$. This gives $[\mathrm{val}_{D_0}] = (-1,-1,-2)$, $[\mathrm{val}_{D_1}] = (1,0,0)$, $[\mathrm{val}_{D_2}] = (0,1,0)$, and $[\mathrm{val}_{D_C}] = (0,0,1)$.

The cone over the image of π constructed in Proposition 6.5.9 depends on the choice of conic C. Consider first the case that C passes through the three coordinate points $\{(1:0:0), (0:1:0), (0:0:1)\}$ of $\mathbb{P}^2$. One such conic is given by $f = x_0x_1 + x_0x_2 + x_1x_2$. The boundary complex $\Delta(\partial\overline{Y})$ is a subdivision of a triangle; there is a simplex for each collection $\{D_i, D_j, D_C\}$ with $0 \leq i < j \leq 2$. The cone over the image of π is then the union of the three cones $\mathrm{pos}((1,0,0),(0,1,0),(0,0,1))$, $\mathrm{pos}((1,0,0),(-1,-1,-2),(0,0,1))$, and $\mathrm{pos}((0,1,0),(-1,-1,-2),(0,0,1))$. This is not two dimensional, so it never equals $\mathrm{trop}(Y)$. For the conic f above, the tropical surface $\mathrm{trop}(Y)$ is the fan with rays $(1,1,1)$, $(-1,0,-1)$, $(0,-1,-1)$, and $(0,0,1)$. It has a two-dimensional cone spanned by any two of these rays. This is properly contained in the cone over the image of π.

Suppose now that C is a general conic that does not pass through any of the three coordinate points. One such conic is $f = x_0^2 + x_1^2 + x_2^2$. The boundary complex is the complete graph K_4 as in Example 6.5.7. The cone over the image of π is the union of the six cones spanned by any two of the rays $(-1,-1,-2), (1,0,0), (0,0,1), (0,0,1)$. As this has the correct dimension and no proper subfan is balanced, we conclude that the support of this fan is $\mathrm{trop}(Y)$. This can be verified using Proposition 3.1.10. ◊

We now explain how assuming that the compactification has simple normal crossings guarantees that the inclusion of Proposition 6.5.9 is an equality. This means that the compactification $\overline{Y}$ actually computes $\mathrm{trop}(Y)$.

Theorem 6.5.11. *Let $(\overline{Y}, \partial\overline{Y})$ be a smooth snc pair compactifying a d-dimensional variety $Y \subset T^n$. Let $\pi : \Delta(\partial\overline{Y}) \to N_{\mathbb{R}}$ be the map defined by sending v_i to $[\mathrm{val}_{D_i}]$ and extending linearly on every simplex. Then the cone over the image of π is equal to the tropical variety* $\mathrm{trop}(Y)$.

Proof. Each component D_i of the boundary $\partial\overline{Y}$ gives rise to a divisorial valuation val_{D_i} on $K(Y)$, and thus to a vector $[\mathrm{val}_{D_i}] \in N_{\mathbb{R}}$. Since $\overline{Y}$ is a smooth variety birational to Y, Proposition 6.5.4 implies that the point $[\mathrm{val}_{D_i}]$ lies in $\mathrm{trop}(Y)$. For $\sigma \in \Delta(\partial\overline{Y})$ with $|\sigma| = d$ and positive integers n_i for $i \in \sigma$, we consider the weighted blow-up of $\overline{Y}$ at the intersection $Z = \bigcap_{i\in\sigma} D_i$ with the weight n_i on D_i. Since $\partial\overline{Y}$ has simple normal crossings, locally this is blowing up the monomial ideal $\langle \prod_{i\in\sigma} x_i^{n_i}\rangle$ in $\mathbb{A}^d$.

The exceptional divisor E of this blow-up gives a divisorial valuation on $K(Y)$. This valuation satisfies $\mathrm{val}_E = \sum_{i\in\sigma} n_i \,\mathrm{val}_{D_i}$. Thus any nonnegative rational combination of the $[\mathrm{val}_{D_i}]$ with $i \in \sigma$ lies in $\mathrm{trop}(Y)$. Since $\mathrm{trop}(Y)$ is closed, the cone $\mathrm{pos}([\mathrm{val}_{D_i}] : i \in \sigma)$ lies in $\mathrm{trop}(Y)$. This shows that the cone over the image of the map π is contained in $\mathrm{trop}(Y)$. As the other inclusion follows from Proposition 6.5.9, we have the required equality. □

Remark 6.5.12. For general $\mathbf{w} \in \mathrm{trop}(Y)$ (those not in the cone over the image of the $(d-1)$-skeleton of $\Delta(\partial\overline{Y})$) the multiplicity of the cell of $\mathrm{trop}(Y)$ containing $\mathbf{w}$ can also be determined from the pair $(\overline{Y}, \partial\overline{Y})$. It equals

$$\mathrm{mult}(\mathbf{w}) \;=\; \sum_{\sigma} (D_{i_1} \cdot \ldots \cdot D_{i_d})[\mathbb{R}\sigma \cap M : \mathbb{Z}\sigma],$$

where the sum is over all simplices $\sigma = \{v_{i_1}, \ldots, v_{i_d}\}$ in the boundary complex $\Delta(\partial\overline{Y})$ with $\mathbf{w} \in \mathrm{relint}(\pi(\sigma))$, the symbols $\mathbb{R}\sigma$ and $\mathbb{Z}\sigma$ denote linear and integer spans of the set $\{[\mathrm{val}_{D_{i_j}}] : 1 \le j \le d\}$, and $D_{i_1} \cdot \ldots \cdot D_{i_d}$ is the intersection number of these divisors on $\overline{Y}$. For a proof, see [**Cue11**].

Example 6.5.13. We continue Example 6.5.3. By part (1) of Example 6.5.7, the boundary complex of $(\overline{Y}, \partial\overline{Y})$ is the graph K_4, so the image of π consists of the six line segments joining any two of the points $(1,0,0), (0,1,0), (0,0,1)$, and $(-1,-1,-1)$. The cone over this is the 2-skeleton of the fan of $\mathbb{P}^3$. ♢

A particularly nice example illustrating Theorem 6.5.11 is the Deligne–Mumford compactification $\overline{M}_{0,n}$ of the moduli space $M_{0,n}$. Recall from Section 6.4 that $M_{0,n}$ is the moduli space of n points on $\mathbb{P}^1$ and $\overline{M}_{0,n}$ is the moduli space of stable genus zero curves with n marked points.

By Theorem 6.4.12, $M_{0,n}$ is a subvariety of the torus $T^{\binom{n-1}{2}-1}$. This description comes from the realization of $M_{0,n}$ as the complement of a hyperplane arrangement in $\mathbb{P}^{n-3}$. For the point of $M_{0,n}$ corresponding to $(x_0 : \cdots : x_{n-3}) \in \mathbb{P}^{n-3}$, we set $p_n = (1:0)$, $p_1 = (0:1)$, $p_2 = (1:1)$, and $p_i = (x_{i-2} : x_0)$ for $3 \leq i \leq n-1$. This differs slightly from the choice made in Section 6.4. With this convention, the coordinate z_{ij} of Theorem 6.4.12 equals $x_{j-2} - x_{i-2}$ for $2 \leq i < j \leq n-1$, and $z_{1i} = x_{i-2}$ for $2 \leq i \leq n-1$.

Theorem 6.5.11 lets us recover the tropicalization of $M_{0,n}$ in $\mathbb{R}^{\binom{n-1}{2}-1}$ from the combinatorics of the boundary $\partial\overline{M}_{0,n}$. The boundary $\partial\overline{M}_{0,n} = \overline{M}_{0,n}\backslash M_{0,n}$ consists of $2^{n-1} - n - 1$ irreducible components δ_I. These are indexed by partitions $I \cup I^c$ of $\{1,\ldots,n\}$ into two parts, each of which has size at least two. The *boundary divisor* δ_I is the closure in $\overline{M}_{0,n}$ of the locus parameterizing stable curves with two components, one containing the points labeled by I and the other containing the points labeled by I^c; we identify δ_I and δ_{I^c}. It contains the limit of any family of points in $M_{0,n}$ where the p_i with $i \in I$ come together. See [**KV07**] for more details.

We now calculate the divisorial valuation determined by each δ_I. For $1 \leq i < j \leq n-1$, $(i,j) \neq (1,2)$, the ratios z_{ij}/z_{12} give a choice of coordinates on the torus $T^{\binom{n-1}{2}-1}$. We compute $[\mathrm{val}_{\delta_I}] \in \mathbb{R}^{\binom{n-1}{2}-1}$ using these coordinates.

Proposition 6.5.14. *Fix a boundary divisor δ_I on $\overline{M}_{0,n}$, with $n \notin I$. The divisorial valuation* val_{δ_I} *on $K(M_{0,n})$ given by the divisor δ_I satisfies*

$$\mathrm{val}_{\delta_I}(z_{ij}/z_{12}) = \begin{cases} 1 & \textit{if } i,j \in I \textit{ and } 1,2 \textit{ are not both in } I, \\ -1 & \textit{if } 1,2 \in I \textit{ and } i,j \textit{ are not both in } I, \\ 0 & \textit{otherwise,} \end{cases}$$

for $2 \leq i \leq j \leq n-1$; and

$$\mathrm{val}_{\delta_I}(z_{1i}/z_{12}) = \begin{cases} 1 & \textit{if } 1,i \in I, 2 \notin I, \\ -1 & \textit{if } 1,2 \in I, i \notin I, \\ 0 & \textit{otherwise,} \end{cases}$$

for $3 \leq i \leq n-1$. Thus the vector $[\mathrm{val}_{\delta_I}] \in \mathbb{R}^{\binom{n-1}{2}-1} \cong \mathbb{R}^{\binom{n-1}{2}}/\mathbb{R}\mathbf{1}$ is

$$[\mathrm{val}_{\delta_I}] = \begin{cases} \sum_{\substack{i\notin I \\ \text{or } j \notin I}} \mathbf{e}_{ij} & \textit{if } 1,2 \in I, \\ \sum_{i,j\in I} \mathbf{e}_{ij} & \textit{otherwise.} \end{cases}$$

Proof. The *cross-ratio* $(p_1,p_2;p_3,p_4)$ of four distinct points p_1,p_2,p_3,p_4 on $\mathbb{P}^1$ is defined by fixing the unique automorphism $\phi\colon \mathbb{P}^1 \to \mathbb{P}^1$ that satisfies

$$\phi(p_1) = (1:0), \quad \phi(p_2) = (0:1), \quad \text{and } \phi(p_3) = (1:1).$$

We then have $\phi(p_4) = (\alpha : 1)$ for some $\alpha \in K\backslash\{0,1\}$, and set $(p_1, p_2; p_3, p_4) = \alpha$. If $p_i = (z_i : 1)$ for $1 \leq i \leq 4$, then

$$(6.5.2) \qquad (p_1, p_2; p_3, p_4) = \frac{(z_1 - z_3)(z_2 - z_4)}{(z_2 - z_3)(z_1 - z_4)}.$$

Note that the cross-ratio $(p_i, p_j; p_k, p_l)$ is a rational function on $M_{0,n}$; each point on $M_{0,n}$ corresponds to a copy of $\mathbb{P}^1$ with n marked labeled points, and this function returns the cross-ratio of the points labeled i, j, k, and l.

Recall that the boundary divisor δ_I contains the limit of any family of points in $M_{0,n}$ where the p_i with $i \in I$ come together. The valuation of the cross-ratio function $(p_i, p_j; p_k, p_l)$ with respect to the divisorial valuation defined by the boundary divisor δ_I is

$$\mathrm{val}_{\delta_I}(p_i, p_j; p_k, p_l) = \begin{cases} 1 & \text{if } p_i, p_k \in I, p_j, p_l \notin I, \text{ or } p_j, p_l \in I, p_i, p_k \notin I, \\ -1 & \text{if } p_i, p_l \in I, p_j, p_k \notin I, \text{ or } p_j, p_k \in I, p_i, p_l \notin I, \\ 0 & \text{otherwise.} \end{cases}$$

This follows from the formula in (6.5.2). For example, if $p_i, p_k \in I$ and $p_j, p_l \notin I$, then there is a family $\mathcal{P}_t = \{p_m(t) = (z_m(t) : 1) : 1 \leq m \leq n\}$ of points in $M_{0,n}$ whose limit lies in δ_I with the limit of $z_i(t)$ equal to the limit of $z_k(t)$ and is distinct from the limits of $z_j(t)$ and $z_l(t)$.

We next note that $z_{1i}/z_{12} = (p_n, p_1; p_2, p_i)$ for $3 \leq i \leq n-1$. Indeed, we have $z_{1i}/z_{12} = x_{i-2}/x_0 = (p_n, p_1; p_2, p_i)$, as $p_i = (x_{i-2}/x_0 : 1)$. In addition we have $(p_n, p_j; p_1, p_i) = (x_{j-2} - x_{i-2})/x_{j-2}$ for $2 \leq i < j \leq n-1$, so $z_{ij}/z_{12} = (x_{j-2} - x_{i-2})/x_0 = (p_n, p_j; p_1, p_i)(p_n, p_1; p_2, p_j)$. Thus $\mathrm{val}_{\delta_I}(z_{ij}/z_{12}) = \mathrm{val}_{\delta_I}(p_n, p_j; p_1, p_i) + \mathrm{val}_{\delta_I}(p_n, p_1; p_2, p_j)$ if $2 \leq i < j \leq n-1$, and $\mathrm{val}_{\delta_I}(z_{1i}/z_{12}) = \mathrm{val}_{\delta_I}(p_n, p_1; p_2, p_i)$ for $3 \leq i \leq n$.

By assumption, $n \notin I$. For $2 \leq i \leq j \leq n-1$, we have

$$\mathrm{val}_{\delta_I}(z_{ij}/z_{12}) = \begin{cases} 1 & \text{if } 1, i, j \in I, 2 \notin I \text{ or } i, j \in I, 1 \notin I, \\ -1 & \text{if } 1, 2 \in I, j \notin I \text{ or } 1, 2, j \in I, i \notin I, \\ 0 & \text{otherwise.} \end{cases}$$

This simplifies to the formula in the proposition. For $3 \leq i \leq n-1$, we get the formula for $\mathrm{val}_{\delta_I}(z_{1i}/z_{12})$ given there. To calculate $[\mathrm{val}_{\delta_I}]$, we break into cases depending on whether $1, 2 \in I$. If $1, 2 \in I$, then $\mathrm{val}_{\delta_I}(z_{ij}/z_{12})$ equals -1 if i and j are not both in I and 0 otherwise, so $[\mathrm{val}_{\delta_I}] = -\sum_{i \in I^c \text{ or } j \in I^c} \mathbf{e}_{ij}$. If at least one of 1 and 2 is in I^c, then $\mathrm{val}_{\delta_I}(z_{ij}/z_{12})$ equals 1 if $i, j \in I$, and 0 otherwise, so $[\mathrm{val}_{\delta_I}(z_{ij}/z_{12})] = \sum_{i,j \in I} \mathbf{e}_{ij}$. □

Proposition 6.5.14 implies that when $n \notin I$, we have $[\mathrm{val}_{\delta_I}] = \sum_{i,j \in I} \mathbf{e}_{ij}$; the expressions in the proposition are the result of choosing representatives for $\mathbb{R}^{\binom{n-1}{2}}/\mathbb{R}\mathbf{1}$ with the coordinate labeled by 12 equal to 0. Thus the ray spanned by $[\text{val}(\delta_I)] \in \mathbb{R}^{\binom{n-1}{2}-1}$ is a ray of the tropicalization of the graphic matroid $M_{K_{n-1}}$ in Example 4.2.14. Combinatorially, this is the same as the space Δ of phylogenetic trees from Chapter 4. The simplicial complex $\Delta(\partial\overline{M}_{0,n})$ agrees with the complex described in Section 4.3. This gives another proof that the tropical variety of $M_{0,n}$ is the fan Δ.

We caution that the previous examples were all misleading in the sense that they embedded the simplicial complex $\Delta(\partial\overline{Y})$ into $N_{\mathbb{R}}$. This is usually not the case. In general, it will happen that interiors of disjoint simplices in $\Delta(\partial\overline{Y})$ intersect in $N_{\mathbb{R}}$. In particular, Theorem 6.5.11 does not necessarily give a fan structure on $\text{trop}(Y)$. We will demonstrate this in Example 6.5.19.

The condition in Theorem 6.5.11 that the pair $(\overline{Y}, \partial\overline{Y})$ be snc is unnecessarily strong. We now relax this condition to cnc, at the expense of assuming that the characteristic is zero to allow for resolution of singularities.

Theorem 6.5.15. *Let K be a field of characteristic zero. Let $(\overline{Y}, \partial\overline{Y})$ be a cnc pair compactifying a smooth variety $Y \subset T^n$. Let $\pi : \Delta(\partial\overline{Y}) \to N_{\mathbb{R}}$ be the map defined by sending v_i to $[\text{val}_{D_i}]$ and extending linearly on every simplex. Then the cone over the image of π equals* $\text{trop}(Y)$.

For a proof of Theorem 6.5.15 see [**Cue11**]. The main idea is to check that resolution of singularities does not change the image of the map from the boundary complex to the tropical variety; the extra vertices corresponding to exceptional divisors give a subdivision of the image. This argument turns Theorem 6.5.15 into a corollary of Theorem 6.5.11.

We now present an application of geometric tropicalization to the *implicitization problem* in computer algebra, which we met in Section 1.5. Consider n Laurent polynomials

$$f_i(\mathbf{t}) \quad = \quad \sum_{\mathbf{a}\in A_i} c_{i,\mathbf{a}} \cdot t_1^{a_1} \cdots t_d^{a_d} \qquad\qquad (i = 1, 2, \ldots, n). \tag{6.5.3}$$

Here each A_i is a finite subset of $\mathbb{Z}^d$, and the $c_{i,a}$ are generic elements of K. Our ultimate aim is to compute the prime ideal $I \subset K[x_1^{\pm 1}, \ldots, x_n^{\pm 1}]$ of relations among $f_1(\mathbf{t}), \ldots, f_n(\mathbf{t})$, or at least, some information about its variety $V(I)$. The ideal I is the kernel of the homomorphism $K[x_1^{\pm 1}, \ldots, x_n^{\pm 1}] \to K[t_1, \ldots, t_d]$ given by $x_i \mapsto f_i$ for $1 \leq i \leq n$.

The tropical approach to this problem is based on the following idea. Rather than computing I by algebraic elimination, we shall compute the tropical variety $\text{trop}(I) \subset \mathbb{R}^n$ by combinatorial means, using Theorem 6.5.11.

Let $\Psi : \mathbb{R}^d \to \mathbb{R}^n$ be the tropicalization of the map $f = (f_1, \ldots, f_n)$. Our notion of generic includes the requirement that all the coefficients $c_{I,\mathbf{a}}$ be nonzero, so the Newton polytope of each f_i equals $P_i = \text{conv}(A_i)$. The ith coordinate of Ψ is $\Psi_i(\mathbf{w}) = \text{trop}(f)(\mathbf{w}) = \min\{\mathbf{w} \cdot \mathbf{v} : \mathbf{v} \in P_i\}$.

The image of Ψ is contained in the tropical variety trop(I), but this containment is usually strict. See the discussion after Remark 3.2.14. In other words, the image of the tropicalization of f is usually a proper subset of the tropicalization of the image of f. The following result characterizes the difference $\mathrm{trop}(I)\backslash\mathrm{image}(\Psi)$. Let $\{\mathbf{e}_1, \ldots, \mathbf{e}_n\}$ be the standard basis in $\mathbb{R}^n$. For any subset J of $[n] = \{1, \ldots, n\}$, we write $\mathbb{R}^J_{\geq 0}$ for the orthant $\mathbb{R}_{\geq 0}\{\mathbf{e}_j : j \in J\}$ and P_J for the Minkowski sum $\sum_{j \in J} P_j$.

Theorem 6.5.16. *Let $f : T^d \to T^n$ be a rational map given by Laurent polynomials $f_1, \ldots, f_n$ that are generic relative to their supports A_i in (6.5.3). Let I be the ideal of the image of f. The following subsets of $\mathbb{R}^n$ coincide:*

(1) *the tropical variety* $\mathrm{trop}(V(I))$*;*
(2) *the union of all sets* $\Psi(\mathrm{trop}(\langle f_j : j \in J\rangle)) + \mathbb{R}^J_{\geq 0}$*, where* $J \subseteq [n]$*;*
(3) *the union of all cones* $\Psi(\mathbf{w}) + \mathbb{R}^J_{\geq 0}$ *such that, for all subsets $L \subseteq J$, the face* $\mathrm{face}_{\mathbf{w}}(P_L)$ *of the polytope P_L has dimension $\geq |L|$.*

The characterization (3) gives a combinatorial recipe for computing the tropical variety trop(I) directly from the given Newton polytopes $P_1, \ldots, P_n$. The ideal $\langle f_j : j \in J\rangle$ in (2) lives in the Laurent polynomial ring in d variables. The contribution of the empty set $J = \emptyset$ in Theorem 6.5.16(2) is precisely the image of the tropicalization Ψ of the given map f, as the ideal generated by the empty set of polynomials is the zero ideal:

$$\Psi(\mathrm{trop}(V(0))) + \mathbb{R}^{\emptyset}_{\geq 0} \;=\; \Psi(\mathbb{R}^d) \;=\; \mathrm{image}(\Psi). \tag{6.5.4}$$

Thus, the contributions of the nonempty subsets J make up the difference between the tropicalization of the image and the image of the tropicalization.

Example 6.5.17. We illustrate Theorem 6.5.16 for the case $d = 1, n = 2$. Consider a plane curve parameterized by two Laurent polynomials $x_1 = f_1(t)$ and $x_2 = f_2(t)$ for $f_1, f_2 \in K[t, t^{-1}]$. The Newton polytopes of f_1 and f_2 are line segments

$$P_1 = [\,\alpha\,,\,\beta\,] \quad \text{and} \quad P_2 = [\,\gamma\,,\,\delta\,] \qquad \text{in } \mathbb{R}^1. \tag{6.5.5}$$

The tropicalization of the parameterization $f = (f_1, f_2)$ is the map

$$\Psi : \mathbb{R} \to \mathbb{R}^2 : \tau \mapsto \big(\min\{\alpha{\cdot}\tau, \beta{\cdot}\tau\}, \min\{\gamma{\cdot}\tau, \delta{\cdot}\tau\}\big) \;=\; \begin{cases} \tau \cdot (\alpha, \gamma) & \text{if } \tau \geq 0, \\ \tau \cdot (\beta, \delta) & \text{if } \tau \leq 0. \end{cases}$$

The desired tropical curve in $\mathbb{R}^2$ is constructed from the contributions of the four subsets J of $\{1, 2\}$. The set $J = \{1, 2\}$ contributes the empty set because the segment $P_J = [\alpha + \gamma, \beta + \delta]$ has no face of dimension $|J| = 2$. For $J = \emptyset$ we get the rays spanned by (α, γ) and $(-\beta, -\delta)$, by (6.5.4). For $J = \{1\}$ and $J = \{2\}$, only $w = 0$ is relevant in Theorem 6.5.16(3), and the contributions $\Psi(0) + \mathbb{R}^J_{\geq 0}$ are the coordinate rays $\mathbb{R}_{\geq 0}\mathbf{e}_1$ and $\mathbb{R}_{\geq 0}\mathbf{e}_2$.

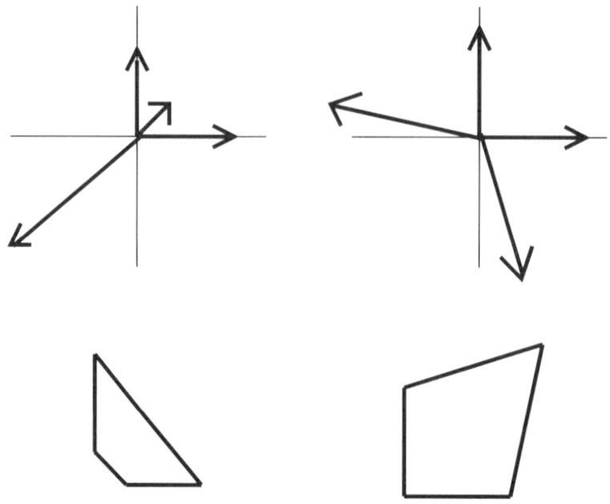

Figure 6.5.2. Tropical plane curves and their Newton polygons.

Assuming that the integers $\alpha, \beta, \gamma, \delta$ are all nonzero, we conclude that the tropical curve consists of four rays and its Newton polygon is a quadrilateral. This is consistent with Theorem 1.5.2. Figure 6.5.2 shows two cases. ◇

Proof of Theorem 6.5.16. The equivalence of the conditions (2) and (3) follows from Corollary 4.6.11 applied to the Laurent polynomials f_j for $j \in J$. To make the connection to condition (1), we use geometric tropicalization. Let $E_i = \{\mathbf{t} \in T^d : f_i(\mathbf{t}) = 0\}$ and $Z = T^d \backslash \bigcup_{i=1}^n E_i$. The Laurent polynomials $f_1, \ldots, f_n$ specify a morphism of very affine algebraic varieties

$$f : Z \to T^n, \quad \mathbf{z} \mapsto \big(f_1(\mathbf{z}), \ldots, f_n(\mathbf{z})\big). \tag{6.5.6}$$

Our goal is to compute the tropicalization of its image $Y = f(Z)$.

We first note that we may assume that f is an isomorphism, so a compactification of Z is a compactification of Y. Otherwise, we augment f by adding some of the coordinate functions z_i to $(f_1, \ldots, f_n)$ and then project the resulting tropical variety into n-space. None of the coordinate functions will contribute to a set J in (2) for the augmented function, and by Corollary 3.2.13 the projection of the tropicalization equals the tropicalization of the projection, so it suffices to prove the theorem for the augmented f.

Let $X_{\tilde{P}}$ be a d-dimensional smooth projective toric variety whose polytope $\tilde{P}$ has the given Newton polytopes $P_1, \ldots, P_n$ as Minkowski summands. This means that the normal fan of $\tilde{P}$ refines the normal fan of each P_i. The smooth toric variety $X_{\tilde{P}}$ is a compactification of T^d, and thus of Z. Since Y is isomorphic to Z, the toric variety $X_{\tilde{P}}$ is thus a compactification of Y. For each $i \in [n]$, there is a canonical morphism $X_{\tilde{P}} \to X_{P_i}$ onto the (generally not smooth) projective toric variety X_{P_i} associated with P_i.

We now consider the boundary $X_{\tilde{P}} \backslash Z$. Here $X_{\tilde{P}}$ is playing the role of $\overline{Y}$, and Z plays the role of Y. The irreducible components of the boundary are of two types. Firstly, we have toric divisors $D_1, \ldots, D_l$ indexed by the facets of $\tilde{P}$. The toric boundary $D_1 \cup \cdots \cup D_l$ of $\overline{Y}$ has simple normal crossings because $\tilde{P}$ is a simple polytope. Secondly, we have divisors $\overline{E}_1, \ldots, \overline{E}_n$ which are the closures in $X_{\tilde{P}}$ of the hypersurfaces $V(f_i) \subset T^d$. By the genericity assumption on the coefficients, each f_i is nondegenerate with respect to its Newton boundary (see the discussion after Definition 6.4.19). Together with Bertini's Theorem this implies that the $\overline{E}_i$ are smooth and irreducible, and that the union of all D_i's and all $\overline{E}_j$'s has simple normal crossings.

Here we are tacitly assuming that each polytope P_i has dimension ≥ 2. If $\dim(P_i) = 1$, then $\overline{E}_i$ is the disjoint union of smooth and irreducible divisors, and the following argument needs to be slightly modified. If $\dim(P_i) = 0$, then E_i is the empty set, and hence so is $\overline{E}_i$. Such indices i will not appear in any index set J which contributes to the union in part (2) of the theorem.

We conclude that Theorem 6.5.11 can be applied to the snc pair $(\overline{Y}, \partial\overline{Y}) = (X_{\tilde{P}}, X_{\tilde{P}} \backslash Z)$, with the boundary having the irreducible decomposition

$$\partial\overline{Y} \quad = \quad D_1 \cup D_2 \cup \cdots \cup D_l \;\cup\; \overline{E}_1 \cup \cdots \cup \overline{E}_n.$$

The simplicial complex $\Delta(\partial\overline{Y})$ has dimension $d-1$. It has $m = l+n$-vertices, one for each of the divisors D_i and $\overline{E}_j$. Its maximal simplices correspond to pairs (C, J) where $C = \{i_1, \ldots, i_{d-r}\} \subseteq [l]$ and $J = \{j_1, \ldots, j_r\} \subseteq [n]$ and

$$D_{i_1} \cap \cdots \cap D_{i_{d-r}} \cap \overline{E}_{j_1} \cap \cdots \cap \overline{E}_{j_r} \quad \neq \quad \emptyset. \tag{6.5.7}$$

There are no larger simplices because the boundary has simple normal crossings. For any $J \subseteq [n]$, let Δ_J denote the subset of $\Delta(\partial\overline{Y})$ consisting of all simplices with fixed J. Note that $\Delta_\emptyset$ is the boundary complex of the simplicial polytope dual to $\tilde{P}$. Moreover, $\Delta_J = \{\emptyset\}$ if $|J| = d$, and $\Delta_J = \emptyset$ if $|J| > d$. The vector $[\mathrm{val}_{\overline{E}_j}] = (\mathrm{val}_{\overline{E}_j} f_1, \ldots, \mathrm{val}_{\overline{E}_j} f_n)$ is the jth basis vector $\mathbf{e}_j$ in $\mathbb{R}^n$. With this, the image of π in Theorem 6.5.11 equals

$$\mathrm{trop}(Y) \quad = \quad \bigcup_{J \subset [n]} \Big(\mathbb{R}^J_{\geq 0} + \bigcup_{C \in \Delta_J} \mathbb{R}_{\geq 0}\big\{[\mathrm{val}_{D_i}] \,:\, i \in C\big\} \Big).$$

Hence to prove the remaining equivalence (1) = (2), it suffices to show that

$$\Psi(\mathrm{trop}(\langle f_j : j \in J \rangle)) \quad = \quad \bigcup_{C \in \Delta_J} \mathbb{R}_{\geq 0}\big\{[\mathrm{val}_{D_i}] \,:\, i \in C\big\}. \tag{6.5.8}$$

Let $\mathbf{v}_i \in \mathbb{R}^d$ be the primitive inner normal vector of the facet of the polytope $\tilde{P}$ corresponding to the divisor D_i. Then $\mathrm{val}_{D_i}(f_j) = \min(\mathbf{v}_i \cdot \mathbf{u} \colon \mathbf{u} \in P_j)$. Hence $[\mathrm{val}_{D_i}] = \Psi(\mathbf{v}_i)$ in $\mathbb{R}^n$, so the right-hand side of (6.5.8) is the image under Ψ of the subfan of the normal fan of $\tilde{P}$ indexed by Δ_J. But, by Corollary 4.6.11, the support of this subfan coincides with the tropical variety defined by $\langle f_j : j \in J \rangle$. This completes our proof of Theorem 6.5.16. □

Remark 6.5.18. Theorem 6.5.16 characterizes the tropical variety of I only as a set. A formula for the multiplicities on $\mathrm{trop}(V(I))$, in terms of mixed volumes, was given in [**STY07**, Theorem 4.1]. This formula can be derived from Theorem 4.6.8 and proved using Remark 6.5.12.

The map π of Theorem 6.5.11 gives an immersion of $\Delta(\partial\overline{Y})$ but generally not an embedding. In Theorem 6.5.16 for $d \geq 2$, image cones of distinct simplices of $\Delta(\partial\overline{Y})$ may intersect in their relative interiors in $\mathbf{R}^n$. In particular, any fan structure on $\mathrm{trop}(Y)$ usually has more cones than $\Delta(\partial\overline{Y})$ has simplices. The following example, due to Hyunsuk Moon, illustrates this.

Example 6.5.19. Let $d = 2$, $n = 3$ and consider the surface parameterized by $f_1 = st^2(1+t)$, $f_2 = s^2t^3(1+st)$, and $f_3 = s^3t(1+s)$. The Newton polytopes of these binomials are the line segments $P_1 = \mathrm{conv}\{(1,2),(1,3)\}$, $P_2 = \mathrm{conv}\{(2,3),(3,4)\}$, and $P_3 = \mathrm{conv}\{(3,1),(4,1)\}$. The hexagon $\tilde{P} = P_1 + P_2 + P_3$ defines a smooth toric surface $\overline{Y} = X_{\tilde{P}}$, namely the blow-up of $\mathbb{P}^2$ at three points, with boundary curves $D_1, D_2, D_3, D_4, D_5, D_6$ and curves $\overline{E_1}, \overline{E_2}, \overline{E_3}$ defined by f_1, f_2, f_3. The boundary complex $\Delta(\partial\overline{Y})$ is the graph that represents the intersections among these nine divisors. It has the 15 edges

$$\begin{gathered}\{D_1,D_2\},\ \{D_2,D_3\},\ \{D_3,D_4\},\ \{D_4,D_5\},\ \{D_5,D_6\},\ \{D_6,D_1\},\\ \{D_1,\overline{E_1}\},\ \{D_2,\overline{E_2}\},\ \{D_3,\overline{E_3}\},\ \{D_4,\overline{E_1}\},\ \{D_5,\overline{E_2}\},\ \{D_6,\overline{E_3}\},\\ \{\overline{E_1},\overline{E_2}\},\ \{\overline{E_1},\overline{E_3}\},\ \{\overline{E_2},\overline{E_3}\}.\end{gathered}$$

The graph $\Delta(\partial\overline{Y})$ is not planar, so it cannot be drawn on the 2-sphere without self-intersections. It is thus impossible for the map π to be an embedding. For our specific choice of f_1, f_2, f_3, the coefficients are generic, and any triple among the nine vectors $[\mathrm{val}_{D_i}]$ and $[\mathrm{val}_{\overline{E_j}}]$ is linearly independent in $\mathbb{R}^3$. The map π creates three new rays by intersections, so $\mathrm{trop}(Y)$ is a fan with 12 rays and 21 two-dimensional cones. The implicit equation has 74 terms. Its Newton polytope has 11 vertices, 21 edges, and 12 facets. ◊

6.6. Degenerations

The tropical variety of $Y \subset T^n$ also determines degenerations of Y, and there is a beautiful interplay between the compactifications of the previous section and these degenerations, which are the topic of this section.

We first describe how a Γ_{val}-rational polyhedral complex gives rise to a degeneration of a toric variety. For this we need toric varieties over the valuation ring R of K. In this section we assume that the field K is algebraically closed and has a nontrivial valuation. We also assume $1 \in \Gamma_{\mathrm{val}}$. As with standard toric varieties over a field, we start with the affine case.

Definition 6.6.1. A Γ_{val}-*admissible* cone is a polyhedral cone of the form

$$\sigma = \{(\mathbf{w}, s) \in N_{\mathbb{R}} \times \mathbb{R}_{\geq 0} : \mathbf{w} \cdot \mathbf{u}_i + sc_i \geq 0 \text{ for } i = 1, 2, \ldots, r\}$$

that does not contain a line, where $\mathbf{u}_i \in M$ and $c_i \in \Gamma_{\text{val}}$ for all i. For $s \in \mathbb{R}_{\geq 0}$, let σ_s denote the convex polyhedron $\{\mathbf{w} \in N_{\mathbb{R}} : (\mathbf{w}, s) \in \sigma\}$. We define a twisted version of the Laurent polynomial ring as

$$K[M]^\sigma = \Big\{ \sum_{\mathbf{u} \in \sigma_0{}^\vee \cap M} c_{\mathbf{u}} x^{\mathbf{u}} : s \operatorname{val}(c_{\mathbf{u}}) + \mathbf{w} \cdot \mathbf{u} \geq 0 \text{ for all } (\mathbf{w}, s) \in \sigma \Big\}. \tag{6.6.1}$$

This ring contains R (since $s \geq 0$) so is an R-algebra.

Example 6.6.2. The orthant $\sigma = (\mathbb{R}_{\geq 0})^n \times \mathbb{R}_{\geq 0}$ is Γ_{val}-admissible. Then $cx^{\mathbf{u}} \in K[M]^\sigma$ implies $\operatorname{val}(c) \geq 0$, using $(\mathbf{w}, s) = (\mathbf{0}, 1)$, and $u_i \geq 0$, using $(\mathbf{w}, s) = (\mathbf{e}_i, 0)$. This argument is reversible. Hence $K[M]^\sigma = R[x_1, \ldots, x_n]$.

Consider now $n=1$ and $\sigma = \operatorname{pos}\{(0,1), (1,1)\} \subset \mathbb{R} \times \mathbb{R}_{\geq 0}$. Then $\sigma_0 = \{0\}$ in $\mathbb{R}$, so $\sigma_0^\vee = \mathbb{R}$. A term cx^j lies in $K[M]^\sigma$ if and only if $\operatorname{val}(c) \geq 0$ and $\operatorname{val}(c) + j \geq 0$. Thus $K[M]^\sigma = R[x, tx^{-1}]$ where $t \in K$ with $\operatorname{val}(t) = 1$. ◊

Definition 6.6.3. For any Γ_{val}-admissible cone σ, we set $\mathcal{U}_\sigma = \operatorname{Spec}(K[M]^\sigma)$. We call $\mathcal{U}_\sigma$ the *affine toric scheme* over R defined by σ.

This definition includes toric varieties over the field K as a special case. If $\sigma \subset N_{\mathbb{R}} \times \{0\}$, then $K[M]^\sigma = K[\sigma_0^\vee \cap M]$, so $\mathcal{U}_\sigma$ is the affine toric variety U_{σ_0}. In general, the scheme $\mathcal{U}_\sigma$ is still well behaved, as the following shows.

Proposition 6.6.4. *Suppose $\sigma \subset N_{\mathbb{R}} \times \mathbb{R}_{\geq 0}$ is not contained in $N_{\mathbb{R}} \times \{0\}$. The affine toric scheme $\mathcal{U}_\sigma$ is integral, normal, of finite type, and flat over* $\operatorname{Spec}(R)$.

Proof. We show that $K[M]^\sigma$ is an integrally closed domain that is finitely generated as an R-algebra and flat as an R-module. Since $K[M]^\sigma$ is a subalgebra of the domain $K[M]$, it is also a domain and a torsion-free R-module. Note that every finitely generated ideal in R is principal, as if $a, b \in R$ with $\operatorname{val}(a) \leq \operatorname{val}(b)$, then $b/a \in R$, so $b \in \langle a \rangle$. Thus $K[M]^\sigma$ is a flat R-module; the proof in Corollary 6.3 of [**Eis95**] that torsion-free is the same as flat for PIDs only uses that finitely generated ideals are principal.

We next show that $K[M]^\sigma$ is finitely generated as an R-algebra. Let $\mathbf{v}_1, \ldots, \mathbf{v}_s$ be the vertices of the Γ_{val}-rational polyhedron σ_1. Since K is algebraically closed, Γ_{val} is divisible, and hence $\mathbf{v}_1, \ldots, \mathbf{v}_s \in \Gamma_{\text{val}}^n$. Write $\tau_i^\vee = \mathcal{N}_{\sigma_1}(\mathbf{v}_i)$ for the inner normal cone to σ_1 at $\mathbf{v}_i$. Then $\tau_1^\vee \cup \cdots \cup \tau_s^\vee = \sigma_0^\vee$ holds in $M_{\mathbb{R}}$. Thus $K[M]^\sigma$ is generated as an R-module by the rings $K[M]^\sigma \cap K[\tau_i^\vee \cap M]$. It suffices to show that these are all finitely generated.

The ring $K[\tau_i^\vee \cap M]$ is the coordinate ring of an affine toric variety. By Gordan's lemma [**CLS11**, Proposition 1.2.17] it is generated by a finite set $\{x^{\mathbf{u}_1}, \ldots, x^{\mathbf{u}_r}\}$. Choose $c_1, \ldots, c_r \in K$ with $\mathrm{val}(c_j) + \mathbf{v}_i \cdot \mathbf{u}_j = 0$. We claim

$$K[M]^\sigma \cap K[\tau_i^\vee \cap M] \;=\; R\big[c_1 x^{\mathbf{u}_1}, \ldots, c_r x^{\mathbf{u}_r}\big]. \tag{6.6.2}$$

Indeed, let $c_{\mathbf{u}} x^{\mathbf{u}}$ be in the left-hand side. Then $\mathbf{u} = \sum_{j=1}^r \lambda_j \mathbf{u}_j$ for some $\lambda_j \in \mathbb{N}$. Let $c = c_{\mathbf{u}} / \prod_{j=1}^r c_j^{\lambda_j}$. We have $\mathrm{val}(\prod_{j=1}^r c_j^{\lambda_j}) = -\mathbf{v}_i \cdot (\sum_{j=1}^r \lambda_j \mathbf{u}_j) = -\mathbf{v}_i \cdot \mathbf{u}$. Since $c_{\mathbf{u}} x^{\mathbf{u}} \in K[M]^\sigma$, and $(\mathbf{v}_i, 1) \in \sigma$, we have $\mathrm{val}(c_{\mathbf{u}}) \geq -\mathbf{v}_i \cdot \mathbf{u}$. Thus $\mathrm{val}(c) \geq 0$, so $c \in R$. This means that $c_{\mathbf{u}} x^{\mathbf{u}} = c \prod_{j=1}^r (c_j x^{\mathbf{u}_j})^{\lambda_j}$ so it lies in the ring on the right of (6.6.2).

It remains to be seen that $K[M]^\sigma$ is integrally closed. Since σ is a polyhedral cone, we can write $\sigma = \sigma_0 + \mathrm{pos}\{(\mathbf{v}_1, 1), \ldots, (\mathbf{v}_r, 1)\}$. With this, the definition of $K[M]^\sigma$ in (6.6.1) is equivalent to

$$K[M]^\sigma \;=\; K[\sigma^\vee \cap M] \;\cap\; \bigcap_{i=1}^{r} \Big\{\sum c x^{\mathbf{u}} : \mathrm{val}(c) + \mathbf{v}_i \cdot \mathbf{u} \geq 0\Big\}.$$

As above, each coordinate v_{ij} of $\mathbf{v}_i$ lies in Γ_{val}. The ring $K[\sigma^\vee \cap M]$ is the coordinate ring of a normal affine toric variety, so it is integrally closed (see, for example, [**CLS11**, Theorem 1.3.5]). Fix $\alpha_j \in K$ with $\mathrm{val}(\alpha_j) = v_{ij}$. The ring $\{\sum c_{\mathbf{u}} x^{\mathbf{u}} : \mathrm{val}(c_{\mathbf{u}}) + \mathbf{v}_i \cdot \mathbf{u} \geq 0\}$ is isomorphic to $R[M]$ by the map that sends x_j to $\alpha_j x_j$ for $1 \leq j \leq n$. The ring R is integrally closed, since it is a valuation ring, so $R[M]$ is integrally closed. The intersection of integrally closed rings with the same field of fractions is again integrally closed. Hence, $K[M]^\sigma$ is integrally closed. □

Remark 6.6.5. The ring $K[M]^\sigma$ is in fact finitely presented as an R-module, as opposed to merely finitely generated. This follows from [**RG71**, Corollary 3.4.7], as $K[M]^\sigma$ is a flat finitely generated R-module. The distinction between "finitely generated" and "finitely presented" arises from the fact that R is not Noetherian when Γ_{val} is divisible; see Remark 2.4.13. For a good treatment of this issue, see [**Vak13**, §13.6].

We will consider $\mathcal{U}_\sigma$ as a family over $\mathrm{Spec}(R)$. Recall that the valuation ring R has exactly two prime ideals: the zero ideal and the maximal ideal $\mathfrak{m}_K$. Hence $\mathrm{Spec}(R)$ has only two points, namely the general point, corresponding to the zero ideal, and the special point, corresponding to $\mathfrak{m}_K$. For our family $\mathcal{U}_\sigma$ over $\mathrm{Spec}(R)$, these give rise to the *general fiber*, which is $\mathrm{Spec}(K[M]^\sigma \otimes_R K)$, and the *special fiber*, which is $\mathrm{Spec}(K[M]^\sigma \otimes_R \Bbbk)$. The next result tells us that the general fiber of $\mathcal{U}_\sigma$ is an affine toric variety over $\mathrm{Spec}(K)$, and the special fiber is a union of affine toric varieties over $\mathrm{Spec}(\Bbbk)$. Thus the family $\mathcal{U}_\sigma$ encodes a degeneration of an affine toric variety.

Proposition 6.6.6. *Let σ be a Γ_{val}-admissible cone in $N_{\mathbb{R}} \times \mathbb{R}_{\geq 0}$ not contained in $N_{\mathbb{R}} \times \mathbb{R}_{\geq 0}$. The translation action of $T = \operatorname{Spec}(K[M])$ on itself extends to an algebraic action of $\operatorname{Spec}(R[M])$ on $\mathcal{U}_\sigma$. The general fiber of the family $\mathcal{U}_\sigma$ over $\operatorname{Spec}(R)$ is the affine toric variety U_{σ_0}. The special fiber of $\mathcal{U}_\sigma$ is a union of affine toric varieties over $\operatorname{Spec}(\Bbbk)$ with one irreducible component for each vertex $\mathbf{v}_i$ of σ_1. The component corresponding to the vertex $\mathbf{v}_i$ is the toric variety over $\Bbbk$ defined by the cone τ_i spanned by $\sigma_1 - \mathbf{v}_i$.*

Proof. To prove that the general fiber $\operatorname{Spec}(K[M]^\sigma \otimes_R K)$ is the affine toric variety U_σ, we must show that $K[M]^\sigma \otimes_R K \cong K[\sigma_0^\vee \cap M]$. The R-algebra $K[M]^\sigma \otimes_R K$ is generated as an R-module by elements of the form $cx^{\mathbf{u}} \otimes a$, where $\mathbf{u} \in \sigma_0^\vee \cap M$ and $s\operatorname{val}(c) + \mathbf{w} \cdot \mathbf{u} \geq 0$ for all $(\mathbf{w}, s) \in \sigma$, $a \in K$. The map $K[M]^\sigma \times K \to K[\sigma_0^\vee \cap M]$ given by sending $(cx^{\mathbf{u}}, a)$ to $acx^{\mathbf{u}}$ and extending linearly is multilinear and compatible with the R-module action, so extends to a homomorphism $\phi\colon K[M]^\sigma \otimes_R K \to K[\sigma_0^\vee \cap M]$. To see that ϕ is surjective, we argue that for any $\mathbf{u} \in \sigma_0^\vee$ there is $c \in K^*$ with $cx^{\mathbf{u}} \in K[M]^\sigma$. Indeed, we can take c with $\operatorname{val}(c) \geq -(1/s)\mathbf{w} \cdot \mathbf{u}$ for all generators $(\mathbf{w}, s)$ of σ. Then for any $ax^{\mathbf{u}} \in K[\sigma_0^\vee \cap M]$, we have $ax^{\mathbf{u}} = \phi(cx^{\mathbf{u}} \otimes a/c)$. To see that ϕ is injective, suppose $\phi(\sum \mu_i c_i x^{\mathbf{u}_i} \otimes a_i) = \sum \mu_i c_i a_i x^{\mathbf{u}_i} = 0$. We may restrict to the case $\mathbf{u}_i = \mathbf{u}$ for all i, so $\sum \mu_i c_i a_i = 0$. Without loss of generality $\operatorname{val}(\mu_1 c_1) \leq \operatorname{val}(\mu_i c_i)$ for all i, so $\mu_i c_i / \mu_1 c_1 \in R$ for all i. Now $\mu_i c_i x^{\mathbf{u}} \otimes a_i = \mu_1 c_1 x^{\mathbf{u}} \otimes (\mu_i c_i / \mu_1 c_1) a_i$, so $\sum \mu_i c_i x^{\mathbf{u}} \otimes a_i = \mu_1 c_1 x^{\mathbf{u}} \otimes (1/\mu_1 c_1)(\sum \mu_i c_i a_i) = 0$.

The special fiber of the family $\mathcal{U}_\sigma$ is isomorphic to the quotient of the R-algebra $K[M]^\sigma$ by the ideal $\mathfrak{m}_\sigma$ generated by those $cx^{\mathbf{u}} \in K[M]^\sigma$ for which $s\operatorname{val}(c) + \mathbf{w} \cdot \mathbf{u} > 0$ for all $(\mathbf{w}, s) \in \sigma$. We claim that this ideal equals

$$\mathfrak{m}_\sigma = \bigcap_{i=1}^{r} \big\langle\, cx^{\mathbf{u}} \in K[M]^\sigma \,:\, \operatorname{val}(c) + \mathbf{v}_i \cdot \mathbf{u} > 0 \big\rangle, \tag{6.6.3}$$

where the intersection is over the vertices $\mathbf{v}_i$ of the slice σ_1. The inclusion $\subseteq$ is immediate since $(\mathbf{v}_i, 1) \in \sigma$ for all i. For the other inclusion, note that every vector $(\mathbf{w}, s) \in \sigma$ has the form $\sum_{i=1}^r \mu_i(\mathbf{v}_i, 1) + (\mathbf{w}', 0)$, for some $\mu_i \geq 0$ and $\mathbf{w}' \in \sigma_0$. Thus if $cx^{\mathbf{u}}$ is in each of the r ideals on the right, then $s\operatorname{val}(c) + \mathbf{w} \cdot \mathbf{u} = \sum_{i=1}^r \mu_i(\operatorname{val}(c) + \mathbf{v}_i \cdot \mathbf{u}) + \mathbf{w}' \cdot \mathbf{u} > 0$, so $cx^{\mathbf{u}} \in \mathfrak{m}_\sigma$.

We next show that each ideal on the right in (6.6.3) is prime, and the quotient of $K[M]^\sigma$ modulo that ideal has the desired form. We reduce to the case that the vertex $\mathbf{v}_i$ is $\mathbf{0}$. Since the polyhedron σ_1 is Γ_{val}-rational, the coordinates v_{ij} of $\mathbf{v}_i$ lie in Γ_{val}. Fix $\alpha_j \in K$ with $\operatorname{val}(\alpha_j) = v_{ij}$, and apply the change of coordinates $\phi\colon x_j \mapsto \alpha_j x_j$. This takes $cx^{\mathbf{u}}$ to $c\alpha^{\mathbf{u}} x^{\mathbf{u}}$, so the condition $\operatorname{val}(c) + \mathbf{v}_i \cdot \mathbf{u} > 0$ becomes $\operatorname{val}(c) > 0$. We have $\phi(K[M]^\sigma) = K[M]^{\phi^*(\sigma)}$, where $\phi^*\colon N_{\mathbb{R}} \times \mathbb{R}_{\geq 0} \to N_{\mathbb{R}} \times \mathbb{R}_{\geq 0}$ is given by $\phi^*((\mathbf{w}, s)) = (\mathbf{w} - s\mathbf{v}_i, s)$. The cone τ_i of σ_1 at $\mathbf{v}_i$ is preserved by this map.

When $\mathbf{v}_i = \mathbf{0}$, the ideal $\mathfrak{m}_{\mathbf{v}_i}$ is the ideal generated by $\mathfrak{m}_K$ in $K[M]^\sigma$, which is prime. Since the choice of which vertex to move to $\mathbf{0}$ was arbitrary, this shows that the ideal $\mathfrak{m}_\sigma$ is radical with primary decomposition (6.6.3). The special fiber has one irreducible component for each vertex of σ_1.

We finish by showing that $K[M]^\sigma/\mathfrak{m}_{\mathbf{v}_i} \simeq \Bbbk[\tau_i^\vee \cap M]$, so each component has the desired form. As before, we suppose $\mathbf{v}_i = \mathbf{0}$. Let $\mathbf{v}_{j_1}, \ldots, \mathbf{v}_{j_s}$ be the vertices of σ_1 which are also generators for the cone τ_i. If $cx^{\mathbf{u}} \in K[M]^\sigma$ with $\mathbf{u} \notin \tau_i^\vee$, then since $\mathrm{val}(c) + \mathbf{v}_{j_\nu} \cdot \mathbf{u} \geq 0$ for all ν, we must have $\mathrm{val}(c) > 0$ and hence $c \in \mathfrak{m}_K$. Hence, we have a well-defined map $cx^{\mathbf{u}} \mapsto \bar{c}x^{\mathbf{u}}$ from $K[M]^\sigma$ onto $\Bbbk[\tau_i^\vee \cap M]$. It is easy to see that the kernel equals $\langle \mathfrak{m}_K \rangle = \mathfrak{m}_{\mathbf{v}_i}$. □

Example 6.6.7. We examine the two toric schemes $\mathcal{U}_\sigma$ from Example 6.6.2. If $\sigma = (\mathbb{R}_{\geq 0})^n \times \mathbb{R}_{\geq 0}$, then $K[M]^\sigma \otimes_R K \cong K[x_1, \ldots, x_n]$ and $K[M]^\sigma \otimes_R \Bbbk \cong \Bbbk[x_1, \ldots, x_n]$. Thus the general fiber of $\mathcal{U}_\sigma$ is $\mathbb{A}^n_K$ and the special fiber is $\mathbb{A}^n_\Bbbk$.

For $\sigma = \mathrm{pos}\{(0,1),(1,1)\}$, we introduce unknowns a and b, and we write

$$K[M]^\sigma \;=\; R[x, tx^{-1}] \;\simeq\; R[a,b]/\langle ab - t\rangle.$$

Then $K[M]^\sigma \otimes_R K \simeq K[x, x^{-1}]$, so the general fiber of $\mathcal{U}_\sigma$ is the torus K^*. To compute the special fiber, we divide by the ideal $\mathfrak{m}_K$. This yields

$$K[M]^\sigma \otimes_R \Bbbk \;=\; R[a,b]/(\langle ab\rangle + \mathfrak{m}_K) \;=\; \Bbbk[a,b]/\langle ab\rangle \;=\; \Bbbk[a,b]/(\langle a\rangle \cap \langle b\rangle).$$

This ideal decomposition is (6.6.3) with $r = 2$; the ideals $\langle a\rangle$ and $\langle b\rangle$ correspond to the two vertices of $\sigma_1 = [0,1] \subset \mathbb{R}$. The special fiber is thus obtained by gluing $\mathbb{A}^1_\Bbbk = \mathrm{Spec}(\Bbbk[a])$ and $\mathbb{A}^1_\Bbbk = \mathrm{Spec}(\Bbbk[b])$. The flat family $\mathcal{U}_P \to \mathrm{Spec}(R)$ is a degeneration of the torus K^* into two copies of $\mathbb{A}^1_\Bbbk$. ◇

Remark 6.6.8. At the start of the section we restricted to the case that K is algebraically closed with a nontrivial valuation. The only consequence of this assumption used in the proofs so far was that $\mathbf{w} \in \Gamma^n_{\mathrm{val}}$ whenever $(\mathbf{w}, 1)$ lies on a ray of σ. This was used to reduce to the case that $\mathbf{w} = \mathbf{0}$. Propositions 6.6.4 and 6.6.6 thus also hold when all vertices of σ_1 lie in Γ^n_{val}.

Example 6.6.9. Here is a degeneration of a toric surface. Let $n = 2$ and

$$\sigma \;=\; \left\{ (w_1, w_2, s) \in \mathbb{R}^2 \times \mathbb{R}_{\geq 0} \;:\; \begin{bmatrix} 1 & 0 & 4 \\ 1 & 1 & 3 \\ 1 & 2 & 1 \\ 1 & 4 & 0 \end{bmatrix} \cdot \begin{bmatrix} w_1 \\ w_2 \\ s \end{bmatrix} \geq \begin{bmatrix} 0 \\ 0 \\ 0 \\ 0 \end{bmatrix} \right\}.$$

The set σ is a pointed three-dimensional cone with four extreme rays:

$$\sigma \;=\; \mathbb{R}_{\geq 0} \left\{ \begin{bmatrix} 0 \\ 1 \\ 0 \end{bmatrix}, \begin{bmatrix} 4 \\ -1 \\ 0 \end{bmatrix}, \begin{bmatrix} -4 \\ 1 \\ 2 \end{bmatrix}, \begin{bmatrix} -8 \\ 3 \\ 2 \end{bmatrix} \right\}.$$

The last two vectors are the vertices of $\sigma_2 = 2 \cdot \sigma_1$, which is an unbounded two-dimensional polyhedron with one bounded edge, so we expect a special fiber with two irreducible components. Algebraically, we find the representation

$$\begin{aligned} K[M]^\sigma &= R\big[\,t^4x,\, t^{5/2}xy,\, txy^2,\, t^{1/2}xy^3,\, xy^4\,\big] \\ &= R[a,b,c,d,e]/I, \end{aligned}$$

where I is the finitely generated ideal promised by Remark 6.6.5. Explicitly,

$$I \;=\; \big\langle\, ac-b^2,\, ad-tbc,\, ae-t^2c^2,\, bd-tc^2, be-tcd, ce-d^2\,\big\rangle.$$

Here, as before, $t \in K$ with $\mathrm{val}(t)=1$. The general fiber is the spectrum of

$$K[M]^\sigma \otimes_R K \;=\; K[a,b,c,d,e]/(I\otimes_R K).$$

This toric surface is the cone over the rational normal curve of degree 4. The special fiber is obtained by replacing I with its initial ideal. Here,

$$K[M]^\sigma \otimes_R \Bbbk \;=\; \Bbbk[a,b,c,d,e]/\mathrm{in}_{\mathbf{0}}(I),$$

$$\text{where} \qquad \begin{aligned} \mathrm{in}_{\mathbf{0}}(I) = I\otimes_R \Bbbk &= \langle ac-b^2,\, ad,\, ae,\, bd, be, ce-d^2\rangle \\ &= \langle a,b,ce-d^2\rangle \,\cap\, \langle ac-b^2, d, e\rangle. \end{aligned}$$

The special fiber of the family $\mathcal{U}_P \to \mathrm{Spec}(R)$ is obtained by gluing two quadric cones over $\Bbbk$ along the common line given by $\langle a,b,d,e\rangle$ in $\mathbb{A}^5_{\Bbbk}$. ◊

This concludes our introduction to affine toric schemes over R. Our next goal is to construct general toric schemes from Γ_{val}-rational polyhedral complexes. As in the standard construction of toric varieties, this requires the ability to relate the affine toric schemes defined by neighboring cells.

Lemma 6.6.10. *Let τ be a face of a Γ_{val}-admissible cone σ. Then $K[M]^\sigma$ is a subalgebra of $K[M]^\tau$, and the morphism $\mathcal{U}_\tau \to \mathcal{U}_\sigma$ is an open immersion.*

Proof. If $c_{\mathbf{u}}x^{\mathbf{u}}$ lies in $K[M]^\sigma$, then $\lambda\,\mathrm{val}(c_{\mathbf{u}}) + \mathbf{w}\cdot\mathbf{u} \geq 0$ for all $(\mathbf{w},\lambda)\in\sigma$, and hence for all $(\mathbf{w},\lambda)\in\tau$. Thus $K[M]^\sigma$ is contained in $K[M]^\tau$.

To prove the second claim, we write $\tau = \sigma \cap \{(\mathbf{w},\lambda) \in N_{\mathbb{R}}\times\mathbb{R}_{\geq 0} : \lambda b + \mathbf{w}\cdot\mathbf{u}' = 0\}$ for some $\mathbf{u}'\in M$ and $b\in\mathbb{R}$. Since σ is Γ_{val}-admissible, $b\in\Gamma_{\mathrm{val}}$, so $b = \mathrm{val}(\alpha)$ for some $\alpha\in K$. We may assume $\lambda b + \mathbf{w}\cdot\mathbf{u}' \geq 0$ for all $(\mathbf{w},\lambda)\in\sigma$, so $\alpha x^{\mathbf{u}'}\in K[M]^\sigma$. For the open immersion property, we prove that $K[M]^\tau$ is the localization $K[M]^\sigma_{\alpha x^{\mathbf{u}'}}$. Consider any $cx^{\mathbf{u}} \in K[M]^\tau$. We must show that $(\alpha x^{\mathbf{u}'})^m c x^{\mathbf{u}}$ is in $K[M]^\sigma$ for some m, or equivalently, that $|(\lambda\,\mathrm{val}(c) + \mathbf{w}\cdot\mathbf{u})|/(\lambda b + \mathbf{w}\cdot\mathbf{u}')$ is bounded above as $(\mathbf{w},\lambda)$ varies over σ. Write $\sigma = \mathrm{pos}\{(\mathbf{w}_1,\lambda_1),\ldots,(\mathbf{w}_s,\lambda_s)\}$, and set $\tilde{m} = \max_{1\leq i\leq s}\big\{(\lambda_i\,\mathrm{val}(c) + \mathbf{w}_i\cdot\mathbf{u})/(\lambda_i b + \mathbf{w}_i\cdot\mathbf{u}')\big\}$. Then

$$-\lambda_i\,\mathrm{val}(c) - \mathbf{w}_i\cdot\mathbf{u} + \tilde{m}(\lambda_i b + \mathbf{w}_i\cdot\mathbf{u}') \geq 0 \quad\text{for}\quad 1\leq i\leq s.$$

Any $(\mathbf{w}, \lambda) \in \sigma$ can be written in the form $(\mathbf{w}, \lambda) = \sum_{i=1}^{s} \mu_i(\mathbf{w}_i, \lambda_i)$, where $\mu_i \geq 0$ for $1 \leq i \leq s$. Then

$$\begin{aligned} & -\lambda \operatorname{val}(c) - \mathbf{w} \cdot \mathbf{u} + \tilde{m}(\lambda b + \mathbf{w} \cdot \mathbf{u}') \\ & \quad = \sum_{i=1}^{s} \mu_i(-\lambda_i \operatorname{val}(c) - \mathbf{w}_i \cdot \mathbf{u} + \tilde{m}(\lambda_i b + \mathbf{w}_i \cdot \mathbf{u}')) \geq 0. \end{aligned}$$

Any integer m greater than $\tilde{m}$ has the required property. □

The construction of a toric scheme over $\operatorname{Spec}(R)$ mimics the construction of a toric variety over a field, with the role of a rational cone σ replaced by a Γ_{val}-rational polyhedron and the role of a polyhedral fan replaced by a polyhedral complex. We associate to each polyhedron $P \subset N_{\mathbb{R}}$ in the complex a cone $\sigma \subset N_{\mathbb{R}} \times \mathbb{R}_{\geq 0}$, and use Lemma 6.6.10 to glue together the schemes U_σ coming from neighboring polyhedra. The choice of allowable polyhedral complexes is subtle. It requires the following definitions.

Definition 6.6.11. Let Σ be a Γ_{val}-rational polyhedral complex in $\mathbb{R}^n$ with the property that Σ is a subcomplex of a polyhedral complex Σ' whose support is all of $\mathbb{R}^n$. We define the cone $C(\Sigma)$ over Σ as follows. For each cell $P \in \Sigma$, let $C(P)$ be the closure of $\{(\lambda \mathbf{x}, \lambda) \in \mathbb{R}^{n+1} : \mathbf{x} \in P, \lambda > 0\}$. Let $C(\Sigma)$ be the collection of all cones $C(P)$ and their faces as P varies over Σ.

Example 6.6.12. Let Σ be the tropical curve in $\mathbb{R}^2$ shown in Figure 3.3.3. The cone $C(\Sigma)$ over Σ is a fan in $\mathbb{R}^2 \times \mathbb{R}_{\geq 0}$ with seven two-dimensional cones; one for each line segment or unbounded edge of the tropical curve Σ. It has seven one-dimensional cones. Three of these (one for each vertex of Σ) intersect the open half-space $\{x_3 > 0\}$. The other four (one for each direction of a ray in Σ) lie in the plane $\{x_3 = 0\}$. For example, the ray $P = \{(1,1) + \lambda(0,1) : \lambda \geq 0\}$ gives $C(P) = \operatorname{pos}\{(0,1,0), (1,1,1)\}$, while the ray $P' = \{(-1,0) + \lambda(0,1) : \lambda \geq 0\}$ gives $C(P') = \operatorname{pos}\{(0,1,0), (-1,0,1)\}$. These intersect in the one-dimensional cone $\operatorname{pos}\{(0,1,0)\}$ of $C(\Sigma)$. ◊

Recall from (3.5.1) that the *recession cone* of a polyhedron P is the largest cone σ for which $P + \sigma \subseteq P$. If P is a Γ_{val}-rational polyhedron in $N_{\mathbb{R}}$, then $C(P)_0$ is the recession cone of P, and the slice $C(P)_1$ equals P.

Lemma 6.6.13. *Let Σ be a Γ_{val}-rational polyhedral complex in $\mathbb{R}^n$ that is a subcomplex of a polyhedral complex Σ' whose support is all of $\mathbb{R}^n$. Then $C(\Sigma)$ is a polyhedral fan, with each cone Γ_{val}-admissible, and the intersection of $C(\Sigma)$ with the hyperplane $\{x_{n+1} = 0\}$ is the recession fan of Σ.*

Proof. We first show that $C(\Sigma')$ is a fan. Suppose that a cell P of Σ' has a nontrivial lineality space, so contains a translate of a linear space L. Then, since Σ' covers $\mathbb{R}^n$, and the intersection of any two polyhedra in Σ' is a face of each, every cell contains a translate of L. We may thus replace $\mathbb{R}^n$ by $\mathbb{R}^n/L$ and assume that no cell of Σ' contains a translate of a linear space. There are two types of cones in $C(\Sigma')$: $C(P)$ and $C(P)_0$ for all $P \in \Sigma'$. If P and Q are cells in Σ', then $F = P \cap Q$ is a common face of each, and

$$C(P) \cap C(Q) = C(F),\ C(P) \cap C(Q)_0 = C(F)_0,\ C(P)_0 \cap C(Q)_0 = C(F)_0.$$

In each of the three cases, the intersection is a face of the two larger cones. Hence $C(\Sigma')$ is a fan with support $\mathbb{R}^n \times \mathbb{R}_{\geq 0}$, and $C(\Sigma)$ is a subfan of it. The recession fan of $C(\Sigma)$ is the set of all of cones $C(P)_0$, where $P \in \Sigma$. So, this is indeed a fan, with support equal to $|C(\Sigma)| \cap \{x_{n+1} = 0\}$. □

Remark 6.6.14. The hypothesis that Σ is a subcomplex of a polyhedral complex Σ' with $|\Sigma'| = \mathbb{R}^n$ is essential. Otherwise, the set $C(\Sigma)$ might not be a well-defined fan. For a simple example in $\mathbb{R}^2$, suppose Σ consists of the point $P = \{(1,0)\}$ and the line $L = \{(0,1) + \lambda(1,0) : \lambda \in \mathbb{R}\}$. The cone over P is $C(P) = \operatorname{pos}\{(1,0,1)\}$, while the cone over L is $C(L) = \operatorname{pos}\{(0,1,1)\} + \operatorname{span}\{(1,0,0)\}$. Their intersection $C(P) \cap C(L) = \{(0,0,0)\}$ is not a face of $C(L)$, as the lineality space of $C(L)$ is one dimensional.

This extra hypothesis in Lemma 6.6.13 is not a topological obstruction; after a subdivision we can assume that Σ is a subcomplex of a polyhedral complex Σ' with $|\Sigma'| = \mathbb{R}^n$. In our example, it suffices to replace L by $(0,1)+\operatorname{pos}((1,0))$ and $(0,1) + \operatorname{pos}((-1,0))$. Subdivision does, however, change the corresponding toric variety. For more on this phenomenon, see [**BGS11**].

We can now finally define a general toric scheme over $\operatorname{Spec}(R)$.

Definition 6.6.15. Let Σ be a Γ_{val}-rational polyhedral complex in $\mathbb{R}^n$ that is a subcomplex of a Σ' with $|\Sigma'| = \mathbb{R}^n$. By Lemma 6.6.13, the cone $C(\Sigma)$ over Σ is a fan in $N_{\mathbb{R}} \times \mathbb{R}$ with each cone Γ_{val}-admissible. The *toric scheme* $\mathcal{X}_\Sigma$ is obtained by gluing together the affine toric schemes $\mathcal{U}_\sigma$ for $\sigma \in C(\Sigma)$ along the open subschemes $\mathcal{U}_\tau$ of Lemma 6.6.10 corresponding to faces τ of σ.

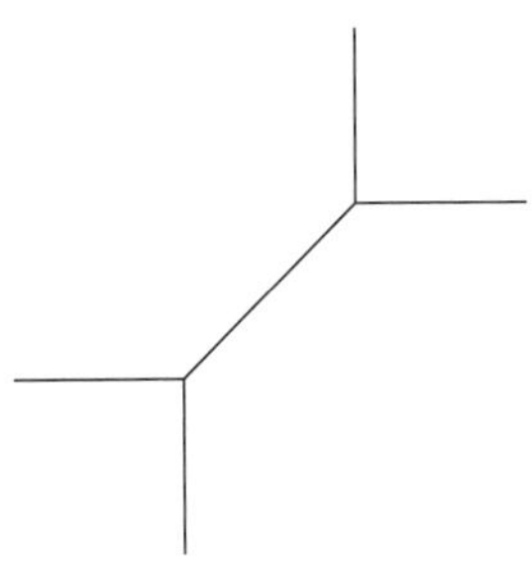

Figure 6.6.1. Example 6.6.18 derives a toric scheme from this tropical curve.

Remark 6.6.16. The construction of toric schemes over $\mathrm{Spec}(R)$ can be slightly more general than presented here; the fan $C(\Sigma)$ associated to a polyhedral complex $\Sigma \subset N_{\mathbb{R}}$ can be replaced by more general fan $\tilde{\Sigma} \subset N_{\mathbb{R}} \times \mathbb{R}$ whose cones are Γ_{val}-admissible. See see [**Gub13**] for details and subtleties.

Here now is our main result about toric schemes over a valuation ring:

Theorem 6.6.17. *Let Σ be a Γ_{val}-rational polyhedral complex in $\mathbb{R}^n$ that is a subcomplex of a polyhedral complex Σ' whose support is all of $\mathbb{R}^n$. The toric scheme $\mathcal{X}_\Sigma$ is integral, separated, normal, of finite type, and flat over* $\mathrm{Spec}(R)$*, with an algebraic action of* $\mathrm{Spec}(R[M])$*. It is proper over* $\mathrm{Spec}(R)$ *if $|\Sigma| = \mathbb{R}^n$. The general fiber of $\mathcal{X}_\Sigma$ is the toric variety $X_{\mathrm{rec}(\Sigma)}$ over K associated to the recession fan of Σ. The special fiber is a union of toric varieties over $\Bbbk$, one for each vertex of Σ. The component corresponding to a vertex $\mathbf{v} \in \Sigma$ is the toric variety $X_{\mathrm{star}_\Sigma(\mathbf{v})}$ over $\Bbbk$.*

Proof. That $\mathcal{X}_\Sigma$ is integral, normal, of finite type, and flat over $\mathrm{Spec}(R)$ all follow from Proposition 6.6.4. These facts were shown there for the affine pieces $\mathcal{U}_\sigma$. The existence of the action of $\mathrm{Spec}(R[M])$ on X_Σ also follows from Definition 6.6.3 and Proposition 6.6.4, which imply that each $\mathcal{U}_\sigma$ has such an action.

To see that $\mathcal{X}_\Sigma$ is separated, we need to show that the image of the diagonal morphism $\mathcal{X}_\Sigma \to \mathcal{X}_\Sigma \times_{\mathrm{Spec}(R)} \mathcal{X}_\Sigma$ is closed. Since $\{\mathcal{U}_\sigma : \sigma \in \Sigma\}$ is an open affine cover of $\mathcal{X}_\Sigma$, the set $\{\mathcal{U}_\sigma \times_{\mathrm{Spec}(R)} \mathcal{U}_{\sigma'} : \sigma, \sigma' \in \Sigma\}$ is an open affine cover of $\mathcal{X}_\Sigma \times_{\mathrm{Spec}(R)} \mathcal{X}_\Sigma$. The image of the diagonal in $\mathcal{U}_\sigma \times_{\mathrm{Spec}(R)} \mathcal{U}_{\sigma'}$ is the image of $\mathcal{U}_\tau$, where $\tau = \sigma \cap \sigma'$. To show that this image is closed, it thus suffices to show that the homomorphism $\phi\colon K[M]^\sigma \otimes_R K[M]^{\sigma'} \to K[M]^\tau$ is a surjection. As in the proof of Lemma 6.6.10, we write $\tau = \sigma \cap \{(\mathbf{w}, \lambda) \in N_{\mathbb{R}} \times \mathbb{R}_{\geq 0} : \lambda \,\mathrm{val}(\alpha) + \mathbf{w} \cdot \mathbf{u}' = 0\}$ for some $\mathbf{u}' \in M$ and $\alpha \in K^*$ with $\mathrm{val}(\alpha) = \lambda$, so we have $K[M]^\tau = K[M]^\sigma_{\alpha x^{\mathbf{u}'}} = K[M]^{\sigma'}_{(1/\alpha)x^{-\mathbf{u}'}}$, with $\alpha x^{\mathbf{u}'} \in K[M]^\sigma$. Then any $f \in K[M]^\tau$ has the form $(\alpha x^{\mathbf{u}'})^m g$ for some $g \in K[M]^{\sigma'}$ and $m \geq 0$. Thus $f = \phi((\alpha x^{\mathbf{u}'})^m \otimes g)$, which shows that ϕ is surjective.

We now consider the description of the general fibers and special fibers. If σ, σ' are cones of the fan $C(\Sigma)$, then by Proposition 6.6.4 the general fibers of $\mathcal{U}_\sigma$ and $\mathcal{U}_{\sigma'}$ are the affine toric varieties U_{σ_0} and $U_{\sigma_0'}$ over K. The intersection of $C(\Sigma)$ with the coordinate hyperplane $x_{n+1} = 0$ is the recession fan $\mathrm{rec}(\Sigma)$ of Σ by Lemma 6.6.13. Let $\tau = \sigma \cap \sigma'$. Then $\sigma_0 \cap \sigma_0' = \tau_0$, so these general fibers are glued together to make the toric variety $X_{\mathrm{rec}(\Sigma)}$ over K.

Similarly, the special fiber is obtained by gluing together the special fibers of the toric schemes $\mathcal{U}_\sigma$ for $\sigma \in C(\Sigma)$. Fix a vertex $\mathbf{v}$ of Σ, and let σ be a cone of $C(\Sigma)$ that contains $(\mathbf{v}, 1)$. Then by Proposition 6.6.4 the special fiber of $\mathcal{U}_\sigma$ has an irreducible component equal to the toric variety over $\Bbbk$ defined by the cone $\tau = \text{pos}(\mathbf{w} - \mathbf{v} : \mathbf{w} \in \sigma_1)$. This is a cone in $\text{star}_\Sigma(\mathbf{v})$. Gluing together the components of the special fibers of the $\mathcal{U}_\sigma$ for $\sigma \in C(\sigma)$ containing $(\mathbf{v}, 1)$, we thus get the toric variety $X_{\text{star}_\Sigma(\mathbf{v})}$.

If the support of Σ is all of $\mathbb{R}^n$, then the recession fan $\text{rec}(\Sigma)$ is a complete fan, so the general fiber of $\mathcal{X}_\Sigma$ is a complete toric variety over $\text{Spec}(K)$. Similarly, each of the irreducible components of the special fiber is a complete toric variety over $\text{Spec}(\Bbbk)$. Both fibers are connected; the general fiber is irreducible, and for the special fiber a path in Σ induces a path between the corresponding $(\Bbbk^*)^n$-orbits. By Proposition 6.6.4 and Remark 6.6.5, the toric scheme $\mathcal{X}_\Sigma$ is separated, flat, and finitely presented over $\text{Spec}(R)$. Corollary 15.7.11 of [**EGAIV**] now implies that $\mathcal{X}_\Sigma$ is proper over $\text{Spec}(R)$. □

Example 6.6.18. Let Σ' be the two-dimensional polyhedral complex shown in Figure 6.6.1, and let Σ be the subcomplex given by the edges and rays. The recession fan of Σ consists of three rays $(\pm 1, 0)$ and $(0, \pm 1)$. The corresponding toric variety is $\mathbb{P}^1 \times \mathbb{P}^1$ with four points removed; this is the general fiber of the toric scheme $\mathcal{X}_\Sigma$. The general fiber of $\mathcal{X}_{\Sigma'}$ is $\mathbb{P}^1 \times \mathbb{P}^1$. The special fiber of $\mathcal{X}_{\Sigma'}$ is a union of two copies of $\mathbb{P}^2$ over $\Bbbk$, one for each vertex of Σ. In the special fiber of $\mathcal{X}_{\Sigma'}$, each of these two copies of $\mathbb{P}^2_\Bbbk$ has three points removed. The incidences of cells in Σ' and Σ determine the gluing. ◊

We now apply Theorem 6.6.17 to study subvarieties of a toric variety. This generalizes the theory of tropical compactifications studied in Section 6.4 to the case that the field K has a nontrivial valuation.

Given a subscheme $Y \subset T^n$ and a polyhedral complex $\Sigma \subset N_\mathbb{R}$ as in Theorem 6.6.17, we can take the closure $\mathcal{Y}$ of Y in the toric scheme $\mathcal{X}_\Sigma$. Explicitly, the subscheme Y is defined by an ideal $I \subset K[M] \cong K[x_1^{\pm 1}, \ldots, x_n^{\pm 1}]$. For any Γ_{val}-admissible cone σ we have an injection $i^* \colon K[M]^\sigma \to K[M]$. The ideal $I_\sigma = (i^*)^{-1}(I)$ of $K[M]^\sigma$ then defines the closure of Y in $\mathcal{U}_\sigma$. The closure $\mathcal{Y}$ of Y in $\mathcal{X}_\Sigma$ is then obtained by gluing together these subvarieties.

Example 6.6.19. Suppose that $\Sigma = \{\mathbf{0}\}$ is just of the origin in $N_\mathbb{R}$. Then $\mathcal{X}_\Sigma$ is $\text{Spec}(R[x_1^{\pm 1}, \ldots, x_n^{\pm 1}])$, which has general fiber T^n_K and special fiber $T^n_\Bbbk$. For a subvariety $Y \subset T^n_K$ defined by an ideal $I \subset K[x_1^{\pm 1}, \ldots, x_n^{\pm 1}]$, the closure $\mathcal{Y}$ is the subvariety of $\mathcal{X}_\Sigma$ defined by the ideal $I_R = I \cap R[x_1^{\pm 1}, \ldots, x_n^{\pm 1}]$.

Next, let Σ be the tropical curve shown in Figure 6.6.1 and discussed in Example 6.6.18. Fix $K = \mathbb{C}\{\{t\}\}$, and consider the curve $Y \subset (K^*)^2$ defined by the polynomial $f = txy + x + y + t$. Let $\mathcal{Y}$ be the closure of Y in $\mathcal{X}_\Sigma$. The general fiber is then the closure of Y in $\mathbb{P}^1 \times \mathbb{P}^1$. The special fiber is two lines, one contained in each of two components of the special fiber of $\mathcal{X}_\Sigma$. ◊

The following is a generalization of Theorem 6.3.4 and Proposition 6.4.7 to the setting of toric schemes. The proof is similar.

Theorem 6.6.20. *Let Σ be a Γ_{val}-rational subcomplex of a polyhedral complex Σ' whose support is $\mathbb{R}^n$. Let $Y \subset T^n$, and let $\mathcal{Y}$ be its closure in $\mathcal{X}_\Sigma$.*

(1) *The special fiber of the scheme $\mathcal{Y}$ intersects the $(\Bbbk^*)^n$ orbit of the special fiber of $\mathcal{X}_\Sigma$ corresponding to a cell $\sigma \in \Sigma$ if and only if* $\mathrm{trop}(Y)$ *intersects the relative interior of σ.*
(2) *The scheme $\mathcal{Y}$ is proper over* $\mathrm{Spec}(R)$ *if and only if* $\mathrm{trop}(Y) \subseteq |\Sigma|$.
(3) *If* $|\Sigma| = \mathrm{trop}(Y)$, *then the intersection in* (1) *has codimension* $\dim(\sigma)$.

Every variety $Y \subset T^n$ has a compactification satisfying the condition in part (3) above. Namely, we can take Σ' to be the Gröbner complex of the homogenization of the ideal of Y, and we take Σ to be the subcomplex with $|\Sigma| = \mathrm{trop}(Y)$.

Remark 6.6.21. Theorem 6.6.20 reveals a fundamental connection between compactifications and degenerations in tropical geometry; the tropical compactification of a variety $Y \subset T^n_K$ over a valued field K gives rise to a degeneration of Y over its residue field $\Bbbk$. This construction has appeared in several different guises in the literature. An important precursor is Viro's patchworking method for constructing real algebraic hypersurfaces with controlled topology; see [**Vir08**]. The combinatorics and commutative algebra of degenerations of toric varieties were explored in [**Stu96**, Chapter 10].

Our exposition most closely follows the description of Gubler [**Gub13**]. Degenerations via tropical geometry also play a pivotal role in the Mirror Symmetry program of Gross and Siebert; see [**Gro11, GS11, NS06**].

As an application, we now finally prove the base case of Theorem 3.5.1. Our proof follows [**CP12**].

Proposition 6.6.22. *If Y is an irreducible curve in the torus T^n, then its tropicalization* $\mathrm{trop}(Y)$ *is connected.*

Proof. The tropical curve $\mathrm{trop}(Y)$ is the support of a unique coarsest polyhedral complex Σ in $\mathbb{R}^n$. This is one dimensional, so consists of vertices, edges, and rays. The complex Σ deformation retracts to the union of the edges and the vertices. Let G be the graph consisting of these vertices and edges. We must show that G is connected.

Let $\mathcal{Y}$ be the closure of Y in the toric scheme $\mathcal{X}_\Sigma$. By Theorem 6.6.20(3), the intersection of the special fiber of $\mathcal{Y}$ with the $T^n_{\Bbbk'}$-orbit corresponding to $\sigma \in \Sigma$ has codimension equal to $\dim(\sigma)$. Explicitly, this means that the special fiber is a curve that intersects the orbit corresponding to a line segment in a collection of points, and intersects the orbit corresponding to a vertex in a one-dimensional component.

If C is an irreducible component of the special fiber, then C intersects the $T^n_{\Bbbk}$-orbit of the special fiber of $\mathcal{X}_\Sigma$ corresponding to a unique vertex $\mathbf{v}$ of Σ. Thus, if C_1, C_2 are irreducible components of the special fiber with $C_1 \cap C_2 \neq \emptyset$, then either the corresponding vertices $\mathbf{v}_1$ and $\mathbf{v}_2$ coincide or the intersection point is in the closure of both $T^n_{\Bbbk}$-orbits, so lies in the $T^n_{\Bbbk}$ orbit corresponding to the edge joining $\mathbf{v}_1$ to $\mathbf{v}_2$. To show that $\operatorname{trop}(Y)$ is connected, it suffices to show that any two vertices $\mathbf{v}_1, \mathbf{v}_2$ in G are connected by a path of edges. For $i = 1, 2$, fix a component C_i of the intersection of the special fiber of $\mathcal{Y}$ with the $T^n_{\Bbbk}$-orbit of $\mathcal{X}_\Sigma$ corresponding the vertex of Σ labeled by $\mathbf{v}_i$. It suffices to show that C_1 and C_2 are connected by a chain of irreducible components $C'_1, \dots, C'_s$ with $C'_i \cap C'_{i+1} \neq \emptyset$ for $1 \leq i \leq s-1$, and $C_1 \cap C'_1 \neq \emptyset$, $C'_s \cap C_2 \neq \emptyset$.

To show this, we pass to a Noetherian subring of R over which $\mathcal{Y}$ is defined. By Theorem 3.2.4 we may assume that K is algebraically closed, so there is a splitting $w \mapsto t^w$ of the valuation map. Only finitely many elements of K occur as coefficients of defining equations for Y and for its closure in each $K[M]^{C(\sigma)}$ with $\sigma \in \Sigma$. In addition, there are only a finite number of elements of K occurring in the gluing isomorphisms when constructing $\mathcal{X}_\Sigma$. Let $\mathcal{C}$ be set of all such elements of K, and let $\mathcal{C}'$ be the set $\{t^{-\operatorname{val}(c)}c, t^{|\operatorname{val}(c)|} : c \in \mathcal{C}\}$. Let $\mathbb{F}$ be the subring of R generated by 1, and let $\tilde{R}$ be the $\mathbb{F}$-algebra generated by $\mathcal{C}'$. Let $\mathfrak{m}' = \tilde{R} \cap \mathfrak{m}_K$, and let R' be the normalization of the localization $\tilde{R}_{\mathfrak{m}'}$. We denote by K' the field of fractions of R'. By construction R' is a Noetherian local ring with two distinguished prime ideals: the zero ideal and $\mathfrak{m}'$. We can form the family $\mathcal{Y}'$ over $\operatorname{Spec}(R')$ as above with R' and K' playing the roles of R and K. This is still irreducible and proper. By choosing $\mathcal{C}'$ large enough we can guarantee that the fiber over $\mathfrak{m}'$ has a component for each component C of the special fiber of $\mathcal{Y}$ by construction. Zariski's connectedness theorem implies that this fiber is connected. See [**EGAIII**, Corollary 4.3.2] for a general form of Zariski's theorem. This means that the components corresponding to C_1 and C_2 are connected by a chain of curves in this fiber. Each curve in this chain corresponds to a curve in the special fiber of $\mathcal{Y}$, so we conclude that C_1 and C_2 are connected by a chain of components of the special fiber of $\mathcal{Y}$. This means that the graph G is connected, and thus $\operatorname{trop}(Y)$ is connected. □

Remark 6.6.23. We have focused in this book on tropicalizing a subvariety of a torus over a valued field K. This gives rise to a flat family $\operatorname{Spec}(R)$ with general fiber the original variety and special fiber the initial ideal. In this section we have seen that this extends to a subvariety $\overline{Y}$ of a toric variety. It is also natural to consider a flat family $\overline{\mathcal{Y}} \to \operatorname{Spec}(R)$ abstractly, with no corresponding embedding of the general fiber $\overline{Y}$ into a toric variety. Such a flat family describes a degeneration of $\overline{Y}$ to the special fiber. This idea has been particularly exploited in the case when $\overline{Y}$ is a curve. The combinatorics of the dual graph of the special fiber then reflects some of the geometry of the curve. Highlights include a tropical Riemann-Roch theorem, a tropical proof of the Brill-Noether theorem, and a tropical understanding of the gonality of a curve. For further reading on these topics we refer to [**ABBR13, Bak08b, BN07, BPR11, CV10, CDPR12, MZ08**].

6.7. Intersection Theory

In this final section we investigate the connection between intersection theory on toric varieties and tropical geometry. In Theorem 6.7.7 we will see that tropical varieties can be understood as cycle classes on toric varieties. Throughout this section we assume that the valuation on K is trivial.

We recall the following basic definitions from intersection theory [**Ful98**]. Let Z be a smooth projective variety of dimension n. The *Chow group* $A_m(Z)$ consists of m-dimensional cycles $\sum a_i Y_i$, where $a_i \in \mathbb{Z}$ and Y_i is a subvariety of dimension m, modulo rational equivalence. The *Chow ring* $A^*(Z)$ is a commutative associative graded ring with identity. Its graded piece $A^r(Z)$ is isomorphic to the group $A_{n-r}(Z)$ of codimension r cycles on Z modulo rational equivalence.

Let Y and Y' be subvarieties of Z of codimension r and r', respectively. We say that Y and Y' *intersect properly* if $Y \cap Y' = \bigcup W_i$, where the W_i are all irreducible of codimension $r + r'$. In that case, their product is

$$Y \cdot Y' \;=\; \sum i(W_i, Y \cdot Y') \cdot W_i,$$

where $i(W_i, Y \cdot Y')$ is the length of W_i in the scheme-theoretic intersection of Y and Y'. In particular, if $Y \cap Y' = \emptyset$, then $Y \cdot Y' = 0$ in $A^*(Z)$.

We now review intersection theory on a smooth projective toric variety X_Σ. The Chow group $A^r(X_\Sigma)$ is generated by the set $\{V(\sigma) : \sigma \in \Sigma(r)\}$ of orbit closures of codimension r. A cone $\sigma \in \Sigma(r)$ has r generators, corresponding to torus invariant divisors $D_1, \ldots, D_r$. The class $[V(\sigma)] \in A^r(X_\Sigma)$ is the intersection $D_1 \cdot \ldots \cdot D_r$. For $\sigma \in \Sigma(r)$ and a divisor D_i corresponding to a ray $\mathrm{pos}(\mathbf{v}_i)$ of Σ not contained in σ, we have $D_i \cdot [V(\sigma)] = [V(\sigma + \mathrm{pos}(\mathbf{v}_i))]$ if $\sigma + \mathrm{pos}(\mathbf{v}_i)$ is a cone of Σ, and $D_i \cdot [V(\sigma)] = 0$ otherwise.

The relations in each Chow group are as follows. For $\tau \in \Sigma(r-1)$ and $\sigma \in \Sigma(r)$ with $\tau \subset \sigma$, the orbit closure $V(\sigma)$ is a codimension-1 subvariety of $V(\tau)$, and thus defines a divisor D_σ on $V(\tau)$. The relations between the D_σ in $A^r(X_\Sigma)$ come from the relations between the torus-invariant divisors in the Picard group of $V(\tau)$. Explicitly, as in Definition 3.3.1, let L be the linear space parallel to τ, let $N(\tau) = N/(L \cap N)$, and let $\mathbf{v}_\sigma$ be the first lattice point of the ray $(\sigma + L)/L$ in $N(\tau)_{\mathbb{R}} \times \mathbb{R}$. Here, $\sigma \in \Sigma$ with $\tau \subset \sigma$ and $\dim(\sigma) = r$. These rays $(\sigma + L)/L$ are the rays of the fan of the orbit closure $V(\tau)$. Note that the lattice dual to $N(\tau)$ is $M(\tau) = \tau^{\perp} \cap M$.

Let $\tau[1]$ be the set of rays of the fan of $V(\tau)$, or, equivalently, the set of r-dimensional cones $\sigma \in \Sigma$ with $\tau \subset \sigma$. For any $\mathbf{u} \in M(\tau)$ we have $[\mathrm{div}(x^{\mathbf{u}})] = \sum_{\sigma \in \tau[1]} (\mathbf{u} \cdot \mathbf{v}_\sigma) D_\sigma$. This expression is 0 in $A^1(V(\tau))$, and thus

$$(6.7.1) \qquad \sum_{\sigma \in \tau[1]} (\mathbf{u} \cdot \mathbf{v}_\sigma)[V(\sigma)] \;=\; 0 \qquad \text{in } A^r(X_\Sigma).$$

These are the only relations on $A^r(X_\Sigma)$; see [**FS97**, Proposition 2.1].

The Chow ring $A^*(X_\Sigma)$ has an explicit description in terms of generators and relations. Let s be the number of rays of Σ. The Chow ring is generated in degree 1 by the classes $[D_1], \ldots, [D_s]$ of the torus invariant divisors. Since $A^1(Z) \cong \mathrm{Pic}(Z)$ for any smooth variety, we have the relations coming from the Picard group, which are given by (6.1.1). If V is the $n \times s$ matrix with columns the vectors $\mathbf{v}_j$, then these relations are encoded in the ideal

$$(6.7.2) \qquad L_\Sigma \;=\; \Big\langle \sum_{j=1}^{s} V_{ij} D_j \colon 1 \leq i \leq n \Big\rangle.$$

The other relations come from the fact that certain divisors do not intersect. If $\{i_1, \ldots, i_l\}$ are such that $\mathrm{pos}(\mathbf{v}_{i_1}, \ldots, \mathbf{v}_{i_l})$ is not a cone of Σ, then the divisors $D_{i_1}, \ldots, D_{i_l}$ do not intersect, so $D_{i_1} \cdot \ldots \cdot D_{i_l} = 0$ holds in $A^*(X_\Sigma)$. These relations correspond to the *Stanley–Reisner ideal* of the fan Σ:

$$\mathrm{SR}(\Sigma) \;=\; \Big\langle \prod_{i \in \sigma} D_i : \sigma \notin \Sigma \Big\rangle \;\subset\; \mathbb{Z}[D_1, \ldots, D_s].$$

This ideal is generated by the *minimal nonfaces*. These are subsets of $\{1, \ldots, s\}$ that do not span a cone of Σ, but every proper subcone does. The Stanley–Reisner ideal is a central character in combinatorial commutative algebra; see, for example, [**Sta96**, Chapter 2] or [**MS05**, Chapter 1].

We summarize the description of the Chow ring of X_Σ in the following theorem. For more details see [**CLS11**, Chapter 12] or [**Ful93**, Chapter 5].

Theorem 6.7.1. *Let X_Σ be a smooth complete toric variety whose fan Σ has s rays, one for each torus-invariant divisor D_i. The Chow ring of X_Σ is*

$$A^*(X_\Sigma) \;\cong\; \mathbb{Z}[D_1, \ldots, D_s]/(\mathrm{SR}(\Sigma) + L_\Sigma).$$

This holds with $\mathbb{Z}$ replaced by $\mathbb{Q}$ when X_Σ is simplicial instead of smooth.

The connection with tropical geometry comes from the following alternative formulation, due to Fulton and Sturmfels [**FS97**]. We mildly extend the notion of a weighted fan here, by allowing the weights on the top-dimensional cones of the fan to be possibly negative integers. The definition of balancing from Definition 3.3.1 still makes sense; there is no need for the multiplicities $m(\sigma)$ to be positive. Balanced fans in this sense are known as *tropical cycles*, or *tropical fan cycles* to emphasize that the underlying set is a fan.

Proposition 6.7.2. *Let X_Σ be a smooth complete toric variety of dimension n, and let Z be a cycle in $A_r(X_\Sigma) \simeq A^{n-r}(X_\Sigma)$. For each $\sigma \in \Delta(r)$, set $m_\sigma = Z \cdot V(\sigma) \in A^n(X_\Sigma) \cong \mathbb{Z}$. Let Δ be the r-dimensional subfan of Σ with maximal cones those σ with $m_\sigma \neq 0$. Then $(\Delta, \mathbf{m})$ is a weighted balanced fan.*

Proof. Fix $\tau \in \Delta(r-1)$. By (6.7.1) we have $\sum_{\sigma\in\Sigma(r),\tau\subset\sigma}(\mathbf{u}\cdot\mathbf{v}_\sigma)V(\sigma) = 0$ in $A^r(X_\Sigma)$ for all $\tau \in \Sigma(r-1)$ and $\mathbf{u} \in M(\tau)$. Here, $\mathbf{v}_\sigma$ is the first lattice point on the ray defined by σ in N_τ. This means that, for all $\mathbf{u} \in M(\tau)$,

$$\begin{aligned}\mathbf{u}\cdot\Bigg(\sum_{\sigma\in\Sigma(r),\tau\subset\sigma} m_\sigma\mathbf{v}_\sigma\Bigg) &= \mathbf{u}\cdot\Bigg(\sum_{\sigma\in\Sigma(r),\tau\subset\sigma}(Z\cdot V(\sigma))\mathbf{v}_\sigma\Bigg)\\ &= Z\cdot\Bigg(\sum_{\sigma\in\Sigma(r),\tau\subset\sigma}(\mathbf{u}\cdot\mathbf{v}_\sigma)[V(\sigma)]\Bigg) = 0.\end{aligned}$$

This implies $\sum m_\sigma\mathbf{v}_\sigma = 0 \in N(\tau)$, and hence $(\Delta, \mathbf{m})$ is balanced at τ. □

Example 6.7.3. Let $X_\Sigma = \mathbb{P}^2$ with $D_i = \{x_i = 0\}$ for $i = 0, 1, 2$. Fix any irreducible curve C of degree d in $\mathbb{P}^2$, and let Z be its class in $A^1(\mathbb{P}^2)$. For $i = 0, 1, 2$ we have $Z \cdot D_i = d$ in $A^2(\mathbb{P}^2) \cong \mathbb{Z}$. Here $(\Delta, \mathbf{m})$ is the standard tropical line but with multiplicity $m_i = d$ on each ray. Note that this fan differs from $\operatorname{trop}(C \cap T^2)$ unless the Newton polygon of C is a triangle. ◊

The next example shows that some weights of $(\Delta, \mathbf{m})$ may be negative.

Example 6.7.4. Let X_Σ be the projective plane $\mathbb{P}^2$ blown up at one point. The fan Σ has four rays $(1, 0)$, $(1, 1)$, $(0, 1)$, $(-1, -1)$ and four two-dimensional cones between them. The Stanley Reisner ideal is

$$\mathrm{SR}(\Sigma) = \langle D_1D_3, D_2D_4\rangle.$$

Take $Z = D_2$, which is the exceptional fiber of the blow-up. We find that

$$Z\cdot D_1 = 1,\;\; Z\cdot D_2 = -1,\;\; Z\cdot D_3 = 1,\;\; Z\cdot D_4 = 0.$$

Hence $(\Delta, \mathbf{m})$ is the one-dimensional subfan of Σ with rays $(1, 0), (1, 1)$ and $(0, 1)$. The weights are $m_1 = 1, m_2 = -1, m_3 = 1$, so this fan is balanced. ◊

So far, we have restricted ourselves to toric varieties X_Σ that are smooth. This means that the fan Σ is simplicial and unimodular. We now relax that hypothesis, and we consider arbitrary toric varieties instead. Their top nonvanishing Chow groups can always be realized as spaces of weights that make a subfan balanced. This is the content of the following theorem:

Theorem 6.7.5. *Let Σ be a rational polyhedral fan that is pure of dimension d. The elements of $A_{n-d}(X_\Sigma)$ are in bijection with choices of weights that make Σ into a balanced fan. This bijection takes a cycle $Z \in A_{n-d}(X_\Sigma)$ to the weight function $m : \Sigma(d) \to \mathbb{Z}$ given by*

$$m(\sigma) = \pi^*(i_*(Z)) \cdot V(\sigma'),$$

where $i : X_\Sigma \to X_{\tilde{\Sigma}}$ is the inclusion of X_Σ into any compactification $X_{\tilde{\Sigma}}$, $\pi : X_{\Sigma'} \to X_{\tilde{\Sigma}}$ is a resolution of singularities induced by a map of fans $\pi : \Sigma' \to \tilde{\Sigma}$, and $\sigma' \in \Sigma'(d)$ is a cone with $\pi(\sigma') \subseteq \sigma$. The multiplicity $m(\sigma)$ is independent of the choice of inclusion i, resolution π, and cone σ'.

Proof. Choose a smooth complete fan Σ' that refines a completion $\tilde{\Sigma}$ of Σ. The toric variety $X_{\Sigma'}$ is then a resolution of a compactification of X_Σ as follows:

$$\begin{array}{ccc} & & X_{\Sigma'} \\ & & \downarrow{\scriptstyle \pi} \\ X_\Sigma & \xrightarrow{\ i\ } & X_{\tilde{\Sigma}} \end{array}$$

Fix $Z \in A_d(X_\Sigma)$, and let $Z' = \pi^*(i_*(Z)) \in A_d(X_{\Sigma'})$. Let $(\Delta, \mathbf{m})$ be the weighted balanced fan associated to Z' by Proposition 6.7.2.

We first note that $m(\sigma') = 0$ for all $\sigma' \in \Sigma'(d)$ that satisfy $\pi(\sigma') \not\subseteq |\Sigma|$. Indeed, by the projection formula [**Ful98**, Proposition 2.3c)] we have

$$\pi_*(Z' \cdot V(\sigma')) = \pi_*(\pi^*(i_*(Z)) \cdot V(\sigma')) = i_*(Z) \cdot \pi_*(V(\sigma')) = i_*(Z) \cdot V(\sigma),$$

where $\sigma \in \tilde{\Sigma}$ is the smallest cone containing $\pi(\sigma')$. If $\sigma \not\subseteq |\Sigma|$, then $i_*(Z) \cdot V(\sigma) = 0$, as required. Next suppose that σ', σ'' are cones in $\Sigma'(d)$ with $\pi(\sigma') = \pi(\sigma'')$. We claim that the balancing condition implies $m(\sigma') = m(\sigma'')$. It suffices to prove this for cones σ', σ'' that share a facet τ. Let $\mathbf{u}_{\sigma'}$ be the first lattice point on the ray $(\sigma' + L)/L$, where L is the linear space parallel to τ. Then $-\mathbf{u}_{\sigma'}$ is the first lattice point in N_τ on the ray $(\sigma'' + L)/L$, and $m(\sigma')\mathbf{u}_{\sigma'} - m(\sigma'')\mathbf{u}_{\sigma'} = 0 \in N_\tau$, since Σ' is balanced with the weights $\mathbf{m}$ by Proposition 6.7.2. Thus $m(\sigma') = m(\sigma'')$.

The previous paragraph shows that the weight function $m\colon \Sigma(d) \to \mathbb{Z}$ given by $m(\sigma) = \pi^*(i_*(Z)) \cdot V(\sigma')$ is well defined and independent of the choice of $\sigma' \in \Sigma'$. The calculation to see if Σ is balanced at a codimension-one cone $\tau \in \Sigma(d-1)$ is the same as checking if Σ' is balanced at any cone $\tau' \in \Sigma'(d-1)$ with $\pi(\tau') \subseteq i(\tau)$. This shows that for every element $Z \in A_d(X_\Sigma)$ we get a weight function $m\colon \Sigma(d) \to \mathbb{Z}$ that makes Σ balanced.

We now show that this correspondence is a bijection. The fact that different Z give different weight functions holds because the classes of the $V(\sigma')$ span $A_{n-d}(X_{\Sigma'})$, and $A_d(X_{\Sigma'})$ is dual to $A_{n-d}(X_{\Sigma'})$ under the intersection pairing. To see that every weight function comes from an element

$Z \in A_d(X_\Sigma)$, choose a collection $\sigma_1, \ldots, \sigma_r \in \Sigma'(d)$ such that the classes $\{[V(\sigma_i)] \colon 1 \leq i \leq r\}$ form a basis for $A_{n-d}(\Sigma')$. We may choose the σ_i as follows. Order the cones $\tau \in \Sigma'(d-1)$ so that for $i > 1$ the cone τ_i is a facet of a cone $\sigma' \in \Sigma'(d)$ that has another facet τ_j for some $j < i$. Starting with τ_1, choose a maximal subset of the $\sigma' \in \Sigma'(d)$ that contain τ_1 for which the classes $[V(\sigma')]$ are linearly independent. For $j > 1$, add to the collection of σ_i a maximal collection of $\sigma' \in \Sigma'(d)$ containing τ_j that keep the entire collection of $[V(\sigma_i)]$ linearly independent. Let Z' in the dual space $A^d(X_\Sigma)$ be defined by $Z' \cdot V(\sigma_i) = m(\sigma_i)$. We set $Z = i^*(\pi_*(Z'))$. It suffices to check that when $\sigma' \in \Sigma'(d)$ with $\pi(\sigma') = \sigma$, then $Z' \cdot V(\sigma') = m(\sigma)$. This follows by induction on $t_{\sigma'} = \min\{j : \tau_i \text{ is a face of } \sigma'\}$. When $t_{\sigma'} = 1$ the balancing condition and (6.7.1) show that $m(\sigma)$ and $Z' \cdot V(\sigma')$ are the same linear combination of the $m(\sigma_i)$. The same holds for the induction hypothesis.

Finally, we argue that the correspondence is independent of the choice of compactification $\tilde{\Sigma}$ and resolution Σ'. If Σ'' is a smooth refinement of Σ', with corresponding morphisms $p \colon X_{\Sigma''} \to X_{\Sigma'}$ and $q = \pi \circ p \colon X_{\Sigma''} \to X_{\tilde{\Sigma}}$, then the multiplicities induced by Σ'' agree with those induced by Σ'. Indeed, if $\sigma'' \in \Sigma''$ with $p(\sigma'') \subseteq \sigma' \in \Sigma'$, and $\pi(\sigma') \subseteq \sigma \in \Sigma$, then

$$m(\sigma) = q^*(i_*(Z)) \cdot V(\sigma'') = p_*(p^*\pi^*(i_*(Z)) \cdot V(\sigma'')) = \pi^*(i_*(Z)) \cdot V(\sigma),$$

which is the multiplicity given by Σ'.

Now suppose Σ_2' is a different smooth complete fan that refines a complete fan $\tilde{\Sigma}_2$ which contains Σ as a subfan. Let $\tilde{\Sigma}_3$ be the common refinement of $\tilde{\Sigma}$ and $\tilde{\Sigma}_2$, and let Σ_3' be a smooth common refinement of Σ' and Σ_2'. We label the corresponding morphisms as in the following diagram.

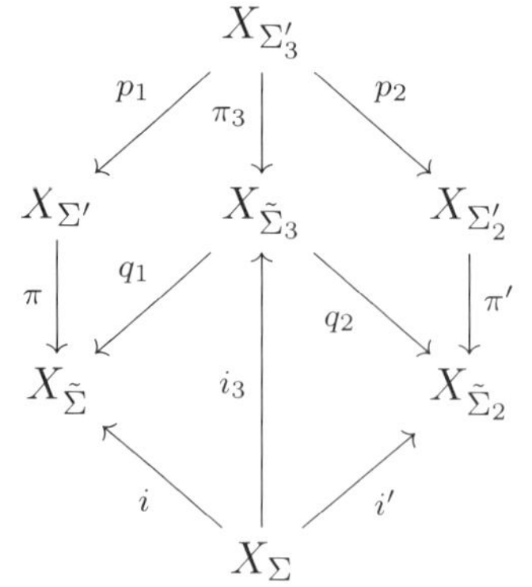

The morphism $i_3 \colon X_\Sigma \to X_{\tilde{\Sigma}_3}$ comes from the fact that Σ is a subfan of both $\tilde{\Sigma}$ and $\tilde{\Sigma}_2$, so is a subfan of its refinement $\tilde{\Sigma}_3$. By the previous paragraph, the multiplicity induced on a cone $\sigma \in \Sigma$ by the inclusion i and the map $q_1 \circ \pi_3 : X_{\Sigma_3'} \to X_{\tilde{\Sigma}}$ equals the multiplicity induced by i and π. Similarly, the multiplicity induced by i' and $q_2 \circ \pi_3$ equals the multiplicity induced by i' and

π. To finish we just need to show that this equals the multiplicity induced by i_3 and π_3. This follows from $q_1^*(i_*(Z)) = i_{3*}(Z)$, as Z is supported on the torus orbit closures corresponding $\sigma \in \Sigma$, and Σ is a subfan of $\tilde{\Sigma}_3$. $\square$

Note that the only hypothesis on Σ in Theorem 6.7.5 is that it is pure of dimension d. In the context of tropical geometry, the fan Σ is typically the tropicalization $\Sigma = \text{trop}(Y)$ of a very affine variety Y of dimension d. One important consequence of Theorem 6.7.5 is that the Chow group $A^d(X_\Sigma)$ depends only on the support $|\Sigma|$ but not on the particular fan structure Σ.

Example 6.7.6. Let Σ be the one-dimensional fan in $\mathbb{R}^2$ with rays spanned by the column vectors of the matrix

$$C = \begin{pmatrix} 1 & -1 & -1 & 0 & 1 \\ 1 & 1 & 0 & -1 & -1 \end{pmatrix}.$$

The Chow group $A_1(X_\Sigma)$ is the cokernel of the map $\mathbb{Z}^2 \xrightarrow{C^T} \mathbb{Z}^5$, so it equals $\mathbb{Z}^3$. This is spanned by the five torus-invariant divisors $D_1, D_2, \ldots, D_5$, subject to the linear relations

$$D_1 - D_2 - D_3 + D_5 = D_1 + D_2 - D_4 - D_5 = 0.$$

This is also the Chow group $A_1(X_{\tilde{\Sigma}})$ of the compactification obtained by taking the unique complete fan $\tilde{\Sigma}$ with the same rays as Σ.

A resolution $X_{\Sigma'}$ of $X_{\tilde{\Sigma}}$ is obtained by taking the stellar subdivision at the rays $(1,0)$ and $(0,1)$. This is shown in the first row of Figure 6.7.1. By

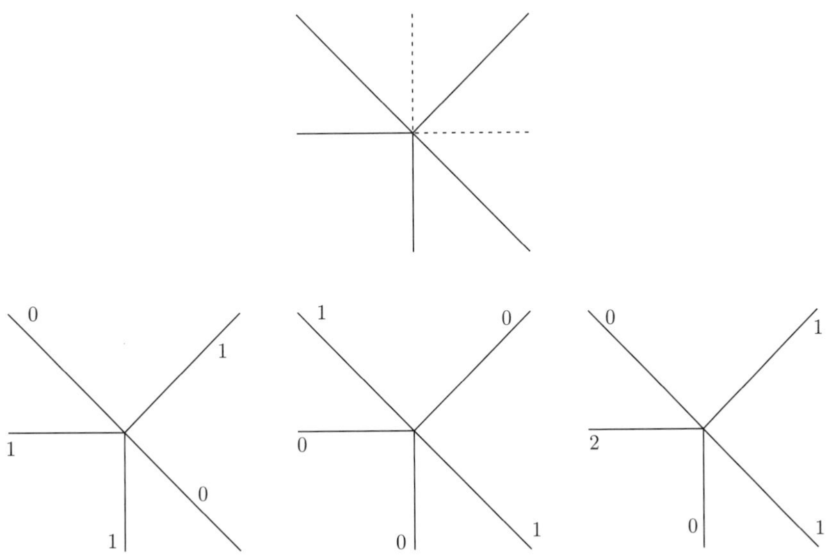

Figure 6.7.1. Effective tropical cycles on a one-dimensional fan in $\mathbb{R}^2$.

computing the normal forms of the products D_iD_j modulo a Gröbner basis of $\mathrm{SR}(\tilde{\Sigma}) + I_\Sigma$, we find the matrix of intersection numbers,

$$(6.7.3) \qquad \big(\pi^*(i_*(D_i)) \cdot D_j\big) \;=\; \frac{1}{2}\begin{pmatrix} 0 & 1 & 0 & 0 & 1 \\ 1 & -1 & 2 & 0 & 0 \\ 0 & 2 & -2 & 2 & 0 \\ 0 & 0 & 2 & -2 & 2 \\ 1 & 0 & 0 & 2 & -1 \end{pmatrix}.$$

Following Theorem 6.7.1, this can also be computed on the resolution $X_{\Sigma'}$. Our 5×5-matrix has rank 3. Its rows span the space of balanced weights $\mathbf{m}$ on the fan Σ. Inside that space lives the cone of effective tropical cycles

$$\big\{\mathbf{m} \in \mathbb{R}^5 : \mathbf{m} \geq 0 \text{ and } C \cdot \mathbf{m} = 0\big\}.$$

This cone is spanned by $\mathbf{m}_1 = (1,0,1,1,0)$, $\mathbf{m}_2 = (0,1,0,0,1)$, and $\mathbf{m}_3 = (1,0,2,0,1)$. These weights are shown in the second row of Figure 6.7.1. ♢

We now consider the following situation: Y is a given subvariety of a torus T^n, and X_Σ is a toric variety such that the closure $\overline{Y}$ of Y in X_Σ happens to be a flat tropical compactification. Let $Z = [\overline{Y}]$ be the class of $\overline{Y}$ in $A_r(X_\Sigma)$. The next theorem tells us that the tropical variety $\mathrm{trop}(Y)$ can be recovered as the balanced fan associated to Z by Proposition 6.7.2.

Theorem 6.7.7. *Let $Y \subset T^n$ be a variety, and let $\overline{Y}$ be any tropical compactification in a toric variety X_Σ. The balanced fan associated to $\overline{Y}$ by Proposition* 6.7.2, *after taking any completion Σ' of Σ, has support* $\mathrm{trop}(Y)$, *and its weights agree with the multiplicities on* $\mathrm{trop}(Y)$. *If X_Σ is smooth and the compactification $\overline{Y}$ is flat tropical, then $m(\sigma) = \overline{Y} \cdot V(\sigma)$ for any maximal cone $\sigma \in \Sigma$. Thus the tropicalization of Y determines the class $[\overline{Y}] \in A^*(X_\Sigma)$.*

Proof. Let $d = \dim(\overline{Y})$. We first assume that $\overline{Y}$ is a flat tropical compactification of Y inside a smooth toric variety X_Σ. By Proposition 6.4.14, the support $|\Sigma|$ of Σ equals $\mathrm{trop}(Y)$. Fix a maximal cone $\sigma \in \Sigma(d)$. By part (2) of Proposition 6.4.7, the scheme $\overline{Y} \cap \mathcal{O}_\sigma$ is zero dimensional. By Proposition 6.4.15, $\overline{Y}$ is Cohen–Macaulay at any point $p \in \overline{Y} \cap \mathcal{O}_\sigma$. The class $[\overline{Y}] \cdot V(\sigma) \in A_0(X_\Sigma)$ equals $i(p, \overline{Y} \cdot V(\sigma))[p]$, where $i(p, \overline{Y} \cdot V(\sigma))$ is the intersection multiplicity of p in $\overline{Y} \cdot V(\sigma)$; see [**Ful98**, Definition 7.1]. By [**Ful98**, Proposition 7.1(b)], $i(p, \overline{Y} \cdot V(\sigma))$ equals the length of $\overline{Y} \cap \mathcal{O}_\sigma$. By Remark 6.4.18, $\overline{Y} \cap \mathcal{O}_\sigma$ equals the quotient of the subscheme of T^n defined by the initial ideal $\mathrm{in}_{\mathbf{w}}(I_Y)$ by the torus $T_\sigma = N_\sigma \otimes K^*$. Here $K = \mathbb{k}$, since K has the trivial valuation. Thus by Lemma 3.4.7 the length equals $m(\sigma)$.

Now let X_Σ be an arbitrary toric variety with $\mathrm{trop}(Y) = |\Sigma|$. Fix a completion Σ' of the fan Σ. By Theorem 6.7.5, the weighted balanced fan associated to $\overline{Y}$ is independent of the chosen resolution of singularities of Σ'.

By Proposition 6.4.14, a refinement of a fan giving a flat tropical compactification also gives a flat tropical compactification. We may thus assume, after taking a resolution Σ'' of the common refinement of Σ' with a completion of the fan of a flat tropical compactification, that Σ'' contains a subfan $\tilde{\Sigma}$ with support $|\Sigma|$ that gives a flat tropical compactification. Write $\pi\colon X_{\Sigma''} \to X_{\Sigma'}$ and $i\colon X_\Sigma \to X_{\Sigma'}$ for the two morphisms. In the fan associated to $\overline{Y}$ by Proposition 6.7.2, the multiplicity of σ is $m(\sigma) = \pi^*(i_*([\overline{Y}])) \cdot [V(\sigma')]$ for any $\sigma' \in \tilde{\Sigma}$ with $\pi(\sigma') \subset \sigma$. Now $\pi^*(i_*([\overline{Y}]))$ is the class of the closure $\overline{Y}''$ of $Y \subset T^n$ in $X_{\Sigma''}$, and $\overline{Y}''$ is a flat tropical compactification, so $m(\sigma)$ equals the multiplicity of σ', and thus also of σ.

We conclude that the multiplicities on $\operatorname{trop}(Y)$ determine the intersection numbers $\pi^*(i_*([\overline{Y}])) \cdot V(\sigma')$, and thus $\pi_*(\pi^*(i_*([\overline{Y}])) \cdot V(\sigma')) = i_*([\overline{Y}]) \cdot V(\sigma)$. Since the $V(\sigma)$ for $\sigma \in \Sigma'(d)$ span $A_{n-d}(X_{\Sigma'})$, the class $i_*([\overline{Y}]) \in A_d(X_{\Sigma'})$ is determined by these intersection numbers, as required. □

Remark 6.7.8. Theorem 6.7.7 lets us give a toric proof of the balancing condition (Theorem 3.4.14). As in the proof given in Chapter 3 we reduce to the case of Proposition 3.4.13 that $C \subset T^n$ is a curve over the residue field $\Bbbk$ (or over a field with the trivial valuation). There is only one choice of fan structure Σ on $\operatorname{trop}(C) \subset \mathbb{R}^n$, so the closure $\overline{C}$ of C in the toric variety X_Σ is a flat tropical compactification. By Theorem 6.7.7, the multiplicity $m(\sigma)$ equals $\overline{C} \cdot V(\sigma)$ for all rays σ, so the balanced fan associated to $\overline{C}$ by Proposition 6.7.2 equals $(\operatorname{trop}(C), \mathbf{m})$. Hence $(\operatorname{trop}(C), \mathbf{m})$ is balanced. To check that this is indeed a proof, note that the proof of Proposition 6.7.2 did not use any tropical geometry (and indeed, predates it), while the proof of Theorem 6.7.7 did not use the fact that $\operatorname{trop}(Y)$ is balanced. This proof trades the delicate commutative algebra of Chapter 3 for a simpler approach, but requires knowledge of intersection theory on toric varieties.

An important point about tropical geometry and toric geometry is that their intersection theories are in harmony. For instance, the tropical concept of *stable intersection*, which was previewed in Theorem 1.3.3 and further developed in Section 3.6, can be interpreted as a toric computation. This is the main theorem of [**FS97**], restated in the language of this book.

Theorem 6.7.9. *Let $Z \in A^r(X_\Sigma)$ and $Z' \in A^s(X_\Sigma)$ be cycles on a smooth complete toric variety X_Σ with $r + s \leq n$. Let $(\Delta, \mathbf{m})$ and $(\Delta', \mathbf{m}')$ be the weighted balanced fans associated to Z and Z', respectively, by Proposition 6.7.2. Then the weighted balanced fan associated to the intersection $Z \cdot Z'$ is the stable intersection $\Delta \cap_{st} \Delta'$ of the two fans.*

Remark 6.7.10. Theorem 6.7.9 gives rise to a combined tropical/toric proof of Bézout's Theorem as follows. Fix homogeneous polynomials $f_1, \dots, f_n$ in $\mathbb{C}[x_0, \dots, x_n]$ of degree $d_1, \dots, d_n$ for which $V(f_1, \dots, f_n)$ is zero dimensional.

After a general change of coordinates, each $\operatorname{trop}(V(f_i))$ is the $(n-1)$-skeleton of the fan of $\mathbb{P}^n$ (see Exercise 6.8(11)), and the multiplicity on each maximal cone in $\operatorname{trop}(V(f_i))$ is d_i. The stable intersection $\operatorname{trop}(V(f_1)) \cap_{st} \cdots \cap_{st} \operatorname{trop}(V(f_n))$ is the origin $\mathbf{0}$ with multiplicity $d_1 \cdots d_n$, so the classical intersection $V(f_1, \ldots, f_n)$ consists of $d_1 \cdots d_n$ points, counted with multiplicity.

The assumption that X_Σ is complete in Theorem 6.7.1 can be relaxed as follows. If X_Σ is complete, then the rth graded piece $A^r(X_\Sigma)$ of the cohomology ring $A^*(X_\Sigma)$ is isomorphic to $\operatorname{Hom}(A_{n-r}(X_\Sigma), \mathbb{Z})$. Now assume that Σ is a simplicial toric variety, but not necessarily complete. We *define* $A^r(X_\Sigma) = A^r(X_\Sigma, \mathbb{Q})$ to be $\operatorname{Hom}(A_{n-r}(X_\Sigma), \mathbb{Q})$. The direct sum $A^*(X_\Sigma) = \bigoplus_{r \geq 0} A^r(X_\Sigma)$ then has a ring structure as follows. Write $\operatorname{mult}(\sigma)$ for the lattice index of the lattice generated by the rays of σ in $N_\sigma = N \cap \operatorname{span}(\sigma)$. For cones $\sigma, \tau \in \Sigma$ with $\sigma \cap \tau = \{\mathbf{0}\}$ and $\sigma + \tau \in \Sigma$, set

$$m_{\sigma\tau} \quad = \quad \frac{\operatorname{mult}(\sigma)\operatorname{mult}(\tau)}{\operatorname{mult}(\sigma+\tau)}.$$

We then have a multiplication defined as follows. If $\sigma \cap \tau = \{\mathbf{0}\}$, then

$$V(\sigma) \cdot V(\tau) \quad = \quad \begin{cases} m_{\sigma\tau} V(\sigma+\tau) & \text{if } \sigma + \tau \in \Sigma, \\ 0 & \text{otherwise.} \end{cases}$$

If $\sigma \cap \tau \neq \{\mathbf{0}\}$, then we use the relations (6.7.1) to rewrite $V(\sigma)$ as a linear combination of $V(\sigma')$ with none of the σ' nontrivially intersecting τ.

The group $A^1(X_\Sigma) = \operatorname{Hom}(A_{n-1}(X_\Sigma), \mathbb{Q})$ is always generated by the torus invariant divisors D_i. If X_Σ is smooth, then we recover the presentation seen in Theorem 6.7.1, namely $A^*(X_\Sigma) \cong \mathbb{Q}[D_1, \ldots, D_s]/(\operatorname{SR}(\Sigma) + L_\Sigma)$.

Example 6.7.11. Let Σ be the standard tropical line in $\mathbb{R}^2$, so X_Σ is $\mathbb{P}^2$ with the three torus-invariant points removed. The Stanley–Reisner ideal of Σ is $\langle D_1D_2, D_1D_3, D_2D_3\rangle$, and L_Σ equals $\langle D_1 - D_3, D_2 - D_3\rangle$. The Chow ring $A^*(X_\Sigma, \mathbb{Q})$ of the toric surface X_Σ is isomorphic to

$$\mathbb{Q}[D_1, D_2, D_3]/\langle D_1D_2, D_1D_3, D_2D_3, D_1 - D_3, D_2 - D_3\rangle \;\cong\; \mathbb{Q}[t]/\langle t^2\rangle. \qquad \diamond$$

Consider a subvariety $Y \subset T^n$ and a tropical compactification $\overline{Y}$, obtained by an embedding $i : \overline{Y} \to X_\Sigma$ into a toric variety X_Σ. This induces a ring homomorphism $i^* : A^*(X_\Sigma) \to A^*(\overline{Y})$, with $A^*(X_\Sigma)$ defined as above.

The homomorphism i^* is generally not surjective. Consider again a d-dimensional projective variety $\overline{Y} \subset \mathbb{P}^m$ that has Picard rank at least 2. Choose generic coordinates on $\mathbb{P}^m$ so that, for $Y = \overline{Y} \cap T^m$, the tropical variety $\operatorname{trop}(Y)$ equals the d-skeleton of the fan of $\mathbb{P}^m$ (see Exercise 6.8(11) for details on this construction). The embedding $\overline{Y}$ is then a tropical compactification of Y, but the induced map $i^* : A^1(\mathbb{P}^m) \to A^1(\overline{Y})$ is not surjective.

The homomorphism i^* need not be injective either. Suppose that $\overline{Y} \subset X_\Sigma$ with $\dim(\overline{Y}) = d$, and consider the cycle $[\overline{Y}] \in A^{n-d}(X_\Sigma)$. For $Z \in A^d(X_\Sigma)$, we have $i^*(Z) = 0$ if $Z \cdot [\overline{Y}] = 0$. By Theorem 6.7.7, the intersection number $Z \cdot [\overline{Y}] \in A^n(X_\Sigma) = \mathrm{Hom}(A_0(X_\Sigma), \mathbb{Q}) \cong \mathbb{Q}$ is determined by the multiplicities on trop(Y). If $Z = \sum_{\sigma \in \Sigma(d)} a_\sigma [V(\sigma)]$, then we have $Z \cdot [\overline{Y}] = \sum_{\sigma \in \Sigma(d)} a_\sigma m(\sigma)$. Thus $\sum_{\sigma \in \Sigma(d)} a_\sigma [V(\sigma)]$ lies in $\ker(i^*)$ whenever $\sum_{\sigma \in \Sigma(d)} a_\sigma m(\sigma) = 0$. Nonzero solutions (a_σ) to this equation often exist.

Example 6.7.12. We continue Example 6.7.11. Fix a general line $\overline{Y}$ in $\mathbb{P}^2$ and set $Y = \overline{Y} \cap T^2$. Then trop$(Y) \subset \mathbb{R}^2$ is the standard tropical line Σ. The cohomology of X_Σ is isomorphic to $\mathbb{Q}[t]/\langle t^2 \rangle$. The map $i^* : A^*(X_\Sigma) \to A^*(\overline{Y})$ sends D_i to the class of the intersection of $\overline{Y}$ with the corresponding line in $\mathbb{P}^2$. This intersection is a point, and its class is nonzero. Since $A^*(\mathbb{P}^1) \cong \mathbb{Q}[t]/\langle t^2 \rangle$, the map $i^* : A^*(X_\Sigma) \to A^*(\overline{Y})$ is an isomorphism. ◊

Example 6.7.13. Let $C = V(x^2 + y^2 + x^3y + xy^3 + x^3y^3)$ in T^2. The tropical curve trop$(C) \subset \mathbb{R}^2$ is the fan Σ of Figure 6.7.1. The multiplicities are one on the rays D_2 and D_5, and two on the others. Let $\overline{Y}$ be the tropical compactification of Y using the toric variety X_Σ. Then $A^1(\overline{Y}, \mathbb{Q}) \cong \mathbb{Q}$, since $\overline{Y}$ is a curve, but $A^1(X_\Sigma, \mathbb{Q}) \cong \mathbb{Q}^3$. Hence the induced map $i^*\colon A^*(X_\Sigma) \to A^*(\overline{Y})$ cannot be injective. Indeed, using the notation of Example 6.7.6, we have $D_2 - D_5 \in \ker(i^*)$, but $D_2 - D_5 \neq 0 \in A^1(X_\Sigma)$. This is because $D_2 \cdot [\overline{Y}] = 1 = D_5 \cdot [\overline{Y}]$, as the multiplicities on these two rays are one. ◊

We now present an important case where the map i^* is an isomorphism. Let $\mathcal{A}$ be an arrangement of $n+1$ hyperplanes in $\mathbb{P}^d$ that do not all pass through one point. Then $Y = \mathbb{P}^d \backslash \cup \mathcal{A}$ is a very affine variety in T^n, as seen in (4.1.1). Fix any building set $\mathcal{G}$ as in Exercise 4.7(10). The choice of $\mathcal{G}$ defines a simplicial fan $\Sigma_\mathcal{G}$ whose support is the tropicalized linear space trop(Y). For instance, the unique minimal building set $\mathcal{G}$ consists of the irreducible flats, and this often (but not always) corresponds to the coarsest fan on trop(Y). If we take $\mathcal{G}$ to be the set of all flats, then $\Sigma_\mathcal{G}$ is the order complex of the geometric lattice of the matroid $M(\mathcal{A})$, as in Theorem 4.1.11.

The rays of the fan $\Sigma_\mathcal{G}$ are the incidence vectors $\mathbf{e}_\sigma = \sum_{i \in \sigma} \mathbf{e}_i$ of the flats σ in $\mathcal{G}$. These vectors live in $\mathbb{R}^{n+1}/\mathbb{R}\mathbf{1}$. The Chow ring $A^*(X_{\Sigma_\mathcal{G}})$ is the quotient of $\mathbb{Z}[x_\sigma : \sigma \in \mathcal{G}]$ modulo the ideal $\mathrm{SR}(\Sigma_\mathcal{G}) + L_{\Sigma_\mathcal{G}}$. The Stanley–Reisner ideal $\mathrm{SR}(\Sigma_\mathcal{G})$ is generated by squarefree monomials that represent nonnested subsets of $\mathcal{G}$, and the linear ideal $L_{\Sigma_\mathcal{G}}$ is generated by the relations

$$\sum_{\substack{\sigma \in \mathcal{G} \\ \sigma \ni i}} x_\sigma = \sum_{\substack{\tau \in \mathcal{G} \\ \tau \ni j}} x_\tau \qquad \text{for } 1 \leq i < j \leq n+1. \tag{6.7.4}$$

Let $\overline{Y}$ denote the closure of Y in the toric variety $X_{\Sigma_\mathcal{G}}$. Since $|\Sigma_\mathcal{G}| =$ trop(Y), this is a tropical compactification of Y. The compactification $\overline{Y}$

predates the development of tropical geometry. First constructed by De Concini and Procesi [**DCP95**], the complete variety $\overline{Y}$ is known as the *wonderful compactification* of the arrangement complement $Y = \mathbb{P}^d \backslash \cup \mathcal{A}$. Feichtner and Yuzvinsky [**FY04**] showed that the cohomology of $\overline{Y}$ agrees with that of $X_{\Sigma_\mathcal{G}}$. Since both varieties are smooth, the cohomology ring is the same object as the Chow ring, and we conclude the following result.

Theorem 6.7.14. *The map i^* is an isomorphism for the wonderful compactification $\overline{Y}$ of the hyperplane arrangement complement $Y = \mathbb{P}^d \backslash \cup \mathcal{A}$ defined by a building set $\mathcal{G}$. In symbols, we have*

$$A^*(\overline{Y}) \simeq A^*(X_{\Sigma_\mathcal{G}}) = \mathbb{Z}[x_\sigma : \sigma \in \mathcal{G}]/(\mathrm{SR}(\Sigma_\mathcal{G}) + L_{\Sigma_\mathcal{G}}). \tag{6.7.5}$$

We wish to stress that the Chow ring in (6.7.5) is not an invariant of the very affine variety Y. It depends on the choice of tropical compactification $\overline{Y}$. In the present context of tropical linear spaces, it depends on our choice of the building set $\mathcal{G}$ for $\mathcal{A}$. Here is a simple example to illustrate this.

Example 6.7.15. Fix $d = 2$, $n = 3$, and let $\mathcal{A}$ consist of four general lines in $\mathbb{P}^2$. Hence $\mathrm{trop}(Y)$ is a plane in $\mathbb{R}^3$, or, combinatorially, the cone over the complete graph K_4. The smallest building set $\mathcal{G}$ consists of just the four lines. Here $A^*(\overline{Y})$ is isomorphic to $\mathbb{Z}[t]/\langle t^3 \rangle$, given by the presentation

$$\mathbb{Z}[x_1, x_2, x_3, x_4]/\langle x_1x_2x_3, x_1x_2x_4, x_1x_3x_4, x_2x_3x_4, x_1 - x_2, x_2 - x_3, x_3 - x_4\rangle.$$

On the other hand, if $\mathcal{G}$ consists of all ten proper flats, then $A^*(\overline{Y})$ is the quotient of $\mathbb{Z}[x_1, x_2, x_3, x_4, x_{12}, x_{13}, x_{14}, x_{23}, x_{24}, x_{34}]$ modulo the monomials

$x_1x_2, x_1x_3, x_1x_4, x_2x_3, x_2x_4, x_3x_4, x_1x_{23}, x_1x_{24}, x_1x_{34}, x_2x_{13}, x_2x_{14}, x_2x_{34}$,
$x_3x_{12}, x_3x_{14}, x_3x_{24}, x_4x_{12}, x_4x_{13}, x_4x_{23}, x_{12}x_{13}, x_{12}x_{14}, x_{12}x_{23}, x_{12}x_{24}, x_{12}x_{34}$,
$x_{13}x_{14}, x_{13}x_{23}, x_{13}x_{24}, x_{13}x_{34}, x_{14}x_{23}, x_{14}x_{24}, x_{14}x_{34}, x_{23}x_{24}, x_{23}x_{34}, x_{24}x_{34}$,

and the linear relations $x_1 + x_{12} + x_{13} + x_{14} = x_2 + x_{12} + x_{23} + x_{24} = x_3 + x_{13} + x_{23} + x_{34} = x_4 + x_{14} + x_{24} + x_{34}$. Here, we have $A^1(\overline{Y}) \simeq \mathbb{Z}^7$. ◊

A generalization of this computation to generic arrangements in arbitrary dimensions appears in [**FY04**, p. 526]. The following example, discussed under the header *partition lattices* in [**FY04**, §7], is of great interest to algebraic geometers.

Example 6.7.16. Consider the embedding $i : \overline{M}_{0,n} \to X_\Delta$ of the moduli space $\overline{M}_{0,n}$ into the toric variety defined by the space of phylogenetic trees. This was discussed in Theorem 6.4.12, where $\overline{M}_{0,n}$ was realized as a tropical compactification of $M_{0,n} = \mathbb{P}^{n-3} \backslash \mathcal{A}$ for a particular arrangement $\mathcal{A}$ of $\binom{n-1}{2}$ hyperplanes. By Theorem 6.7.14, the pull-back morphism $i^* \colon A^*(X_\Delta) \to A^*(\overline{M}_{0,n})$ is an isomorphism, and we obtain a combinatorial recipe.

The Chow ring of $\overline{M}_{0,n}$ was computed by Keel in [**Kee92**] using his description of $\overline{M}_{0,n}$ as a blow-up of $(\mathbb{P}^1)^{n-3}$. The boundary divisors of $\overline{M}_{0,n}$

are indexed by partitions $\{1,\ldots,n\} = I \cup I^c$ with $|I|, |I^c| \geq 2$. We write δ_I for the boundary divisor indexed by (I, I^c), and identify δ_I and δ_{I^c}. Then

$$A^*(\overline{M}_{0,n}) = \mathbb{Z}\big[\,\delta_I : I \subset \{1,\ldots,n\} \text{ with } |I|, |I^c| \geq 2\big]/J_n, \tag{6.7.6}$$

where J_n is the ideal corresponding to the relations

$$\sum_{i,j\in I,k,l\notin I} \delta_I = \sum_{i,k\in I,j,l\notin I} \delta_I = \sum_{i,l\in I,j,k\notin I} \delta_I, \tag{6.7.7}$$

$$\text{and} \qquad \delta_I\delta_J = 0 \qquad \text{unless } I \subset J, J \subset I, \text{ or } I \cap J = \emptyset. \tag{6.7.8}$$

For example, for $n = 4$ we have $\overline{M}_{0,4} \cong \mathbb{P}^1$. The rules above give $A^*(\overline{M}_{0,4}) = \mathbb{Z}[\delta_{12}, \delta_{13}, \delta_{23}]/\langle \delta_{12} - \delta_{13}, \delta_{12} - \delta_{23}, \delta_{12}\delta_{13}, \delta_{12}\delta_{23}, \delta_{13}\delta_{23}\rangle \simeq \mathbb{Z}[t]/t^2$. For general n, the relations on $A^*(\overline{M}_{0,n})$ of the form (6.7.7) come from pulling back the relations on $A^*(\overline{M}_{0,4})$ under the forgetful morphism π_{ijkl} that forgets all marked points except for those labeled i, j, k, l, and then stabilizing. See [**KV07**, §1.3] for details. The relations (6.7.8) express the fact that the corresponding boundary divisors do not intersect. Keel [**Kee92**] shows directly that there are no other relations other than these natural ones.

In our approach, we consider the simplicial fan structure Δ on the space of phylogenetic trees with n leaves, as in Proposition 4.3.10. Its rays correspond to the boundary divisors δ_I. The ideal $\langle \delta_I\delta_J : I \cap J \neq \emptyset, I \not\subseteq J, J \not\subseteq I\rangle$ in (6.7.8) is the Stanley–Reisner ideal of the simplicial complex given by Δ.

We derive the presentation (6.7.6) directly from Theorem 6.7.14, using our realization (in Theorem 6.4.12) of $\overline{M}_{0,n}$ as the complement of $\binom{n-1}{2}$ hyperplanes in $\mathbb{P}^{n-3}$. The building set $\mathcal{G}$ consists of $2^{n-1} - n - 1$ flats in that arrangement, one for each boundary divisor δ_I. These flats are $\{x_{i-1} = x_{j-1} : i, j \in I\}$, where $I \subset \{1,\ldots,n-1\}$, $n \in I^c$ and $x_{n-2} = 0$. ◊

Remark 6.7.17. Another connection between tropical varieties and cohomology comes from the consideration of the Hodge structure on the cohomology of Y. In [**Hac08**] Hacking proves the following result: if $\overline{Y}$ is a smooth projective variety compactifying a d-dimensional variety $Y \subset T^n$ for which the boundary $\overline{Y}\backslash Y$ has simple normal crossings, then the reduced ith homology of the boundary Delta-complex $\Delta(\partial\overline{Y})$ from Section 6.5 equals the top graded piece of the weight filtration on the cohomology of Y:

$$\tilde{H}_i(\Delta(\partial(\overline{Y})), \mathbb{C}) = \mathrm{Gr}^W_{2d} H^{2d-(i+1)}(Y, \mathbb{C}).$$

This implies $\tilde{H}_i(\Delta(\partial(\overline{Y}))) = 0$ for $i \neq d-1$. By Theorem 6.5.11, the cone over $\Delta(\partial(\overline{Y}))$ maps surjectively onto the tropical variety $\mathrm{trop}(Y)$. If this map is injective (for which a sufficient, but not necessary, condition is that all multiplicities on $\mathrm{trop}(Y)$ equal one, or that $\overline{Y}$ is a *schön* compactification of Y in the sense of Definition 6.4.19), then this shows that the link of $\mathrm{trop}(Y)$ at its lineality space has only top-dimensional homology.

Helm and Katz [**HK12a**] studied a generalization to fields with nontrivial valuation. This was further developed by Payne [**Pay13**] whose results imply that the homotopy type of $\Delta(\partial(\overline{Y}))$ is independent of choices.

The focus of this section was understanding the intersection theory of subvarieties $\overline{Y}$ of a toric variety X_Σ via the tropicalization of $Y = \overline{Y} \cap T$. An intersection theory has also been developed [**AR10**] for all tropical cycles (balanced weighted polyhedral complexes), regardless of whether or not they are the tropicalization of some subvariety of the torus. This includes tropicalized linear spaces that are not tropical linear spaces, as in Chapter 4. This situation was studied in [**FR13, Sha13**]. An extensive exposition of this intersection theory can be found in the book by Mikhalkin and Rau [**MR**].

6.8. Exercises

(1) (For toric novices) Check that $A_{n-1}(\mathbb{P}^n) \cong \mathbb{Z}$ and $A_2((\mathbb{P}^1)^3) \cong \mathbb{Z}^3$.

(2) Show that $\operatorname{trop}(\mathbb{P}^1)$ is homeomorphic to the closed interval $[0, 1]$ in the usual topology on $\mathbb{R}$. Show explicitly that the two definitions given for tropical $\mathbb{P}^1$ are homeomorphic.

(3) Verify the claim of Definition 6.2.1 that every $\phi \in U_\sigma^{\text{trop}}$ satisfies $\{\mathbf{u} : \phi(\mathbf{u}) \neq \infty\} = (\sigma^\vee \cap \tau^\perp) \cap M$ for some face τ of σ.

(4) Verify the claim of Remark 6.2.3 that the topology on U_σ^{trop} is the induced topology coming from regarding U_σ^{trop} as a subset of $(\overline{\mathbb{R}})^m$ for *any* choice of m generators for the semigroup $\sigma^\vee \cap M$.

(5) Let X_Σ be the toric surface obtained by blowing up $\mathbb{P}^2$ at the three coordinate points (1:0:0), (0:1:0), and (0:0:1). Draw a picture of some tropical curves on $\operatorname{trop}(X_\Sigma)$.

(6) The 3×3-determinant is a polynomial in nine variables with six terms. Its tropical hypersurface in $\mathbb{R}^9$ was described in Example 3.1.11. Using Definition 6.2.10, compute its extended tropical hypersurface in $\overline{\mathbb{R}}^9$. What is the f-vector of this polyhedral complex?

(7) Let $Y = V(x + 3y + 7x^2y - 8xy^2 - x^2y^2) \subset T^2$. Do the following for each of the following toric varieties X_Σ listed with bullets below:
   (a) Compute the closure $\overline{Y}$ of Y in X_Σ.
   (b) For each torus orbit $\mathcal{O}_\sigma$ of X_Σ compute $\overline{Y} \cap \mathcal{O}_\sigma$.
   (c) Compare your answer with that predicted by Theorem 6.3.4.
   - $X_\Sigma = \mathbb{P}^2$, with $i : T^2 \to \mathbb{A}^2$ given by $i(x, y) = (x, y)$;
   - $X_\Sigma = \mathbb{P}^1 \times \mathbb{P}^1$ with $i : T^2 \to \mathbb{P}^1 \times \mathbb{P}^1$ given by

$$i(x, y) = (x : 1) \times (y : 1);$$

- X_Σ is the toric surface with five rays. The first lattice points on these are $\{(1,1),(-1,1),(-1,0),(0,-1),(1,-1)\}$. The five maximal cones in Σ are generated by adjacent rays.

(8) Let M be a matroid on $n+1$ elements, and let Σ be the Bergman fan structure on the tropical linear space $\mathrm{trop}(M)$, as described in Theorem 4.2.6. Show that the toric variety X_Σ is smooth.

(9) Compute the tropicalization of the curve Y of part (4) of Example 6.4.3 in both its embedding into $(\mathbb{C}^*)^2$ and $(\mathbb{C}^*)^3$. Verify Proposition 6.4.4 for this example.

(10) Give an *explicit* example of a subvariety $Y \subseteq T^n$ for which the closure $\overline{Y}$ of Y in a toric variety X_Σ with $|\Sigma| = \mathrm{trop}(Y)$ is not a flat tropical compactification. *Hint*: Example 6.4.16.

(11) Let $\overline{Y}$ be a d-dimensional subvariety of $\mathbb{P}^n$.
   (a) Show that there is a Zariski open set $U \subset \mathrm{PGL}(n+1)$ for which the change of coordinates coming from $g \in U$ has the property that $g\overline{Y}$ does not intersect any coordinate subspace of dimension less than $n-d$.
   (b) Conclude that, for a generic choice of coordinates on $\mathbb{P}^n$, the tropical variety $\mathrm{trop}(\overline{Y} \cap T^n)$ equals the d-skeleton of the fan of $\mathbb{P}^n$. This is Theorem 1.1 of [**RS12**].

(12) Suppose that a group G acts on a very affine variety Y.
   (a) Show that the action of G extends to the intrinsic torus T_{in} of Y so that the embedding $i\colon Y \to T_{\mathrm{in}}$ is G-equivariant.
   (b) Show that an automorphism of the algebraic torus T^n induces an automorphism of $\mathbb{R}^n$ via tropicalization.
   (c) Deduce that G acts on the tropicalization $\mathrm{trop}(Y)$ of Y embedded into its intrinsic torus. Give examples to show that this action need not be faithful even if the original action is.
   (d) Did we need the assumption on the intrinsic torus here?

(13) Fix n points in $\mathbb{P}^2$ with no three on a line and such that, for any six points and any partition of these into three pairs, the three lines through these pairs do not share a common intersection point. Let $\mathcal{A}$ be the line arrangement in $\mathbb{P}^2$ consisting of all $\binom{n}{2}$ lines joining pairs of points, and let $Y = \mathbb{P}^2 \setminus \mathcal{A}$ be the complement.
   (a) Describe the embedding of Y into its intrinsic torus $(K^*)^{\binom{n}{2}-1}$.
   (b) Describe $\mathrm{trop}(Y) \subseteq \mathbb{R}^{\binom{n}{2}-1}$. Show that there is a unique coarsest fan Σ with $|\Sigma| = \mathrm{trop}(Y)$.
   (c) Show that the tropical compactification $\overline{Y}$ of Y using this coarsest fan Σ is the blow-up of $\mathbb{P}^2$ at the original n points. For $n \leq 8$, this is a del Pezzo surface of degree $9-n$.

(14) Fix an arrangement $\mathcal{A}$ of five planes in $\mathbb{P}^3$. Describe three different building sets $\mathcal{G}$, and determine the corresponding tropical compactifications of $Y = \mathbb{P}^3 \backslash \cup \mathcal{A}$.

(15) Compute the Chow ring $A^*(\overline{Y})$ for each of the three threefolds $\overline{Y}$ in the previous exercise. *Hint*: Theorem 6.7.14 and Example 6.7.15.

(16) Let D_1, D_2, D_3 be three lines in $\mathbb{P}^2$ that do not intersect in a common point. Let D_4 be a conic in $\mathbb{P}^2$ that intersects the lines D_1, D_2, D_3 transversely at six distinct points. Choose concrete equations and determine the embedding of $Y = \mathbb{P}^2 \backslash \bigcup_{i=1}^4 D_i$ into $(K^*)^3$. Compute the prime ideal I_Y and the tropical surface $\operatorname{trop}(Y)$.

(17) Let Y be a cubic surface in T^3. Describe the set of all divisorial valuations on the function field $K(Y)$. Explain and verify the statement in Proposition 6.5.4 for this example.

(18) Let $D_1, D_2, \ldots, D_{27}$ be the 27 lines on a smooth cubic surface $\overline{Y}$ in $\mathbb{P}^3$. Set $\partial \overline{Y} = \bigcup_{i=1}^{27} D_i$, and show that $Y = \overline{Y} \backslash \partial \overline{Y}$ is a very affine variety. Determine the corresponding boundary complex $\Delta(\partial \overline{Y})$.

(19) Some smooth cubic surfaces have *Eckhart points*. What are these points, and what do they mean for the previous exercise?

(20) Show that the formula given in (6.6.1) of Definition 6.6.1 of the affine toric scheme $K[M]^\sigma$ did not need the restriction $\mathbf{u} \in \sigma_0^\vee$ in the summation. In other words, show that if σ is a Γ_{val}-admissible cone in $N_\mathbb{R} \times \mathbb{R}_{\geq 0}$ and $\lambda \mathbf{c} + \mathbf{w} \cdot \mathbf{u} \geq 0$ for all $(\mathbf{w}, \lambda) \in \sigma$, then $\mathbf{u} \in \sigma_0^\vee$.

(21) Let $P \subset \mathbb{R}^2$ be the polyhedron $\{(x, y) \in \mathbb{R}^2 : x \geq 0, y \geq 0, x + y \geq 1\}$, and let $\sigma = C(P) \subset \mathbb{R}^2 \times \mathbb{R}_{\geq 0}$ be the cone over P. Describe explicitly the affine toric scheme $\mathcal{U}_\sigma$. What is the general fiber? What is the special fiber?

(22) Let X_Σ be the toric surface given by the fan Σ in Example 6.7.6. Consider the two curves in T^2 given in parts (1) and (4) of Example 6.4.3, and let Z and Z' be their closures in X_Σ.
  (a) Write the equations of Z and Z' in Cox homogeneous coordinates on X_Σ.
  (b) Compute the intersection of Z and Z' with each torus invariant boundary stratum. Thus compute the associated weighted balanced fan for Z and Z'.
  (c) Write $[Z]$ and $[Z']$ as linear combinations of the (classes of) toric boundary divisors on X_Σ.
  (d) Compute the product $[Z] \cdot [Z']$ in the Chow ring of X_Σ.

(23) Determine the Chow ring of the smooth toric surface $X_{\Sigma'}$ in Example 6.7.6. Find a Gröbner basis for the defining ideal of $A^*(X_{\Sigma'})$. Use this to recompute the intersection numbers in (6.7.3).

(24) A pure weighted balanced fan $(\Sigma, \mathbf{m})$ of dimension d is *tropically reducible* if there is a refinement Σ' of Σ and two subfans Σ_1, Σ_2 of Σ and weightings $\mathbf{m}_i : \Sigma_i(d) \to \mathbb{Z}_{>0}$ for $i = 1, 2$ that make Σ_1 and Σ_2 into balanced fans, with the property that for all $\sigma \in \Sigma'$ we have $\mathbf{m}(\sigma) = \mathbf{m}_1(\sigma) + \mathbf{m}_2(\sigma)$, where we set $\mathbf{m}_i(\sigma) = 0$ if $\sigma \notin \Sigma_i$, and for each $i = 1, 2$ there is $\sigma_i \in \Sigma$ with $\mathbf{m}_i(\sigma_i) < \mathbf{m}(\sigma_i)$.

(a) Show that the weighted balanced fan in Figure 3.4.1 is tropically reducible.

(b) What does it mean for a one-dimensional fan in $\mathbb{R}^2$ to be tropically irreducible?

(c) What does it mean for the Chow group $A^d(X_\Sigma)$ if a d-dimensional weighted balanced fan $(\Sigma, \mathbf{m})$ is tropically reducible?

(25) Let Σ be a fan in $\mathbb{R}^n$ with $|\Sigma| = \operatorname{trop}(M)$ for a rank $d+1$ matroid M on a ground set of size $n+1$. Let $\mathbf{m} : \Sigma(d) \to \mathbb{N}$ given by $\mathbf{m}(\sigma) = 1$ for all $\sigma \in \Sigma(d)$. Is $(\Sigma, \mathbf{m})$ always tropically irreducible?

(26) Given an example of a curve Y in T^2 such that the intrinsic torus T_{in} of Y is isomorphic to T^{13}. Determine the defining equations of your very affine curve Y in its intrinsic embedding into $T_{\text{in}} = T^{13}$.

Bibliography

[ABBR13] Omini Amini, Matthew Baker, Erwan Brugallé, and Joseph Rabinoff, *Lifting harmonic morphisms I: metrized complexes and Berkovich skeleta*, arXiv:1303.4812, 2013. 334

[AD09] Federico Ardila and Mike Develin, *Tropical hyperplane arrangements and oriented matroids*, Math. Z. **262** (2009), no. 4, 795–816. 239, 266

[AK06] Federico Ardila and Caroline J. Klivans, *The Bergman complex of a matroid and phylogenetic trees*, J. Combin. Theory Ser. B **96** (2006), no. 1, 38–49. 167

[AKW06] Federico Ardila, Caroline Klivans, and Lauren Williams, *The positive Bergman complex of an oriented matroid*, European J. Combin. **27** (2006), no. 4, 577–591. 275

[AR10] Lars Allermann and Johannes Rau, *First steps in tropical intersection theory*, Math. Z. **264** (2010), no. 3, 633–670. 145, 346

[AZ04] Martin Aigner and Günter M. Ziegler, *Proofs from The Book*, third ed., Springer-Verlag, Berlin, 2004. 247

[Bak08a] Matthew Baker, *An introduction to Berkovich analytic spaces and non-Archimedean potential theory on curves*, *p*-adic geometry, Univ. Lecture Ser., vol. 45, Amer. Math. Soc., Providence, RI, 2008, pp. 123–174. 311

[Bak08b] ———, *Specialization of linear systems from curves to graphs*, Algebra Number Theory **2** (2008), no. 6, 613–653. 334

[BCOQ92] François Louis Baccelli, Guy Cohen, Geert Jan Olsder, and Jean-Pierre Quadrat, *Synchronization and linearity: An algebra for discrete event systems*, John Wiley & Sons Ltd., Chichester, 1992. ix, 4, 228

[Ber71] George M. Bergman, *The logarithmic limit-set of an algebraic variety*, Trans. Amer. Math. Soc. **157** (1971), 459–469. 17, 20, 21, 28

[Ber75] D. N. Bernstein, *The number of roots of a system of equations*, Funkcional. Anal. i Priložen. **9** (1975), no. 3, 1–4. 210

[BG84] Robert Bieri and J. R. J. Groves, *The geometry of the set of characters induced by valuations*, J. Reine Angew. Math. **347** (1984), 168–195. 25, 28

[BGS11] José Ignacio Burgos Gil and Martín Sombra, *When do the recession cones of a polyhedral complex form a fan?*, Discrete Comput. Geom. **46** (2011), no. 4, 789–798. 132, 329

[BHV01] Louis Billera, Susan Holmes, and Karen Vogtmann, *Geometry of the space of phylogenetic trees*, Adv. in Appl. Math. **27** (2001), no. 4, 733–767. 178

[BJS$^+$07] Tristram Bogart, Anders N. Jensen, David Speyer, Bernd Sturmfels, and Rekha R. Thomas, *Computing tropical varieties*, J. Symbolic Comput. **42** (2007), no. 1-2, 54–73. 129

[BJWW98] Alexander Barvinok, David S. Johnson, Gerhard Woeginger, and Russell Woodroofe, *The maximum traveling salesman problem under polyhedral norms*, Integer programming and combinatorial optimization, Lecture Notes in Comput. Sci., vol. 1412, Springer, Berlin, 1998, pp. 195–201. 244

[BN07] Matthew Baker and Serguei Norine, *Riemann-Roch and Abel-Jacobi theory on a finite graph*, Adv. Math. **215** (2007), no. 2, 766–788. 334

[BNS87] Robert Bieri, Walter D. Neumann, and Ralph Strebel, *A geometric invariant of discrete groups*, Invent. Math. **90** (1987), no. 3, 451–477. 25, 26

[Bos] Siegfried Bosch, *Lectures on formal and rigid geometry*, Available at `www.math.uni-muenster.de/sfb/about/publ/heft378.pdf`. 50

[BPR06] Saugata Basu, Richard Pollack, and Marie-Françoise Roy, *Algorithms in real algebraic geometry*, second ed., Algorithms and Computation in Mathematics, vol. 10, Springer-Verlag, Berlin, 2006. 230

[BPR11] Matthew Baker, Sam Payne, and Joseph Rabinoff, *Nonarchimedean geometry, tropicalization, and metrics on curves*, arXiv:1104.0320, 2011. 334

[BS80] Robert Bieri and Ralph Strebel, *Valuations and finitely presented metabelian groups*, Proc. London Math. Soc. (3) **41** (1980), no. 3, 439–464. 25, 26, 27, 28, 29, 30

[BS89] Jürgen Bokowski and Bernd Sturmfels, *Computational Synthetic Geometry*, Lecture Notes in Mathematics, vol. 1355, Springer, Heidelberg, 1989. 183

[But10] Peter Butkovič, *Max-linear Systems: Theory and Algorithms*, Springer Monographs in Mathematics, Springer-Verlag, London, 2010. ix, 4, 228

[CDPR12] Filip Cools, Jan Draisma, Sam Payne, and Elina Robeva, *A tropical proof of the Brill-Noether theorem*, Adv. Math. **230** (2012), no. 2, 759–776. 334

[CGQS05] Guy Cohen, Stéphane Gaubert, Jean-Pierre Quadrat, and Ivan Singer, *Max-plus convex sets and functions*, Idempotent mathematics and mathematical physics, Contemp. Math., vol. 377, Amer. Math. Soc., Providence, RI, 2005, pp. 105–129. 228

[CJR11] Melody Chan, Anders Jensen, and Elena Rubei, *The 4×4 minors of a $5\times n$ matrix are a tropical basis*, Linear Algebra Appl. **435** (2011), no. 7, 1598–1611. 244, 255

[CLO07] David Cox, John Little, and Donal O'Shea, *Ideals, varieties, and algorithms*, third ed., Undergraduate Texts in Mathematics, Springer, New York, 2007. ix, 22, 52, 53, 55, 65, 67, 74, 90, 104, 105, 106, 118, 124, 146, 177, 209

[CLS11] David A. Cox, John B. Little, and Henry K. Schenck, *Toric varieties*, Graduate Studies in Mathematics, vol. 124, American Mathematical Society, Providence, RI, 2011. 277, 280, 283, 299, 300, 305, 324, 335

[CM13] Andrew J. Chan and Diane Maclagan, *Gröbner bases over fields with valuations*, `arXiv:1303.0729`, 2013. 67, 124

[Cox95] David A. Cox, *The homogeneous coordinate ring of a toric variety*, J. Algebraic Geom. **4** (1995), no. 1, 17–50. 284

[CP12] Dustin Cartwright and Sam Payne, *Connectivity of tropicalizations*, Math. Res. Lett. **19** (2012), no. 5, 1089–1095. 332

[Cue11] Maria Angelica Cueto, *Implicitization of surfaces via geometric tropicalization*, `arXiv:1105.0509`, 2011. 315, 318

[CV10] Lucia Caporaso and Filippo Viviani, *Torelli theorem for graphs and tropical curves*, Duke Math. J. **153** (2010), no. 1, 129–171. 334

[DCP95] Corrado De Concini and Claudio Procesi, *Wonderful models of subspace arrangements*, Selecta Math. (N.S.) **1** (1995), no. 3, 459–494. 215, 344

[DFS07] Alicia Dickenstein, Eva Maria Feichtner, and Bernd Sturmfels, *Tropical discriminants*, J. Amer. Math. Soc. **20** (2007), no. 4, 1111–1133. 270, 273

[DK86] Vladimir I. Danilov and Askold G. Khovanskiĭ, *Newton polyhedra and an algorithm for calculating Hodge-Deligne numbers*, Izv. Akad. Nauk SSSR Ser. Mat. **50** (1986), no. 5, 925–945. 309

[Dol03] Igor Dolgachev, *Lectures on invariant theory*, London Mathematical Society Lecture Note Series, vol. 296, Cambridge University Press, 2003. 280

[DRS10] Jesús De Loera, Jörg Rambau, and Francisco Santos, *Triangulations*, Springer-Verlag, 2010. 62, 239

[DS04] Mike Develin and Bernd Sturmfels, *Tropical convexity*, Doc. Math. **9** (2004), 1–27. 228, 236

[DSS05] Mike Develin, Francisco Santos, and Bernd Sturmfels, *On the rank of a tropical matrix*, Combinatorial and computational geometry, Math. Sci. Res. Inst. Publ., vol. 52, Cambridge Univ. Press, 2005, pp. 213–242. 244, 255

[DW92] Andreas Dress and Walter Wenzel, *Valuated matroids*, Adv. Math. **93** (1992), 214–250. 184

[DY07] Mike Develin and Josephine Yu, *Tropical polytopes and cellular resolutions*, Experiment. Math. **16** (2007), no. 3, 277–291. 230

[EGAIII] Alexandre Grothendieck, *Éléments de géométrie algébrique. III. Étude cohomologique des faisceaux cohérents. I*, Inst. Hautes Études Sci. Publ. Math. (1961), no. 11, 167. 333

[EGAIV] ______, *Éléments de géométrie algébrique. IV. Étude locale des schémas et des morphismes de schémas. III*, Inst. Hautes Études Sci. Publ. Math. (1966), no. 28, 255. 331

[Eis95] David Eisenbud, *Commutative algebra: With a view toward algebraic geometry*, Graduate Texts in Mathematics, vol. 150, Springer-Verlag, New York, 1995, 49, 53, 65, 67, 71, 72, 104, 118, 293, 309, 323

[EKL06] Manfred Einsiedler, Mikhail Kapranov, and Douglas Lind, *Non-Archimedean amoebas and tropical varieties*, J. Reine Angew. Math. **601** (2006), 139–157. 95

[EP05] Antonio J. Engler and Alexander Prestel, *Valued fields*, Springer Monographs in Mathematics, Springer-Verlag, Berlin, 2005. 51

[Ewa96] Günter Ewald, *Combinatorial convexity and algebraic geometry*, Graduate Texts in Mathematics, vol. 168, Springer-Verlag, New York, 1996. 300

[Fei05] Eva Maria Feichtner, *De Concini-Procesi wonderful arrangement models: a discrete geometer's point of view*, Combinatorial and computational geometry, MSRI Publ., vol. 52, Cambridge Univ. Press, 2005, pp. 333–360. 215

[Fin13] Alex Fink, *Tropical cycles and Chow polytopes*, Beitr. Algebra Geom. **54** (2013), no. 1, 13–40. 191

[FR13] Georges François and Johannes Rau, *The diagonal of tropical matroid varieties and cycle intersections*, Collect. Math. **64** (2013), no. 2, 185–210. 346

[FS97] William Fulton and Bernd Sturmfels, *Intersection theory on toric varieties*, Topology **36** (1997), no. 2, 335–353. 145, 335, 341

[FS05] Eva Maria Feichtner and Bernd Sturmfels, *Matroid polytopes, nested sets and Bergman fans*, Port. Math. (N.S.) **62** (2005), no. 4, 437–468. 39, 167, 215, 262, 303

[Ful93] William Fulton, *Introduction to toric varieties*, Annals of Mathematics Studies, vol. 131, Princeton University Press, Princeton, NJ, 1993, The William H. Roever Lectures in Geometry. 277, 279, 283, 284, 335

[Ful98] ———, *Intersection theory*, second ed., Ergebnisse der Mathematik und ihrer Grenzgebiete. 3. Folge. A Series of Modern Surveys in Mathematics, vol. 2, Springer-Verlag, Berlin, 1998. 334, 337, 340

[FY04] Eva Maria Feichtner and Sergey Yuzvinsky, *Chow rings of toric varieties defined by atomic lattices*, Invent. Math. **155** (2004), no. 3, 515–536. 344

[GGMS87] I. M. Gel′fand, R. M. Goresky, R. D. MacPherson, and V. V. Serganova, *Combinatorial geometries, convex polyhedra, and Schubert cells*, Adv. in Math. **63** (1987), no. 3, 301–316. 167

[GJ00] Ewgenij Gawrilow and Michael Joswig, *polymake: a framework for analyzing convex polytopes*, Polytopes — Combinatorics and Computation (Gil Kalai and Günter M. Ziegler, eds.), Birkhäuser, 2000, pp. 43–74. 65

[GKZ08] Israel M. Gel′fand, Mikhail M. Kapranov, and Andrei V. Zelevinsky, *Discriminants, resultants and multidimensional determinants*, Birkhäuser, Boston, 2008. 17, 32, 62, 191, 268, 271

[GM07a] Andreas Gathmann and Hannah Markwig, *The Caporaso-Harris formula and plane relative Gromov-Witten invariants in tropical geometry*, Math. Ann. **338** (2007), no. 4, 845–868. 34

[GM07b] ———, *The numbers of tropical plane curves through points in general position*, J. Reine Angew. Math. **602** (2007), 155–177. 34

[GM08] ———, *Kontsevich's formula and the WDVV equations in tropical geometry*, Adv. Math. **217** (2008), no. 2, 537–560. 34

[GM10] Angela Gibney and Diane Maclagan, *Equations for Chow and Hilbert quotients*, Algebra Number Theory **4** (2010), no. 7, 855–885. 302

[Gro11] Mark Gross, *Tropical geometry and mirror symmetry*, CBMS Regional Conference Series in Mathematics, vol. 114, Published for the Conference Board of the Mathematical Sciences, Washington, DC, 2011. ix, 332

[GS11] Mark Gross and Bernd Siebert, *An invitation to toric degenerations*, Surveys in differential geometry. Volume XVI. Geometry of special holonomy and related topics, vol. 16, Int. Press, Somerville, MA, 2011, pp. 43–78. 332

[Gub13] Walter Gubler, *A guide to tropicalizations*, Algebraic and Combinatorial Aspects of Tropical Geometry, Contemporary Mathematics, vol. 589, Amer. Math. Soc., Providence, RI, 2013, pp. 125–189. 330, 332

[Hac08] Paul Hacking, *The homology of tropical varieties*, Collect. Math. **59** (2008), no. 3, 263–273. 257, 313, 345

[Har77] Robin Hartshorne, *Algebraic geometry*, Springer-Verlag, New York, 1977, Graduate Texts in Mathematics, No. 52. 198

[Har95] Joe Harris, *Algebraic geometry*, Graduate Texts in Mathematics, vol. 133, Springer-Verlag, New York, 1995. 53

[Hat02] Allen Hatcher, *Algebraic topology*, Cambridge University Press, 2002. 313

[HJ08] Sven Herrmann and Michael Joswig, *Splitting polytopes*, Münster J. Math. **1** (2008), 109–141. 259

[HJJS09] Sven Herrmann, Anders Jensen, Michael Joswig, and Bernd Sturmfels, *How to draw tropical planes*, Electron. J. Combin. **16** (2009), no. 2, R6. 255, 257, 258, 266, 268

[HJS14] Sven Herrmann, Michael Joswig, and David Speyer, *Dressians, tropical Grassmannians, and their rays*, Forum Mathematicum, **6** (2014). 1853–1882. 257

[HK12a] David Helm and Eric Katz, *Monodromy filtrations and the topology of tropical varieties*, Canad. J. Math. **64** (2012), no. 4, 845–868. 346

[HK12b] June Huh and Eric Katz, *Log-concavity of characteristic polynomials and the Bergman fan of matroids*, Math. Ann. **354** (2012), no. 3, 1103–1116. 169

[Hol01] Jan E. Holly, *Pictures of ultrametric spaces, the p-adic numbers, and valued fields*, Amer. Math. Monthly **108** (2001), no. 8, 721–728. 176

[HS95] Birkett Huber and Bernd Sturmfels, *A polyhedral method for solving sparse polynomial systems*, Math. Comp. **64** (1995), no. 212, 1541–1555. 210

[HS04] Mark Haiman and Bernd Sturmfels, *Multigraded Hilbert schemes*, J. Algebraic Geom. **13** (2004), no. 4, 725–769. 308

[HT09] Kerstin Hept and Thorsten Theobald, *Tropical bases by regular projections*, Proc. Amer. Math. Soc. **137** (2009), no. 7, 2233–2241. 86

[Hun80] Thomas W. Hungerford, *Algebra*, Graduate Texts in Mathematics, vol. 73, Springer-Verlag, New York, 1980, Reprint of the 1974 original. 50

[IMS07] Ilia Itenberg, Grigory Mikhalkin, and Eugenii Shustin, *Tropical algebraic geometry*, Oberwolfach Seminars, vol. 35, Birkhäuser Verlag, Basel, 2007. ix

[Jen] Anders N. Jensen, *Gfan, a software system for Gröbner fans and tropical varieties*, Available at `http://home.imf.au.dk/jensen/software/gfan/gfan.html`. 37, 40, 65, 107, 129, 308

[JK10] Michael Joswig and Katja Kulas, *Tropical and ordinary convexity combined*, Adv. Geom. **10** (2010), no. 2, 333–352. 236

[Jon14] Mattias Jonsson, *Degenerations of amoebae and Berkovich spaces*, `arXiv:1406.1430`, 2014. 20

[Jos] Michael Joswig, *Essentials of tropical combinatorics*, Book in preparation. ix, 228

[Jou83] Jean-Pierre Jouanolou, *Théorèmes de Bertini et applications*, Progress in Mathematics, vol. 42, Birkhäuser, Boston, MA, 1983. 129

[JSY07] Michael Joswig, Bernd Sturmfels, and Josephine Yu, *Affine buildings and tropical convexity*, Albanian J. Math. **1** (2007), no. 4, 187–211. 284

[JY13] Anders Jensen and Josephine Yu, *Stable intersections of tropical varieties*, `arXiv:1309.7064`, 2013. 145

[Kaj08] Takeshi Kajiwara, *Tropical toric geometry*, Toric topology, Contemp. Math., vol. 460, Amer. Math. Soc., Providence, RI, 2008, pp. 197–207. 283

[Kap93] Mikhail M. Kapranov, *Chow quotients of Grassmannians. I*, I. M. Gel′fand Seminar, Adv. Soviet Math., vol. 16, Amer. Math. Soc., Providence, RI, 1993, pp. 29–110. 262

[Kar78] Richard M. Karp, *A characterization of the minimum cycle mean in a digraph*, Discrete Math. **23** (1978), no. 3, 309–311. 223

[Ked01] Kiran S. Kedlaya, *The algebraic closure of the power series field in positive characteristic*, Proc. Amer. Math. Soc. **129** (2001), no. 12, 3461–3470. 49

[Kee92] Sean Keel, *Intersection theory of moduli space of stable n-pointed curves of genus zero*, Trans. Amer. Math. Soc. **330** (1992), no. 2, 545–574. 344, 345

[Kou76] A. G. Kouchnirenko, *Polyèdres de Newton et nombres de Milnor*, Invent. Math. **32** (1976), no. 1, 1–31. 309

[KSZ92] Mikhail Kapranov, Bernd Sturmfels, and Andrei Zelevinsky, *Chow polytopes and general resultants*, Duke Math. J. **67** (1992), no. 1, 189–218. 191

[Kuh55] Harold W. Kuhn, *The Hungarian method for the assignment problem*, Naval Res. Logist. Quart. **2** (1955), 83–97. 10

[Kuš76] A. G. Kušnirenko, *Newton polyhedra and Bezout's theorem*, Funkcional. Anal. i Priložen. **10** (1976), no. 3, 82–83. 210

[KV07] Joachim Kock and Israel Vainsencher, *An invitation to quantum cohomology*, Progress in Mathematics, vol. 249, Birkhäuser, Boston, MA, 2007. 316, 345

[LQ11] Mark Luxton and Zhenhua Qu, *Some results on tropical compactifications*, Trans. Amer. Math. Soc. **363** (2011), no. 9, 4853–4876. 309

[M2] Daniel R. Grayson and Michael E. Stillman, *Macaulay2, a software system for research in algebraic geometry*, Available at http://www.math.uiuc.edu/Macaulay2/. 53, 179

[Mac01] Diane Maclagan, *Antichains of monomial ideals are finite*, Proc. Amer. Math. Soc. **129** (2001), no. 6, 1609–1615. 75

[Mar10] Thomas Markwig, *A field of generalised Puiseux series for tropical geometry*, Rend. Semin. Mat. Univ. Politec. Torino **68** (2010), no. 1, 79–92. 49

[Mik04] Grigory Mikhalkin, *Decomposition into pairs-of-pants for complex algebraic hypersurfaces*, Topology **43** (2004), no. 5, 1035–1065. 20

[Mik05] ———, *Enumerative tropical algebraic geometry in $\mathbb{R}^2$*, J. Amer. Math. Soc. **18** (2005), no. 2, 313–377. 31, 33, 34

[MR] Grigory Mikhalkin and Johannes Rau, *Tropical geometry*, Book in preparation. ix, 34, 346

[MR01] Grigory Mikhalkin and Hans Rullgård, *Amoebas of maximal area*, Internat. Math. Res. Notices (2001), no. 9, 441–451. 20

[MS05] Ezra Miller and Bernd Sturmfels, *Combinatorial commutative algebra*, Graduate Texts in Mathematics, vol. 227, Springer-Verlag, New York, 2005. 53, 57, 152, 279, 335

[MZ08] Grigory Mikhalkin and Ilia Zharkov, *Tropical curves, their Jacobians and theta functions*, Curves and abelian varieties, Contemp. Math., vol. 465, Amer. Math. Soc., Providence, RI, 2008, pp. 203–230. 334

[NS06] Takeo Nishinou and Bernd Siebert, *Toric degenerations of toric varieties and tropical curves*, Duke Math. J. **135** (2006), no. 1, 1–51. 332

[NS13] Mounir Nisse and Frank Sottile, *The phase limit set of a variety*, Algebra Number Theory **7** (2013), no. 2, 339–352. 21

[Oda88] Tadao Oda, *Convex bodies and algebraic geometry*, Ergebnisse der Mathematik und ihrer Grenzgebiete (3), vol. 15, Springer-Verlag, Berlin, 1988. 277

[OP13] Brian Osserman and Sam Payne, *Lifting tropical intersections*, Doc. Math. **18** (2013), 121–175. 124

[Oxl11] James Oxley, *Matroid theory*, Graduate Texts in Mathematics, vol. 21, Oxford University Press, 2011. 163

[Pas85] Donald S. Passman, *The algebraic structure of group rings*, Robert E. Krieger Publishing Co. Inc., Melbourne, FL, 1985. 48

[Pay09a] Sam Payne, *Analytification is the limit of all tropicalizations*, Math. Res. Lett. **16** (2009), no. 3, 543–556. 311

[Pay09b] ______, *Fibers of tropicalization*, Math. Z. **262** (2009), no. 2, 301–311. 109

[Pay12] ______, *Erratum to: Fibers of tropicalization*, Math. Z. **272** (2012), no. 3-4, 1403–1406. 109

[Pay13] ______, *Boundary complexes and weight filtrations*, Michigan Math. J. **62** (2013), no. 2, 293–322. 313, 346

[Pin98] Jean-Eric Pin, *Tropical semirings*, Idempotency (Bristol, 1994), Publ. Newton Inst., vol. 11, Cambridge Univ. Press, 1998, pp. 50–69. 1

[Poo93] Bjorn Poonen, *Maximally complete fields*, Enseign. Math. (2) **39** (1993), no. 1-2, 87–106. 48, 51

[PR04] Mikael Passare and Hans Rullgård, *Amoebas, Monge-Ampère measures, and triangulations of the Newton polytope*, Duke Math. J. **121** (2004), no. 3, 481–507. 18, 19

[PS04] Lior Pachter and Bernd Sturmfels, *Tropical geometry of statistical models*, Proc. Natl. Acad. Sci. USA **101** (2004), no. 46, 16132–16137. 11

[PS05] Lior Pachter and Bernd Sturmfels (eds.), *Algebraic statistics for computational biology*, Cambridge University Press, 2005. ix, 11, 171, 173, 174

[PT05] Mikael Passare and August Tsikh, *Amoebas: their spines and their contours*, Idempotent mathematics and mathematical physics, Contemp. Math., vol. 377, Amer. Math. Soc., Providence, RI, 2005, pp. 275–288. 18

[Pur08] Kevin Purbhoo, *A nullstellensatz for amoebas*, Duke Math. J. **141** (2008), no. 3, 407–445. 18

[Rab12] Joseph Rabinoff, *Tropical analytic geometry, Newton polygons, and tropical intersections*, Adv. Math. **229** (2012), no. 6, 3192–3255. 283

[Rad42] Richard Rado, *A theorem on independence relations*, Quart. J. Math., Oxford Ser. **13** (1942), 83–89. 204

[Ram02] Jörg Rambau, *TOPCOM: triangulations of point configurations and oriented matroids*, Mathematical software (Beijing, 2002), World Sci. Publ., River Edge, NJ, 2002, pp. 330–340. 241

[Rei88] Miles Reid, *Undergraduate algebraic geometry*, London Mathematical Society Student Texts, vol. 12, Cambridge University Press, 1988. 198

[RG71] Michel Raynaud and Laurent Gruson, *Critères de platitude et de projectivité. Techniques de "platification" d'un module*, Invent. Math. **13** (1971), 1–89. 324

[Rib99] Paulo Ribenboim, *The theory of classical valuations*, Springer Monographs in Mathematics, Springer-Verlag, New York, 1999. 48

[Rin13] Felipe Rincón, *Computing tropical linear spaces*, J. Symbolic Comput. **51** (2013), 86–98. 161

[RS12] Tim Römer and Kirsten Schmitz, *Generic tropical varieties*, J. Pure Appl. Algebra **216** (2012), no. 1, 140–148. 208, 347

[RSS14] Qingchun Ren, Steven V. Sam, and Bernd Sturmfels, *Tropicalization of classical moduli spaces*, Math. Comput. Sci. **8** (2014), no. 2, 119–145. 268

[Sam66] Pierre Samuel, *À propos du théorème des unités*, Bull. Sci. Math. (2) **90** (1966), 89–96. 297

[San05] Francisco Santos, *The Cayley trick and triangulations of products of simplices*, Integer points in polyhedra—geometry, number theory, algebra, optimization, Contemp. Math., vol. 374, Amer. Math. Soc., Providence, RI, 2005, pp. 151–177. 239, 266

[Sch93] Rolf Schneider, *Convex bodies: the Brunn-Minkowski theory*, Encyclopedia of Math. and its Applications, vol. 44, Cambridge University Press, 1993. 204

[Sha13] Kristin M. Shaw, *A tropical intersection product in matroidal fans*, SIAM J. Discrete Math. **27** (2013), no. 1, 459–491. 346

[Shi13] Yaroslav Shitov, *When do the r-by-r minors of a matrix form a tropical basis?*, J. Combin. Theory Ser. A **120** (2013), no. 6, 1166–1201. 244, 253, 255, 274

[Sim88] Imre Simon, *Recognizable sets with multiplicities in the tropical semiring*, Mathematical foundations of computer science, 1988 (Carlsbad, 1988), Lecture Notes in Comput.Sci., vol. 324, Springer, Berlin, 1988, pp. 107–120. 1

[SS04] David Speyer and Bernd Sturmfels, *The tropical Grassmannian*, Adv. Geom. **4** (2004), no. 3, 389–411. 190

[ST08] Bernd Sturmfels and Jenia Tevelev, *Elimination theory for tropical varieties*, Math. Res. Lett. **15** (2008), no. 3, 543–562. 269

[ST10] Reinhard Steffens and Thorsten Theobald, *Combinatorics and genus of tropical intersections and Ehrhart theory*, SIAM J. Discrete Math. **24** (2010), no. 1, 17–32. 212

[Sta96] Richard P. Stanley, *Combinatorics and commutative algebra*, second ed., Progress in Mathematics, vol. 41, Birkhäuser, Boston, MA, 1996. 335

[Sta12] ———, *Enumerative combinatorics. Volume 1*, second ed., Cambridge Studies in Advanced Mathematics, vol. 49, Cambridge University Press, 2012. 158

[Str84] Ralph Strebel, *Finitely presented soluble groups*, Group theory, Academic Press, London, 1984, pp. 257–314. 27, 30

[Stu96] Bernd Sturmfels, *Gröbner bases and convex polytopes*, University Lecture Series, vol. 8, American Mathematical Society, Providence, RI, 1996. 67, 81, 152, 157, 178, 201, 332

[Stu02] ———, *Solving systems of polynomial equations*, CBMS Regional Conference Series in Mathematics, vol. 97, Published for the Conference Board of the Mathematical Sciences, Washington, DC, 2002. 118, 211

[STY07] Bernd Sturmfels, Jenia Tevelev, and Josephine Yu, *The Newton polytope of the implicit equation*, Mosc. Math. J. **7** (2007), no. 2, 327–346. 25, 110, 117, 322

[SY08] Bernd Sturmfels and Josephine Yu, *Tropical implicitization and mixed fiber polytopes*, Software for algebraic geometry, IMA Vol. Math. Appl., vol. 148, Springer, New York, 2008, pp. 111–131. 25, 110, 117

[Tev07] Jenia Tevelev, *Compactifications of subvarieties of tori*, Amer. J. Math. **129** (2007), no. 4, 1087–1104. 36, 302, 303, 304, 309, 312

[Tev14] ———, *On a question of B. Teissier*, Collect. Math. **65** (2014), no. 1, 61–66. 309

[The02] Thorsten Theobald, *Computing amoebas*, Experiment. Math. **11** (2002), no. 4, 513–526 (2003). 18

[Tyo12] Ilya Tyomkin, *Tropical geometry and correspondence theorems via toric stacks*, Math. Ann. **353** (2012), no. 3, 945–995. 34

[Vak13] Ravi Vakil, *Foundations of algebraic geometry*, Available at `http://math216.wordpress.com/`, 2013. 303, 304, 324

[Vig10] Magnus Dehli Vigeland, *Smooth tropical surfaces with infinitely many tropical lines*, Ark. Mat. **48** (2010), no. 1, 177–206. 200

[Vir08] Oleg Viro, *From the sixteenth Hilbert problem to tropical geometry*, Japan J. Math. **3** (2008), no. 2, 185–214. 332

[Wel76] D. J. A. Welsh, *Matroid theory*, Academic Press, London, 1976, L.M.S. Monographs, No. 8. 163

[Whi86] Neil White (ed.), *Theory of matroids*, Encyclopedia of Mathematics and its Applications, vol. 26, Cambridge University Press, 1986. 163

[Whi87] Neil White (ed.), *Combinatorial geometries*, Encyclopedia of Mathematics and its Applications, vol. 29, Cambridge University Press, 1987. 163

[Whi92] Neil White (ed.), *Matroid applications*, Encyclopedia of Mathematics and its Applications, vol. 40, Cambridge University Press, 1992. 163, 165

[Zie95] Günter M. Ziegler, *Lectures on polytopes*, Graduate Texts in Mathematics, vol. 152, Springer-Verlag, New York, 1995. x, 58, 59

Index